JAGUAR

3·4 'S' and 3·8 'S'

MODELS

SERVICE MANUAL

JAGUAR CARS LIMITED, COVENTRY, ENGLAND

Publication No. E/133/3

INDEX TO SECTIONS

GENERAL INFORMATION

ADDENDA

The following corrections should be noted when using SECTION A.

Page	Correction
A.16	Electrically Heated Backlight. On later cars a separate heater switch is provided. An amber warning light, situated on the facia panel, indicates that the system is in operation. A resistance through the side and headlamp switch automatically dims the warning light for night driving.
A.17	Note that on later cars the nave plate is removed with the aid of the tool provided in the tool kit.
A.20	Note that on later cars, a red warning light in the quadrant also is illuminated when the overdrive is in operation.

INDEX

SEAT BELTS

Anchorage points for seat belts are incorporated in the construction of the car. If it is required to fit seat belts, contact your usual Jaguar dealer.

GENERAL INFORMATION

CAR IDENTIFICATION

Fig. 1. *Location of car number.*

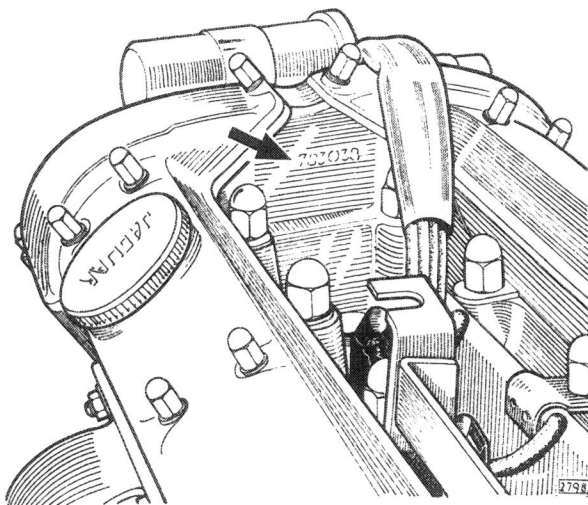

Fig. 2. *Location of engine number.*

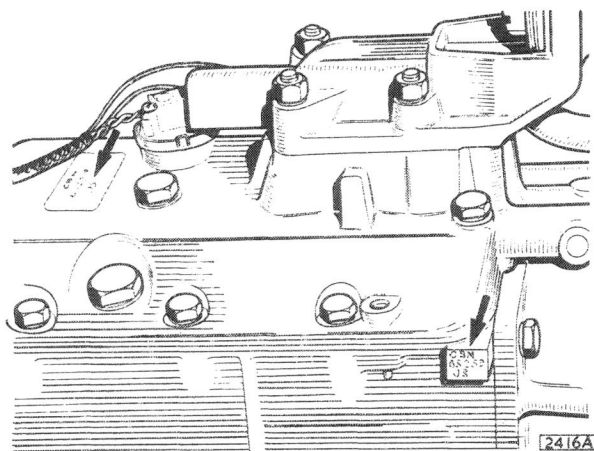

Fig. 3. *Location of gearbox numbers.*

It is imperative that the Car and Engine numbers, together with any prefix or suffix letters, are quoted in any correspondence concerning this vehicle. If the unit in question is the Gearbox or Overdrive, the Gearbox number and any prefix or suffix letters must also be quoted. This also applies when ordering spare parts.

Car Number

Stamped in the bonnet catch channel, forward of the radiator header tank.

Suffix "DN" to the car number indicates that an overdrive unit is fitted.

Engine Number

Stamped on the right-hand side of the cylinder block above the oil filter and at the front of the cylinder head casting.

/7, /8 or /9 following the engine number denotes the compression ratio.

Gearbox Number

Stamped on a shoulder at the left-hand rear corner of the gearbox casing and on the top cover.

Letter "N" at the end of the prefix letters indicates that an overdrive unit is fitted.

The above numbers are stamped on an identification plate mounted on either the left-hand wheel valance or on the bulkhead.

Key Numbers

Two different types of key are provided to enable the car to be left with the luggage boot and cubby locker locked, on the occasions when it is required to leave the ignition key with the car.

(a) The round headed key operates the ignition switch and door locks.

(b) The rectangular headed key operates the locks for the luggage boot lid and the cubby locker.

GENERAL DATA

Dimensions and Weights

Wheelbase	8′ 11⅜″ (2.727 m.)
Track, Front—Disc Wheels	4′ 7¼″ (1.403 m.)
Track, Front—Wire Wheels	4′ 7¼″ (1.403 m.)
Track, Rear—Disc Wheels	4′ 6¼″ (1.373 m.)
Track, Rear—Wire Wheels	4′ 4⅞″ (1.343 m.)
Overall length	15′ 7 13/16″ (4.770 m.)
Overall width	5′ 6¾″ (1.695 m.)
Overall height	4′ 6½″ (1.38 m.)
Weight (dry) approximate	30.7 cwts. (1560 kg.)
Turning circle	33′ 6″ (10.21 m.)
Ground Clearance	7″ (178 mm.)

Capacities

	Imperial	U.S.	Litres
Engine (refill, including filter)	12 pints	14½ pints	6.75
Gearbox (without overdrive)	2½ ,,	3 ,,	1.5
Gearbox (with overdrive)	4 ,,	4¾ ,,	2.25
Automatic transmission unit	15 ,,	18 ,,	8.5
Rear axle	2¾ ,,	3¼ ,,	1.5
Cooling system (including heater)	22 ,,	26½ ,,	12.5
Petrol tanks—left hand	7 galls.	8¼ galls.	31.75
Petrol tanks—right hand..	7 ,,	8¼ ,,	31.75

PERFORMANCE DATA

The following table gives the relationship between engine revolutions per minute and road speed in miles and kilometres per hour.

It is recommended that engine revolutions **in excess of 5,000 per minute** should not be exceeded for long periods. Therefore, if travelling at sustained high speed on motorways, the accelerator should be released occasionally to allow the car to overrun for a few seconds.

AXLE RATIO 3.54 : 1

ROAD SPEED		ENGINE REVOLUTIONS PER MINUTE			
Kilometres per hour	Miles per hour	First and Reverse Gears 11.954	Second Gear 6.584	Third Gear 4.541	Top Gear 3.54
16	10	1569	864	596	465
32	20	3138	1728	1192	930
48	30	4708	2593	1789	1395
64	40		3456	2385	1860
80	50		4320	2981	2325
96	60		5186	3577	2790
112	70			4174	3255
128	80			4770	3720
144	90			5367	4185
160	100				4650
176	110				5115
192	120				5580

Note: The figures in these tables are theoretical and actual figures may vary slightly from those quoted due to such factors as tyre wear, pressures etc.

The following table gives the relationship between engine revolutions per minute and road speed in miles and kilometres per hour.

It is recommended that engine revolutions **in excess of 5,000 per minute** should not be exceeded for long periods. Therefore, if travelling at sustained high speed on motorways, the accelerator should be released occasionally to allow the car to overrun for a few seconds.

AXLE RATIO 3.77 : 1

ROAD SPEED		ENGINE REVOLUTIONS PER MINUTE				
Kilometres per hour	Miles per hour	First and Reverse Gears 12.731	Second Gear 7.012	Third Gear 4.836	Top Gear 3.77	Overdrive 2.933
16	10	1672	921	636	495	385
32	20	3344	1842	1272	990	770
48	30	5016	2762	1907	1485	1154
64	40		3684	2544	1980	1540
80	50		4605	3180	2475	1925
96	60			3814	2970	2310
112	70			4449	3465	2693
128	80			5088	3960	3078
144	90				4455	3463
160	100				4950	3848
176	110				5445	4232
192	120					4617

Note: The figures in these tables are theroetical and actual figures may vary slightly from those quoted due to such factors as tyre wear, pressures etc.

OPERATING INSTRUCTIONS

INSTRUMENTS

1. Ammeter.
2. Fuel gauge.
3. Lighting switch.
4. Oil pressure gauge.
5. Water temperature gauge.
6. Revolution counter.
7. Speedometer.
8. Brake fluid level/Hand-brake warning light.
9. Interior/Map light switch.
10. Panel light switch.
11. Heater fan switch.
12. Ignition switch.
13. Cigar lighter.
14. Starter switch.
15. Fuel tank change-over switch.
16. Windscreen wiper switch.
17. Windscreen washer switch.
18. Flashing direction indicator and headlamp flashing switch.
19. Clock adjuster.
20. Horn switch ring.
21. Speedo trip control.

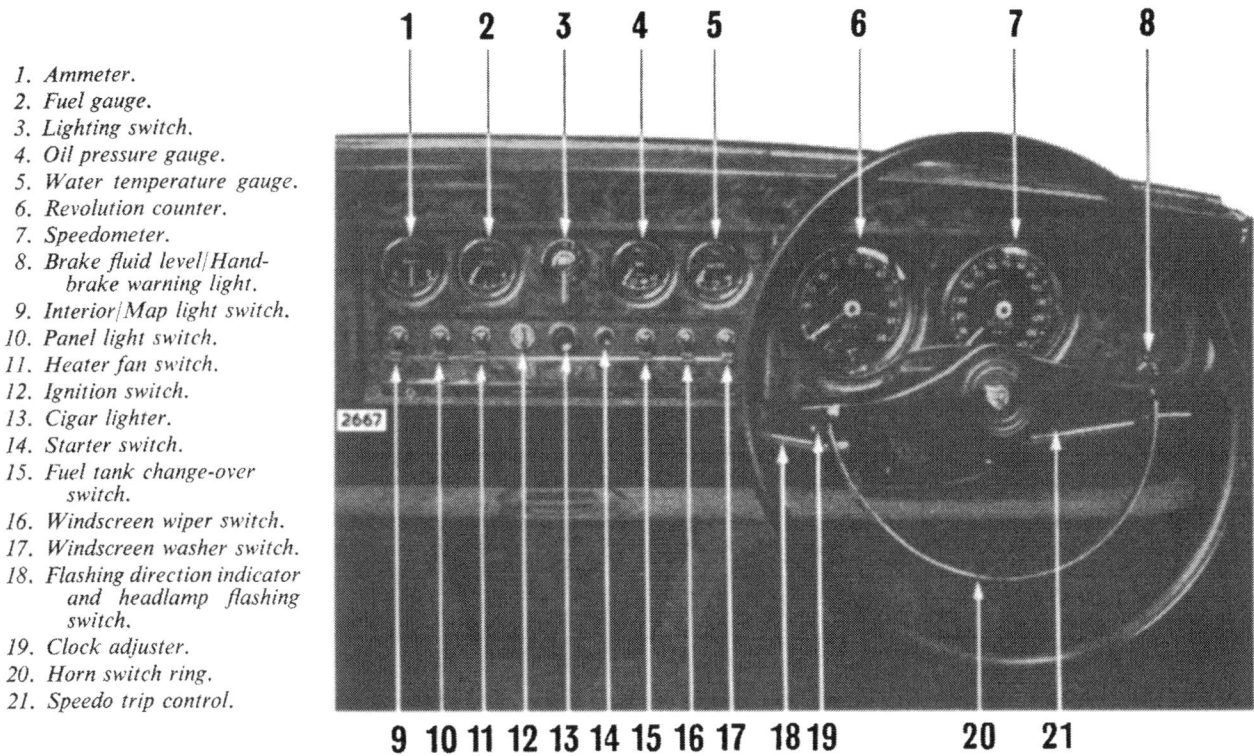

Fig. 4. Instruments and controls—Right hand drive.

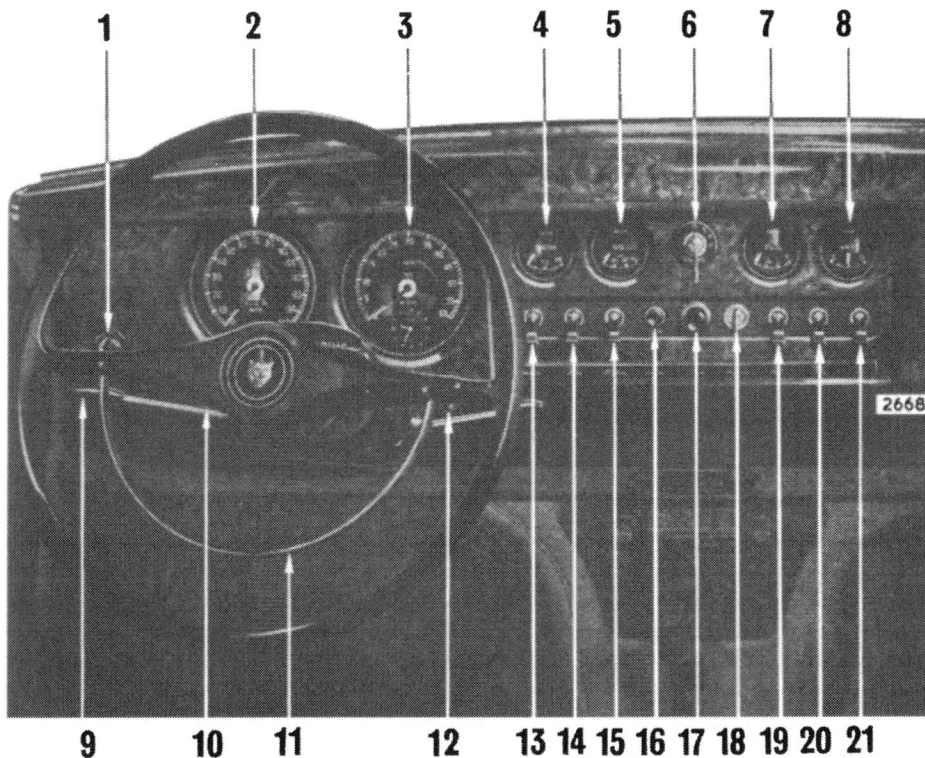

1. Brake fluid level/Hand-brake warning light.
2. Speedometer.
3. Revolution counter.
4. Water temperature gauge.
5. Oil pressure gauge.
6. Lighting switch.
7. Fuel gauge.
8. Ammeter.
9. Flashing direction indicator and headlamp flashing switch.
10. Speedometer trip control.
11. Horn switch ring.
12. Clock adjuster.
13. Windscreen washer switch.
14. Windscreen wiper switch.
15. Fuel tank change-over switch.
16. Starter switch.
17. Cigar Lighter.
18. Ignition switch.
19. Heater fan switch.
20. Panel light switch.
21. Interior/Map light switch.

Fig. 5. Instruments and controls—Left-hand drive.

Ammeter

Records the flow of current into or out of the battery. Since compensated voltage control is incorporated, the flow of current is adjusted to the state of charge of the battery; thus when the battery is fully charged the dynamo provides only a small output and therefore little charge is registered on the ammeter, whereas when the battery is low a continuous high charge is shown.

Oil Pressure Gauge

The electrically operated pressure gauge records the oil pressure being delivered by the oil pump to the engine; it does not record the quantity of oil in the sump. The minimum pressure at 3,000 r.p.m. when hot should not be less than 40 lbs. per square inch.

Note : On starting up, a period of approximately 20 seconds will elapse before the correct reading is obtained.

Water Temperature Gauge

The electrically operated water temperature gauge records the temperature of the coolant by means of a bulb screwed into the inlet manifold water jacket.

Fuel Level Gauge

Records the quantity of fuel in the supply tank in use. To obtain readings for the opposite tank operate the fuel change-over switch on the instrument panel. Readings will only be obtained when the ignition is switched "ON".

Note : Lift the switch lever for the left-hand tank, lower for the right-hand tank as shown on the switch indicator strip.

Electric Clock

The clock is built in the revolution counter instrument and is powered by the battery. The clock hands may be adjusted by pushing up the winder and rotating. Starting is accomplished in the same manner.

Revolution Counter

Records the speed of the engine in revolutions per minute.

Speedometer

Records the vehicle speed in miles per hour, total mileage and trip mileage (kilometres on certain export models). The trip figures can be set to zero by pushing the winder upwards and rotating anti-clockwise.

Fig. 6. Warning lights. A—Ignition. B—Headlamp main beam.

Fig. 7. Flashing direction indicator warning lights.

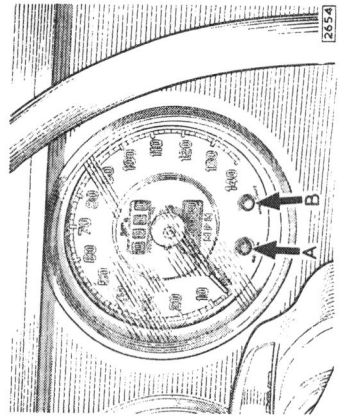

Flashing Direction Indicator-Warning Lights

The warning lights are in the form of green arrows, one at each side of the quadrant, situated behind the steering wheel.

When the flashing indicators are in operation one of the arrows light up on the side selected.

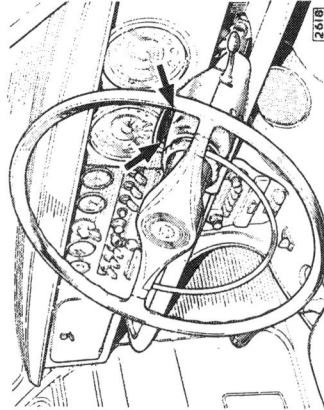

Headlamp Warning Light

A warning light marked " Headlamps " situated in the speedometer, lights up when the headlamps are in full beam position and is automatically extinguished when the lamps are in the dipped beam position.

Ignition Warning Light

A red warning light (marked "Ignition") situated in the speedometer lights up when the ignition is switched " on " and the engine is not running, or when the engine is running at a speed insufficient to charge the battery. The latter condition is not harmful, but always switch " off " when the engine is not running.

Brake Fluid Level and Handbrake Warning Light

A warning light (marked "Brake Fluid-Handbrake") situated on the facia behind the steering wheel, serves to indicate if the level in the brake fluid reservoir has become low, provided the ignition is "on." As the warning light is also illuminated when the handbrake is applied, the handbrake must be fully released before it is assumed that the fluid level is low. If with the ignition "on" and the handbrake fully released the warning light is illuminated the brake fluid must be "topped up" immediately.

As the warning light is illuminated when the handbrake is applied and the ignition is "on" a two-fold purpose is served. Firstly, to avoid the possibility of driving away with the handbrake applied. Secondly, as a check that the warning light bulb has not "blown"; if on first starting up the car with the handbrake fully applied, the warning light does not become illuminated the bulb should be changed immediately.

Fig. 8. Brake fluid level and handbrake warning light.

CONTROLS AND ACCESSORIES

Accelerator Pedal

Controls the speed of the engine.

Brake Pedal

Operates the vacuum-servo assisted disc brakes on all four wheels.

Clutch Pedal

On overdrive and standard transmission cars, connects and disconnects the engine and the transmission. Never drive with the foot resting on the pedal and do not keep the pedal depressed for long periods in traffic. Never coast the car with a gear engaged and clutch depressed.

Headlight Dipper

Situated on the toe boards to the left of the clutch pedal. The switch is of the change over type and if the headlights are in the full beam position a single pressure on the control will switch the lights to the dipped beam position and they will remain so until another single pressure switches them to the full beam position again.

Gear Lever (Overdrive and Standard Transmission Models)

Centrally situated and with gear positions indicated on the control knob. To engage reverse gear first press the gear lever against the spring pressure before pushing the lever forward. Always engage neutral and release the clutch when the car is at rest.

On later cars, to engage reverse gear press the gear lever against the spring pressure and pull the lever rearward.

Overdrive Switch Lever

For full instructions on the operation of the overdrive, see Section F.

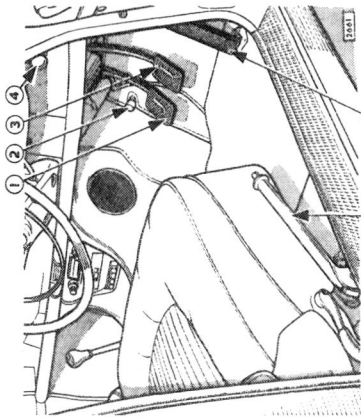

Fig. 9. *The foot and hand controls. 1. Clutch pedal. 2. Head-lamp dipper. 3. Brake pedal. 4. Bonnet lock control. 5. Accelerator pedal. 6. Handbrake.*

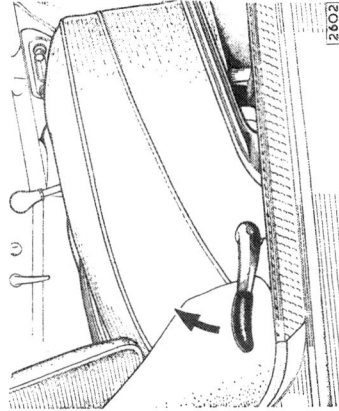

Fig. 10. *Gear positions (early cars).*

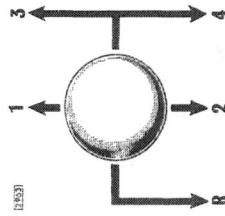

Fig. 10A. *Gear positions (later cars).*

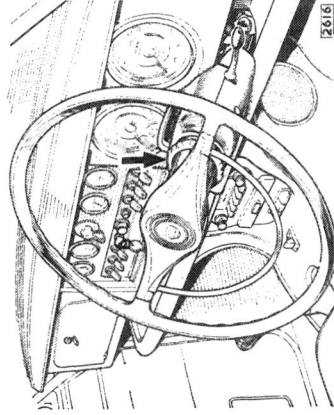

Fig. 11. *Reclining seat control lever.*

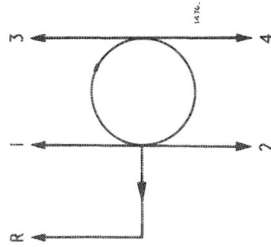

Fig. 12. *Steering wheel adjustment.*

Automatic Transmission Selector Lever

For full instructions on the operation of the automatic transmission, see page A.20.

Handbrake Lever

Positioned at the outside of the driver's seat. The handbrake operates mechanically on the rear discs only and is provided for parking, driving away on a hill and when at a standstill in traffic. To apply the brake, pull the lever upward and the trigger will automatically engage with the ratchet. The handbrake is released by pressing in the knob and pushing the lever downward.

Seat Adjustment

Both front seats are adjustable for reach. Push the lock bar, situated beside the inside runner, towards the inside of the car and slide into the required position. Release the lock bar and slide until the mechanism engages with a click.

Adjustable Front Seat Back

The seats are adjustable from the vertical to the fully reclined position.

To adjust, lift the lever located on the outside of the seat cushion and adjust the seat back to the desired position.

Release the lever to lock in position.

Steering Wheel Adjustment

Rotate the knurled ring at the base of the steering wheel hub in an anti-clockwise direction when the steering wheel may be slid into the desired position. Turn the knurled ring clockwise to lock the steering wheel.

Front Door Locks

The front doors are opened from the outside by pressing the button incorporated in the door handle. The doors are opened from the inside by pulling the interior handles rearwards.

Both front doors can be locked from the interior by turning the lock escutcheon, marked "LOCK", in the direction of the arrow. Return the escutcheon to the original position to unlock.

Both front doors can be locked from the outside by means of the ignition key; the locks are incorporated in the push buttons of the door handles.

To lock the right hand door insert the key in the lock, rotate clockwise as far as possible and allow the lock to return to its original position—the door is now locked. To unlock, turn the key anti-clockwise as far as possible and allow the lock to return to its original position.

To lock the left hand door rotate the key anti-clockwise; to unlock, rotate the key clockwise.

KEYLESS LOCKING is obtainable by first turning the escutcheon in the direction of the arrow and closing the door with the push button in the door fully depressed.

Warning.—If the doors are to be locked by this method the ignition key should be removed beforehand (or the spare key kept on the driver's person) as the only means of unlocking the front door is with the key.

Rear Door Locks

The rear doors are opened from the outside by pressing the button incorporated in the door handle. The doors are opened from the inside by pushing the interior door handle forward and are locked by turning the escutcheon marked "LOCK" in the direction of the arrow.

Horn

Depress the semi-circular ring attached to the steering wheel or press the button in the centre to operate the twin horns.

Ignition Switch

Inserting the key provided in the switch and turning clockwise will switch on the ignition.

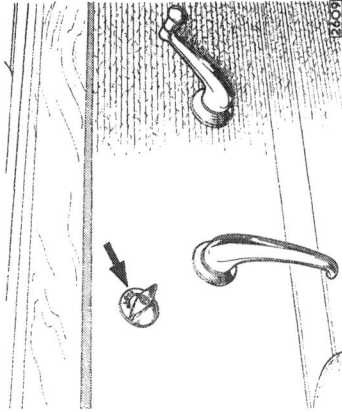

Fig. 13.   *The door lock control. Turn in the direction of the arrow to lock.*

Never leave the ignition on when the engine has stopped, a reminder of such circumstances is provided by the ignition warning light situated in the speedometer.

Interior/Map Light Switch

The map and interior lights are controlled by a three-position switch marked "Interior/ Map" on the indicator strip. Lift the switch lever to the second position to operate the map light situated above the instrument panel. For interior lights lift the switch lever to the third position. To provide ease of entry into the car at night the interior lights are automatically switched on when any one of the doors is opened and are extinguished when the door is closed.

Lighting Switch

From "Off" can be rotated clockwise into three positions, giving in the first location, side and tail, in the second location, head, side and tail, and in the third position, fog, side and tail lamps.

(Fog lamps are not fitted on cars for U.S.A.).

Panel Light Switch

Lift the switch lever (marked "Panel" on the indication strip) to enable the instruments to be read at night and to provide illumination of the switch markings. The switch has two positions "Dim" and "Bright" to suit the driver's requirements. The panel lights will only operate when the side lights are switched on.

Starter Switch

Press the button (marked "Starter" on the indicator strip) with the ignition switched on, to start the engine. Release the switch immediately the engine fires and never operate the starter when the engine is running.

Flashing Direction Indicators

The "Flashers" are operated by a lever behind the steering wheel. To operate the flashing direction indicators on the right-hand side of the car, move the lever clockwise; to operate the left-hand side indicators move the lever anti-clockwise. While the flashing indicators are in operation one of the warning lights in the quadrant behind the steering wheel lights up on the side selected.

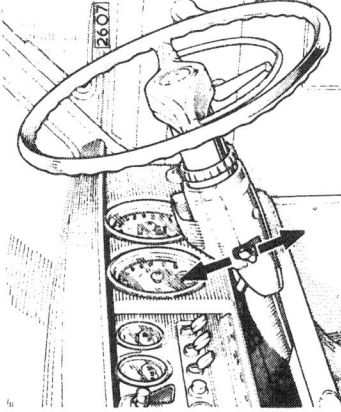

Fig. 14.   *Flashing direction indicator control.*

Headlamp Flasher

To "flash" the headlamps as a warning signal, lift and release the flashing indicator switch lever to the second position. The headlamps can be "flashed" when the lights are "off" or when they are in the dipped beam position; they will not "flash" in the main beam position.

Glovebox Light

A lamp in the glovebox is automatically illuminated when the lid is opened and the sidelights are "on".

Braking Lights

Twin combined tail and brake lights automatically function when the footbrake is applied.

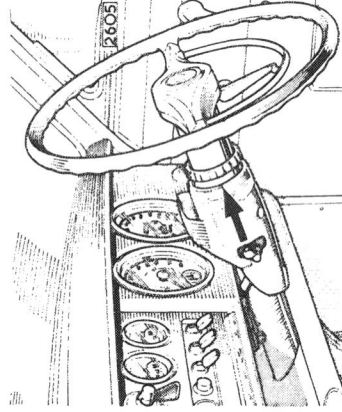

Fig. 15.   *Method of "flashing" the headlamps.*

Reversing Light

The reversing light is automatically brought into operation when reverse gear is engaged and the ignition is switched on.

Luggage Compartment Illumination

The luggage compartment is automatically illuminated by a lamp when the lid is opened. The lamp operates only when the sidelights are switched on.

Cigar Lighter

To operate, press holder into the socket (marked "Cigar" on the indicator strip) and remove the hand. On reaching the required temperature, the holder will return to the extended position. Do not hold the lighter in the "pressed in" position.

Windscreen Wipers

The wipers are controlled by a three-position switch (marked "Wiper" on the indicator strip). Lift the switch to the second position (Slow) which is recommended for all normal adverse weather conditions and snow.

For conditions of very heavy rain and for fast driving in rain lift the switch to the third position (Fast). This position should not to be used in heavy snow or with a drying windscreen, that is, when the load on the motor is in excess of normal; the motor incorporates a protective cut-out switch which under conditions of excessive load cuts off the current supply until normal conditions are restored.

When the switch is placed in the "Off" position the wipers will automatically return to the parked position along the lower edge of the screen.

Windscreen Washer

For full instructions on the use of Windscreen Washing Equipment see Section "O".

Heating and Ventilating Equipment

For full instructions on the use of the Heating and Ventilating Equipment see Section "O"

Scuttle Ventilator

The scuttle ventilator can be operated only by the heating and ventilating controls (see Section "O").

Bonnet Lock Control

The bonnet lock is controlled from the driving compartment. To open the bonnet pull the control knob, situated under the facia on the right-hand side. This will release the bonnet which will now be retained by the safety catch.

Insert the fingers under the nose of the bonnet and lift the safety catch upwards when the bonnet may be raised. The bonnet is automatically retained in the fully open position by the action of the hinge springs. The bonnet is self-locking when pushed down firmly into the closed position.

Fig. 16. Bonnet lock control.

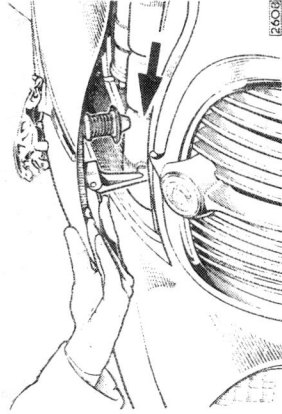

Fig. 17. Bonnet safety catch.

Two-Pin Plug Socket

A two-pin plug socket is provided, situated under the bonnet on the left-hand wing valance. The socket may be used to "trickle charge" the battery, or, provide an inspection lamp point.

Spare Wheel and Jacking Equipment

The spare wheel is housed in a compartment underneath the luggage boot floor and is accessible after raising the hinged panel.

The jacking equipment is retained in clips also in the spare wheel compartment and is accessible after raising the panel.

Tools

The container for the hand tools is housed in the spare wheel compartment.

No-Draught Ventilation

All doors are fitted with no-draught ventilation windows incorporating quick-locking catches.

To open the window, release the locking catch and set the window to the desired position.

It will be observed that initial opening of the front window gives extraction of air from the body. When the window is opened further, air is forced into the body due to the angle of the ventilator and forward motion of the car. It should be observed that using the N.D.V. windows as extractors (that is, partially open) has, to a minor degree, the effect of demisting the windscreen.

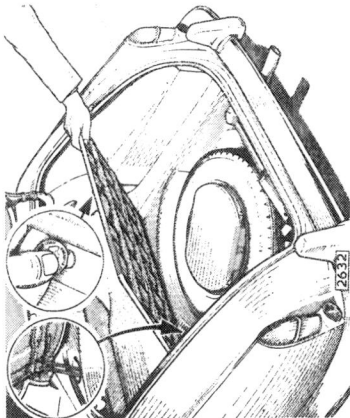

Fig. 18. The spare wheel is housed in a compartment underneath the luggage boot floor. The container for the hand tools is also stored in this compartment.

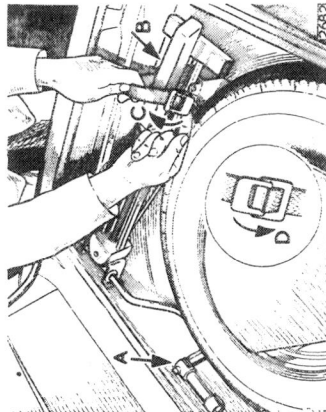

Fig. 19. "A" Location of the wheel brace. "B" Location of the jack. "C" Pull the top strap and lift the buckle tongue to tighten. "D" Press tongue downwards to release the strap.

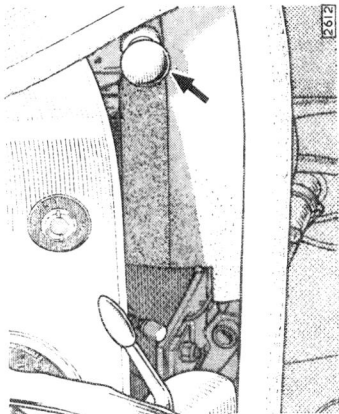

Fuel Tank Fillers

The two fuel tank fillers are situated in recesses in the rear wings and are accessible after lifting the hinged covers.

Electrically Heated Backlight (Optional Extra)

An electrically heated backlight to provide demisting or defrosting of the rear window is available as an optional extra.

A heating element consisting of a fine wire mesh between the laminations of glass is connected to the wiring harness and will come into operation when the ignition is switched on. The current consumption is 5 amps. and a 15 amp. fuse is located below the fuse blocks.

Luggage Compartment

To open the luggage compartment, insert the square-headed key in the lock situated between the twin reversing lamps and unlock by rotating anti-clockwise through a quarter of a turn. Release the catch lever located beneath the lock and raise the lid. The lid is retained in the fully open position by means of torsion bar springs.

Interior Driving Mirror

This is of the dipping type. Move the lever, situated under the mirror, to the left for night driving to avoid being dazzled by the lights of a following car.

The mirror is adjustable for height by releasing the hexagon nut on the stem and sliding the mirror up or down. Re-tighten the nut when the correct position is obtained.

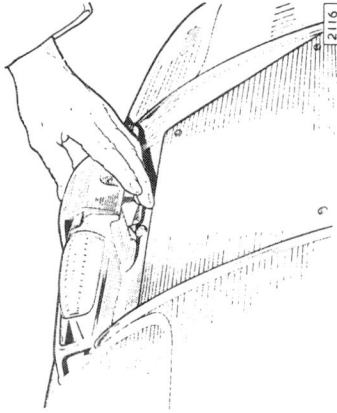

Fig. 20. Opening the luggage compartment.

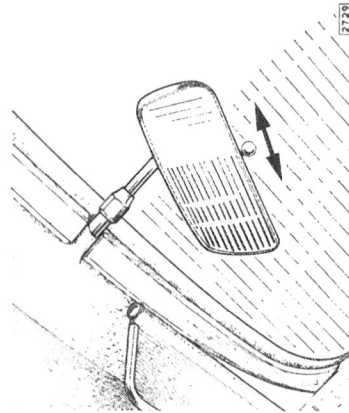

Fig. 21. Interior driving mirror dipping operation.

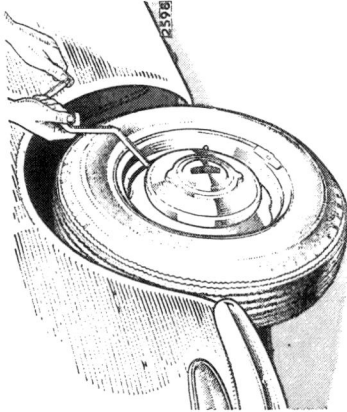

WHEEL CHANGING

(Disc Wheels)

Whenever possible, the wheel changing should be carried out with the car standing on level ground and in all cases with the handbrake fully applied.

The spare wheel and wheel changing equipment is housed in a compartment underneath the luggage boot floor.

Unlock the luggage compartment by turning the key in the lock, situated between the twin reverse lamps, through half-a-turn. Lift the catch lever located beneath the lock and raise the lid as far as possible where it will be retained by the action of the hinge torsion bars.

Insert a screwdriver or coin in the slot in the floor panel fastener. Rotate anti-clockwise until the catch is released.

Raise the panel, which will automatically be retained in the open position by means of the self-locking stay.

Unscrew the spare wheel clamping handle and lift out the wheel. Remove the jack and wheel brace.

Remove the wheel nave plate by levering off with the blade end of the wheel brace. Using the wheel brace loosen, but do not remove, the five wheel nuts; all wheel nuts have right-hand threads, that is, they are unscrewed anti-clockwise.

Remove the rubber plug from the jacking socket adjacent to the wheel to be removed, insert the square portion of the jack well home into the socket and rotate the handle attached to the jack clockwise. Raise the car until the wheel is clear of the ground. Remove the wheel nuts and withdraw the road wheel.

Mount the spare wheel on the fixing studs and start all five nuts on the threads by rotating clockwise. Apply the wheel brace and run up all the nuts until they are tight.

Fig. 22. Removal of the wheel nave plate. To avoid damaging the nave plate do not allow it to fall on the ground.

Fig. 23. Method of using the wheel brace for releasing and also removing the wheel nuts.

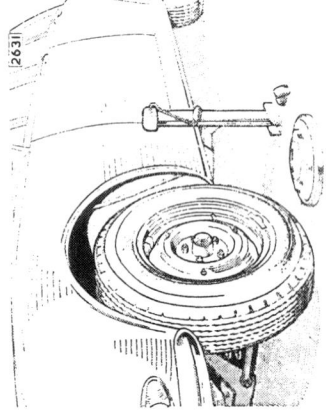

Fig. 24. The jack in position for raising the left-hand front wheel.

(d) The synchromesh gearbox provides a synchronized change into second, third and top. When changing gear the movement should be slow and deliberate.

When changing down a smoother gear change will be obtained if the accelerator is left depressed to provide the higher engine speed suitable to the lower gear. Always fully depress the clutch pedal when changing gear.

(e) Gear changing may be slightly stiff on a new car but this will disappear as the gearbox becomes "run-in".

(f) Always apply the footbrake progressively; fierce and sudden application is bad for the car and tyres. The handbrake is for use when parking the car, when driving away on a hill and when at a standstill in traffic.

STARTING AND DRIVING

Prior to Starting

Before starting the engine, the new owner should be familiar with the location and function of the instruments and controls.

Ensure that the water level in the radiator and the oil level in the sump are correct. Check for sufficient petrol in the tanks.

Place the gear lever in the neutral position and check that the handbrake is applied.

On cars equipped with automatic transmission the selector lever must be in the "P" or "N" position.

Starting from Cold

The auxiliary starting carburettor is entirely automatic and controls the mixture strength without assistance from the driver. The starting carburetter automatically cuts out when the temperature of the water in the cylinder head reaches 35° C.

When starting from cold do not depress the accelerator until the engine has fired and run for a few seconds.

Warming up

Do not operate the engine at a fast speed when first started but allow time for the engine to warm up and the oil to circulate. A thermostat is incorporated in the cooling system to assist rapid warming up. In very cold weather run the engine at 1,500 r.p.m. with the car stationary until a rise in temperature is indicated on the temperature gauge.

Driving

(a) Careful adherence to the "Running-in" Instructions given in the next column will be amply repaid by obtaining the best performance and utmost satisfaction from the car.

(b) The habit should be formed of reading the oil pressure gauge, water temperature gauge and ammeter occasionally as a check on the correct functioning of the car. Should an abnormal reading be obtained an investigation should be made immediately.

(c) Always start from rest in first gear. To start in a higher gear will cause excessive clutch slip and premature wear. Never drive with a foot resting on the clutch pedal and do not keep the clutch depressed for long periods in traffic.

"Running-in" Instructions

Only if the following important recommendations are observed will the high performance and continued good running of which the Jaguar is capable be obtained.

During the "running-in" period do not allow the engine to exceed the following speeds and particularly do not allow the engine to labour on hills; it is preferable to select a lower gear and use a higher speed rather than allow the engine to labour at low speed:—

First 1,000 miles (1,600 km.) .. .. 2,500 r.p.m.
From 1,000—2,000 miles (1,600—3,200 km.) .. 3,000 r.p.m.

Have the engine sump drained and refilled and the oil filter attended to as recommended at the free service, that is, after the first 1,000 miles (1,600 km.).

Rotate the jack handle anti-clockwise and lower the jack until the full weight of the car is on the wheel. Finally, tighten all wheel nuts.

Fit the nave plate over two of the three mounting posts (or raised bosses) and secure by a sharp tap from the hand at a point in line with the third mounting post.

WHEEL CHANGING

(Wire Spoke Wheels)

Remove the copper and hide mallet from the tool kit. Using the mallet, slacken but do not remove the hub cap; the hub caps are marked "Right (off) side" or "Left (near) side" and the direction of rotation to remove, that is, clockwise for the right-hand side and anti-clockwise for the left-hand side.

Remove the rubber plug from the front or rear jacking socket on the side to be raised. Insert the square portion of the jack well home into the socket and rotate the handle attached to the jack clockwise. Raise the car until the wheel is clear of the ground.

Remove the hub cap and withdraw the road wheel.

On cars for certain continental countries it will be necessary to attach the hub cap removal tool contained in the tool kit (see Fig. 28) before the hub cap can be removed.

Mount the spare wheel on the splined hub. Refit the hub and tighten as much as possible by rotating the hub cap in the required direction, that is, anti-clockwise for the right-hand side and clockwise for the left-hand side.

Rotate the jack handle anti-clockwise and lower the jack until the full weight of the car is on the wheel.

Finally, tighten the hub cap fully with the copper and hide mallet.

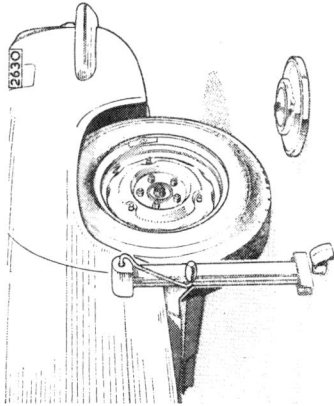
Fig. 25. *The jack in position for raising the left-hand rear wheel.*

Fig. 26. *Hub cap—right hand side.*

Fig. 27. *Hub cap—left-hand side (continental type).*

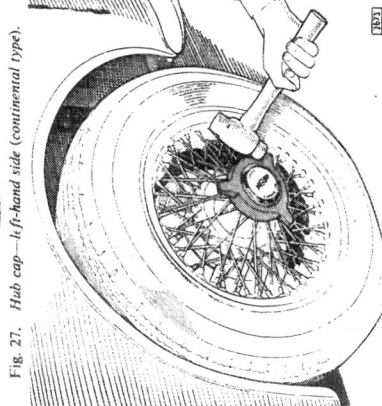
Fig. 28. *Removing the continental type hub cap with the removal tool.*

OVERDRIVE OPERATION

The Laycock de Normanville overdrive unit (fitted as an optional extra) comprise a hydraulically-controlled epicyclic gear housed in a casing which is directly attached to an extension at the rear of the gearbox.

When brought into operation, the overdrive reduces the engine speed in relation to the road speed. This permits high road speeds with low engine revolutions resulting in fuel economy and reduced engine wear.

Operation

The overdrive will operate in top gear only and is brought into action by means of the lever behind the steering wheel on the right-hand side of the column. Operate the lever clockwise to engage the overdrive and anti-clockwise to bring the drive into top (4th) gear.

When the overdrive is in operation the word "Overdrive" in the quadrant behind the steering wheel becomes illuminated. When the sidelights are switched on, the light is automatically dimmed.

Use of the clutch pedal when changing into or out of overdrive is unnecessary, but to ensure maximum smoothness of operation, particularly when changing down from overdrive to top gear, the accelerator pedal should be slightly depressed.

Do NOT bring the overdrive into operation at high speed with a wide throttle opening; release the accelerator momentarily when engaging overdrive.

For driving in towns, heavy traffic or hilly country when the maximum flexibility and low speed performance is required the overdrive manual switch should be placed in the "Out" position which will bring the drive into the normal top gear ratio.

For normal driving in open country the overdrive should be brought into operation when the required cruising speed has been obtained.

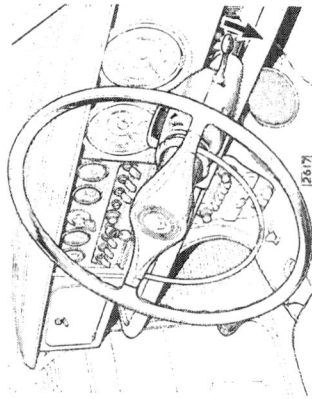

Fig. 29. *Overdrive control lever.*

The manual control lever allows selection of the following conditions:—

P (Park). A pawl is mechanically engaged with teeth on the main shaft. A hydraulic interlock prevents engagement at speeds above 3 to 5 m.p.h. (5 to 8 k.p.h.).

N (Neutral). All clutches are disengaged and there is no drive beyond the torque converter.

D (Drive). Automatic changes between the low gear and intermediate gear and between the intermediate gear and direct drive.

Manual L (Low). A low gear train and the torque converter are operative and no automatic change can occur. Manual changes between L and D may be made while the car is in motion but changes into L should be avoided at speeds above 45 m.p.h. (72 k.p.h.).

R (Reverse). A reverse-gear train and the torque converter are operative. A hydraulic interlock prevents engagement of the reverse clutch at forward speeds above 5 m.p.h. (8 k.p.h.).

Electrical connection to the starter is made only when N and P are selected. An anti-creep device traps brake fluid pressure when the car is stationary after the brakes have been applied. Opening the throttle releases the fluid.

Selector

The operation of the automatic transmission is controlled by the position of the selector lever which is indicated by the quadrant pointer. The quadrant is situated in front of the steering wheel and is marked P, N, D, L and R. The lever must be raised when selecting P, L or R and when moving from P to any other position.

When the ignition is switched on the letters P, N, D, L, R, in the quadrant behind the steering wheel become illuminated; when the sidelights are switched on the illumination is automatically dimmed.

To start the engine the selector lever must be in the P or N position.

P or Park provides a safe, positive lock on the rear wheels when the car is stopped. Movement of the selector lever to the P position actuates a mechanical locking device in the transmission which prevents the rear wheels from turning in either direction. For this reason,

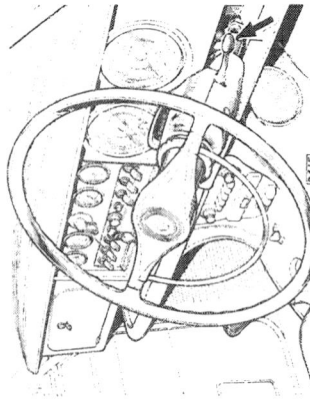

Fig. 30. *Automatic transmission selector lever.*

AUTOMATIC TRANSMISSION

Operation

The transmission assembly consists of a three-element hydraulic torque converter followed by two planetary gear sets which permit the elimination of the clutch pedal and normal gear-shift lever. The planetary gear sets incorporate free-wheels and are controlled by hydraulically-operated band and disc clutches.

Warning

The P (Park) position must **not** be selected whilst the car is moving in Reverse. Always bring the car to a stop and apply the handbrake firmly before selecting "P".

Changes from low to intermediate gear and intermediate to direct drive depend upon the combination of road speed and throttle position; the larger the throttle opening the higher the speed at which the change occurs. This is achieved by mechanically combining the motions of a mechanical centrifugal governor and the throttle linkage. The resultant motion operates a hydraulic valve.

Depression of the accelerator pedal beyond normal travel causes a "kick-down" change from direct to intermediate gear. Below 52 m.p.h. (84 k.p.h.) a downshift from direct to intermediate gear can be obtained by depressing the accelerator to the full throttle position short of "kick-down". No "kick-down" downshift is possible for intermediate to low gear.

The torque converter and a gear reduction are operative in the low intermediate gears. Direct drive is obtained by coupling the engine directly to the main shaft by a disc clutch.

should the car be pushed from front or rear with sufficient force, the car will skid on the rear tyres. This condition is quite similar to that encountered when a car with conventional transmission is parked in gear or with the handbrake applied firmly. The fact that the engine may be started with the selector in P position is convenient when parked on an incline.

Warning

The P (Park) position must **not** be selected whilst the car is moving in Reverse. Always bring the car to a stop and apply the handbrake firmly before selecting "P".

When the car is stopped on a hill and the P (Park) position is selected, the parking mechanism may become very firmly engaged due to the load on the pawl. To disengage the parking pawl under these conditions the following procedure should be adopted:—

To release transmission from P (Park) when facing UP HILL.

1. Start the engine.
2. Release the handbrake.
3. Select D and **hold** lever in this position (irrespective of the direction in which it is desired to move off).
4. Depress accelerator slowly until the car moves forward, indicating the release of the parking pawl.
5. The car is now "free" and can be driven away in the desired direction.

To release transmission from P (Park) when facing DOWN HILL.

1. Start the engine.
2. Release the handbrake.
3. Select R and **hold** lever in this position (irrespective of the direction in which it is desired to move off).

4. Depress the accelerator slowly until the car moves backward, indicating the release of the parking pawl.
5. The car is now "free" and can be driven away in the desired direction.

N or Neutral position permits idling the engine without the possibility of setting the car into motion by pressure on the accelerator and may be used when starting the engine. It is inadvisable to engage neutral for coasting.

D or Drive provides the normal forward driving range and includes automatic shifting between the low, intermediate and direct drive ranges. Virtually all forward driving, accelerating and stopping can be done with the lever in the D position. Once the engine is started and the lever is moved to D it can be left in this position for all normal driving. When accelerating, the transmission shifts automatically from low to intermediate between 11 and 40 m.p.h. (18 and 64 k.p.h.) and from intermediate to direct between 23 and 64 m.p.h. (37 and 103 k.p.h.) depending on the position of the accelerator pedal. On deceleration, it will shift automatically from direct drive to intermediate at approximately 16 m.p.h. (26 k.p.h.) and from intermediate to low at approximately 4 m.p.h. (6 k.p.h.).

L or Low is an emergency engine power range for use on unusually long and steep grades or for braking on descents, for extra heavy pulling, and for rocking the car out of mud, sand or snow.

R or Reverse position of the selector lever provides reverse driving range.

Fig. 31. *Intermediate speed hold switch.*

intermediate range; the drive will continue in the intermediate range until release of the accelerator or approximately 78 m.p.h. (126 k.p.h.) is reached.

Hard Pulling, such as encountered in deep snow, mud or other adverse driving conditions, is best accomplished in the L range.

Rocking out of Mud, Sand or Snow is accomplished with the accelerator pedal slightly depressed and held steady while making quick alternate selections of L and R ranges.

Anti-Creep is a special braking feature which prevents the car from creeping forward when stopped on level ground or slight grades, as long as the ignition key is turned on. Apply the footbrake to stop the car and then remove the foot from the brake pedal. The car will not creep forward or backward. Any movement of the accelerator pedal, or turning off the ignition key, releases the anti-creep action.

Push Starting may sometimes be necessary, as in the case of a flat battery. Turn ignition key ON, place the selector lever in the N position. The car may now be pushed and when it has reached 15 to 20 m.p.h. (24 to 32 k.p.h.) move the selector lever to D or L position. **Do not tow the car to start the engine—it may overtake the tow car.**

Engine Braking, for descending long mountainous grades, is easily secured by bringing the car speed below 45 m.p.h. (72 k.p.h.) and momentarily depressing the selector lever while placing the selector lever in the L position.

Prolonged Idling is sometimes inavoidable. In such cases, as a safety precaution, move the selector lever to the P or N position.

Towing of the car should be done with the selector in the N position. When towing the car for a short distance do not exceed 30 m.p.h. (48 k.p.h.) If the car is to be towed for an appreciable distance the propeller shaft must be disconnected.

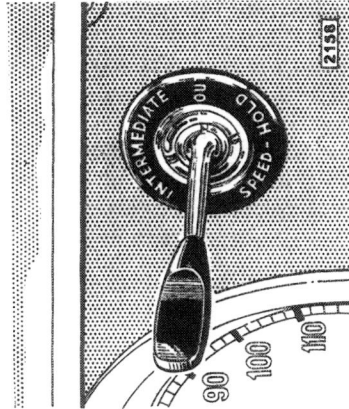

Intermediate Speed Hold. A switch mounted on the facia provides a means for the driver to obtain a downshift from direct to intermediate without depressing the accelerator pedal (as advised under the heading "Additional Power and Acceleration") and to retain the drive in the intermediate range. This will be found convenient for overtaking or when hill climbing.

With the switch in the "IN" position no upshift will take place between intermediate and direct drive; placing the switch lever in the "OUT" position will cause the transmission to shift to direct drive, provided the normal upshift speed has been obtained.

Warning. Do NOT allow the maximum permitted engine revolutions to be exceeded through allowing the "Intermediate Speed Hold" to remain in operation longer than necessary, or by switching in the "Hold" at speeds in excess of 75 m.p.h. (121 k.p.h.).

Additional Power and Acceleration in D range can be obtained as follows:—

(a) Below 52 m.p.h. (84 k.p.h.) depress the accelerator pedal to the full throttle position to effect a change into the intermediate range; the drive will continue in the intermediate range until the release of the accelerator or approximately 64 m.p.h. (103 k.p.h.) is reached.

(b) Between 52 m.p.h. and 68 m.p.h. (84 k.p.h. and 109 k.p.h.) depress the accelerator pedal all the way to the floorboard to effect a "kickdown" change into

ROUTINE MAINTENANCE

The fluid necessary for the operation of the torque converter is common with that used in the transmission. The total capacity of the transmission assembly is approximately 15 Imperial pints (18 U.S. pints; 8.5 litres), but when draining the transmission a small quantity of fluid will remain in the unit and the amount required to refill it will be that needed to bring the fluid level to the FULL mark on the dipstick as described in "Drain and Refill Transmission".

Every 3,000 miles (5,000 km.)

Check Transmission Fluid Level

1. Raise the bonnet. The dipstick will be found forward of the carburetter adjacent to the radiator top water hose.

2. With the car on a level floor, set the handbrake firmly. Set the selector lever in the P position and start the engine. With the footbrake applied move the selector lever to L and raise the transmission fluid temperature by running the engine at 800 r.p.m. for 2 or 3 minutes.

3. Clean the end of the filler tube. Remove the dipstick and wipe it dry. With the foot still on the brake and the selector lever at L run the engine at its normal idling speed and check the fluid level. Add sufficient fluid to bring the level up to the "Full" mark on the dipstick. DO NOT OVERFILL. The space between the "Full" and "Low" marks on the dipstick represents approximately one pint.

Every 12,000 miles (20,000 km.)

Drain and Refill Transmission

1. Raise the bonnet. The dipstick will be found forward of the front carburetter adjacent to the radiator top water hose.

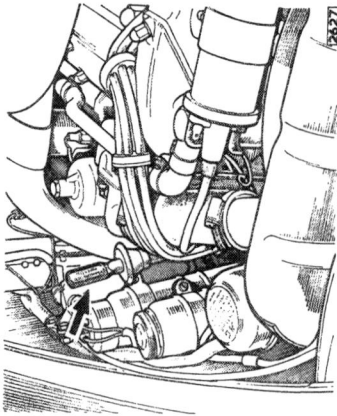

Fig. 32. *Automatic transmission dipstick.*

2. With the car on a level floor, set the handbrake firmly. Set the selector lever in the P position and start engine. With the footbrake applied move the selector lever to L and raise the transmission fluid temperature by running the engine at 800 r.p.m. for two or three minutes.

3. Stop the engine. Clean the end of the filler tube.

4. Remove the transmission oil pan drain plug. (A, Fig. 33).

5. Remove the converter housing cover plate and rotate the converter until the drain plug is in position for draining. Remove the converter drain plug. (B).

6. To facilitate draining, remove the square-headed converter pressure take-off plug from the bottom of the housing attached to the left-hand side of the transmission casing (C).

7. After fluid has drained, refit and tighten the drain plugs in the transmission oil pan and converter. Refit the converter housing cover plate. Refit and tighten the converter pressure take-off plug.

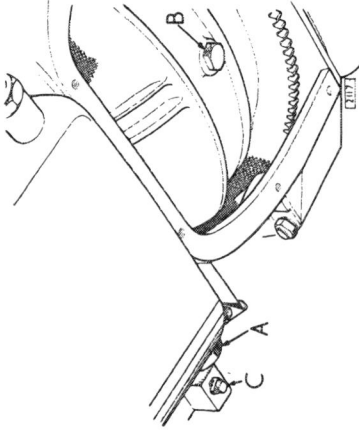

Fig. 33. *Automatic transmission drain plug. (The converter housing cover plate has been removed).*

8. Pour 10 Imperial pints (12 U.S. pints; 5.7 litres) of the recommended grade of fluid into the transmission through the filler tube.

9. Set the selector lever in the P position and start engine. With the footbrake applied move the selector lever to L and run the engine at 800 r.p.m. for two or three minutes to transfer fluid from the transmission case to the converter.

10. With the foot still on the brake and the selector lever at L run the engine at its normal idling speed and add additional fluid (approximately 5 Imperial pints; 6 U.S. pints or 2.8 litres) to bring the level up to the "Full" mark on the dipstick. DO NOT OVERFILL.

RECOMMENDED LUBRICANTS

Component	MOBIL	CASTROL	SHELL	ESSO	B.P.	DUCKHAM	REGENT Caltex/Texaco
Multigrade engine oils	Mobiloil Special*	Castrol GTX	Shell Super Oil	Uniflo	Super Visco-Static	Q Motor Oil	Havoline 20W/40 or 10W/30*
Upper cylinder lubrication	Mobil Upperlube	Castrollo	Shell U.C.I. or Donax U.	Esso U.C.I.	Energol U.C.I.	Adcoid Liquid	Regent U.C.I.
Gearbox (Early cars), Distributor oil can points, Oil can lubrication	Mobiloil A	Castrol GTX	Shell X-100 30	Uniflo	Energol SAE 30	Q Motor Oil	Havoline 30
Rear axle, Gearbox (All synchromesh)	Mobilube GX 90	Castrol Hypoy	Spirax 90 EP	Esso Gear Oil GP 90/140	Gear Oil SAE 90 E.P.	Hypoid 90	Multigear Lubricant EP 90
Steering box (standard steering)	Mobilube C140	Castrol D	Spirax 140 E P	Esso Gear Oil GP90/140	Gear Oil SAE140E.P.	NOL EP 140	Multigear Lubricant EP 140
Front wheel bearings, Rear wheel bearings, Distributor cam	Mobilgrease MP	Castrolease LM	Retinax A	Esso Multi-purpose Grease H	Energrease L.2	LB 10	Marfak All-purpose
Steering idler housing, Steering tie-rods, Wheel swivels, Door hinges	Mobilgrease MP	Castrolease LM	Retinax A	Esso Multi-purpose Grease H	Energrease L.2	LB 10	Marfak All-purpose
Automatic transmission unit, Power steering system	ATF 210	Castrol T.Q.F.	Shell Donax T7	Esso Glide	Autran B	Q-matic	Texamatic Type F

*These oils should NOT be used in worn engines requiring overhaul.
If an SAE 30 or 40 has previously been used in the engine a slight increase in oil consumption may be noticed but this will be compensated by the advantages gained.

RECOMMENDED HYDRAULIC FLUID.

Braking System and Clutch Operation.

Castrol/Girling Crimson Clutch/Brake Fluid is recommended. This conforms to S.A.E. 70 R3 specification modified to give a higher boiling point for additional safety.

Where this is not available, only fluid guaranteed to conform to S.A.E. 70 R3 specification, may be used as an alternative.

SUMMARY OF MAINTENANCE

Daily
Check radiator water level.
Check engine oil level.

Weekly
Check tyre pressures (including spare wheel).

Monthly
Check battery electrolyte level and connections.

Every 3,000 miles (5,000 km.)
Drain engine sump and refill.
Check radiator water level.
Clean oil filter element and replace seal.
Check tyre pressures (including spare wheel).
Check battery electrolyte level and connections.
Check gearbox oil level.
Check fluid level in automatic transmission unit (if fitted).
Check rear axle oil level.
Check fluid level in power-assisted steering reservoir (if fitted).
Check fluid level in brake and clutch master cylinder reservoirs.
Clean, adjust and test sparking plugs.
Lubricate distributor and check contact points.
Top up carburetter hydraulic piston dampers.
Check clutch pedal free travel (early cars).
Check carburetter slow running.

Every 6,000 miles (10,000 km.)
Carry out 3,000 mile (5,000 km.) service.
Top up steering unit (standard steering).
Lubricate all grease nipples (excluding wheel bearings).
Renew oil filter element and seal.
Tune carburetters.
Clean carburetter filters.
Clean fuel feed line filter.
Clean and lubricate brake servo air cleaner (use brake fluid).
Examine brake friction pads for wear.
Clear drain holes in bottoms of doors.
Adjust top timing chain (if necessary).
Check front wheel alignment.
Check fan belt for wear.

Carry out oil can lubrication of (a) seat runner and adjusting mechanism (b) handbrake lever ratchet (c) door locks (d) luggage compartment hinges and lock (e) bonnet hinges and catches (f) windscreen wiper arms (g) accelerator and carburetter linkage (h) fuel filler cover hinges (i) generator end bush.

Every 12,000 miles (20,000 km.)
Carry out 3,000 and 6,000 miles (5,000 and 10,000 km.) service.
Drain and refill gearbox (and overdrive if fitted).
Clean overdrive oil pump filter (if overdrive fitted).
Drain and refill rear axle.
Drain and refill automatic transmission unit (if fitted).
Renew air cleaner element.
Renew steering oil reservoir filter (power-assisted steering only).
Renew sparking plugs.
Lubricate front and rear wheel bearing hubs.
Check front and rear wheel bearings for end-float (additional charge for adjustment).
Check exhaust system for leaks.
Check and tighten all chassis and body nuts, bolts and screws.
Drain engine sump and refill.
Renew oil filter element and seal.

SPECIAL SERVICE TOOLS

Home distributors and dealers should obtain the Churchill Service Tools (that is, tools which have no asterisk preceding the number) mentioned in this manual direct from Messrs. V. L. Churchill & Co. Limited at the address given below. Overseas distributors should order their requirements through the Spares Division, Jaguar Cars Limited, Coventry, England.

> Messrs. V. L. Churchill & Co. Limited,
> London Road,
> Daventry,
> Northants.

* Tools with an asterisk preceding the number are available only from the Jaguar Spares Division.

Tool No.

ENGINE

Upper timing chain adjuster	J.2
Engine lifting plate	J.8
Crankshaft rear oil seal sizing tool	J.17
Valve guide bore reamer	J.18
Valve spring compressor	J.6118
Valve timing gauge (supplied in tool kit)	*C.4015
Piston ring compressor	38.U.3

CLUTCH

Universal checking fixture	99

OVERDRIVE

Drive shaft oil seal remover	L.176
Drive shaft oil seal replacer and cone clutch and spring thrust housing dismantling tool	L.177
Freewheel assembly ring	L.178
Piston and ring fitting tool (3.4 litre model)	L.179
Piston and ring fitting tool (3.8 litre model)	L.180
Accumulator 'O' ring fitting tool (3.4 litre model)	L.181
Accumulator 'O' ring fitting tool (3.8 litre model)	L.217
Accumulator piston housing remover (3.4 litre model)	L.182
Accumulator piston housing remover (3.4 litre model)	L.182
Pump barrel remover	L.183
Pump barrel replacer	L.184
Dummy drive shaft	L.185
Mainshaft bearing replacer	L.186
Hydraulic test equipment	L.188
Drain plug remover	J.3

SPECIAL SERVICE TOOLS

REAR AXLE

Pinion bearing cone removal/replacing adaptor	SL.14-1†
Pinion bearing outer bearing cup replacing adaptor	SL.550-4‡
Pinion bearing inner bearing cup replacing adaptor	SL.550-5‡
Differential bearing cone removal adaptor	SL.14-3†
Differential bearing cone replacing adaptor	SL.550-1‡
Pinion cone setting gauge	SL.3
Hub end-float spacer	J.15
Hub end-float dial gauge	J.13
Hub extractor (disc wheel hubs)	J.1C
Hub extractor (wire wheel hubs)	J.7
Hub outer bearing cup removal adaptor	J.16B†
Hub bearing cup replacing adaptor	J.20A-1§

STEERING

Ball joint separator	JD.24

FRONT SUSPENSION

Front coil spring compressor	JD.6A

REAR SUSPENSION

Rear coil spring compressor adaptor	J.11A†
Rear wishbone dummy shaft	J.14 (2 off)
Radius arm bush removal/replacing tools	J.21
Rear suspension setting link (for camber checking)	J.24

BRAKES

Piston resetting tool	*10416

† Use with main tool SL.14 — Multi-purpose hand press.
‡ Use with main tool 550 — Multi-purpose handle.
§ Use with main tool J.20A — Bearing remover.

SERVICE DEPARTMENTS

Factory:

THE SERVICE DIVISION

JAGUAR CARS LIMITED, COVENTRY

Telephone No. Allesley 2121 (P.B.X.)

U.S.A.

The Technical Service Department
Jaguar Cars Inc.
54-20 Broadway, Woodside,
New York 11377, U.S.A.

Canada:

The Technical Service Department
Jaguar Cars (Canada) Ltd.
8505 Delmeade Road, Montreal 9
Quebec, Canada

London:

Messrs. Henlys Ltd.,
The Hyde,
Hendon,
London, N.W.9.
Telephone No. Colindale 6565

CONVERSION TABLES

METRIC INTO ENGLISH MEASURE

1 millimetre is approximately 1/25″, and is exactly ·03937″.
1 centimetre is approximately ⅜″, and is exactly ·3937″.
1 metre is approximately 39⅜″, and is exactly 39·37″ or 1·0936 yards.
1 kilometre is approximately ⅝ mile, and is exactly ·6213 miles.
1 kilomgramme is approximately 2¼ lbs., and is exactly 2·21 lbs.
1 litre is approximately 1¾ pints, and is exactly 1·76 pints.
To convert metres to yards, multiply by 70 and divide by 64.
To convert kilometres to miles, multiply by 5 and divide by 8 (approx.).
To convert litres to pints, multiply by 88 and divide by 50.
To convert grammes to ounces, multiply by 20 and divide by 567.
To find the cubical contents of a motor cylinder, square the diameter (or bore), multiply by 0·7854, and multiply the result by the stroke.
1 M.P.G.—0·3546 kilometres per litre or 2·84 litres per kilometre.

MILES INTO KILOMETRES

Kilo.	Miles	Kilo.	Miles	Kilo.	Miles	Kilo.	Miles	Kilo.	Miles
1	⅝	16	10	31	19¼	46	28⅝	60	37¼
2	1¼	17	10⅝	32	19⅞	47	29¼	70	43½
3	1⅞	18	11¼	33	20½	48	29⅞	80	49¾
4	2½	19	11¾	34	21⅛	49	30½	90	55⅞
5	3⅛	20	12⅜	35	21¾	50	31⅛	100	62⅛
6	3¾	21	13	36	22⅜	51	31¾	200	124¼
7	4⅜	22	13⅝	37	23	52	32¼	300	186⅜
8	5	23	14¼	38	23⅝	53	32⅞	400	248½
9	5⅝	24	14⅞	39	24¼	54	33½	500	310¾
10	6¼	25	15½	40	24⅞	55	34⅛	600	372⅞
11	6⅞	26	16⅛	41	25½	56	34¾	700	435
12	7½	27	16¾	42	26⅛	57	35⅜	800	497½
13	8⅛	28	17⅜	43	26¾	58	36	900	559¼
14	8¾	29	18	44	27⅜	59	36⅝	1000	621⅜
15	9⅜	30	18⅝	45	28				

PINTS AND GALLONS TO LITRES

Pints	Gallons	Litres Approx.	Litres Exact	Pints	Gallons	Litres Approx.	Litres Exact
1	⅛	½	·57	40	5	23	22·75
2	¼	1	1·14	48	6	27	27·30
3	⅜	1½	1·71	56	7	32	31·85
4	½	2¼	2·27	64	8	36½	36·40
8	1	4½	4·54	72	9	41	40·95
16	2	9	9·10	80	10	45½	45·50
24	3	13½	13·65	88	11	50	50·05
32	4	18	18·20	96	12	54½	54·60

CONVERSION TABLES

RELATIVE VALUE OF MILLIMETRES AND INCHES

mm.	Inches	mm.	Inches	mm.	Inches	mm.	Inches
1	0·0394	26	1·0236	51	2·0079	76	2·9922
2	0·0787	27	1·0630	52	2·0473	77	3·0315
3	0·1181	28	1·1024	53	2·0866	78	3·0709
4	0·1575	29	1·1417	54	2·1260	79	3·1103
5	0·1968	30	1·1811	55	2·1654	80	3·1496
6	0·2362	31	1·2205	56	2·2047	81	3·1890
7	0·2756	32	1·2598	57	2·2441	82	3·2284
8	0·3150	33	1·2992	58	2·2835	83	3·2677
9	0·3543	34	1·3386	59	2·3228	84	3·3071
10	0·3937	35	1·3780	60	2·3622	85	3·3465
11	0·4331	36	1·4173	61	2·4016	86	3·3859
12	0·4724	37	1·4567	62	2·4410	87	3·4252
13	0·5118	38	1·4961	63	2·4803	88	3·4646
14	0·5512	39	1·5354	64	2·5197	89	3·5040
15	0·5906	40	1·5748	65	2·5591	90	3·5433
16	0·6299	41	1·6142	66	2·5984	91	3·5827
17	0·6693	42	1·6536	67	2·6378	92	3·6221
18	0·7087	43	1·6929	68	2·6772	93	3·6614
19	0·7480	44	1·7323	69	2·7166	94	3·7008
20	0·7874	45	1·7717	70	2·7559	95	3·7402
21	0·8268	46	1·8110	71	2·7953	96	3·7796
22	0·8661	47	1·8504	72	2·8347	97	3·8189
23	0·9055	48	1·8898	73	2·8740	98	3·8583
24	0·9449	49	1·9291	74	2·9134	99	3·8977
25	0·9843	50	1·9685	75	2·9528	100	3·9370

RELATIVE VALUE OF INCHES AND MILLIMETRES

Inches	0	1/16	1/8	3/16	1/4	5/16	3/8	7/16
0	0·0	1·6	3·2	4·8	6·4	7·9	9·5	11·1
1	25·4	27·0	28·6	30·2	31·7	33·3	34·9	36·5
2	50·8	52·4	54·0	55·6	57·1	58·7	60·9	61·9
3	76·2	87·8	79·4	81·0	82·5	84·1	85·7	87·3
4	101·6	103·2	104·8	106·4	108·0	109·5	111·1	112·7
5	127·0	128·6	130·2	131·8	133·4	134·9	136·5	138·1
6	152·4	154·0	155·6	157·2	158·8	160·3	161·9	163·5

Inches	1/2	9/16	5/8	11/16	3/4	13/16	7/8	15/16
0	12·7	14·3	15·9	17·5	19·1	20·6	22·2	23·8
1	38·1	39·7	41·3	42·9	44·4	46·0	47·6	49·2
2	63·5	65·1	66·7	68·3	69·8	71·4	73·0	74·6
3	88·9	90·5	92·1	93·7	95·2	96·8	98·4	100·0
4	114·3	115·9	117·5	119·1	120·7	122·2	123·8	125·4
5	139·7	141·3	142·9	144·5	146·1	147·6	149·2	150·8
6	165·1	166·7	168·3	166·9	171·5	173·0	174·6	176·2

SECTION B

ENGINE

ADDENDA

The following corrections should be noted when using SECTION B.

Page	Correction
B.15, B.17	"ensure that the circular rubber seal in the filter head has not become displaced" should read "renew the circular rubber seal in the filter head".
B.39, B.40, B.52, B.56	"Remove sump as described on page B.28" should read "Remove sump as described on page B.55".
B.61	"Remove camshafts as described on page B.33." should read "Remove camshafts as described on page B.40."

INDEX

INDEX (continued)

INDEX (continued)

ENGINE

Both the 3.4 litre and 3.8 litre 'S' models are fitted with twin overhead camshaft ($\frac{3}{8}$" lift) XK type engine with the 'B' type cylinder head.

The engine number will be observed stamped on the right-hand side of the cylinder block above the oil filter and at the front of the cylinder head casting.

Model	Compression Ratio	Commencing Engine Number and Prefix	Colour of Cylinder Head
3.4 litre	7 : 1, 8 : 1 or 9 : 1	7B 1001	Light Blue
3.8 litre	7 : 1, 8 : 1 or 9 : 1	7B 50001	Dark Blue

Compression ratios of 7 to 1, 8 to 1 or 9 to 1 are specified for both the 3.4 and 3.8 litre engines, the differences in compression ratio being provided by varying the crown design of the piston.

The compression ratio of an engine is indicated by /7, /8 or /9 following the engine number.

DATA

CAMSHAFT

Number of journals	..	..	..	..	Four per shaft
Journal diameter	..	..	..	..	$1.00''$—$.0005''$
					—$.001''$
					(25.4 mm.—.013 mm.
					—.025 mm.)
Thrust taken	..	..	..	..	Front end
Number of bearings	..	..	..	..	Four per shaft (eight half bearings)
Type of bearing	..	..	..	..	White metal steel backed shell
Diameter clearance	..	..	..	..	.0005" to .002"
					(.013 to .05 mm.)
Permissible end float	..	..	..	..	.0045" to .008"
					(.11 to .20 mm.)
Tightening torque—Bearing cap nuts	..	..	..	15 lb. ft. (175 lb. ins.)
					(2.0 kg./m.)

CONNECTING ROD

Length centre to centre	7¾" (19.68 cm.)
Big end—Bearing type	Lead bronze steel backed shell, lead indium coated
Bore for big end bearing	2.233" to 2.2335" (56.72 to 56.73 mm.)
Big end—Width	$1\frac{3}{16}"$ $+.006"$ $-.008"$ (30.16 mm. —.15 mm. —.20)
Big end—Diameter clearance	.0015"—.0033" (.04 mm. to .08 mm.)
Big end—Side clearance	.0058" to .0087" (.15 mm. to .22 mm.)
Bore for small end bush	1.00" ±.0005" (25.4 mm. ±.013 mm.)
Small end bush—Type	Phosphor bronze—steel backed
Small end—Width	$1\frac{5}{64}"$ (27.4 mm.)
Small end bush—Bore diameter	.875" $+.0002"$ $-.0000"$ (22.22 mm. +.005 mm. —.000 mm.)
Tightening torque—Con rod bolts	37 lb. ft. (450 lb. ins.) 5.1 kg./m.

CRANKSHAFT

Number of main bearings	Seven
Main bearing—Type	White metal steel backed shell
Journal diameter	2.750" to 2.7505" (69.85 to 69.86 mm.)
Journal length	
Front	$1\frac{11}{16}"$ ±.005" (42.86 mm. ±.13 mm.)
Centre	$1\frac{3}{4}"$ $+.0005"$ $+.001"$ (44.45 mm. +.013 mm. +.025)
Rear	$1\frac{7}{8}"$ (47.63 mm.)
Intermediate	$1\frac{7}{32}"$ ±.002" (30.96 mm. ±.05 mm.)
Thrust taken	Centre bearing thrust washers
Thrust washer—Thickness	.092" ±.001" and .096" ±.001" (2.33 mm. ±.025 mm. and 2.43 mm. ±.025 mm.)
End clearance	.004" to .006" (.10 to .15 mm.)
Main bearing—Length	
Front	$1\frac{1}{2}"$ ±.005" (38.1 mm. ±.13 mm.)
Centre	
Rear	
Intermediate	1" ±.005" (25.4 mm. ±.13 mm.)
Diameter clearance	.0015" to .003" (.04 to .08 mm.)
Crankpin—Diameter	2.086" $+.0006"$ $-.000"$ (52.98 mm. +.015 mm. —.000 mm.)
Length	$1\frac{3}{16}"$ $+.0007"$ $-.0002"$ (30.16 mm. ±.018 mm. —.006 mm.)
Regrind undersize	.010", .020", .030" and .040" (.25, .51, .76 and 1.02 mm.)
Minimum diameter for regrind	—.040" (1.02 mm.)
Tightening torque—main bearing bolts	83 lb. ft. (1,000 lb. ins.) (11.5 kg./m.)

CYLINDER BLOCK

Material—3.4 litre	.. ..	Chromium iron
—3.8 litre	.. ..	"Brivadium" dry liners
Cylinder bores—Nominal—3.4 litre	..	83 mm. +.0127 mm. (3.2677" +.0005") −.0064 mm. −.00025"
—3.8 litre	.. ..	87 mm. +.0127 mm. (3.4252" +.0005") −.0064 mm. −.00025"
Maximum rebore size	.. .. ..	+.030" (.76 mm.)
Bore size for fitting liners—3.4 litre	..	3.391" to 3.392" (86.13 to 86.16 mm.)
—3.8 litre	..	3.561" to 3.562" (90.45 mm. to 90.47)
Outside diameter of liner—3.4 litre	..	3.3945" to 3.3955" (86.22 to 86.25 mm.)
—3.8 litre	..	3.563" to 3.566" (90.50 to 90.58 mm.)
Interference fit	.. ..	.001" to .005" (.02 to .125 mm.)
Overall length of liner—3.4 and 3.8 litre	..	(6 31/32") (17.7 cm.)
Outside diameter of lead-in—3.4 litre	..	3.389" to 3.391" (86.08 to 86.13 mm.)
—3.8 litre	..	3.558" to 3.560" (90.37 to 90.42 mm.)
Size of bore honed after assembly—3.4 litre ..		83 mm. (3.2677")
in cylinder block—Nominal—3.8 litre ..		87 mm. (3.4252")
Main line bore for main bearings	.. ..	2.9165" +.0005" −.0000" (74.08 +.013 mm. −.000 mm.)

CYLINDER HEAD

Type	.. ..	'B ' Type
Material	.. ..	Aluminium Alloy
Valve seat angle—Inlet	.. ..	45°
—Exhaust	.. ..	45°

Valve throat diameter—Inlet	.. ..	1½" (38.1 mm.)
—Exhaust	..	1⅜" (34.9 mm.)
Tightening torque—Cylinder head nuts	..	54 lb. ft. (650 lb. ins.) (7.5 kg./m.)
Firing order	.. ..	1, 5, 3, 6, 2, 4 No. 1 cylinder being at the rear of the engine unit

GUDGEON PIN

Type	.. ..	Fully floating
Length	.. ..	2.840" to 2.845" (72.14 to 72.26 mm.)
Inside diameter	.. .. ..	⅝" (15.87 mm.)
Outside diameter	.. ..	.8750" to .8752" (22.22 to 22.23 mm.)

LUBRICATING SYSTEM

Oil pressure (hot)	.. .. ..	40 lb. per sq. in. at 3,000 r.p.m.
Oil pump—Type	.. .. ..	Eccentric rotor
—Clearance at end of lobes	..	.010" maximum (.25 mm.)
—End clearance	.. ..	.004" maximum (.10 mm.)
—Clearance between outer rotor and body	.. ..	.010" maximum (.25 mm.)

PISTON AND PISTON RINGS

Make	.. ..	Brico
Type	.. ..	Semi-split skirt
Piston		
Skirt clearance (measured at bottom of skirt at 90° to gudgeon pin axis)	.. ..	.0011" to .0017" (.028 to .043 mm.)
Gudgeon pin bore	.. .. ..	.8749" to .8751" (2.223 to 2.227 mm.)

COMPRESSION HEIGHT

	3.4 litre	3.8 litre
7 : 1 compression ratio	1.690" to 1.695" (42.93 to 45.05 mm.)	1.846" to 1.841" (46.76 to 46.89 mm.)
8 : 1 compression ratio	2.163" to 2.168" (54.94 to 55.067 mm.)	2.069" to 2.064" (52.42 to 52.55 mm.)
9 : 1 compression ratio	2.258" to 2.263" (57.35 to 57.48 mm.)	2.247" to 2.242" (56.94 to 57.07 mm.)

Piston rings—Number

Compression	2
Oil control	1

Piston rings—Width

Compression	.0772" to .0777"
Oil Control	.150" to .154"

Piston rings—Thickness

Compression	.132" to .139"
Oil control	.119" to .127" (3.02 to 3.23 mm.)

Piston rings—Side clearance in groove

Compression	.001" to .003" (.02 to .07 mm.)
Oil Control	.001" to .003" (.02 to .07 mm.)

Piston rings—Gap when fitted to cylinder bore

Compression	.015" to .020" (.38 to .51 mm.)
Oil control	.015" to .033" (.38 to .83 mm.)

SPARKING PLUGS

Make	Champion

Type

7 : 1 compression ratio	UN.12.Y
8 : 1 compression ratio	UN.12.Y
9 : 1 compression ratio	UN.12.Y
Gap	.025" (.64 mm.)

TAPPETS AND TAPPET GUIDES

Tappet—Material	Cast iron (chilled)
—Outside diameter	1.3738" to 1.3742" (34.89 to 34.90 mm.)
—Diameter clearance	.0008" to .0019" (.02 to .048 mm.)
Tappet guide—Material	Austenitic iron
—Inside diameter (before reaming)	1.353" to 1.357" (34.37 to 34.48 mm.)
—Reaming size (when fitted to cylinder head)	1.375" +.0007" −.0000" (34.925 mm. +.018 mm. −.000 mm.)
—Interference (shrink) fit in head	.003" (.07 mm.)

TIMING CHAINS AND SPROCKETS

Type	Duplex
Pitch	⅜" (9.5 mm.)
Number of pitches—Top chain	100
—Bottom chain	82
Crankshaft sprocket—Teeth	21
Intermediate sprocket, outer—Teeth	28
Intermediate sprocket, inner—Teeth	20
Camshaft sprocket—Teeth	30
Idler Sprocket	21

VALVE TIMING

Inlet valve opens	15° B.T.D.C.
Inlet valve closes	57° A.B.D.C.
Exhaust valve opens	57° B.B.D.C.
Exhaust valve closes	15° A.T.D.C.

(with valve clearances set at .010" [.25 mm.])

VALVES AND VALVE SPRINGS

Valves—Material, Inlet	Silicon chrome steel
Exhaust	21 - 4 - NS
Valve head diameter, Inlet	1¾" ± .002" (44.45 mm. ± .05 mm.)
Exhaust	1⅝" ± .002" (41.27 mm. ± .05 mm.)
Valve stem diameter, Inlet and Exhaust	5/16" −.0025"/−.0035" (7.95 mm. −.06 mm./−.09 mm.)
Valve lift	3/8" (9.5 mm.)
Valve clearance—Inlet	.004" (.10 mm.)
—Exhaust	.006" (.15 mm.)
Valve seat angle—Inlet	45°
—Exhaust	45°
Valve spring—Free length. Inner	1⅝" (.42 mm.)
Outer	1 15/16" (49.2 mm)
Valve spring—Fitted length. Inner	1⅞" (30.96 mm.)
Outer	1 5/16" (33.34 mm.)
Valve spring—Fitted load. Inner	30.33 lb. (13.76 kg.)
Outer	48.375 lb. (21.94 kg.)
Valve spring—Solid length (max.) Inner	.810" (20.57 mm.)
Outer	.880" (22.35 mm.)
Number of free coils. Inner	6
Outer	5
Diameter of wire. Inner	12 SWG (.104") (2.64 mm.)
Outer	10 SWG (.128") (3.25 mm.)

VALVE GUIDE AND VALVE SEAT INSERT

Valve Guides—Material	Cast iron
Valve Guide—Length. Inlet	1 13/16" (46.04 mm.)
Exhaust	1 15/16" (49.21 mm.)
Valve Guide—Inside diameter. Inlet	5/16" −.0005"/−.0015" (7.94 mm. −.013 mm./−.038 mm.)
Exhaust	5/16" ± .0005" (7.94 mm. ± .01 mm.)
Interference fit in head	.0005" to .0022" (.013 to .055 mm.)
Valve Seat inserts—Material	Cast iron (centrifugally cast)
Inside diameter. Inlet	1½" +.003"/−.001" (38.1 +.076 mm./−.025 mm.)
Exhaust	1.379" to 1.383" (35.03 to 35.13 mm.)
Interference (shrink) fit in head	.003" (.076 mm.)

RECOMMENDED LUBRICANTS

Component	MOBIL	CASTROL	SHELL	ESSO	B.P.	DUCKHAM	REGENT (Caltex/Texaco)
Multigrade engine oils	Mobiloil Special	Castrol GTX	Shell Super Oil	Uniflo	Super Visco-Static	Q Motor Oil	Havoline 20W/40 or 10W/30
Upper cylinder lubrication {	Mobil Upperlube	Castrollo	Shell U.C.L. or Donax U.	Esso U.C.L.	Energol U.C.L.	Adcoid Liquid	Regent U.C.L.

*These oils should NOT be used in worn engines requiring overhaul.

If an SAE 30 or 40 has previously been used in the engine a slight increase in oil consumption may be noticed but this will be compensated by the advantages gained.

ROUTINE MAINTENANCE

Daily

CHECKING THE ENGINE OIL LEVEL

Check the oil level with the car standing on level ground otherwise a false reading will be obtained.

Remove the dipstick and wipe it dry. Replace and withdraw the dipstick; if the oil level is on the knurled patch, with the engine hot or cold, no additional oil is required. If the engine has been run immediately prior to making an oil level check, wait one minute after switching off before checking the oil level.

Note: Almost all modern engine oils contain special additives, and whilst it is permissible to mix the recommended brands it is undesirable. If it is desired to change from one brand to another this should be done when the sump is drained, and the oil company's recommendation in regard to flushing procedure should be followed.

Every 3,000 miles (5,000 km.)

CHANGING THE ENGINE OIL

Note: Under certain adverse operating conditions, conducive to oil dilution and sludge formation, more frequent oil changing than the normal 3,000 mile (5,000 km.) period is advised. Where the car is used mainly for low-speed city driving, stop-start driving particularly in cold weather or in dusty territory the oil should be changed at least every 1,000 miles (1,600 km.).

The draining of the sump should be carried out at the end of a run when the oil is hot and therefore will flow more freely. The drain plug is situated at the right-hand rear corner of the sump. When the engine oil is changed, the oil filter which is situated on the right-hand side of the engine, must also receive attention.

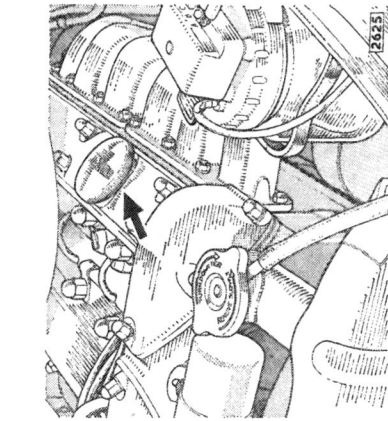

Fig. 1. *Engine oil filler.*

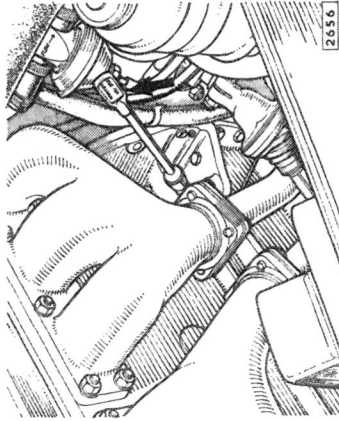

Fig. 2. *Engine oil dipstick.*

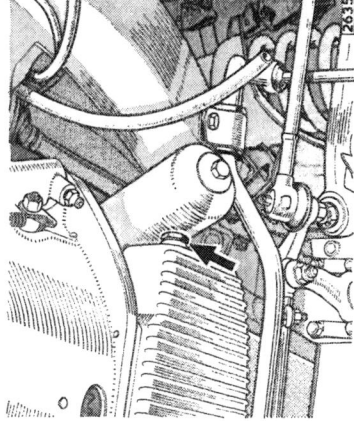

Fig. 3. *Engine drain plug.*

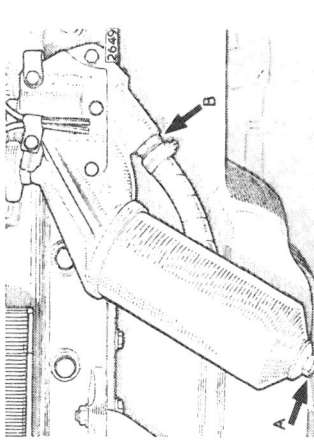

Fig. 4. *The engine oil filter is removed by unscrewing the bolt "A" and withdrawing the canister and element. The oil pressure relief valve is situated behind the outlet "B".*

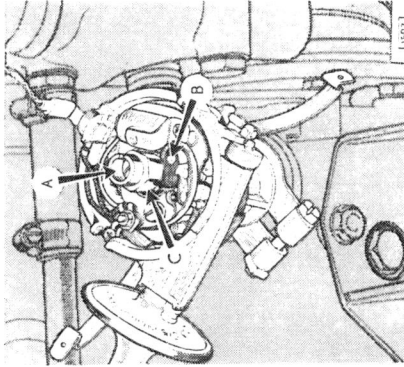

Fig. 5. *Distributor lubrication points.*

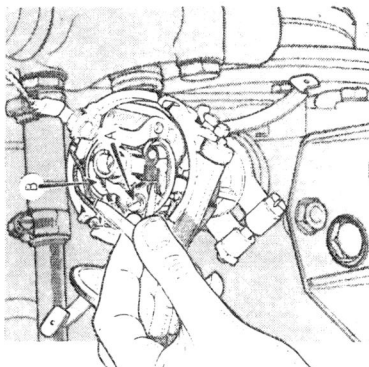

Fig. 6. *Checking the gap between the distributor contact points. The screw "A" secures the fixed contact point, the contact gap is adjusted by turning a screwdriver in the slot "B" in the contact plate.*

Unscrew the central bolt and remove the canister and element. Thoroughly wash these parts in petrol and allow to dry out. When replacing the canister ensure that the circular rubber seal in the filter head has not become displaced. (Attention is drawn to the importance of renewing the filter element at 6,000 miles [10,000 km.] intervals).

Note: Almost all modern engine oils contain special additives, and whilst it is permissible to mix the recommended brands it is undesirable. If it is desired to change from one brand to another this should be done when the sump is drained, and the oil company's recommendation in regard to flushing procedure should be followed.

DISTRIBUTOR—LUBRICATION

Take great care to prevent oil or grease from getting on or near the contact breaker points.

Remove the moulded cap at the top of the distributor by springing back the two clips. Lift off the rotor arm and apply a few drops of engine oil around the screw (A. Fig. 5) now exposed. It is not necessary to remove the screw as it has clearance to permit the passage of oil.

Apply **one** drop of oil to the post (B) on which the contact breaker pivots. Lightly smear the cam (C) with grease. Lubricate the centrifugal advance mechanism by injecting a few drops of engine oil through the aperture at the edge of the contact breaker base plate.

DISTRIBUTOR CONTACT BREAKER POINTS

Check the gap between the contact points with feeler gauges when the points are fully opened by one of the cams on the distributor shaft. A combined screwdriver and feeler gauge is provided in the tool kit.

The correct gap is .014″—.016″ (.36—.41 mm.).

CONTACT BREAKER ADJUSTMENT

If the gap is incorrect, slacken (very slightly) the contact plate securing screw (A. Fig. 6) and adjust the gap by turning a screwdriver in the nick in the contact plate and the slot in the base plate. (B). Turn clockwise to decrease the gap and anti-clockwise to increase. Tighten the securing screw and plate.

Examine the contact breaker points. If the contacts are burned or blackened, clean them with a fine carborundum stone or very fine emery cloth. Afterwards wipe away any trace of grease or metal dust with a petrol moistened cloth.

Cleaning of the contacts is made easier if the contact breaker lever carrying the moving contact is removed. To do this, remove the nut, insulating piece and connections from the post to which the end of the contact breaker spring is anchored. The contact breaker lever can now be lifted off its pivot post.

SPARKING PLUGS

Every 3,000 miles (5,000 km.) or more often if operating conditions demand, withdraw, clean and reset the plugs.

The only efficient way to clean sparking plugs is to have them properly serviced on machines specially designed for this purpose. These machines operate with compressed air and utilise a dry abrasive material specially graded and selected to remove harmful deposits from the plug insulator without damaging the insulator surface. In addition the majority of the machines incorporate electrical testing apparatus enabling the plugs to be pressure tested to check their electrical efficiency and gas tightness.

The gap between the points should be .025" (.64 mm.). When adjusting the gap always move the side wire—never bend the centre wire.

The Champion Sparking Plug Co. supply a special combination gauge and setting tool, the use of which is recommended.

Every 12,000 miles (20,000 km.) a new set of plugs of the recommended type should be fitted. To save petrol and to ensure easy starting, the plugs should be cleaned and tested regularly,

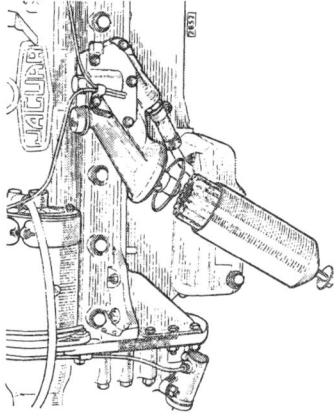

Fig. 7. *Normal condition.*

Fig. 8. *Oil fouling.*

Fig. 9. *Petrol fouling.*

Fig. 10. *Badly burned sparking plug.*

Every 6,000 Miles (10,000 km.)

OIL FILTER ELEMENT (Fig. 11)

It is most important to renew the oil filter element every 6,000 miles (10,000 km.) as after this mileage it will have become choked with impurities.

To guard against the possibility of the filter being neglected to the extent where the element becomes completely choked, a balance valve is incorporated in the filter head which allows **unfiltered** oil to by-pass the element and reach the bearings. This will be accompanied by a drop in the normal oil pressure of some 10 lb. per sq. in. and if this occurs the filter element should be renewed as soon as possible.

To gain access to the element, unscrew the central bolt (A) when the canister complete with the element can be removed. Thoroughly wash out the canister with petrol and allow to dry before inserting the new element.

When replacing the canister ensure that the circular rubber seal (B) in the filter head has not become displaced.

FAN BELT—CHECK FOR WEAR

Periodically check the condition of the fan belt. The belt is automatically tensioned correctly by means of a spring-loaded jockey pulley and routine adjustment is therefore unnecessary.

If the belt has to be replaced carry out the following procedure:—

Slacken the two bolts securing the dynamo to the mounting bracket. Remove the nut and unscrew the bolt securing the top link to the dynamo. Slacken the bolt securing the top link to the engine and press the dynamo as far as possible towards the engine. Remove the belt.

Place the new belt in position on the fan, jockey and crankshaft pulleys and by pressing the jockey pulley against the spring pass the belt over the dynamo pulley.

Pass the dynamo top securing bolt through the link and screw into the dynamo lug. Pull the dynamo away from the engine as far as possible, tighten the dynamo top securing bolt and replace the nut. Tighten the bolt securing the dynamo link to the engine and the two dynamo mounting bolts.

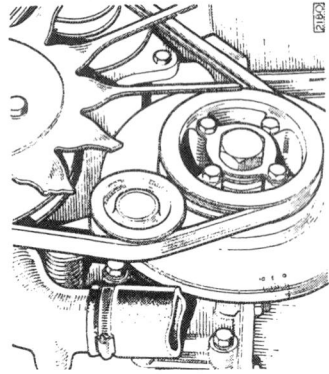

Fig. 11. *Removal of oil filter element.*

Fig. 12. *The automatic fan belt tensioner.*

TOP TIMING CHAIN TENSION

If the top timing chain is audible adjust the tension as follows:—

Slacken the locknut securing the serrated adjuster plate. Tension the chain by pressing the locking plunger inwards and rotating the adjuster plate in an anti-clockwise direction.

When correctly tensioned there should be slight flexibility on both outer sides of the chain below the camshaft sprockets, that is, the chain must not be dead tight. Release locking plunger, and securely tighten lock nut. Refit the breather housing.

This operation requires the use of a special tool to enable the adjuster plate to be rotated. To gain access to the adjuster plate remove the breather housing attached to the front face of the cylinder head.

Every 12,000 Miles (20,000 km.)

AIR CLEANER

The air cleaner is of the paper element type and is fitted on top of the cylinder head.

No maintenance is necessary but the element should be renewed every 12,000 miles (20,000 km.) or more frequently in dusty territories.

Roll back the sealing rubber between the carburetter elbow and the air cleaner. Slacken the two wing nuts securing the air cleaner to the bracket on the cylinder head. Release the air cleaner by pulling it towards the left hand wing valance. Release the two clips securing the end cover to the air cleaner. Withdraw the end cover and element. Remove the wing nut, washer, end cap and rubber ring securing the element to the end cover. When refitting the element ensure that the two rubber sealing rings are in their correct positions.

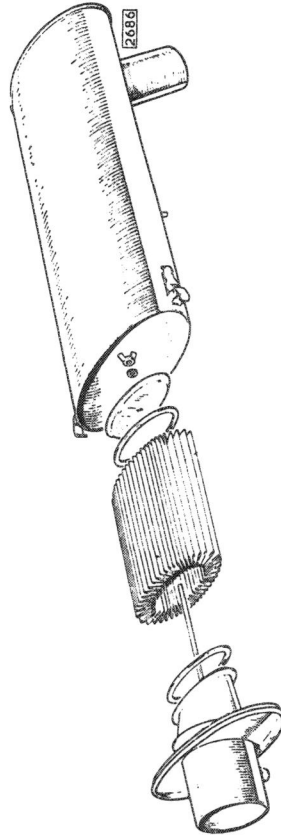

Fig. 13. *The air cleaner element.*

ENGINE DETAILS
(CYLINDER HEAD)

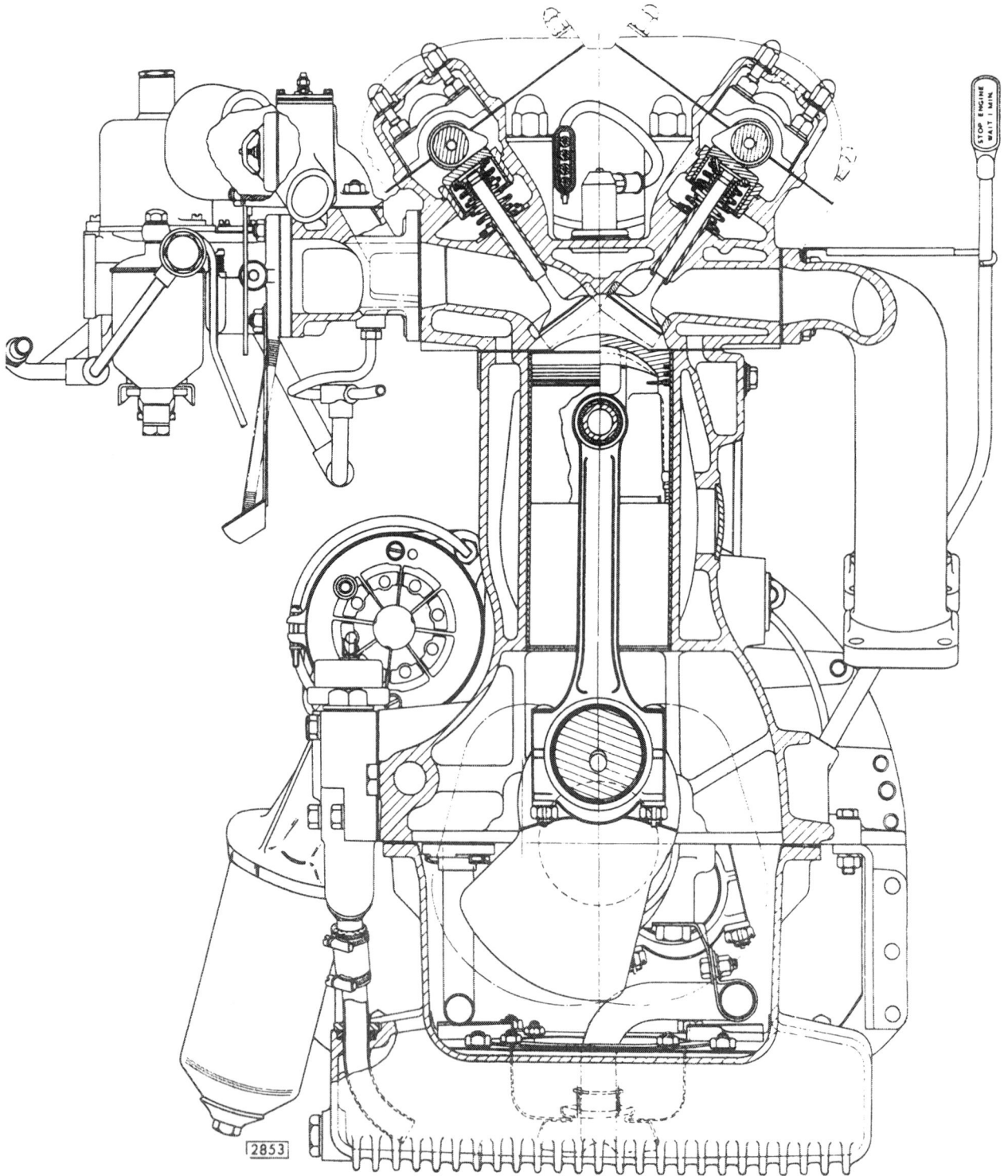

Fig. 15.   *Cross sectional view of engine.*

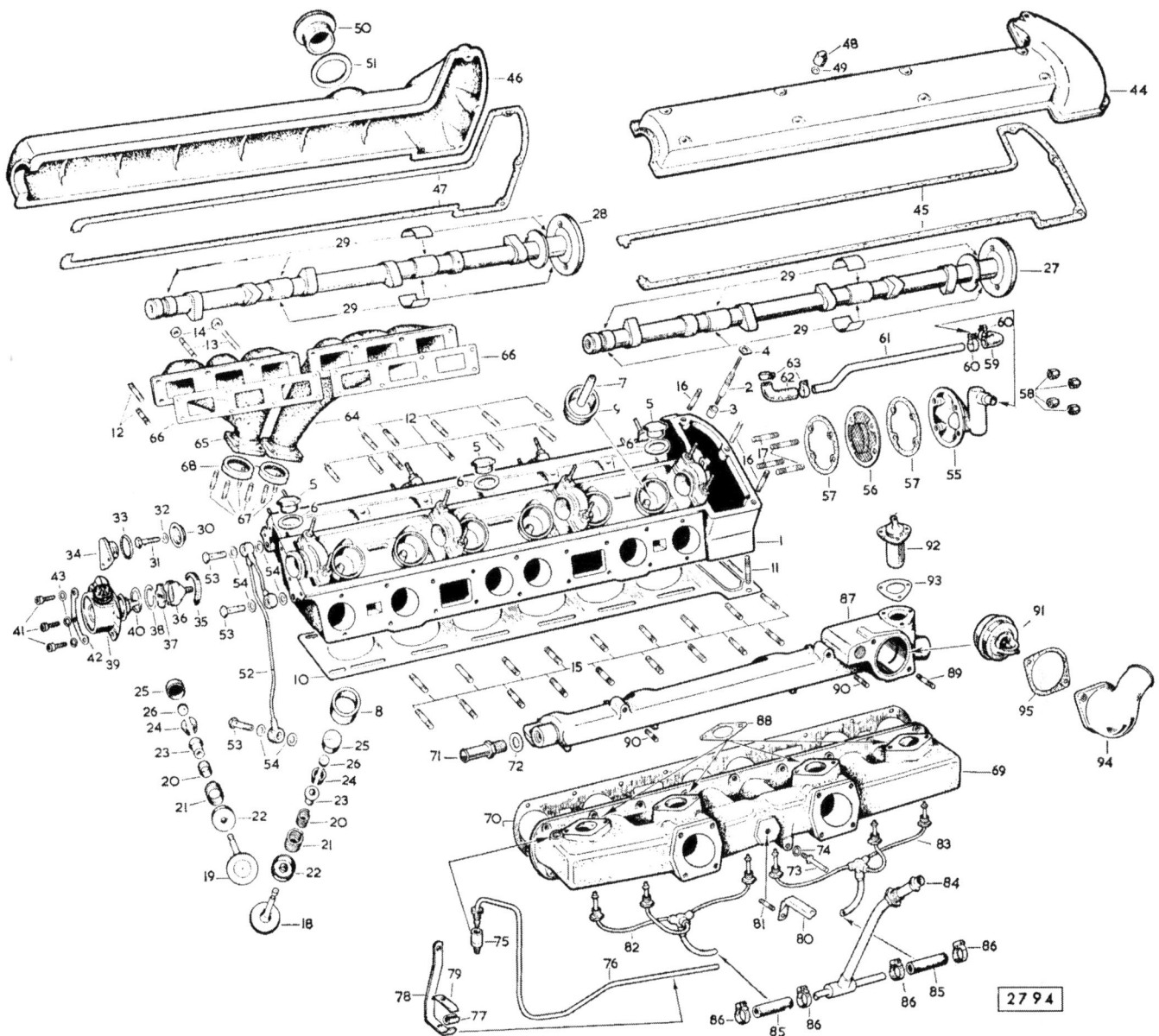

Fig. 14. *Cylinder head.*

1. Cylinder head	25. Tappet	49. Copper washer	73. Pivot pin
2. Stud	26. Adjusting pad	50. Oil filler cap	74. Spring washer
3. Ring dowel	27. Inlet camshaft	51. Fibre washer	75. Vacuum servo adaptor
4. "D" washer	28. Exhaust camshaft	52. Oil pipe	76. Vacuum servo pipe
5. Core plug	29. Bearing	53. Banjo bolt	77. Rubber sleeve
6. Copper washer	30. Oil thrower	54. Copper washer	78. Hanger bracket
7. Guide	31. Screw	55. Front cover and breather housing	79. Clamp
8. Insert	32. Copper washer	56. Gauze filter	80. Anti-creep throttle switch bracket
9. Guide	33. Sealing ring	57. Gasket	81. Stud
10. Gasket	34. Sealing plug	58. Dome nut	82. Starting pipe (L.H. Manifold)
11. Stud	35. Seal	59. Hose elbow	83. Starting pipe (R.H. Manifold)
12. Stud	36. Adaptor	60. Clip	84. Carburetter starting pipe
13. Stud	37. Driving dog	61. Breather pipe	85. Neoprene tube
14. Distance piece	38. Circlip	62. Hose elbow	86. Clip
15. Stud	39. Revolution counter generator	63. Clip	87. Water outlet pipe
16. Stud	40. "O" ring	64. Front exhaust manifold	88. Gasket
17. Stud	41. Screw	65. Rear exhaust manifold	89. Stud
18. Inlet valve	42. Plate washer	66. Exhaust gasket	90. Stud
19. Exhaust valve	43. Lock washer	67. Stud	91. Thermostat (water outlet pipe)
20. Valve spring (inner)	44. R.H. camshaft cover	68. Sealing ring	92. Thermostat (carburetter automatic
21. Valve spring (outer)	45. Gasket	69. Inlet manifold	choke)
22. Seat	46. L.H. camshaft cover	70. Gasket	93. Gasket
23. Collar	47. Gasket	71. Adaptor	94. Water outlet elbow
24. Cotter	48. Dome nut	72. Copper washer	95. Gasket

ENGINE DETAILS
(CYLINDER BLOCK)

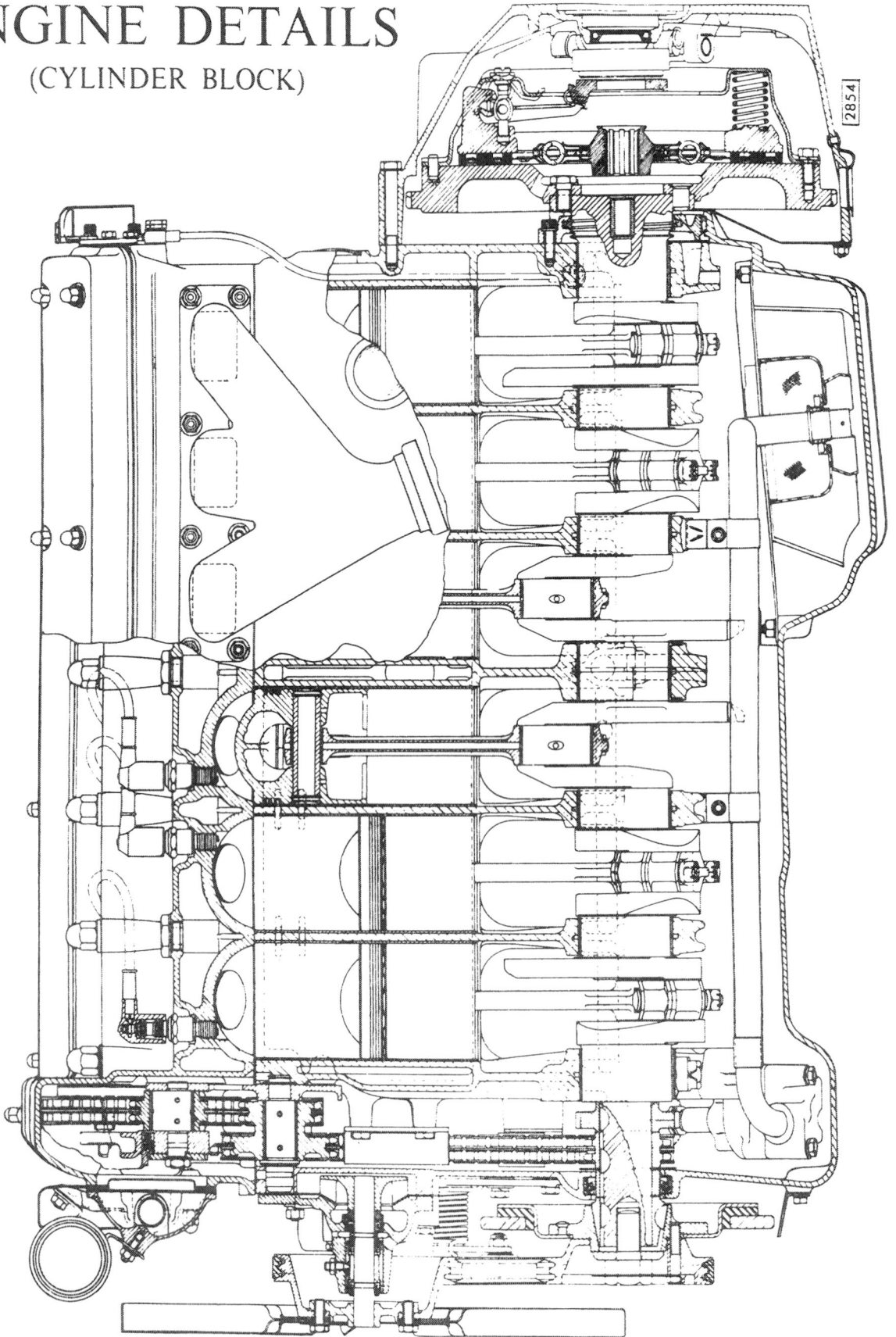

2854

Fig. 16. *Longitudinal section view of engine.*

1. Cylinder block
2. Core plug
3. Plug
4. Washer
5. Front timing cover
6. Plug
7. Headed plug
8. Washer
9. Dowel
10. Stud
11. Stud
12. Cover assembly
13. Oil seal
14. Ring dowel
15. Cup screw
16. Union
17. Union
18. Filter gauze
19. Water drain tap
20. Washer
21. Washer
22. Crankshaft
23. Screwed plug
24. Bush
25. Thrust washer
26. Main bearing
27. Main bearing
28. Crankshaft damper assembly
29. Cone
30. Distance piece
31. Oil thrower
32. Gear
33. Gear
34. Key
35. Pulley
36. Bolt
37. Washer
38. Bolt
39. Washer
40. Tab washer
41. Connecting rod
42. Bearing
43. Flywheel
44. Dowel
45. Dowel

46. Screw
47. Plate
48. Piston
49. Pressure ring (upper)
50. Pressure ring (lower)
51. Scraper ring (Maxiflex)
52. Gudgeon pin
53. Circlip
54. Oil sump
55. Gasket
56. Seal
57. Seal
58. Drain plug
59. Washer
60. Baffle assembly
61. Filter assembly
62. Stud
63. Pipe assembly
64. "O" ring
65. Stud
66. Hose
67. Clip
68. Dipstick assembly
69. Bracket
70. Ignition timing pointer
71. Dynamo and l.h. front engine mounting bracket
72. Bracket assembly for r.h. front engine mounting
73. Front engine mounting (rubber)
74. L.H. flange support bracket
75. R.H. flange support bracket
76. Stabilising link
77. Bush
78. Stepped washer
79. Stepped bush
80. Rubber mounting for stabilising link
81. R.H. bearing bracket
82. L.H. bearing bracket
83. Bolt
84. Nut
85. Channel support
86. Spring seat
87. Centre bush
88. Spring retainer assembly
89. Coil spring
90. Packing block

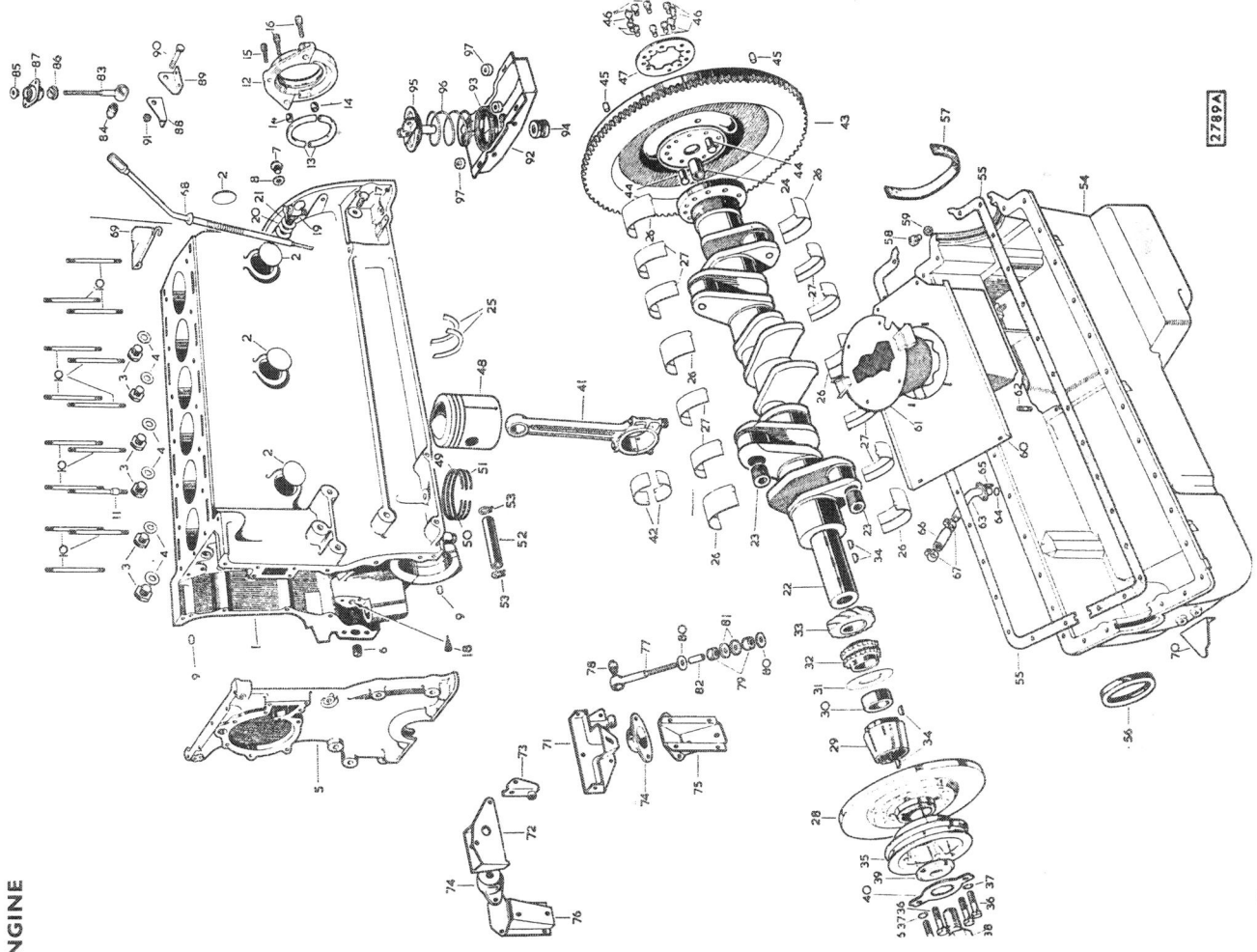

2789A

Fig. 17. Cylinder block.

Page B.22

ENGINE REMOVAL AND FITTING

It is recommended that the engine be removed from above, with the vehicle standing on the workshop floor, using overhead lifting tackle and a trolley jack.

Fig. 18. *Engine removal.*

REMOVAL. (Figs. 19 and 20).

Raise the bonnet, mark the hinge positions and remove bonnet (A).

Remove the air cleaner (B).

Remove the battery (C).

Drain the engine sump (D) and remove the dipstick (E).

Drain the cooling system by removing the radiator filler cap (F), turning the radiator drain tap remote control (G) and the cylinder block drain tap (H). Conserve the coolant if anti-freeze is in use.

Remove engine breather pipe by disconnecting the clip securing the flexible pipe to the breather housing (I).

Remove washer bottle (J).

Slacken the clips securing the top and bottom water hoses (K). Remove hoses.

Remove dynamo connections (L) and ensure that, when reconnecting, the brown/yellow wire is connected to the large terminal (if a radio is installed, the radio suppressor is also connected to the large terminal). Remove the two mounting bolts and nuts underneath the dynamo. Remove the adjusting bolt situated at the top of the dynamo, disengage the fan belt and lift out the dynamo.

In order to remove the radiator proceed as follows:—

(a) Remove the two setscrews securing the sides of the radiator to the body (M).

(b) Remove the two securing nuts at the bottom of the radiator (N).

(c) Lift out the radiator taking care not to damage the matrix with the fan blades.

Note: Unscrew the four nuts securing the cowl and allow the cowl to rest on the water pump housing behind the fan until the radiator is removed.

Disconnect the exhaust system (O).

Disconnect the clips securing the two heater pipes at the rear of the engine and remove the pipes (P).

Detach the revolution counter leads from the Lucar tags on the A.C. generator situated at the rear end of the right hand camshaft cover (Q).

Disconnect the clutch fluid pipe at the bracket at the rear of the cylinder head (R).

Disconnect the oil pressure gauge cable at the oil filter.

Detach the flexible rubber vacuum hose from the rigid pipe beneath the inlet manifold adjacent to the ignition distributor vacuum unit (T).

Disconnect the cable from the starter motor (U).

Disconnect the two snap connectors from the gearbox harness situated at the rear of the exhaust manifold (V).

Withdraw the split pin from the top pin of the accelerator linkage (W). Disconnect the ball-joint at the throttle spindle lever (X).

Remove the lead from the head of the temperature gauge indicator unit situated in the side of the water outlet pipe (Y).

Remove the wire from the S.W. terminal of the ignition coil (AA).

Remove the locknuts and washers from the engine stabilisers at the front and rear of the cylinder block (BB).

Remove the two setscrews from the front engine mounting rubbers (CC).

Detach the two S.U. carburetters from the inlet manifold together with the lead to the electrical thermostat unit situated in the top front face of the water outlet pipe (DD).

Remove the gear lever knob, air distribution pipe cover and the rubber grommet (EE).

Disconnect the earth strap from body at rear of inlet manifold (FF).

Disconnect the speedometer cable from the rear of the gearbox (GG).

Remove cylinder head securing nuts numbers 3, 6, 8 and 9 and fit engine lifting plate (Figs. 18 and 21).

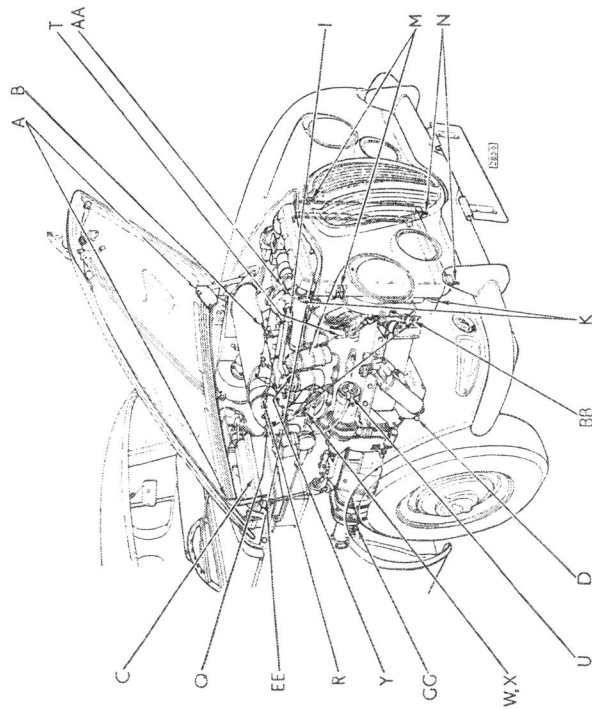

Fig. 19. *Engine removal sequence (1).*

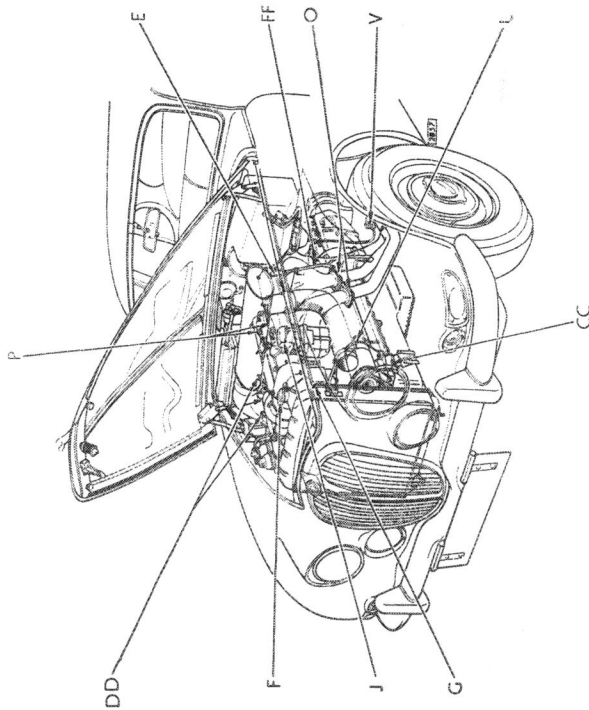

Fig. 20. *Engine removal sequence (2).*

Support the engine on the lifting tackle and remove the eight setscrews from the rear engine mounting member at the rear of the gearbox or overdrive. Remove the propeller shaft.

Note: In the case of a vehicle fitted with automatic transmission:—

Remove the six setscrews securing the rear mounting to the body floor. Remove the two nuts and spring washers securing the mounting plate to the two rubber mountings attached to the rear of the transmission. Remove the mounting plate. Disconnect the propeller shaft from the gearbox flange. Remove the two setscrews securing the centre bearing. Disconnect the propeller shaft from the rear axle flange and remove the propeller shaft. Disconnect the control rod from the selector lever at the left side of the transmission.

Remove the selector cable clamp from the reverse servo cylinder on the left front side of transmission. Disconnect the governor control rod from the governor lever at the rear of the transmission. Remove the leads from the "anti-creep" pressure switch and disconnect the intermediate speed hold solenoid feed wire at the snap connector.

Remove the engine from the engine compartment.

REFITTING

Refitting is the reverse of the removal procedure.

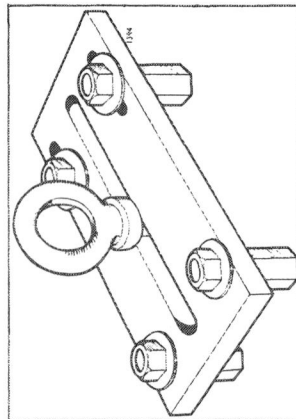

Fig. 21. *The engine lifting plate, Churchill Tool No. J.8.*

ENGINE — TO DISMANTLE

GENERAL

When the power unit is removed from the chassis, the following instructions are applicable for the removal of the engine components.

Dismantling of sub-assemblies and the removal of individual components when the engine is in the chassis frame are dealt with later in this section.

All references made in this section to the top or bottom of the engine assume that the engine is in the normal upright position. References to the left- or right-hand side assume that the engine is upright and viewed from the rear.

REMOVE STARTER

Remove the electrical cable from the terminal on the starter. Unscrew the two nuts securing the starter to the clutch housing and withdraw the starter.

REMOVE GEARBOX

Remove the set bolts and nuts securing the clutch housing to the engine and withdraw the gearbox unit. The gearbox must be supported during this operation in order to avoid straining the clutch driven plate and constant pinion shaft.

REMOVE DISTRIBUTOR

Spring back clips and remove the cover complete with high tension leads. Disconnect the electrical cable from the distributor. Slacken the clamp plate bolt and withdraw distributor. Remove the setscrew and remove the clamp plate. Note the cork seal in recess at the top of the distributor drive hole.

REMOVE CYLINDER HEAD (Fig. 14)

Disconnect the distributor vacuum feed pipe from the front carburetter. Remove the high tension leads from the sparking plugs and lead carrier from the cylinder head studs. Remove the sparking plugs. Disconnect the camshaft oil feed pipe (52) from the rear of the cylinder head. Remove the eleven dome nuts (48) from each camshaft cover and lift off the covers.

Remove the four dome nuts (58) securing the breather housing (55) and withdraw housing. Release the tension on the camshaft chain by slackening the nut on the eccentric idler sprocket shaft, depressing the spring-loaded stop peg and rotating serrated adjuster plate clockwise. Anti-clockwise rotation of the serrated adjuster viewed from the front of the engine tightens the chain.

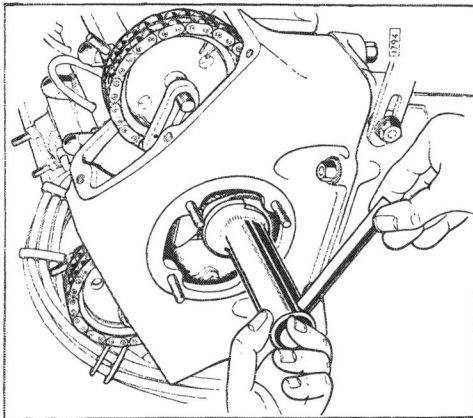

Fig. 22. *The top timing chain adjuster, Churchill Tool No. J.2 in position.*

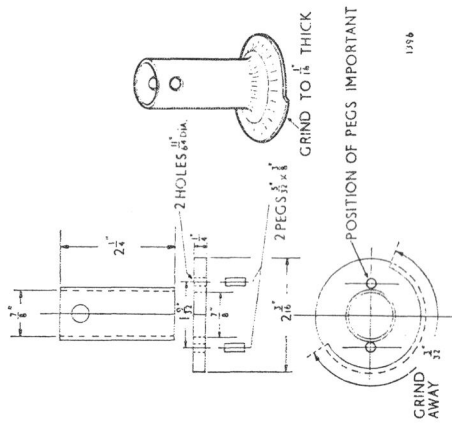

Fig. 23. *Top timing chain adjusting tool.*

Break the locking wire on the two setscrews securing the camshaft sprockets to their respective camshafts. Remove the setscrews and withdraw the sprockets from the camshafts with chain in position. Having once disconnected the camshaft sprockets do NOT rotate the engine or camshafts.

Slacken the fourteen cylinder head dome nuts and six nuts securing the front of the cylinder head a part of a turn at a time in the order shown in Fig. 30 until the nuts become free. Lift off the cylinder head complete with exhaust manifold and inlet manifolds. Remove and scrap the cylinder head gasket.

REMOVE CLUTCH AND FLYWHEEL (Fig. 17)

Unscrew the six setscrews securing the flange of the clutch cover to the flywheel and remove the clutch cover. Note the balance marks "B" stamped on the clutch cover and on the edge of the flywheel.

Knock back the tabs of locking plate securing the ten flywheel bolts. Unscrew the flywheel bolts and remove the locking plate. Remove flywheel from the crankshaft flange by gently tapping with a rawhide mallet.

REMOVE FAN

Remove the fan and fan pulley from the hub by unscrewing the four set bolts fitted with shakeproof washers.

REMOVE CRANKSHAFT DAMPER

Knock back the tab washers and remove the two bolts securing the locking washer to the pulley.

Unscrew the large nut and remove the plain washer.

Insert two levers behind the damper and ease it off the split cone—a sharp tap on the edge of the cone will assist removal.

REMOVE WATER PUMP

Unscrew the set bolts and three nuts, and remove the water pump from the timing cover. Note the gasket between the pump and timing cover.

REMOVE OIL FILTER

Detach the short length of flexible pipe between the oil filter and the oil sump.

Unscrew the five set bolts securing the oil filter to the cylinder block and remove filter.

REMOVE SUMP (Fig. 17)

Remove the twenty-six setscrews securing the sump to the crankcase and the four nuts securing the sump to the timing cover. The sump can now be removed.

REMOVE OIL PUMP AND PIPES

Tap back the tab washers and unscrew the two set bolts securing the oil feed pipe from the oil pump to the bottom face of the crankcase. Withdraw the pipe from the pump.

Remove the nut and bolt securing the oil pump inlet pipe clip to the bracket on the main bearing cap.

Remove the nut and bolt securing the oil pump inlet pipe clip in the bracket on the oil pump.

Withdraw the pipe from the pump.

Tap back the tab washers from the three bolt heads securing the oil pump to the front main bearing cap. The oil pump can now be withdrawn.

REMOVE PISTONS AND CONNECTING RODS

As the pistons will not pass the crankshaft it will be necessary to withdraw the pistons and connecting rods from the top.

Remove the split pins from the connecting rod bolt nuts and unscrew nuts. Remove the connecting rod cap, noting that the corresponding cylinder numbers on the connecting rod and cap are together.

Withdraw the piston and connecting rod from top of cylinder block.

Note: Split skirt pistons MUST be fitted with the split opposite to the thrust side, that is, with the split on the left-hand or exhaust side of the engine. To facilitate correct fitting the pistons' crowns are marked "Front".

REMOVE TIMING COVER

Remove the set bolts securing the timing cover to the front face of the cylinder block. Remove the timing cover, noting that the cover is located to the cylinder block by two dowels.

REMOVE TIMING GEAR ASSEMBLY

When removing the bottom timing chain tensioner from the engine, remove the hexagon head plug and tab washer from the end of the body. Insert an Allen key into the hole until it registers in the end of the restraint cylinder. Turn the Allen key clockwise until the restraint cylinder can be felt to be fully retracted within the body. The adjuster head will then be free of the chain.

Knock back the tab washers on the two set bolts securing the chain tensioner to the cylinder block.

Withdraw the bolts and remove the tensioner together with the conical gauze filter fitted in the tensioner oil feed hole in the cylinder block. This should be cleaned in petrol.

Unscrew the four set bolts securing the front mounting bracket to the cylinder block. Release the tabs of the tab washers and remove the two screwdriver slotted setscrews from the rear mounting bracket.

Remove the coupling shaft from the squared end of the distributor and oil pump drive shaft.

The timing gear can now be removed.

REMOVE DISTRIBUTOR DRIVE GEAR

Tap back the tab washer securing the distributor drive gear nut and remove the nut and washer. Tap the squared end of the distributor drive shaft through the gear, noting that the gear is keyed to the shaft. Remove the gear and thrust washer and withdraw the drive shaft.

REMOVE CRANKSHAFT

Knock back the tab washers securing the fourteen main bearing cap bolts. Unscrew the bolts and the main bearing caps, noting the corresponding numbers stamped on the caps and bottom face of crankcase and also the thrust washers fitted in the recesses in the centre main bearing caps.

Detach the bottom half of the rear oil seal housing from the top half by unscrewing the two Allen screws. Note that the two halves are located by hollow dowels.

The crankshaft can now be lifted out from the crankcase.

ENGINE — TO ASSEMBLE

GENERAL

All references in this section to the top or bottom of the engine assume the engine to be upright, irrespective of the position of the unit when the reference is made. References to the left- or right-hand side assume the engine to be upright and looking from the rear.

FIT DISTRIBUTOR DRIVE SHAFT BUSH

If a new bush is to be fitted, press the bush into the bore of the lug at front of cylinder block.

Ream the bush in position to a diameter of

$$\frac{3}{4}\ {}^{+.0005''}_{-.00025''}\ (19.05\ \text{mm.}\ {}^{+.012\ \text{mm.}}_{-.006\ \text{mm.}})$$

FIT CRANKSHAFT

Fit the asbestos seals in the rear oil seal housing as described under "Replacing the Seal" on page B.43. Fit the rear main bearing cap to the block without the bearings and fit the rear oil seal housing to the cylinder block. Size the rear oil seal as described under "Refitting" on page B.43.

Remove the bottom half of the rear oil seal housing and rear main bearing cap.

Fit the main bearing shells to the top half of the main line bore in the cylinder block. Lay the crankshaft in the bearing shells. Fit the bottom half to the top half of the rear oil seal housing.

FIT DISTRIBUTOR AND OIL PUMP DRIVE GEAR

Fit the distributor drive shaft to the bush on front face of the cylinder block with the offset slot in the top of the shaft as in Fig. 26. Fit the thrust washer and drive gear to the drive shaft, noting that the gear is keyed to the shaft.

Fit the pegged tab washer with the peg in the keyway of the drive gear.

Fully tighten nut and secure with the tab washer. Check the end float of shaft which should be .004" to .006" (.10 to .15 mm.).

If no clearance exists fit a new oil pump/distributor driving gear which will restore the clearance.

FIT OIL PUMP AND PIPES

Fit the coupling shaft between the squared end of the distributor drive shaft and the driving gear of the oil pump. Secure the oil pump to the front main bearing cap by the three dowel bolts and tab washers. Check that there is appreciable end-float of the short coupling shaft. Fit the oil delivery pipe from the oil pump to the bottom face of the crankcase with a new "O" ring and gasket. Fit the suction pipe with a new "O" ring at the oil pump end.

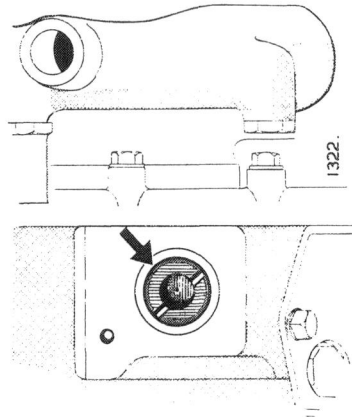

Fig. 26. *Showing the position of the distributor drive shaft offset when No. 6 (front) piston is on Top Dead Centre.*

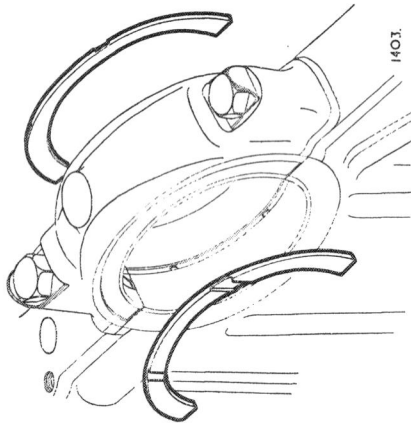

Fig. 24. *The crankshaft thrust washers.*

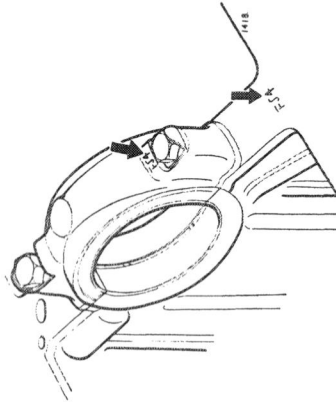

Fig. 25. *Showing the corresponding numbers marked on the main bearing cap and crankcase.*

FIT CRANKSHAFT GEAR AND SPROCKET

Fit the Woodruff key and drive on the crankshaft gear with the widest part of boss to the rear.

Fit the Woodruff key and drive on the crankshaft sprocket. Fit oil thrower, washer and distance piece.

Turn the engine until Nos. 1 and 6 pistons are on T.D.C.

The two halves of the oil seal housing are supplied only as an assembly together with the dowels and screws.

Fit the centre main bearing cap with a thrust washer, white metal side outward, to the recess in each side of cap. Tighten down the cap and check the crankshaft end float, which should be .004" to .006" (.10 to .15 mm.). The thrust washers are supplied in two thicknesses, standard and .004" (.10 mm.) oversize and should be selected to bring the end float within permissible limits. The oversize thrust washers are stamped +.004" (.10 mm.) on the steel face.

Fit the main bearing caps with the numbers stamped on the caps with the corresponding numbers stamped on the bottom face of the crankcase.

Fit the main bearing cap bolts and tab washers and tighten to a torque of 83 lb. ft. (11.5 kg./m.).

Test the crankshaft for free rotation.

The tab washers for the rear main bearing bolts are longer than the remainder and the plain ends should be tapped down around the bolt hole bosses.

FIT PISTONS AND CONNECTING RODS

Turn the engine on its side. Remove the connecting rod caps and fit the pistons and connecting rods to their respective bores from the top of the cylinder block, using a suitable piston ring compressor. The cylinder number is stamped on the connecting rod and cap. No. 1 cylinder being at rear.

Note: Semi-split skirt pistons MUST be fitted with the split opposite the thrust side, that is, with the split on the left-hand or exhaust side of the engine. To facilitate correct fitting the piston crowns are marked "Front".

Fit the connecting rod caps to the connecting rods with the corresponding numbers together. Fit the castellated nuts and tighten to a torque of 37 lb. ft. (5.1 kg./m.). Secure nut with split pins.

TO ASSEMBLE TIMING GEAR (Fig. 63.)

Fit the eccentric shaft to the hole in front mounting bracket. Insert the spring and locking plunger for the serrated plate to the hole in the front mounting bracket. Fit the serrated plate and secure with the shakeproof washer and nut. Fit the idler sprocket (21 teeth) to the eccentric shaft.

Fit the two intermediate sprockets (20 and 28 teeth) to their shaft with the larger sprocket forward and press the shaft through lower central hole in rear mounting bracket. Secure with the circlip at the rear of the bracket.

Fit the top timing chain (longer chain) to the small intermediate sprocket and the bottom timing chain (shorter chain) to the large intermediate sprocket.

Loop upper timing chain under the idler sprocket and offer up the front mounting bracket to the rear mounting bracket with the two chain dampers interposed between the brackets.

Fit the intermediate damper to the bottom of the rear mounting bracket with two screwdriver slotted setscrews and tab washers.

FIT TIMING GEAR

Turn the engine upside down. Fit the lower timing chain damper and bracket to the front face of the cylinder block with two set bolts and locking plate.

Pass the four securing bolts through the holes in the brackets, chain dampers and spacers noting that shakeproof washers are fitted under the bolt heads. Secure the two mounting brackets together with four stud nuts and shakeproof washers.

Turn the timing gear assembly upside down and offer it up to the cylinder block. Loop the bottom timing chain over the crankshaft sprocket and secure the mounting brackets to the front face of the cylinder block with the four long securing bolts.

TIMING CHAIN TENSIONER

Place the timing chain tensioner, backing plate and filter in position so that the spigot on the tensioner aligns with the hole in the cylinder block. Fit shims as necessary, between the backing plate and cylinder block so that the timing chain runs centrally along the rubber slipper. Fit the tab washer and two securing bolts. Tighten the bolts and tap the tab washers against the bolt heads.

It is important that no attempt is made to release the locking mechanism until the adjuster has been finally mounted in the engine WITH THE TIMING CHAIN IN POSITION.

Remove the hexagon head plug and tab washer from the end of the body. Insert the Allen key into the hole until it registers in the end of the restraint cylinder. Turn the key clockwise until the tensioner head moves forward under spring pressure against the chain. Do not attempt to turn the key anti-clockwise, nor force the tensioner head into the chain by external pressure.

Refit the plug and secure with the tab washer.

FIT TIMING COVER

Fit the circular oil seal to the recess in the bottom face of timing cover, ensuring that seal is well bedded in its groove.

Fit the timing cover gasket with good quality jointing compound and secure the timing cover to the front face of the cylinder block with the securing bolts. Do not forget to fit the dynamo adjusting link and distance piece, with the distance piece interposed between the link and the timing cover.

FIT OIL SUMP

Fit a new sump gasket to the bottom face of the crankcase. Fit the cork seal to the recess in the rear main bearing cap. (Fig. 27).

Fit the sump to the crankcase and secure with the twenty-six set screws, four nuts and washers.

Note: The short setscrew must be fitted to the right-hand front corner of the sump.

FIT FLYWHEEL AND CLUTCH

Turn the engine upright.

Check that the crankshaft flanges and the holes for the flywheel bolts and dowels are free from burrs.

Turn the engine until Nos. 1 and 6 pistons are on T.D.C. and fit the flywheel to the crankshaft flange so that the "B" stamped on the edge of the flywheel is at approximately the B.D.C. position. (This will ensure that the balance mark "B" on the flywheel is in line with the balance mark on the crankshaft which is a group of letters stamped on the crank throw just forward of the rear main journal.)

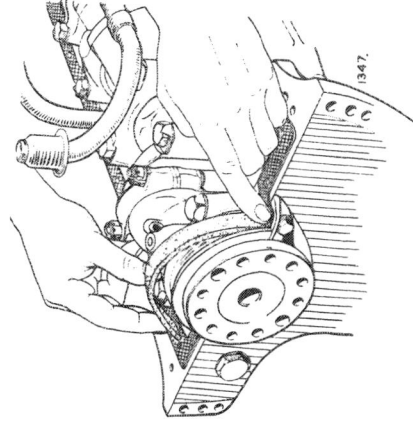

Fig. 27. *Fitting the rear oil seal.*

Tap the two mushroom-headed dowels into position, fit the locking plate and flywheel securing set screws. Tighten the set screws to a torque of 67 lb. ft. (9.2 kg. m.) and secure with the locking plate tabs. Assemble the clutch driven plate to the flywheel, noting that one side of the plate is marked "Flywheel Side". Centralise the driven plate by means of a dummy shaft which fits the splined bore of the driven plate and the spigot bush in the crankshaft. (A constant pinion shaft may be used for this purpose.) Fit clutch cover assembly so that the "B" stamped adjacent to one of the dowel holes coincides with the "B" stamped on the periphery of the flywheel. Secure the clutch assembly with the six set screws and spring washers, tightening the screws at a turn at a time by diagonal selection. Remove the dummy shaft.

FIT CYLINDER HEAD

Before refitting the cylinder head it is important to observe that if the camshafts are out of phase with piston position fouling may take place between the valves and pistons. It is, therefore, essential to adhere to the following procedure before fitting the cylinder head:—

Check that the grooves in the front flanges of the camshafts are vertical to the camshaft housing face and accurately position by engaging the valve timing gauge. If it is found necessary to rotate one of the camshafts the other camshaft must either be removed or the bearing cap nuts slackened to their fullest extent to allow the valves to be released.

Turn No. 6 (front) piston to the top dead centre position with the widest portion of the distributor drive shaft offset positioned as shown in Fig. 26.

Do NOT rotate the engine or camshafts until the camshaft sprockets have been connected to the camshafts. Fit the two camshaft sprockets complete with adjuster plates and circlips to the top timing chain and enter the guide pins in the slots in the front mounting bracket.

FIT CYLINDER HEAD OIL FEED PIPE AND OIL FILTER

Fit the cylinder head oil feed pipe from the tapped holes in the main oil gallery to the two tapped holes in the rear of the cylinder head. Secure the pipe with the three banjo bolts with a copper washer fitted to both sides of each banjo.

Fit the oil filter to the cylinder block with the four setscrews and copper washers. New gasket(s) must always be fitted between the filter and cylinder block.

Fit the short length of flexible hose between the oil filter head and the oil sump and tighten two hose clips.

FIT CRANKSHAFT DAMPER AND PULLEY

Fit a Woodruff key to the crankshaft and the split cone. Fit the split cone to the crankshaft with the widest end towards the timing cover. Fit the damper to the cone and secure with the flat washer, chamfered side outwards, and large nut. Retain the large nut with the locking plate and secure with two setscrews. Secure the setscrews with the tabs at each end of the locking plate.

FIT WATER PUMP

Fit the water pump to the timing cover with a new gasket and secure with six bolts, three nuts and spring washers.

FIT FAN

Fit the fan and pulley and secure with four setscrews and washers.

FIT DYNAMO AND FAN BELT

Slacken the setscrew securing the dynamo adjusting link to the timing cover and swing link upwards.

Place the new belt in position on the fan jockey and crankshaft pulleys and, by pressing the jockey pulley against the spring, pass the belt over the dynamo pulley.

Pass the dynamo top securing bolt through the link and screw into the dynamo lug. Pull the dynamo away from the engine as far as possible, tighten the dynamo top securing bolt and replace the nut. Tighten the bolt securing the dynamo link to the engine and the two dynamo mounting bolts.

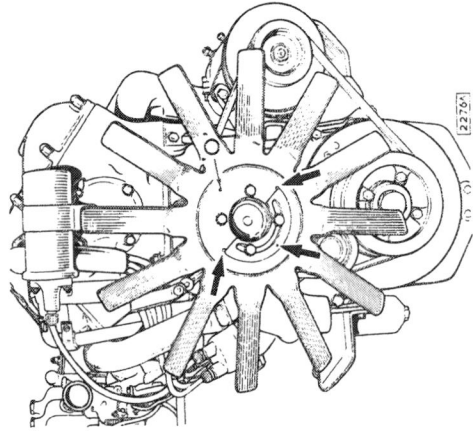

Fig. 29. *The arrows indicate the balance piece location marks.*

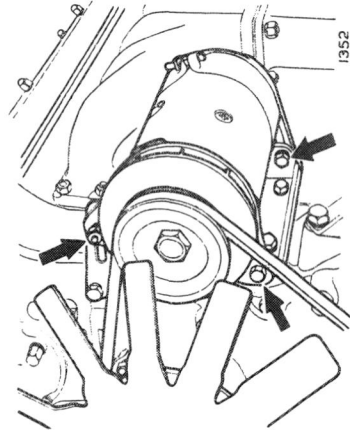

Fig. 30. *The dynamo mounting bolts.*

VALVE TIMING

Check that the No. 6 (front) piston is exactly in the T.D.C. position.

Through the breather aperture in the front of the cylinder head slacken the lock nut securing the serrated plate.

With the camshaft sprocket on the flanges off the camshafts, tension chain by pressing locking plunger inwards and rotating serrated plate by the two holes in an anti-clockwise direction.

When correctly tensioned there should be slight flexibility on both outer sides of the chain below the camshaft sprockets, that is the chain must not be dead tight. Release the locking plunger and securely tighten the locknut. Tap the camshaft sprockets off the flanges of the camshafts.

Fit the cylinder head gasket, taking care that the side marked "Top" is uppermost. Fit the cylinder head complete with manifolds to the cylinder block. Note that the second cylinder head stud from the front on the left-hand side is a dowel stud.

Fit the sparking plug lead carrier to the 3rd and 5th stud on the right-hand side. Fit plain washers to these and the two front stud positions and "D" washers to the remaining studs. Tighten the fourteen large cylinder head dome nuts a part of a turn at a time to a torque of 54 lb. ft. (7.5 kg./m.) in the order shown in Fig.

28. Also tighten the six nuts securing the front end of the cylinder head.

Accurately position the camshaft with the valve timing gauge and check that the T.D.C. marks are in exact alignment.

Withdraw the circlips retaining the adjusting plates to the camshaft sprockets and pull the adjusting plates forward until the serrations disengage. Replace the sprockets on to the flanges of camshafts and align the two holes in the adjuster plate with the two tapped holes in each camshaft flange. Engage the serrations of the adjuster plates with the serrations in the sprockets.

Note: It is most important that the holes are in exact alignment, otherwise when the setscrews are fitted the camshafts will be moved out of position. If difficulty is experienced in aligning the holes exactly, the adjuster plates should be turned through 180°, which, due to the construction of the plate, will facilitate alignment.

Fit the circlips to the sprockets and one setscrew to the accessible hole in each adjuster plate. Turn the engine until the other two holes are accessible and fit the two remaining setscrews.

Finally, recheck the timing chain tension and timing in this order. Secure the four setscrews for camshaft sprockets with new lock wire.

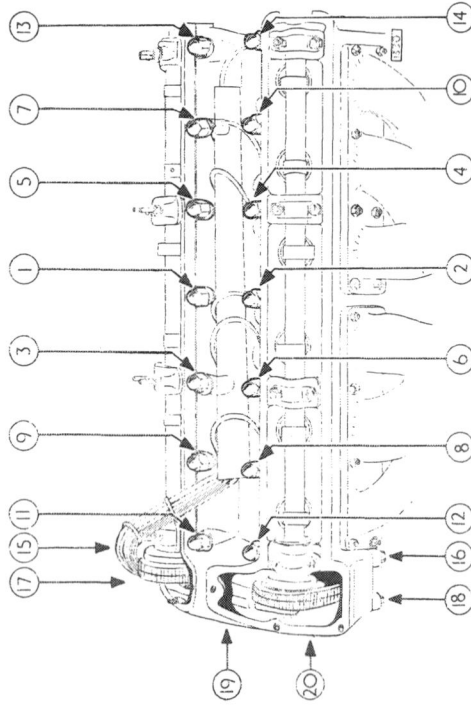

Fig. 28. *Tightening sequence for the cylinder head nuts.*

FIT DISTRIBUTOR AND SPARKING PLUGS

Fit the cork seal to the recess at the top of the hole for the distributor. Secure the distributor clamping plate to the cylinder block with the set-screw. Slacken the clamping plate bolt.

Set the micrometer adjustment in the centre of the scale.

Enter the distributor into the cylinder block with the vacuum advance unit connection facing the cylinder block.

Rotate the rotor-arm until the driving dog engages with the distributor drive shaft.

Rotate the engine until the rotor-arm approaches the No. 6 (front) cylinder segment in the distributor cap.

Slowly rotate the engine until the ignition timing scale on the crank-shaft damper is the appropriate number of degrees before the pointer on the sump.

Slowly rotate the distributor body until the points are just breaking.

Tighten the distributor plate pinch bolt.

A maximum of six clicks on the vernier adjustment from this setting, to either advance or retard, is allowed.

Fit the vacuum advance pipe from the distributor to the union on the front carburetter.

Fit the distributor cover and secure with the two spring clips. Fit the sparking plugs with new copper washers and attach high tension leads.

FIT CAMSHAFT COVERS

Fit each camshaft cover to the cylinder head using a new gasket. Fit the eleven copper washers and dome nuts to the cover retaining studs but do not tighten fully.

Fit the revolution counter generator and flanged plug to the rear of right-hand and left-hand camshaft covers respectively with the rubber sealing rings seated in the recesses provided. Tighten fully the dome nuts securing the camshaft covers.

FIT STARTER

Fit the starter motor to the clutch housing by means of the starter strap assembly.

FIT GEARBOX

Fit the gearbox and clutch housing to the rear of the crankcase with setscrews and shakeproof washers.

Fit the support brackets to each side, at the bottom face of the crankcase with two bolts, nuts and spring washers, and to the clutch housing with three bolts, nuts and shakeproof washers.

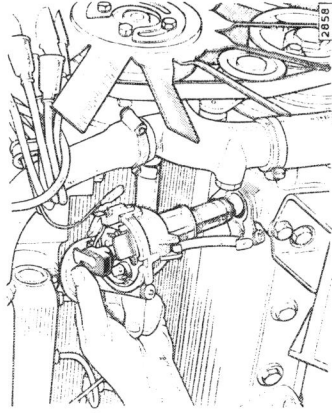
Fig. 31. Fitting distributor.

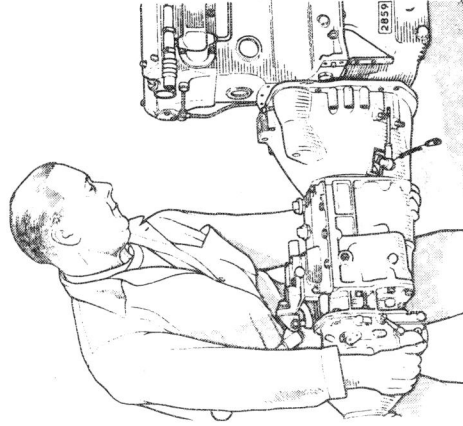
Fig. 32. Fitting gearbox.

DECARBONISING AND GRINDING VALVES

REMOVE CYLINDER HEAD

Remove the cylinder head as described on page B.45.

REMOVE VALVES

With the cylinder head on the bench remove the inlet manifold, and the revolution counter generator.

Remove the four bearing caps from each camshaft and lift out the camshaft (note mating marks on each bearing cap).

Remove the twelve tappets and adjusting pads situated between tappets and valve stems. Lay out the tappets and pads in order, to ensure that they can be replaced in their original guides.

Obtain a block of wood the approximate size of the combustion chambers and place this under the valve heads in No. 1 cylinder combustion chamber. Press down the valve collars and extract the split cotters. Remove the collars, valve springs and spring seats. Repeat for the remaining five cylinders. Valves are numbered and must be replaced in their original locations, No. 1 cylinder being at the rear, that is the flywheel end.

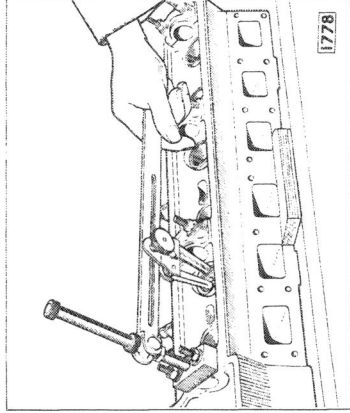
Fig. 33. Fitting the valve springs utilising the valve spring compressing tool Churchill Tool No. J.6118.

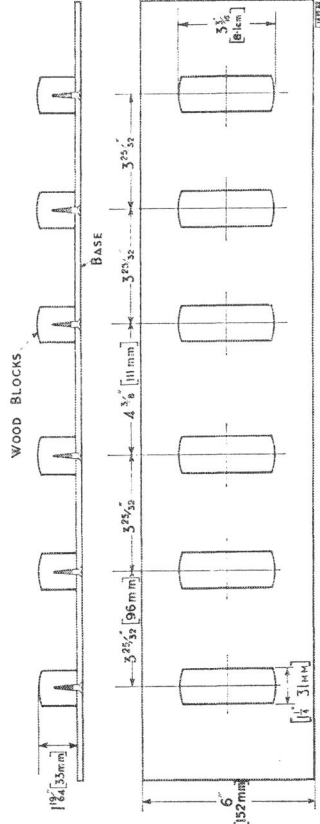
Fig. 34. Combustion chamber blocks for valve removal.

DECARBONISE AND GRIND VALVES

Remove all traces of carbon and deposits from the combustion chambers from the induction and exhaust ports. The cylinder head is of aluminium alloy and great care should be exercised not to damage this with scrapers or sharp pointed tools. Use worn emery cloth and paraffin only. Thoroughly clean the water passages in the cylinder head. Clean the carbon deposits from the piston crowns and ensure that the top face of the cylinder block is quite clean particularly round the cylinder head studs. Remove any pitting in the valve seats, using valve seat grinding equipment. Reface the valves if necessary using valve grinding equipment; grind the valves to the seats, using a suction valve grinding tool.

Clean the sparking plugs and set gaps; if possible use approved plug cleaning and testing equipment. Clean and adjust distributor contact breaker points.

VALVE CLEARANCE ADJUSTMENT

Thoroughly clean all traces of valve grinding compound from the cylinder head and valve gear. Assemble the valves to the cylinder head. When **checking the valve clearances the camshafts must be fitted one at a time as if one camshaft is rotated when the other camshaft is in position, fouling is likely to take place between the inlet and exhaust valves. Obtain and record all valve clearances by using a feeler gauge between the back of each cam and the appropriate valve tappet.**

Correct valve clearances are:—

    Inlet   ..   .004″ (.10 mm.).
    Exhaust ..   .006″ (.15 mm.).

Adjusting pads are available rising in .001″ (.03 mm.) sizes from .085″ to .110″ (2.16 to 2.79 mm.) and are etched on the surface with the letter "A" to "Z", each letter indicating an increase in size of .001″ (.03 mm.). Should any valve clearance require correction, remove the camshaft, tappet and adjusting pad. Observe the letter etched on the existing adjusting pad and should the recorded clearance for this valve have shown say .002″ (.05 mm.) excessive clearance, select a new adjusting pad bearing a letter two lower than the original pad.

As an example, assume that No. 1 inlet valve clearance is tested and recorded as .007″ (.18 mm.). On removal of the adjusting pad, if this is etched with the letter "D" then substitution with a pad bearing the letter "G" will correct the clearance for No. 1 inlet valve.

When fitting the camshafts prior to fitting the cylinder head to the engine it is most important that the keyway in the front bearing flange of each camshaft is perpendicular (at 90°) to the adjacent camshaft cover face (using valve timing gauge) before tightening down the camshaft bearing cap nuts.

Tighten the camshaft bearing cap nuts to a torque of 15 lb./ft. (2.0 kg./m.).

REFIT CYLINDER HEAD

Before attempting to refit the cylinder head refer to the instructions given on page B.46.

COMPRESSION PRESSURES

Pressures must be taken with all the sparking plugs removed, carburettor throttles wide open and the engine at its normal operating temperature (70 C approximately).

Note: When taking compression pressures ensure that the ignition switch is "off"; rotate the engine by operating the push button on the starter solenoid.
On automatic transmission models it will first be necessary to remove the rubber and metal cover from the end of the solenoid to enable the switch to be operated. Check that the selector lever is in the P (Park) position before operating the starter. Replace the solenoid push button cover after the pressure tests have been taken.

The compression pressures for all the six cylinders should be even and should approximate to the figures given below.

If one or more compressions are weak it will most probably be due to poor valve seatings when the cylinder head must be removed and the valves and valve seats refaced and reground.

COMPRESSION PRESSURES

7 to 1 compression ratio: 125 lb. per sq. in (8.79 kg./cm.²).

8 to 1 compression ratio: 155 lb. per sq. in. (10.90 kg./cm.³).

9 to 1 compression ratio: 180 lb. per sq. in. (12.65 kg./cm.³).

THE CONNECTING ROD AND BEARINGS

The connecting rods are steel stampings and are provided with precision shell big-end bearings and steel backed phosphor-bronze small end bushes. A longitudinal drilling through the connecting rod provides an oil feed from the big end to the small end bush.

REMOVAL

As the pistons will not pass the crankshaft it will be necessary to withdraw the pistons and connecting rods from the top.

Proceed as follows:—

Remove Cylinder Head

Remove the cylinder head as described on page B.28.

Remove Sump

Remove the sump as described on page B.28.

Remove Piston and Connecting Rod

Remove the split pins from the connecting rod bolt nuts and unscrew the nuts. Remove the connecting rod cap, noting that the corresponding cylinder numbers on the connecting rod and cap are on the same side. Remove the connecting rod bolts and withdraw the piston and connecting rod from the top of the cylinder block.

OVERHAUL

If connecting rods have been in use for a very high mileage, or if bearing failure has been experienced, it is desirable to renew the rod(s) owing to the possibility of fatigue.

The connecting rods fitted to an engine should not vary one with another by more than 2 drams (3.5 grammes). The alignment should be checked on an approved connecting rod alignment jig. Correct any misalignment as necessary. The big end bearings are of the precision shell type and under no circumstances should they be hand scraped or the bearing caps filed.

The small ends are fitted with steel-backed phosphor-bronze bushes which are a press fit in the connecting rod. After fitting, the bush should be reamed or honed to a diameter of .875″ to .8752″ (22.225 to 22.23 mm.). Always use new connecting bolts and nuts at overhauls.

REFITTING

Refitting is the reverse of the removal procedure. Pistons and connecting rods must be fitted to their respective cylinders (pistons and connecting rods are stamped with their cylinder number, No. 1 being at the rear) and the same way round in the bore.

The pistons must be fitted with split on the left-hand or exhaust side of the engine. To facilitate correct fitting the piston crowns are marked "Front", see Fig. 57.

The cap must be fitted to the connecting rod so that the cylinder numbers stamped on each part are on the same side.

Tighten the connecting rod nuts to a torque of 37 lb./ft. (5.1 kg./m.).

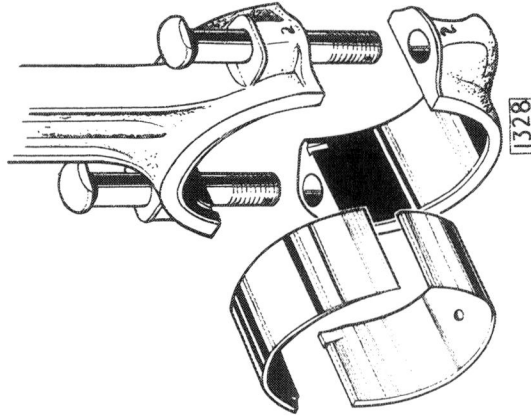

Fig. 35. The connecting rod and cap are stamped with the cylinder number.

BIG-END BEARING REPLACEMENT

The big-end bearings can be replaced without removing the engine from the car but before fitting the new bearings the crankpin must be examined for damage or for the transfer of bearing metal. The oilway in the crankshaft must also be tested for blockage.

Remove the sump as described on page B.28.

Turn the engine until the big-end is approximately at the bottom dead centre position.

Remove the split pins from the connecting rod bolt nuts and unscrew the nuts. Remove the connecting rod cap, noting that the corresponding cylinder numbers on the connecting rod and cap are on the same side.

Lift the connecting rod off the crankpin and detach the bearing shell.

If all the big-end bearings are to be replaced they are most easily replaced in pairs, that is, in pairs of connecting rods having corresponding crank throws.

THE CAMSHAFTS

The camshafts are manufactured of cast iron and each shaft is supported in four white metal steel backed bearings. End float is taken on the flanges formed at each side of the front bearing. Oil is fed from the main oil gallery to the camshaft rear bearing housings through an external pipe. Oil then passes through the rear bearing into a longitudinal drilling in the camshaft; cross drillings which break into this oilway feed the three remaining bearings. On later engines a drilling is made through the base of each cam into the oilway to reduce tappet noise when starting from cold.

Warning: Before carrying out any work on the camshafts the following points must be observed to avoid possible fouling between (a) the inlet and exhaust valves and (b) the valves and pistons.

(1) Do NOT rotate the engine or the camshafts with the camshafts sprockets disconnected. If, with the cylinder head removed from the engine, it is required to rotate a camshaft, the other camshaft must either be removed or the bearing cap nuts slackened to their fullest extent to allow the valves to be **released**.

(2) When fitting the camshafts to the cylinder head ensure that keyway in the front bearing flange of each camshaft is perpendicular (at 90°) to the adjacent camshaft cover face (use valve timing gauge) before tightening down the camshaft bearing cap nuts.

If this operation is being carried out with the cylinder head fitted to the engine, rotate the engine until No. 6 (front) piston is on Top Dead Centre in the firing position, that is with the distributor rotor opposite No. 6 cylinder segment, before fitting the camshafts.

REMOVAL

Remove the eleven dome nuts and copper washers securing each camshaft cover and lift off the cover.

Unscrew the three Allen setscrews attaching the revolution counter generator to the right-hand side of the cylinder head and the sealing plug from the left-hand side (note the copper washers under the heads of the setscrews and the half gaskets between the sealing plug and the cylinder head). Remove the circular rubber sealing rings.

Break the wire locking the camshaft adjuster plate setscrews.

Rotate the engine until No. 6 (front) piston is approximately on Top Dead Centre on compression stroke (firing position), that is, when the keyway in the front bearing flange of each camshaft is at 90° to the adjacent cover face (see Fig. 36).

Note the positions of the **inaccessible** adjuster plate setscrews and rotate the engine until they can be removed.

Turn back the engine to the T.D.C. position with No. 6 firing and remove the two remaining setscrews.

Tap the sprockets off their respective camshaft flanges. Release the eight nuts securing the bearing caps a turn at a time. Remove the nuts, spring washers and "D" washers from the bearing studs.

Remove the bearing caps, noting that the caps and cylinder head are marked with corresponding numbers. Also note that the bearing caps are located to the lower bearing housings with hollow dowels.

If the same bearing shells are to be replaced they should be refitted to their original positions.

The camshaft can now be lifted out from the cylinder head.

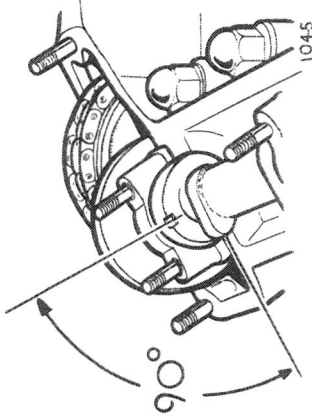

REFITTING

Check that No. 6 (front) piston is exactly on T.D.C. on the compression stroke (firing position), that is, with the distributor rotor opposite No. 6 cylinder segment.

Replace the shell bearings—in their original positions if the same bearings are being refitted.

Replace each camshaft with the keyways in the front bearing flange at 90° to the adjacent cover face (using the valve timing gauge).

Refit the bearing caps to their respective positions and the "D" washers, spring washers and nuts.

Tighten down the bearing caps evenly a turn at a time. Finally, tighten the nuts to a torque of 15 lb./ft. (2.0 kg./m.).

Set the valve timing as described on page B.69.

OVERHAUL

It is unlikely, except after very high mileages, to find wear in the camshafts and camshaft bearings. The camshaft bearings are of the precision shell type and under no circumstances should these be hand scraped or the bearing caps filed. Undersize bearings are not supplied.

CRANKSHAFT DAMPER AND PULLEY

A torsional vibration damper is fitted at the front end of the crankshaft.

The damper consists of a malleable iron ring bonded to a thick rubber disc. An inner member also bonded to the disc is attached to a hub which is keyed to a split cone on the front extension of the crankshaft.

The crankshaft damper and pulley are balanced as an assembly and if they are to be separated mark each part before dismantling so that they can be refitted in their original positions.

REMOVAL

In order to remove the crankshaft damper it will first be necessary to remove the radiator. Remove the two setscrews securing the radiator at the sides and the nuts from the two mountings at bottom of the radiator. On cars fitted with a fan cowl remove the four nuts and hang the cowl on the fan. Lift out the radiator taking care not to damage the matrix against the fan blades.

Remove the fan belt after slackening the dynamo and pushing towards the engine.

Fig. 36. *When fitting a camshaft the keyway must be at 90° to the camshaft cover face.*

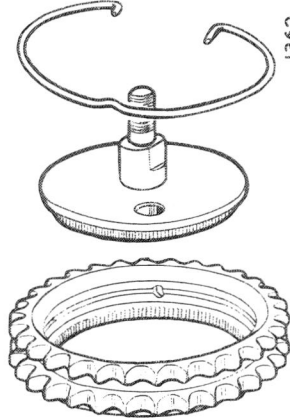

Fig. 37. *Exploded view of the camshaft sprocket assembly.*

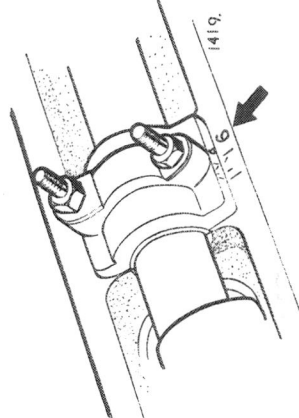

Fig. 38. *Showing the corresponding numbers on the bearing cap and cylinder head.*

Remove the locking washer securing the damper bolt by knocking back the tabs and unscrewing the two setscrews. Unscrew the large damper securing bolt and remove the flat washer. Insert two levers behind the damper and ease it off the split cone—a sharp tap on the end of the cone will assist removal.

OVERHAUL

Examine the rubber portion of the damper for signs of deterioration and if necessary fit a new one. Also examine the crankshaft pulley for signs of wear and renew if necessary.

REFITTING

Refitting is the reverse of the removal procedure.

THE CRANKSHAFT

The counterbalanced crankshaft is of manganese molybdenum steel and is supported in seven precision shell bearings. End thrust of the crankshaft is taken on two semi-circular white metal faced steel thrust washers fitted in recesses in the centre main bearing cap. A torsional vibration damper is fitted at the front end of the crankshaft

Initially, the crankshaft is itself balanced both statically and dynamically and is then re-balanced as an assembly with the flywheel and clutch unit attached.

REMOVAL

Proceed as detailed under "Engine—To Dismantle" on page B.27.

OVERHAUL

Regrinding of the crankshaft journals is generally recommended when wear or ovality in excess of .003" (.08 mm.) is found. Factory reconditioned crankshafts are available on an exchange basis, subject to the existing crankshaft being fit for satisfactory reconditioning, with undersize main and big end bearings —.010" (.25 mm.), —.020" (.51 mm.), —.030" (.76 mm.), and —.040" (1.02 mm.).

Grinding beyond the limits of .040" (1.02 mm.) is not recommended and under such circumstances a new crankshaft should be obtained.

New crankshaft thrust washers should be fitted, these being in two halves located in recesses in the centre main bearing cap. Fit the main bearing cap with a thrust washer, white metal side outwards, to the recess in each side of cap. Tighten down the cap and check the crankshaft end float, which should be .004" to .006" (.10 to .15 mm.). The thrust washers are supplied in two thicknesses, standard and .004" (.10 mm.) oversize and should be selected to bring the end float within the required limits. It is permissible to fit a standard size thrust washer to one side of the main bearing cap and on oversize washer to the other. Oversize thrust washers are stamped .004" on the steel face.

Ensure that the oil passages in the crankshaft are clear and perfectly clean before re-assembling. If the original crankshaft is to be refitted remove the Allen headed plugs in the webs (which are secured by staking) and thoroughly clean out any accumulated sludge with a high pressure jet followed by blowing out with compressed air.

After refitting the plugs, secure by staking with a blunt chisel.

REFITTING

Proceed as detailed under "Engine—To Assemble" on page B.29.

CRANKSHAFT REAR OIL SEAL

The crankshaft rear oil seal consists of a cast iron housing in two halves and an asbestos seal also in two halves, fitted to the housing. The two halves of the housing are located by hollow dowels and secured by Allen screws. The top half of the housing is secured to the cylinder block by three Allen screws and is located by two hollow dowels.

REMOVAL

Having removed the lower half of the oil seal and the crankshaft as described on page B.29, remove the three Allen screws securing the upper half to the oil seal noting the hollow locating dowels at the two outer holes.

Prise out the asbestos seal from its groove and discard it.

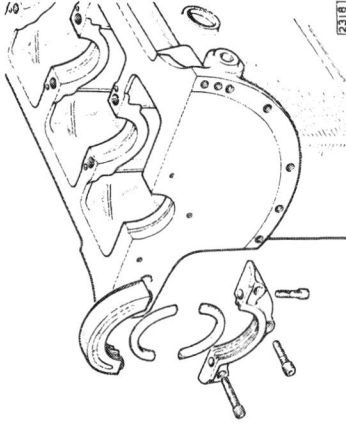

Fig. 39. Exploded view of the crankshaft rear oil seal.

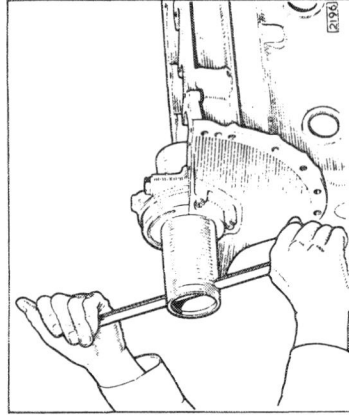

Fig. 40. Sizing the rear oil seal using the special tool (Churchill Tool No. J.17).

REPLACING THE SEAL

Take the new asbestos seals and carefully tap them on the side face to narrow the section of the seal. Fit the seals to the housing and press into the groove using a hammer handle until the seal does not protrude from the ends of the housing. Do NOT cut the ends off the seal if they protrude from the housing but continue pressing the seal into the groove until both ends are flush.

Using a knife or similar tool, press all loose ends of asbestos into the ends of the groove so that they will not be trapped between the two halves of the housing when assembled.

REFITTING

Assemble the two halves of the rear seal and secure with the two Allen screws. Fit the rear main bearing cap to the block without the bearings and tighten to a torque of 83 lb. ft. (11.5 kg./m.). Fit the seal and housing to the cylinder block and secure with the three Allen screws. Smear a small quantity of colloidal graphite around the inner surface of the asbestos seal and insert the sizing bar (Churchill Tool No. J.17). Ensure that the pilot end of the sizing bar enters the bore of the rear main bearing then press the bar inwards and rotate at the same time until the bar is fully home. Remove the bar by pulling and twisting at the same time. Remove the three Allen screws securing the oil seal housing to the cylinder block and remove the Allen screws securing the two halves of the seal. Separate the two halves of the seal and remove the rear main bearing. The crankshaft may now be refitted as described on page B.29.

THE CYLINDER BLOCK

The cylinder block is of chromium iron and is integral with the crankcase. The main bearing housings are line bored and the caps are not interchangeable, corresponding numbers being stamped on the caps and the bottom face of the crankcase for identification purposes. In the case of the 3.8 litre cylinder block, pressed-in dry liners are fitted.

OVERHAUL

Check the top face of the cylinder block for truth. Check that the main bearing caps have not been filed and that the bores for the main bearings are in alignment. If the caps have been filed or if there is mis-alignment of the bearing housings the caps must be re-machined and the bearing housings line bored.

After removal of the cylinder head studs prior to reboring, check the area around the stud holes for flatness. When the edges of the stud holes are found to be raised they must be skimmed flush with the surrounding joint face, to ensure a dead flat surface on which to mount the boring equipment.

Reboring is normally recommended when the bore wear exceeds .006" (.15 mm.). Reboring beyond the limit of .030" (.76 mm.) is not recommended and when the bores will not clean out at .030" (.76 mm.), liners and standard size pistons should be fitted.

In the instance of the 3.8 litre cylinder block the worn liners must be pressed out from below utilizing the illustrated stepped block.

Before fitting the new liner, lightly smear the cylinder walls with jointing compound to a point half-way down the bore and also smear the top outer surface of the liner.

Press the new liners in from the top and lightly skim the tops of the liners flush with the top face of the cylinder block.

Bore out and hone the liners to suit the grade (or grades) of pistons to be fitted. (See piston grades on page B.57.)

The following oversize pistons are available: +.010" (.25 mm.), +.020" (.51 mm.) and .030" (.76 mm.).

Following reboring the blanking plugs in the main oil gallery should be removed and the cylinder block oilways and the crankcase interior thoroughly cleaned. After cleaning, paint the crankcase interior with heat and oil resisting paint.

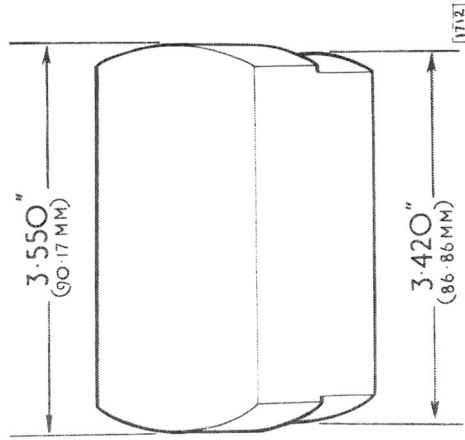

Fig. 41. *Stepped block for 3.8 litre cylinder liner removal.*

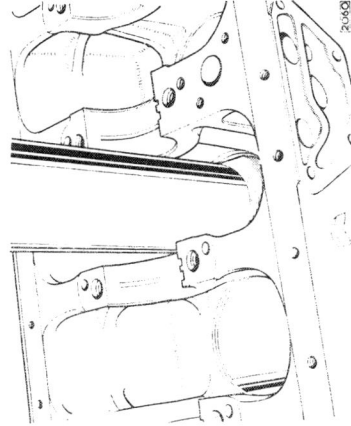

Fig. 42. *Pressing out a cylinder liner using a stepped block.*

THE CYLINDER HEAD
(Fig. 14)

The cylinder head is manufactured of aluminium alloy and has machined hemispherical combustion chambers. Cast iron valve seat inserts, tappet guides and valve guides are shrunk into the cylinder head castings.

Warning: Before carrying out any work on the cylinder head the following points should be observed to avoid possible fouling between (a) the inlet and exhaust valves, and (b) the valves and pistons.

(1) Do NOT rotate the engine or the camshafts with the camshaft sprockets disconnected.

If, with the cylinder head removed from the engine, it is required to rotate a camshaft, the other camshaft must either be removed or the bearing cap nuts slackened to their fullest extent to allow the valves to be released.

(2) When fitting the camshafts to the cylinder head ensure that the keyway in the front bearing flange of each camshaft is perpendicular (at 90°) to the adjacent camshaft cover face before tightening down the camshaft bearing cap nuts. If this operation is being carried out with the cylinder head fitted to the engine, rotate the engine until No. 6 (front) piston is on Top Dead Centre in the firing position, that is with the distributor rotor opposite No. 6 cylinder segment, before fitting the camshafts.

Note: As the valves in the fully open position protrude below the cylinder head joint face, the cylinder head must not be placed joint face downwards directly on a flat surface; support the cylinder head on wooden blocks, one at each end.

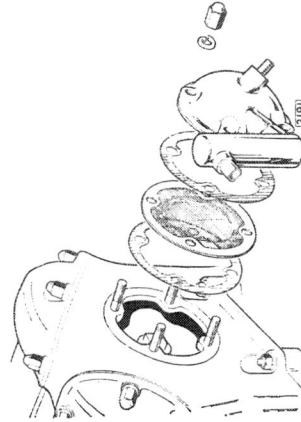

Fig. 43. *Removal of the engine breather.*

REMOVAL.

Drain the cooling system by turning the radiator drain tap remote control and opening the cylinder block drain tap. Conserve water if anti-freeze is in use.

Remove the bonnet by unscrewing the four setscrews, having previously marked the position of the hinges to facilitate adjustment on re-assembly.

Remove the battery and battery platform.

Remove the air cleaner and air intake pipe.

Disconnect the accelerator linkage at the throttle spindle and at the attachment to the inlet manifold.

Disconnect the distributor vacuum advance pipe from the front carburetter.

Disconnect the petrol feed pipe at the float chamber unions.

Disconnect the leads from auxiliary starting carburetter solenoid. Remove the pipe between the auxiliary starting carburetter and the inlet manifold.

Disconnect the revolution counter lead from the generator.

Disconnect the top water hose and by-pass hose from the front of the inlet manifold water jacket.

Remove the high tension leads from the sparking plug and the lead carrier from the cylinder head studs.

Remove the clutch flexible pipe bracket from the rear of the cylinder head.

Disconnect the wires from the ignition coil and remove the coil.

Remove the sparking plugs.

Disconnect the engine breather pipe from the front of the cylinder head.

Disconnect the exhaust manifolds from the engine.

Disconnect the two camshaft oil feed pipe unions from the rear of the cylinder head.

Disconnect the heater hose from the rear of the inlet manifold water jacket.

Disconnect the heater pipe clips from the inlet manifold lower securing nuts.

Remove the Lucar connector from the water temperature transmitter in the inlet manifold water jacket.

Slacken the clip and disconnect the metal vacuum servo pipe from the rubber hose connection to the inlet manifold.

Remove the eleven dome nuts from each camshaft cover and lift off the covers.

Remove the four nuts securing the breather housing to the front of the cylinder head and withdraw housing observing the position of the baffle plate with the two holes vertical.

Release the tension on the top timing chain by slackening the nut on the eccentric idler sprocket shaft, depressing the spring-loaded stop peg and rotating serrated adjuster plate clockwise.

Break the locking wire on the two setscrews, securing camshaft sprockets to respective camshafts.

Remove one setscrew only from each of the camshaft sprockets; rotate the engine until the two remaining setscrews are accessible and remove these screws.

Do NOT rotate the engine or the camshaft after having disconnected the sprockets.

The two camshaft sprockets may now be slid up the support brackets.

Slacken the fourteen cylinder head dome nuts a part at a time in the order shown in Fig. 28 until the nuts become free. Remove the six nuts securing the front of the cylinder head.

Lift off the cylinder head complete with the inlet manifolds. Remove and scrap the cylinder head gasket.

Check the bottom face of the cylinder head for truth.

OVERHAUL

As the cylinder head is of aluminium alloy, great care should be exercised when carrying out overhaul work, not to damage or score the machined surfaces. When removing carbon do not use scrapers or sharply pointed tools—use worn emery cloth and paraffin only.

Remove all traces of carbon and deposits from the combustion chambers and the inlet and exhaust ports and regrind the valve and seats if necessary, as described under "Decarbonising and Grinding Valves," on page B.37.

If it is required to replace the valve guides, valve seat inserts or tappet guides, only the special replacement parts must be used. The replacement parts must be shrunk into the cylinder head in accordance with the instructions given under the appropriate headings in this section.

REFITTING
Fit Cylinder Head

Before refitting the cylinder head it is important to observe that if the camshafts are out of phase with piston position fouling may take place between the valves and pistons. It is, therefore, essential to adhere to the following procedure before fitting the cylinder head:—

Check that the keyways in the front flanges of the camshafts are vertical to the camshaft housing face and accurately position by engaging the valve timing gauge. If it is found necessary to rotate one of the camshafts the other camshaft must either be removed or the bearing cap nuts slackened to their fullest extent to allow the valves to be released.

Turn No. 6 (front) piston to the Top Dead Centre position with the distributor rotor arm opposite No. 6 cylinder segment.

Do NOT rotate the engine or camshafts until the camshaft sprockets have been connected to the camshafts.

Fit the cylinder head gasket, taking care that the side marked "Top" is uppermost. Fit the cylinder head complete with manifolds to the cylinder block. Note that the second cylinder head stud from the front on the left-hand side is a dowel stud.

Fit the sparking plug lead carrier to the 3rd and 6th stud from the front on the right-hand side. Fit plain washers to these and the two front stud positions. Fit the clutch flexible pipe bracket to the two studs at the rear of the cylinder head. Fit "D" washers to the remaining studs.

Tighten the fourteen large cylinder head dome nuts a part of a turn at a time to a torque of 54 lb./ft. (7.5 kg./m.) in the order shown in Fig. 28. Also tighten the six nuts securing the front end of the cylinder head.

Valve Timing

Check that No. 6 (front) piston is exactly in the T.D.C. position.

Through the breather aperture in the front of the cylinder head slacken the locknut securing the serrated plate.

With the camshaft sprocket on the flanges of the camshafts, tension chain by pressing locking plunger inwards and rotating serrated plate by two holes in an anti-clockwise direction.

When correctly tensioned there should be slight flexibility on both outer sides of the chain below the camshaft sprockets, that is, the chain must not be dead tight. Release the locking plunger and securely tighten the locknut. Tap the camshaft sprockets off the flanges of the camshafts.

Accurately position the camshafts with the valve timing gauge and check that the T.D.C. marks are in exact alignment.

Withdraw the circlips retaining the adjusting plates to the camshaft sprockets and pull the adjusting plates forward until the serrations disengage. Replace the sprockets on to the flanges of camshafts and align the two holes in each adjuster plate with the two tapped holes in each camshaft flange. Engage the serrations of the adjuster plates with the serrations in the sprockets.

Note: It is most important that the holes are in exact alignment, otherwise when the setscrews are fitted the camshafts will be moved out of position. If difficulty is experienced in aligning the holes exactly, the adjuster plates should be turned through 180", which, due to the construction of the plate, will facilitate alignment.

Fit the circlips to the sprockets and one setscrew to the accessible hole in each adjuster plate. Turn the engine until the other two holes are accessible and fit the two remaining setscrews.

Finally, recheck the timing chain tension and valve timing in this order. Secure the four setscrews for camshaft sprockets with new locking wire.

Fit Cylinder Head Oil Feed Pipe

Fit the cylinder head oil feed pipe from the tapped hole in the main oil gallery to the two tapped holes in the rear of the cylinder head. Secure the pipe with the three banjo bolts with a copper washer fitted to both sides of each banjo.

Fit Camshaft Covers

Fit each camshaft cover to the cylinder head using a new gasket. Fit the eleven copper washers and dome nuts to the cover retaining studs but do not tighten fully.

Fit the revolution counter generator and flanged plug to the rear of left-hand and right-hand camshaft covers respectively with the rubber sealing rings seated in the recesses provided and secure with the setscrews and copper washers. Tighten fully the dome nuts securing the camshaft covers.

The remainder of the re-assembly is the reverse of the removal procedure.

THE EXHAUST MANIFOLDS

REMOVAL

Remove the eight nuts and spring washers securing the exhaust pipe flanges to the exhaust manifolds.

Remove the sixteen nuts and spring washers securing the exhaust manifolds to the cylinder head when the manifolds can be detached.

REFITTING

Refitting is the reverse of the removal procedure. Use new gaskets between the manifolds and the cylinder head and new sealing rings between the exhaust pipe and manifold flanges.

THE FLYWHEEL

The flywheel is a steel forging and has integral starter gear teeth. The flywheel is located to the crankshaft by two mushroom-headed dowels and is secured by ten setscrews retained by a circular locking plate.

REMOVAL

Remove the engine as described on page B.25. Unscrew the four setscrews and remove the cover plate from the front face of the clutch housing.

Remove the bolts and nuts securing the clutch housing to the engine and withdraw the gearbox unit.

Unscrew the six setscrews securing the flange of clutch cover to the flywheel and remove clutch assembly. Note the balance marks "B" stamped on the clutch cover and on the periphery of the flywheel.

Knock back the tabs of locking plate securing the ten flywheel bolts. Unscrew the flywheel bolts and remove the locking plate. Remove flywheel from the crankshaft flange by gently tapping with a rawhide mallet.

THE INLET MANIFOLD

The inlet manifold is an aluminium casting and is heated by the coolant from the cylinder head through cast-in passages. A water outlet pipe attached to the inlet manifold houses the thermostat and has the top water hose and by-pass hose connected at the front end.

REMOVAL.

Drain the radiator.

Remove the carburetters as described in Section C. Slacken the clips and disconnect the top water hose and by-pass hoses from the inlet manifold water outlet pipe.

Disconnect the cable from water temperature gauge indicator unit situated in the side of the water outlet pipe.

Disconnect the cable from the auxiliary starting carburetter switch. Detach the flexible rubber vacuum hose from the rigid pipe beneath the inlet manifold adjacent to the ignition distributor vacuum unit.

Disconnect the heater hose from the connection at the rear of the manifold. Remove the split pin and detach the accelerator linkage from the pin in the manifold.

Remove the eighteen nuts and spring washers, detach the heater pipe clips from the lower studs when the inlet manifold can be withdrawn.

REFITTING

Refitting is the reverse of the removal procedure.

IGNITION TIMING

Set the micrometer adjustment in the centre of the scale.

Rotate the engine until the rotor-arm approaches the No. 6 (front) cylinder segment in the distributor cap.

Slowly rotate the engine until the ignition timing scale on the crankshaft damper is the appropriate number of degrees before the pointer on the sump.

Ignition Settings

Connect a 12 volt test lamp with one lead to the distributor terminal (or the CB terminal of the ignition coil) and the other to a good earth.

Slacken the distributor plate pinch bolt.

Switch on the ignition.

Slowly rotate the distributor body until the points are just breaking, that is, when the lamp lights up with the fibre heel leading the appropriate cam lobe in the normal direction of rotation.

Tighten the distributor plate pinch bolt.

A maximum of six clicks on the vernier adjustment from this setting, to either advance or retard, is allowed.

Static Ignition Timing

7 : 1 compression ratio	TDC
8 : 1 compression ratio	7 BTDC
9 : 1 compression ratio	5 BTDC

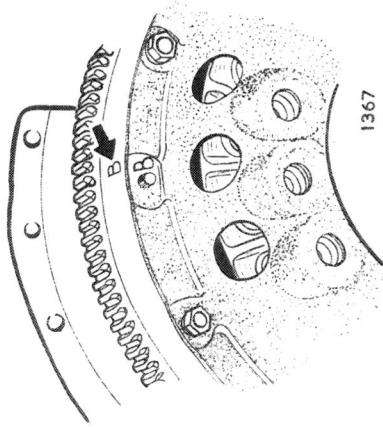

1367

Fig. 44. *Showing the balance marks "B" on the clutch and flywheel*

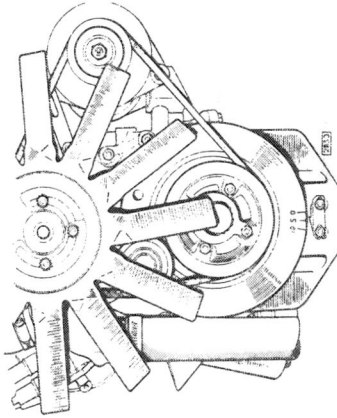

Fig. 45. *Showing the timing scale marked on the crankshaft damper The scale is marked in crankshaft degrees from 0 (top dead centre) to 10 advance (before top dead centre).*

OVERHAUL

If the starter gear is badly worn a new flywheel should be used, since the starter gear teeth are integral with the flywheel, and in this case it will be necessary to balance the flywheel and clutch as an assembly.

If a new flywheel is being fitted, check the flywheel and clutch balance as an assembly by mounting on a mandrel and setting up on parallel knife edges. Mark the relative position of clutch and flywheel. If necessary, remove the clutch and drill $\frac{3}{8}''$ (9.5 mm.) balance holes not more than $\frac{1}{2}''$ (12.7 mm.) deep at a distance of $\frac{3}{8}''$ (9.5 mm.) from the edge of the flywheel.

REFITTING

Turn the engine upright.

Check that the crankshaft flange and the holes for the flywheel bolts and dowels are free from burrs.

Turn the engine until Nos. 1 and 6 pistons are on T.D.C. and fit the flywheel to the crankshaft flange so that the "B" stamped on the edge of the flywheel is at approximately the B.D.C. position. (This will ensure that the balance mark "B" on the flywheel is in line with the balance mark on the crankshaft which is a group of letters stamped on the crank throw just forward of the rear main journal.)

Tap the two mushroom-headed dowels into position, fit the locking plate and flywheel securing setscrews. Tighten the setscrews to a torque of 67 lb./ft. (9.2 kg./m.) and secure with the locking plate tabs. Assemble the clutch driven plate to the flywheel, noting that one side of the plate is marked "Flywheel Side". Centralise the driven plate by means of a dummy shaft which fits the splined bore of the driven plate and the spigot bush in the crankshaft. (A constant pinion shaft may be used for this purpose.) Fit clutch cover assembly so that the "B" stamped adjacent to one of the dowel holes coincides with the "B" stamped on the periphery of the flywheel. Secure the clutch assembly with the six setscrews and spring washers, tightening the screws a turn at a time by diagonal selection. Remove the dummy shaft.

THE OIL FILTER

The oil filter is of the full flow type and has a renewable felt element.

The oil from the oil pressure relief valve is returned to the engine sump via an external rubber hose. The oil pressure relief valve is retained by the outlet adaptor to which the hose to the sump is secured.

A balance valve fitted in the filter head which opens at a pressure differential of 10 to 15 lb. per sq. in. (0.703 to 1.055 kg./cm.²) provides a safeguard against the possibility of the filter element becoming so choked that oil is prevented from reaching the bearings.

REMOVAL OF OIL FILTER (Fig. 46)

It is advisable to catch any escaping oil in a drip-tray. Detach the oil pressure gauge indicator leads from the oil pressure unit situated in the head of the oil filter. Remove the rubber hose (24) from below the filter head by slackening the hose clip (25). Detach the oil filter assembly from the side face of the cylinder block by withdrawing the five bolts.

Collect the gasket fitted between the filter head and the cylinder block.

REFITTING THE OIL FILTER

Refitting is the reversal of the removal procedure, but a new gasket must be fitted between the oil filter head and cylinder block.

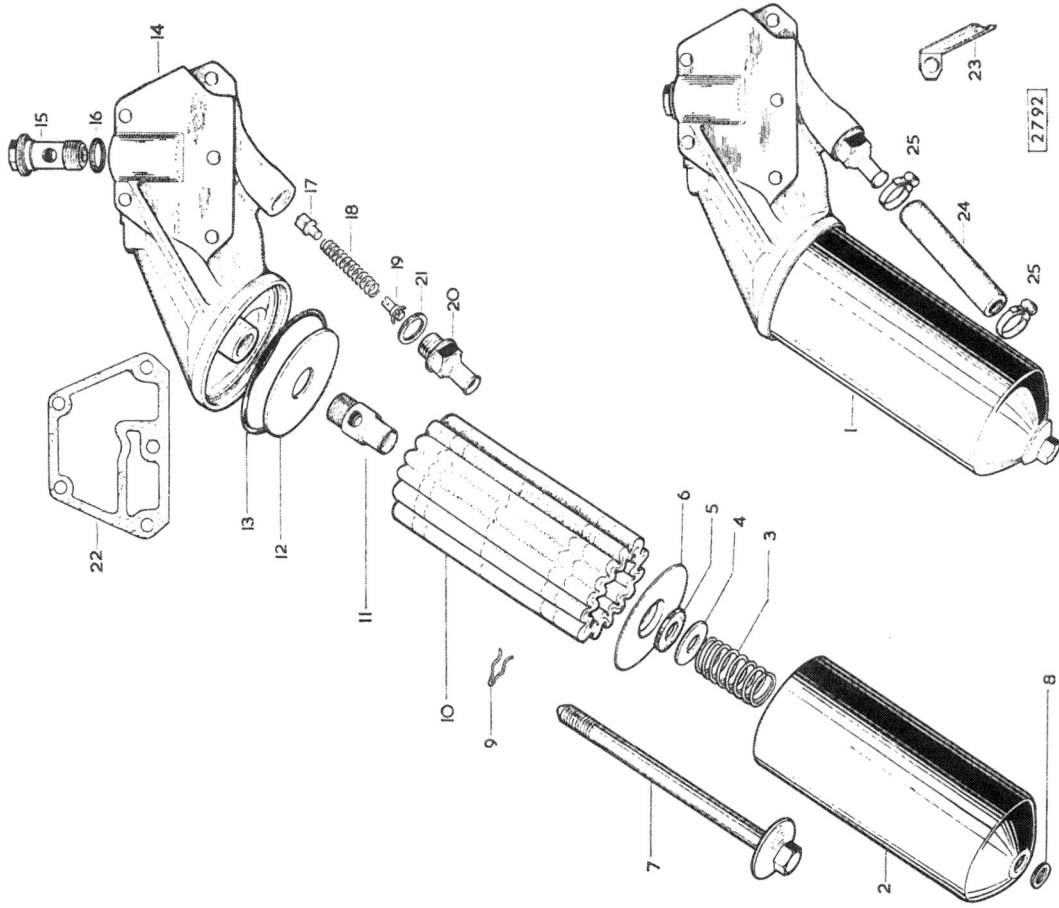

Fig. 46. *Exploded view of oil filter.*

Key to Fig. 46.

1. Oil filter
2. Canister
3. Spring
4. Washer
5. Felt washer
6. Pressure plate
7. Bolt
8. Rubber washer
9. Spring clip
10. Element
11. Anchor insert
12. Clamping plate
13. Sealing ring
14. Filter head
15. Balance valve
16. Seal
17. Relief valve
18. Spring
19. Spider and pin
20. Adaptor
21. Seal
22. Gasket
23. Bracket
24. Hose
25. Clip

ELEMENT REPLACEMENT

It is most important to renew the oil filter element at the recommended periods as after this mileage it will have become choked with impurities.

To guard against the possibility of the filter being neglected to the extent where the element becomes completely choked, a balance valve is incorporated in the filter head which allows unfiltered oil to by-pass the element and reach the bearings. This will be accompanied by a drop in the normal oil pressure of some 10 lb. per sq. in. and if this occurs the filter element should be renewed as soon as possible.

The oil filter is situated on the right-hand side of the engine and in the instance of the upward pointing oil filter drain the element by withdrawing the flat headed drain plug from the bottom of the unit, but with the downward pointing oil filter this is not necessary although it is advisable to catch any escaping oil in a driptray. To gain access to the element, unscrew the centre bolt when the canister and element can be removed. Empty out the oil, thoroughly wash out the canister with petrol and allow to dry before inserting new element.

THE OIL PUMP

The oil pump is of the eccentric rotor type and consists of five main parts: the body, the driving spindle with the inner rotor pinned to it, the outer rotor and the cover, which is secured to the main body by four bolts, finally being secured to the engine with additional dowel bolts. The inner rotor has one lobe less than the number of internal segments in the outer rotor. The spindle centre is eccentric to that of the bore in which the outer rotor is located, thus the inner rotor is able to rotate within the outer, and causes the outer rotor to revolve. The inlet connection is the space in which it is contained decreasing in size as it passes over the port.

REMOVAL

Remove the sump as described on page B.28.

Detach the suction and delivery pipe brackets and withdraw the pipes from the oil pump.

Tap back the tab washers and remove the three bolts which secure the oil pump to the front main bearing cap.

Withdraw the oil pump and collect the coupling sleeve at the top of the drive shaft.

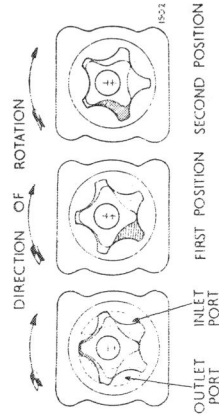

DISMANTLING

Unscrew the four bolts and detach the bottom cover from the oil pump.

Withdraw the inner and outer rotors from the oil pump body. The inner rotor is pinned to the drive shaft and must not be dismantled.

OVERHAUL

Check the clearance between lobes of the inner and outer rotors which should be .006" (.15 mm.) maximum (Fig. 48).

Check the clearance between the outer rotor and the pump body which should not exceed .010" (.25 mm.) (Fig. 49).

Check the end-float of the rotors by placing a straight edge across the joint face of the body and measuring the clearance between the rotors and straight edge (Fig. 50). This clearance should be .0025" (.06 mm.) and in an emergency can be restored by lapping the pump body and outer rotor on a surface plate to suit the inner rotor.

Examine the pump body and bottom cover for signs of scoring and the drive shaft bores for signs of wear; fit new parts as necessary.

Place the drive shaft in a vice fitted with soft jaws and check that the inner rotor is tight on the securing pin.

Note that the drive shaft, inner and outer rotors are supplied only as an assembly.

RE-ASSEMBLING

Re-assembly is the reverse of the dismantling procedure but it is important when fitting the outer rotor to the pump body to insert the chamfered end of the rotor foremost.

Always fit new "O" rings to the suction and delivery pipe bores.

REFITTING

Refitting is the reverse of the removal procedure.

Do not omit to fit the coupling sleeve to the squared end of the drive shaft before offering up the oil pump.

After fitting of the oil pump, check that there is appreciable end-float of the coupling sleeve.

DIRECTION OF ROTATION

OUTLET PORT   INLET PORT   FIRST POSITION   SECOND POSITION

Fig. 47. *Operation of eccentric rotor type oil pump.*

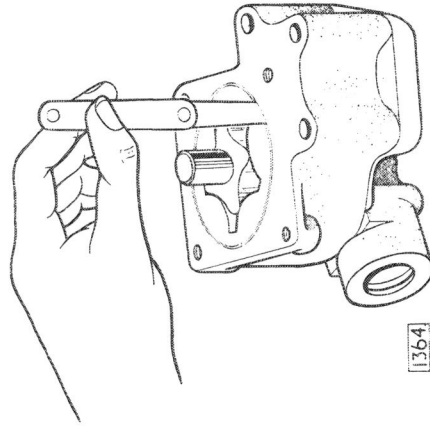

Fig. 48. *Measuring the clearance between the inner and outer rotors.*

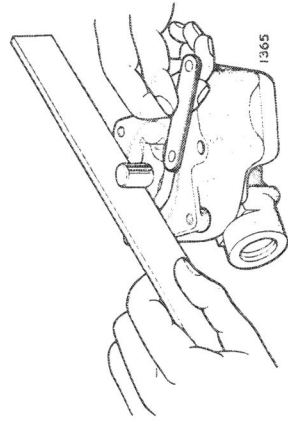

Fig. 49. *Measuring the clearance between the outer rotor and the pump body.*

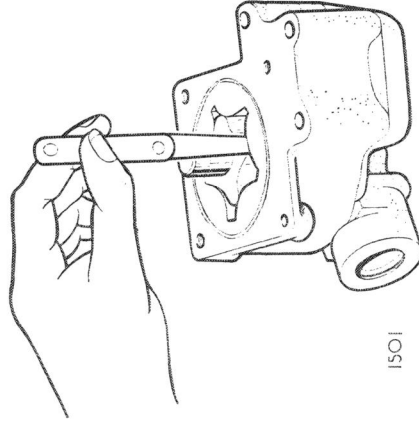

Fig. 50. *Measuring the end float of the rotors.*

THE OIL SUMP

All engine units are fitted with aluminium sumps which have an external connection for a rubber hose, the second end of which is attached to the oil filter.

A gauze bowl-type strainer is fitted in the sump.

REMOVAL

Drain the oil sump.

Remove the front suspension unit as described in Section "J".

Unscrew the twenty-six setscrews and four nuts and detach the sump from the cylinder block, noting that a short setscrew is fitted at the right-hand front corner of the sump.

REFITTING

Scrape off all traces of the old gaskets or sealing compound from the joint faces of the sump and crankcase.

Always fit new gasket(s) and a rear oil seal when refitting the sump. If time permits, roll the rear oil seal into a coil and retain with string for a few hours. This will facilitate the fitting of the seal to its semi-circular recess.

Ensure that the short setscrew is fitted to the right-hand front corner of the sump.

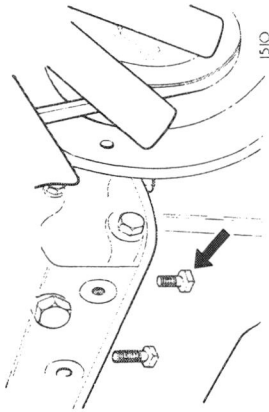

Fig. 52. Showing the location of the short setscrew.

1. Body
2. Rotor assembly
3. Cover
4. Screw
5. Screw
6. Washer
7. "O" ring
8. "O" ring
9. Drive shaft
10. Bush
11. Washer
12. Helical gear
13. Key
14. Nut
15. Special washer
16. Shaft
17. Dowel bolt
18. Tab washer
19. Oil delivery pipe
20. Gasket
21. Oil suction pipe
22. Clip
23. Strut
24. Strut
25. Strut
26. Plate
27. Spring
28. Split pin

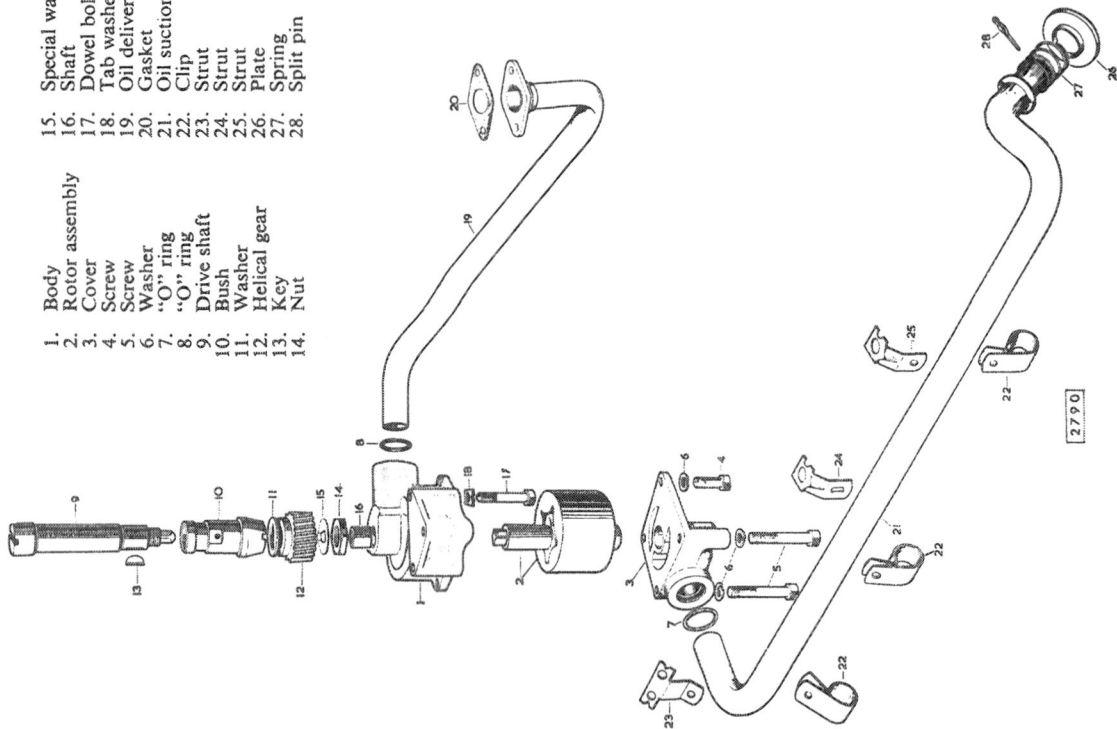

Fig. 51. Exploded view of oil pump.

PISTONS AND GUDGEON PINS

The pistons are made from low expansion aluminium alloy and are of the semi-split skirt type.

The pistons have three rings each, two compression and one oil control. The top compression ring only is chromium plated: both the top and second compression rings have a tapered periphery.

The fully floating gudgeon pin is retained in the piston by a circlip at each end.

REMOVAL

As the pistons will not pass the crankshaft it will be necessary to withdraw the pistons and connecting rods from the top. The connecting rod bolts should, however, be removed to allow the big end to pass easily through the bore. Proceed as follows:—

Remove Cylinder Head

Remove cylinder head as described on page B.45.

Remove Sump

Remove the sump as described on page B.28.

Remove Piston and Connecting Rod

Remove the split pins from the connecting rod bolt nuts and unscrew nuts. Remove the connecting rod cap, noting the corresponding cylinder numbers on the connecting rod and cap. Remove the connecting rod bolts and withdraw the piston and connecting rod from the top of cylinder block.

OVERHAUL

Pistons are supplied complete with gudgeon pins which have been selectively assembled and are, therefore, not interchangeable one with another.

The pistons fitted to an engine should not vary one with another by more than 2 drams (3.5 grammes).

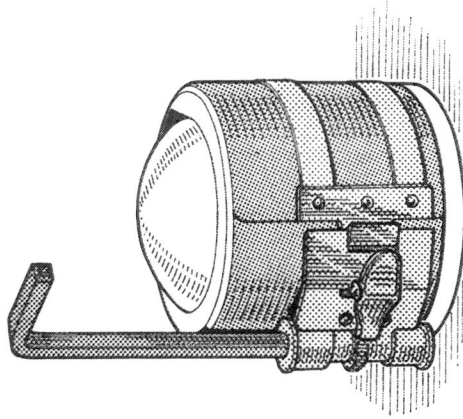

Fig. 53. *Exploded view of piston and connecting rod.*

Fig. 54. *3.8 litre piston.*

Fig. 55. *Piston ring compressor. (Churchill Tool No. 3843)*

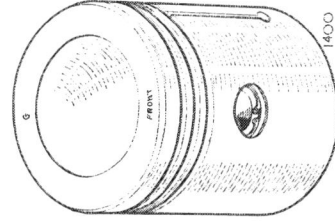

Fig. 56. *Showing the markings on the piston crown.*

Gudgeon Pin Fitting

Gudgeon pins are a finger push fit in the piston at normal room temperature 68°F. (20°C.).

When actually removing or refitting the gudgeon pin, the operation should be effected by immersing the piston, gudgeon pin and connecting rod little end in a bath of hot oil. When the piston and little end have reached a sufficient temperature (230°F. 110°C.) the gudgeon pin can be moved into position. Always use new circlips on assembly.

When assembling the engine, centralise the small end of the connecting rod between the gudgeon pin bosses in the piston and ensure that the connecting rod mates up with the crankshaft journal without any pressure being exerted on the rod.

Piston Grades

The following selective grades are available in standard size pistons only. When ordering standard size pistons the identification letter of the selective grade should be clearly stated. Pistons are stamped on the crown with the letter identification and the cylinder block is also stamped on the top face adjacent to the bores.

Grade Identification Letter	Cylinder bore size for 3.4 litre engine units
F	3.2673" to 3.2676" (82.990 to 82.997 mm.)
G	3.2677" to 3.2680" (83.000 to 83.007 mm.)
H	3.2681" to 3.2684" (83.010 to 83.017 mm.)
J	3.2685" to 3.2688" (83.020 to 83.027 mm.)
K	3.2689" to 3.2692" (83.030 to 83.037 mm.)

Grade Identification Letter	Cylinder bore size for 3.8 litre engine units
F	3.4248" to 3.4251" (86.990 to 86.997 mm.)
G	3.4252" to 3.4255" (87.000 to 87.007 mm.)
H	3.4256" to 3.4259" (87.010 to 87.017 mm.)
J	3.4260" to 3.4263" (87.020 to 87.027 mm.)
K	3.4264" to 3.4267" (87.030 to 87.037 mm.)

Oversize Pistons

Oversize pistons are available in the following sizes:

+.010" (.25 mm.) +.020" (.51 mm.) +.030" (.76 mm.).

There are no selective grades in oversize pistons as grading is necessarily purely for factory production methods.

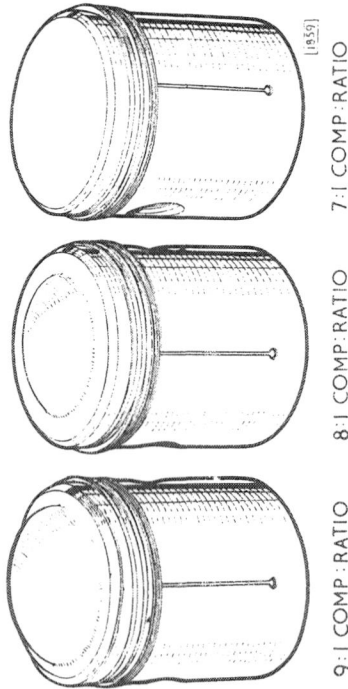

9:1 COMP:RATIO   8:1 COMP:RATIO   7:1 COMP:RATIO

[1858]

Fig. 57. *3.4 litre pistons.*

Piston Rings

Check the piston ring gap with the ring as far down the cylinder bore as possible. Push the ring down the bore with a piston to ensure that it is square and measure the gap with a feeler gauge. The correct gaps are as follows:—

Compression rings:

.015″ to .020″ (.38 to .51 mm.)

Oil control rings:

.015″ to .033″ (.38 to .83 mm.)

With the compression rings fitted to the piston check the side clearance in the grooves which should be .001″ to .003″ (.025 to .076 mm.).

One of the compression rings is hard chrome plated and this ring must be fitted to the top groove in the piston.

Tapered Periphery Rings

All engine units are fitted with tapered periphery piston rings in at least one position and these must be fitted the correct way up.

The narrowest part of the ring must be fitted uppermost; to assist in identifying the narrowest face a letter "T" or "Top" is marked on the side of the ring to be fitted uppermost.

The oil control ring consists of two steel rails with a spacer between the two. These rails are held together on assembly with an adhesive. The expander, which is fitted inside the oil control ring, should be assembled with the two lugs positioned in the hole directly above the gudgeon pin bore.

REFITTING

Pistons and connecting rods must be fitted to their respective cylinders (piston and connecting rods are stamped with their cylinder number, No. 1 being at the rear) and the same way round in the bore.

The pistons must be fitted with split on the left-hand or exhaust side of the engine. To facilitate correct fitting the piston crowns are marked "Front", see Fig. 56.

Use a piston ring clamp when entering the rings into the cylinder bore.

The cap must be fitted to the connecting rod so that the cylinder numbers stamped on each part are on the same side.

Tighten the connecting rod nuts to a torque of 37 lb./ft. (5.1 kg./m.).

SPARKING PLUGS

SERVICE PROCEDURE

To maintain peak sparking plug performance, plugs should be inspected, cleaned and re-gapped at regular intervals of 3,000 miles. Under certain fuel and operating conditions, particularly extended slow speed town driving, sparking plugs may have to be serviced at shorter intervals.

Disconnect the ignition cables from all sparking plugs.

Loosen the sparking plugs about two turns anti-clockwise using the proper sized deep-socket wrench.

Blow away the dirt from around the base of each plug.

Remove the sparking plugs and place them in a suitable holder, preferably in the order they were in the engine.

ANALYSING SERVICE CONDITIONS

Examine the gaskets to see if the sparking plugs were properly installed. If the gaskets were excessively compressed, installed on dirty seats or distorted, leakage has probably occurred during service which would tend to cause overheating of the sparking plugs. Gaskets properly installed will have flat, clean surfaces. Gaskets which are approximately one-half their original thickness will be satisfactory but thinner ones should be renewed.

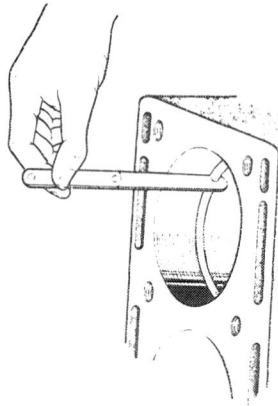

Fig. 58. *Checking the piston ring gap.*

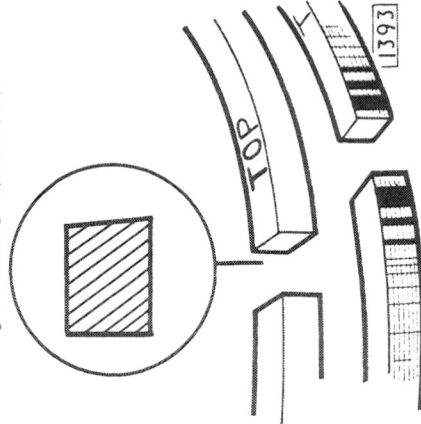

Fig. 59. *Showing the identification marks on tapered periphery compression rings.*

Examine the firing ends of the sparking plugs, noting the type of the deposits and the degree of electrode erosion. The typical conditions illustrated may indicate the use of a sparking plug with an incorrect heat range or faulty engine and ignition-system operation. Remember that if sufficient voltage is not delivered to the sparking plug, no type of plug can fire the mixture in the cylinder properly.

Normal Condition (Fig. 7, page 16)

Look for powdery deposits ranging from brown to greyish tan. Electrodes may be worn slightly. These are signs of a sparking plug of the correct heat range used under **normal** conditions, that is mixed periods of high speed and low speed driving. Cleaning and re-gapping of the sparking plugs is all that is required.

Normal Condition

Watch for white to yellowish powdery deposits. This usually indicates long periods of constant speed driving or a lot of slow speed city driving. These deposits have no effect on performance if the sparking plugs are cleaned **thoroughly** at approximately 3,000 miles intervals. Remember to "wobble" the plug during abrasive blasting in the Champion Service Unit. Then file the sparking surfaces vigorously to expose bright, clean metal.

Oil Fouling (Fig. 8, page 16)

This is usually indicated by wet, sludgy deposits traceable to excessive oil entering the combustion chamber through worn cylinders, rings and pistons, excessive clearances between intake valve guides and stems, or worn and loose bearings, etc. Hotter sparking plugs may alleviate oil fouling temporarily, but in severe cases engine overhaul is called for.

Petrol Fouling (Fig. 9, page 16)

This is usually indicated by dry, fluffy black deposits which result from incomplete combustion. Too rich an air-fuel mixture, excessive use of the mixture control or a faulty automatic choke can cause incomplete burning. In addition, a defective coil, contact breaker points, or ignition cable, can reduce the voltage supplied to the sparking plug and cause misfiring. If fouling is evident in only a few cylinders, sticking valves may be the cause. Excessive idling, slow speeds, or stop-and-go driving, can also keep the plug temperatures so low that normal combustion deposits are not burned off. In the latter case, hotter plugs may be installed.

(3) Grind down the 1.643" (41.73 mm.) outside diameter of tappet guide to a diameter of .003" (.08 mm.) larger than the tappet guide bore dimension, that is to give an interference fit of .003" (.08 mm.).

(4) Also grind off the same amount from the "lead-in" at the bottom of tappet guide. The reduction in diameter from the adjacent diameter should be .0032" to .0057" (.08 to .14 mm.).

(5) Heat the cylinder head in an oven for half an hour from cold at a temperature of 300 F. (150 C.).

(6) Fit the tappet guide, ensuring that the lip at top of guide beds evenly in the recess.

(7) After fitting, ream tappet guide bore to a diameter of $1\frac{3}{8}$" $+.0007"$ (34.925 $+.018$ mm.) $-.0003"$ $-.000$ mm.).

Note: It is essential that, when reamed, the tappet guide bore is concentric with the bore of the valve guide.

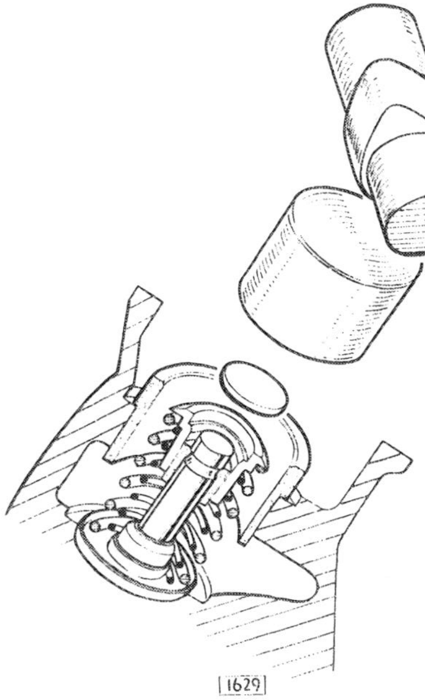

[1629]

Fig. 61. Showing the tappet and adjusting pad.

TAPPETS, TAPPET GUIDES AND ADJUSTING PADS

The chilled cast iron tappets are of cylindrical form and run in guides made of austenitic iron which are shrunk into the cylinder head. A steel pad for adjustment of the valve clearance is sandwiched between the underside of the tappet and top of the valve stem. The pads are available in a range of thicknesses, rising in .001" (.025 mm.) steps, from .085" to .110" (2.16 to 2.79 mm.) and are etched on the surface with the letter "A" to "Z", each letter indicating an increase in size of .001" (.025 mm.).

REMOVAL OF TAPPETS AND ADJUSTING PADS

Remove the camshafts as described on page 33. The tappets can now be withdrawn with a suction valve grinding tool.

Remove the adjusting pads. If valve clearance adjustment is not being carried out the adjusting pads must be refitted to their original positions.

OVERHAUL

Examine the tappets and tappet guides for signs of wear. The diametrical clearance between the tappet and tappet guide should be .0008" to .0019" (.02 to .05 mm.).

Examine the adjusting pads for signs of indentation. Renew if necessary with the appropriate size when making valve clearance adjustment on re-assembly.

TAPPET GUIDE REPLACEMENT

If it is found necessary to replace the tappet guides they must be fitted in accordance with the following instructions and only genuine factory replacement parts used.

(1) Remove the old tappet guide by boring out until the guide collapses. Take care not to damage the bore for the guide in the cylinder head.

(2) Carefully measure the diameter of the tappet guide bore in the cylinder head at room temperature - 68 F. (20 C.).

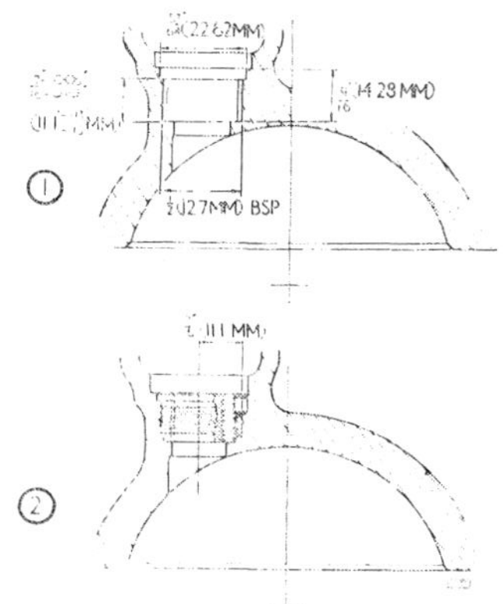

Fig. 60. Fitting dimensions for sparking plug inserts.

SPARKING PLUG INSERTS (Fig. 60)

When it becomes necessary to fit a sparking plug insert in the event of a stripped thread proceed as detailed below.

Bore out the stripped thread to .75" (19.05 mm.) diameter and tap $\frac{7}{8}$" B.S.P.

Counterbore 57/64" (22.62 mm.) diameter to accommodate the larger diameter of the insert.

Fit the screwed insert ensuring that it sits firmly on the face at the bottom of the thread.

Drill and ream a $\frac{1}{8}$" (3.17 mm.) diameter hole $\frac{3}{16}$" (4.76 mm.) deep between the side of the insert and the cylinder head as shown. Drive in the locking pin and ensure that the pin is below the surface. To secure peen over the aluminium on the chamfered portion of the insert and also peen over the locking pin.

Burned or Overheated Condition (Fig. 10, page 16)

This condition is usually identified by a white, burned or blistered insulator nose and badly eroded electrodes, inefficient engine cooling and improper ignition timing can cause general overheating. Severe service, such as sustained high speed and heavy loads, can also produce abnormally high temperatures in the combustion chamber which necessitate the use of colder sparking plugs.

File the sparking surfaces of the electrodes by means of a point file. If necessary, open the gaps slightly and file vigorously enough to obtain bright, clean, parallel surfaces. For best results, hold the plug in a vice.

Reset the gaps using the bending fixture of the Champion Gap Tool. Do not apply pressure on the centre electrode as insulator fractures may result. Use the bending fixture to obtain parallel sparking surfaces for maximum gap life.

Visually inspect all sparking plugs for cracked or chipped insulators. Discard all plugs with insulator fractures.

Test the sparking ability of a used sparking plug on a comparator.

Clean the threads by means of wire hand or power-driven brush. If the latter type is used, wire size should not exceed .005" diameter. Do not wire brush the insulator nor the electrodes.

Clean gasket seats on the cylinder head before installing sparking plugs to assure proper seating of the sparking plug gasket. Then, using a new gasket, screw in the plug by hand fingertight.

Note: If the sparking plug cannot be seated on its gasket by hand, clean out the cylinder head threads with a clean-out tap or with another used sparking plug having three or four vertical flutes filed in its threads.

Tighten the sparking plugs to a torque of 27 lb. ft.

STANDARD GAP SETTING

The sparking plug gap settings recommended in this Service Manual have been found to give the best overall performance under all service conditions. They are based on extensive dynamometer testing and experience on the road, and are generally a compromise between the wide gaps necessary for best idling performance and the small gaps required for the best high-speed performance.

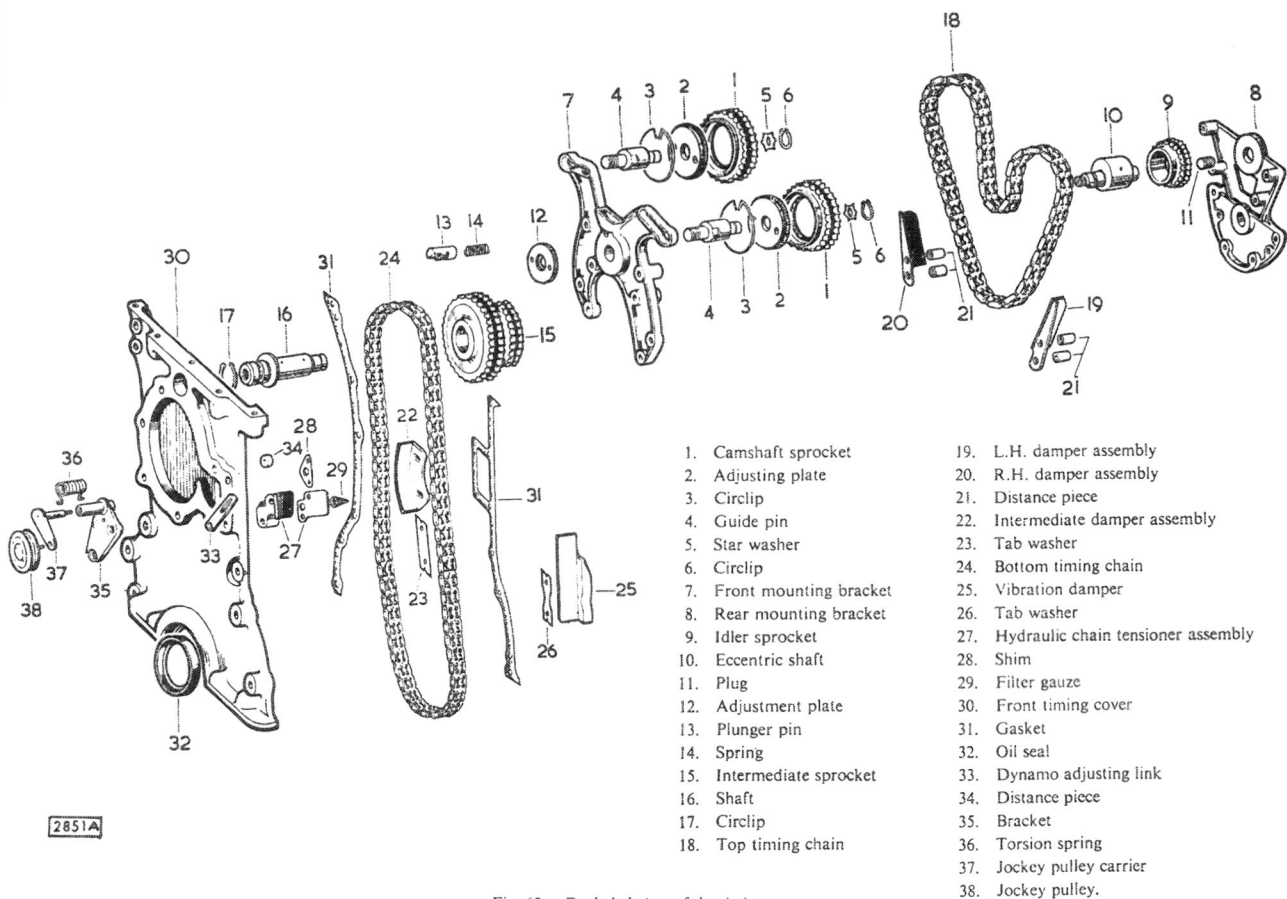

1. Camshaft sprocket
2. Adjusting plate
3. Circlip
4. Guide pin
5. Star washer
6. Circlip
7. Front mounting bracket
8. Rear mounting bracket
9. Idler sprocket
10. Eccentric shaft
11. Plug
12. Adjustment plate
13. Plunger pin
14. Spring
15. Intermediate sprocket
16. Shaft
17. Circlip
18. Top timing chain

19. L.H. damper assembly
20. R.H. damper assembly
21. Distance piece
22. Intermediate damper assembly
23. Tab washer
24. Bottom timing chain
25. Vibration damper
26. Tab washer
27. Hydraulic chain tensioner assembly
28. Shim
29. Filter gauze
30. Front timing cover
31. Gasket
32. Oil seal
33. Dynamo adjusting link
34. Distance piece
35. Bracket
36. Torsion spring
37. Jockey pulley carrier
38. Jockey pulley.

Fig. 63.   *Exploded view of the timing gear.*

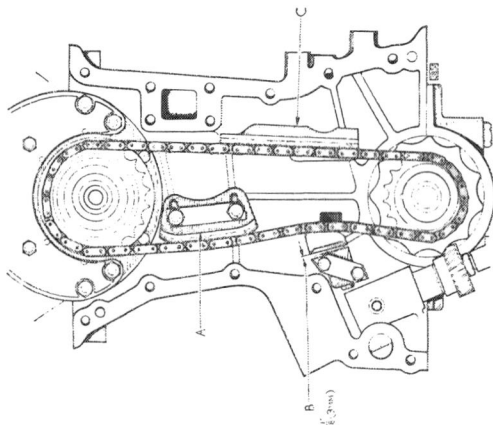

Fig. 62.   *When fitting a new lower timing chain, set the intermediate damper (A) in light contact with the chain when there is ¼" (3 mm.) gap between the rubber slipper and the tensioner body. In the case of a worn chain the gap (B) may have to be increased to avoid fouling between the chain and the cylinder block. Set the lower damper (C) in light contact with the chain.*

THE TIMING GEAR

The camshafts are driven by Duplex endless roller chains in two stages.

The first stage or bottom timing chain drives the larger wheel of a double intermediate sprocket; the second stage or top timing chain passes round the smaller wheel of the intermediate sprocket, both camshaft sprockets and is looped below an idler sprocket.

The idler sprocket has an eccentric shaft for top timing chain tension adjustment and the bottom chain is automatically tensioned by an hydraulic tensioner bolted to the cylinder block. Rubber vibration dampers are located at convenient points around the chains.

REMOVAL

Unscrew the four set bolts and remove the bonnet. Care should be taken to mark the original position of the bonnet to facilitate replacement.

Remove the battery, battery tray and drain tube. Remove the windscreen washer bottle from the wing valance.

Drain the water from the radiator and cylinder block.

Disconnect the top and bottom water hoses from the radiator.

Remove the two set bolts securing the sides of the radiator to the body.

Remove the two bottom radiator mounting nuts, washers and rubber mountings.

Remove the nuts and washers securing the fan cowl and servo air cleaner to the radiator. Place the cowl over the fan and withdraw the radiator.

Remove the cylinder head as described on page B.45.

Remove the front suspension as described in Section J.

Remove the damper as described on page B.41.

Withdraw the split cone.

Remove the sump as described on page B.28.

Unscrew the set bolts and nuts, and remove the water pump from the timing cover.

Note the gasket between the pump and the timing cover.

Remove the timing cover as described on page B.28.

Remove the bottom timing chain tensioner as described on page B.64.

Unscrew the four setscrews securing the front mounting bracket to the cylinder block.

Remove the two screwdriver slotted setscrews securing the rear mounting bracket;

The timing gear assembly can now be removed.

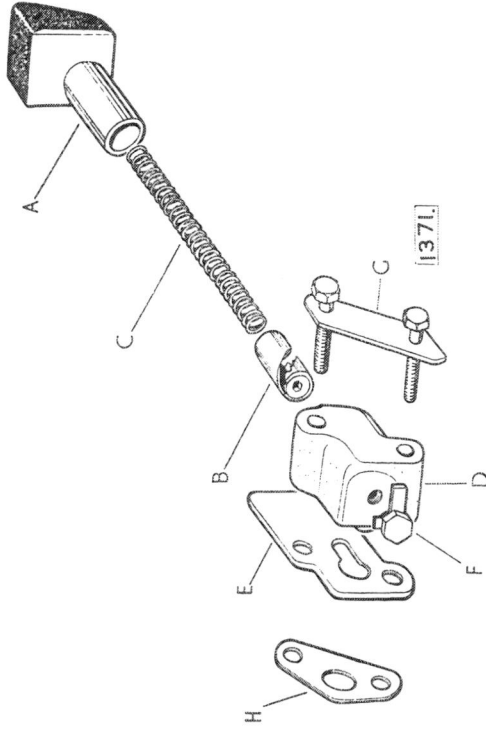

[1371.]

A. Plunger
B. Restraint cylinder
C. Spring
D. Adjuster body
E. Backing plate
F. End plug and tab washer
G. Body securing bolts and tab washer
H. Shim

Fig. 64. *Exploded view of the bottom timing chain tensioner.*

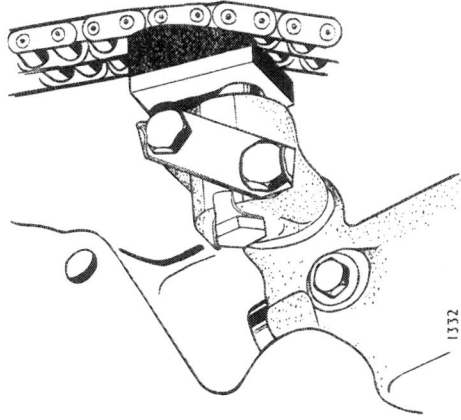

1332

65 *Showing the bottom timing chain tensioner in position.*

DISMANTLING

Remove the nut and serrated washer from the front end of the idler shaft, and withdraw the plunger and spring.

Remove the four nuts securing the front mounting bracket to the rear bracket. Withdraw the front bracket from the studs.

Remove the bottom timing chain from the large intermediate sprocket.

To remove the intermediate sprockets, remove the circlip from the end of the shaft in the mounting bracket. Press the shaft out of the bracket, and withdraw the sprockets from the shaft.

OVERHAUL

If the chain shows signs of stretching or wear, new ones should be fitted. Replace any sprockets and dampers that show signs of wear.

ASSEMBLING

Fit the eccentric shaft to the hole in front mounting bracket. Insert the spring and locking plunger for the serrated plate to the hole in the front mounting bracket. Fit the serrated plate and secure with the shakeproof washer and nut. Fit the idler sprocket (21 teeth) to the eccentric shaft.

Fit the intermediate sprocket (20 and 28 teeth) to shaft with the larger sprocket forward and press the shaft through lower central hole in rear mounting bracket. Secure with the circlip at the rear of bracket.

Fit the top timing chain (longer chain) to the small intermediate sprocket and the bottom timing chain (shorter chain) to the large intermediate sprocket.

Loop the upper timing chain under the idler sprocket and offer up the front mounting bracket to the rear mounting bracket with the two chain dampers interposed between the brackets.

Pass the four securing bolts through the holes in the brackets, chain dampers and spacers noting that shakeproof washers are fitted under the bolt heads. Secure the two mounting brackets together with four stud nuts and shakeproof washers.

REFITTING

Refitting the remainder of the assembly is the reverse of the removal procedure.

When refitting the timing chain tensioner refer to page B.65.

THE BOTTOM CHAIN TENSIONER

The bottom timing chain tensioner is of hydraulic type and consists of an oil resistant rubber slipper mounted on a plunger (A. Fig. 64) which bears on the outside of the chain. The light spring (C) cased by the restraint cylinder (B) and the plunger. in combination with oil pressure holds the slipper head against the chain keeping it in correct tension.

Return movement of the slipper head is prevented by the limit peg at the bottom end of the plunger bore engaging the nearest tooth on the helical slot of the restraint cylinder. The oil is introduced into the adjuster body (D) via a small drilling in the locating spigot and passing through a hole in the slipper head lubricates the chain. The backing plate (E) provides a suitable face along which the slipper head can work.

REMOVAL.

Proceed as described on page B.62 until the bottom chain tensioner is accessible.

Remove the bottom plug which provides access to the hexagonal hole in the end of the restraint cylinder. Insert an Allen key (.125" A.F.) into this and turn the key in a *clockwise* direction until the slipper head remains in the retracted position. Remove the securing bolts and detach the adjuster. If a conical filter is fitted in the oil feed hole in the cylinder block this should be removed and cleaned in petrol.

REFITTING

Fit the conical filter to oil feed hole in the cylinder block.

Fit shims as necessary. between the backing plate and cylinder block so that the timing chain runs centrally along the rubber slipper.

Fit the tab washer and two securing bolts. Tighten the bolts and tap the tab washers against the bolt heads.

It is important that no attempt is made to release the locking mechanism until the adjuster has been finally mounted in the engine WITH THE TIMING CHAIN IN POSITION.

Remove the hexagon head plug and tab washer from the end of the body. Insert the Allen key into the hole until it registers in the end of the restraint cylinder. Turn the key clockwise until the tensioner head moves forward under spring pressure against the chain. Do not attempt to turn the key anti-clockwise, nor force the tensioner head into the chain by external pressure.

Refit the plug and secure with the tab washer.

THE VALVES AND SPRINGS

The inlet valves are of silicon chrome steel and the exhaust valves are of austenitic steel. Double coil valve springs are fitted and are retained by a valve collar with split cotters.

Warning: As the valves in the fully open position protrude below the cylinder head joint face, the cylinder head must not be placed joint face downwards directly on a flat surface; support the cylinder head on wooden blocks, one at each end.

REMOVAL

Remove the cylinder head as described on page B.45.

Remove Valves

With the cylinder head on the bench remove the inlet manifold and the revolution counter generator.

Remove the four bearing caps from each camshaft and lift out the camshafts (note mating marks on each bearing cap).

Remove the twelve tappets and adjusting pads situated between tappets and valve stems. Lay out the tappets and pads in order, to ensure that they can be replaced in their original guides.

Obtain a block of wood the approximate size of the combustion chambers and place this under the valve heads in No. 1 cylinder combustion chamber. Press down the valve collars and extract the split cotters. Remove the collars, valve springs and spring seats. Repeat for the remaining five cylinders. Valves are numbered and must be replaced in the original locations, No. 1 cylinder being at the rear, that is the flywheel end.

OVERHAUL

Valves

Examine the valves for pitting, burning or distortion and reface or renew the valves as necessary. Also reface the valve seats in the cylinder head and grind the valves to their seats using a suction valve tool. When refacing the valves or seat inserts do not remove more metal than is necessary to clean up the facings.

The valve seat angles are as follows: inlet and exhaust, 45°.

Renew valves where the stem wear exceeds .003″ (.08 mm.). The clearance of the valve stem in the guide when new is .001″ to .004″ (.025 to .10 mm.).

Valve Springs

Test the valve springs for pressure, either by comparison with the figures given in the "Valve Spring Data" or by comparison with a new valve spring.

To test against a new valve spring, insert both valve springs end to end between the jaws of a vice or under a press with a flat metal plate interposed between the two springs. Apply a load partly to compress the springs and measure their comparative lengths.

When fitting valve springs to the cylinder head compress the springs using Churchill Tool No. J.6118.

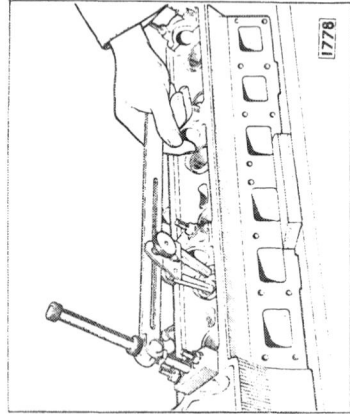

Fig. 66. *Fitting the valve springs, utilising the valve spring compressing tool. Churchill Tool No. J.6118.*

VALVE CLEARANCE ADJUSTMENT

When checking the valve clearances, the camshafts must be fitted one at a time as if one camshaft is rotated when the other camshaft is in position, fouling is likely to take place between the inlet and exhaust valves. Obtain and record all valve clearances by using a feeler gauge between the back of each cam and the appropriate valve tappet.

Correct valve clearances are:

Inlet	..	..	.004″ (.10 mm.)
Exhaust	..	..	.006″ (.15 mm.)

Adjusting pads are available rising in .001″ (.03 mm.) sizes from .085″ to .110″ (2.16 to 2.79 mm.) and are etched on the surface with the letter "A" to "Z", each letter indicating an increase in size of .001″ (.03 mm.). Should any valve clearance require correction, remove the camshaft, tappet and adjusting pad. Observe the letter etched on the existing adjusting pad and should the recorded clearance for this valve have shown say .002″ (.05 mm.) excessive clearance select a new adjusting pad bearing a letter two lower than the original pad.

As an example, assume that No. 1 inlet valve clearance is tested and recorded as .007″ (.18 mm.). On removal of the adjusting pad, if this is etched with the letter "D" then substitution with a pad bearing the letter "G" will correct the clearance for No. 1 inlet valve.

When fitting the camshafts prior to fitting the cylinder head to the engine it is most important that the keyway in the front bearing flange of each camshaft is perpendicular (at 90°) to the adjacent camshaft cover face before tightening down the camshaft bearing cap nuts.

Tighten the camshaft bearing cap nuts to a torque of 15 lb./ft. (2.0 kg. m.).

REFITTING

Before attempting to refit the cylinder head refer to the instructions given on page B.46.

THE VALVE GUIDES

The valve guides are of cast iron and are chamfered at the upper ends. The outside diameter of the guide is reduced at the upper end to provide a "lead-in" when fitting the guide to the cylinder head. The inlet and exhaust guides are of different lengths; the inlet being the shorter of the two.

REPLACEMENT

Examine the valve guides for evidence of wear in the bore. The clearance between the valve stem and the guide when new is .001 to .004″ (.025 to .10 mm.).

If it is found necessary to replace worn valve guides they must be fitted in accordance with the following instructions and only genuine factory replacement parts used.

(1) Press out, or drive out with a piloted drift, the old valve guide from the top of the cylinder head.

(2) Ream the valve guide bore in the cylinder head to a diameter of .505″ +.0005″ +.012 mm.) −.0002″ (12.83 mm. −.005 mm.).

(3) Heat the cylinder head by immersing in boiling water for 30 minutes.

(4) Coat the valve guide with graphite grease and press in, or drive in with a piloted drift, from the combustion chamber end. The correct fitted position for both inlet and exhaust guides is with the top of the guide (chamfered end) 5/16″ (8 mm.) above the spot facing for the valve spring seat. (See Fig. 67).

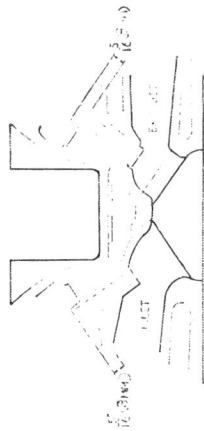

Fig. 67. *Showing the fitted position of the valve guide.*

THE VALVE SEAT INSERTS

The valve seat inserts are centrifugally cast iron and are shrunk into the cylinder head.

REPLACEMENT

If it is found necessary to replace the valve seat inserts they must be fitted in accordance with the following instructions and only genuine factory replacement parts used.

(1) Remove the old valve seat insert by boring out until the insert collapses. Take care not to damage the recess for insert in the cylinder head.

(2) Carefully measure diameter of insert recess in cylinder head at room temperature 68 F. (20°C.).

(3) Grind down outside of insert to a diameter of .003" (.08 mm.) larger than recess dimension, that is, to give an interference fit of .003" (.08 mm.).

(4) Heat the cylinder head in an oven for one hour from cold at a temperature of 300 F. (150°C.).

(5) Fit insert, ensuring that it beds evenly in its recess.

(6) After the valve seat insert has been fitted the following instructions should be carried out to ensure that the valve clearance can be obtained within the range of the adjusting pads, that is, .085" to .110" (2.16 to 2.79 mm.).

(a) Assemble the camshafts to the cylinder head. Fit the appropriate valve to the insert in question and, with the valve seat faces touching, check the distance between the top of valve stem and the **back** of the cam. This should be .320" (8.13 mm.) **plus** the appropriate valve clearance. (The figure of .320" (8.13 mm.) includes an allowance for an adjusting pad thickness of .095" (2.41 mm.) to .097" (2.46 mm.) which will, if necessary, permit the fitting of thicker or thinner adjusting pads when making the final valve clearance adjustment).

(b) If the distance is greater than the figure of .320" (8.13 mm.), plus the appropriate valve clearance, grind the valve seat of the insert with suitable valve grinding equipment until the correct distance is obtained.

Example: Assume that the valve insert in question is an exhaust and the distance between the top of the valve stem and the back of the cam is found to be .344" (8.74 mm.).

Adding the exhaust valve clearance of .006" (.15 mm.) to .320" (8.13 mm.) equals .326" (8.28 mm.). In this case the valve seat of the insert will have to be ground down to reduce the distance between the top of the valve stem and the back of the cam by .018" (.46 mm.) that is, .344" minus .326" (8.74 mm. minus 8.28 mm.).

(c) After assembling the cylinder head, check and adjust the valve clearances in the normal manner.

VALVE TIMING

Turn the engine so that No. 6 (front) piston is exactly in the T.D.C. position on compression stroke (firing position) that is, with the distributor rotor arm opposite No. 6 cylinder segment.

See Figs. 45, 71 or 72 for location of T.D.C. marks.

It is important to tension the top timing chain before attempting to check or set the valve timing. Proceed as follows:

Through the breather aperture in the front of the cylinder head slacken the locknut securing the serrated plate (Fig. 68).

Tension the chain by pressing locking plunger inwards and rotating serrated plate by the two holes in an anti-clockwise direction. Turn the engine each way slightly and recheck the chain tension. When correctly tensioned there should be slight flexibility on both outer sides below the camshaft sprockets, that is, the chain must not be dead tight. Release the locking plunger and securely tighten the locknut.

Remove the locking wire from the setscrews securing the camshaft sprockets. Note the positions of the **inaccessible** setscrews and rotate the engine until they can be removed. Remove the setscrew from each sprocket and turn the engine back to the T.D.C. position with the No. 6 firing and remove the remaining screws. Tap the camshaft sprockets off the flanges of the camshafts.

Accurately position the camshafts with the valve timing gauge, and check that the T.D.C. marks are in exact alignment.

Withdraw the circlips retaining the adjusting plates to the camshaft sprockets and press the adjusting plates forward until the serrations disengage. Replace the sprockets on the flanges of camshafts and align the two holes in the adjuster plate with the two tapped holes in each camshaft flange. Engage the serrations of the adjuster plates with the serrations in the sprockets.

Note: It is most important that the holes are in exact alignment, otherwise when the setscrews are fitted, the camshafts will be moved out of position. If difficulty is experienced in aligning the holes exactly the adjuster plates should be turned through 180°, which due to the construction of the plate will facilitate alignment.

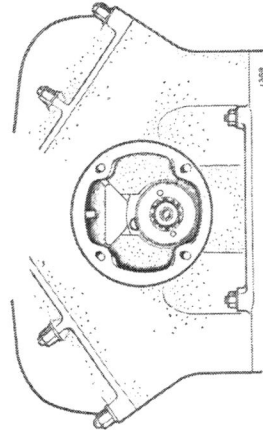

Fig. 68. *Showing the serrated plate for adjustment of the top timing chain tension.*

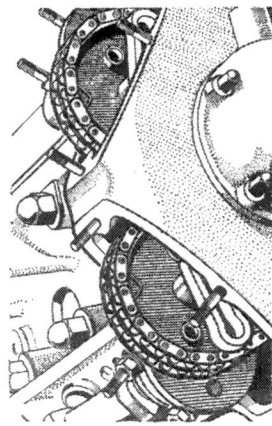

Fig. 69. *Showing the camshaft sprockets disconnected from the camshafts.*

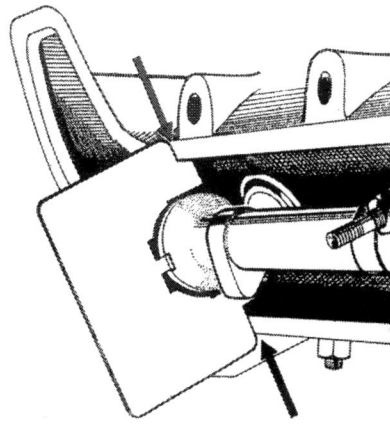

Fig. 70. *The valve timing gauge in position. Ensure that the gauge is seated at the points indicated by the arrows.*

Fit the circlips to the sprockets and one setscrew to the accessible holes in each adjuster plate. Turn the engine until the other two holes are accessible and fit the two remaining setscrews.

Finally, recheck the timing chain tension and valve timing in this order. Secure the four setscrews for camshaft sprockets with new locking wire.

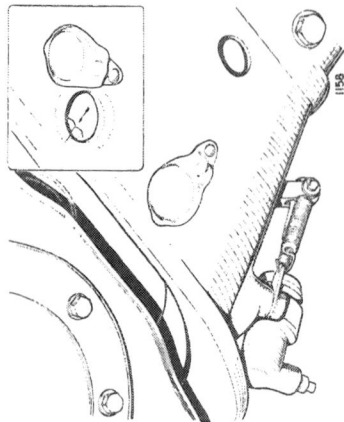

Fig. 71. *Showing the location of the Top Dead Centre marks on standard transmission cars as viewed from below with the dust cover moved to one side.*

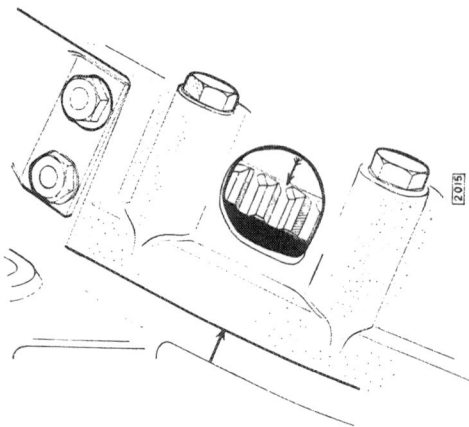

Fig. 72. *Showing the location of the Top Dead Centre marks on automatic transmission cars as viewed from the left-hand side of the combined engine and transmission unit.*

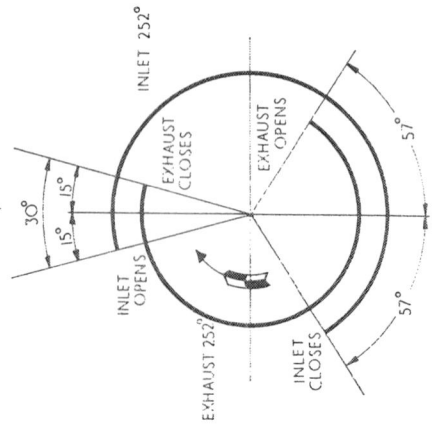

INLET OPENS
EXHAUST CLOSES
EXHAUST OPENS
INLET CLOSES
INLET 252°
EXHAUST 252°
30° 15° 15° 57° 57°

Fig. 73. *Valve timing diagram.*

ENGINE MOUNTINGS

The engine is supported at the front on two rubber mountings which are attached to brackets on the body underframe.

On standard transmission and overdrive cars the rear of the power unit is supported on a coil spring, mounted in a channel support which is bolted to the body floor. An extension of the spring retainer passes through a rubber bush in the channel support.

On automatic transmission cars the rear of the power unit is supported on two rubber mountings fitted between the rear extension case and a mounting bracket bolted to the body floor.

FRONT ENGINE MOUNTINGS
Removal

Either place a sling round the front of the engine or attach a lifting plate to the cylinder head. Unscrew the large set bolt, spring washer and plain washer securing the front engine mounting bracket to the mounting rubber. Repeat for the other side.

Raise the engine so that the front mounting brackets are just clear of the mounting rubbers.

Remove the two nuts, washers and bolts securing the front engine mounting rubber to the support bracket to the body underframe. Repeat for the other side.

Refitting

Refitting is the reverse of the removal procedure.

REAR ENGINE MOUNTING
(Standard Transmission and Overdrive Models)

The rear mounting coil springs fitted to standard transmission and overdrive models are of different lengths and must be fitted to their respective models: the free lengths are as follows:

	Standard transmission models	Overdrive models
Free length	3⅛ (80 mm.)	3½ (90 mm.)

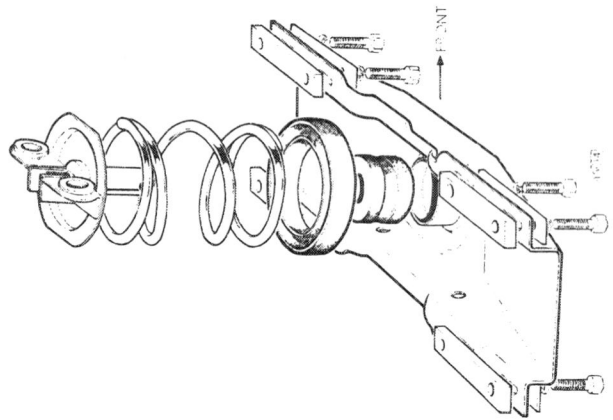

Fig. 74. *Exploded view of the rear mounting assembly (standard transmission and overdrive models).*

Removal

Place a jack under the gearbox to take the weight of the engine.

Place a large washer over the stem of the spring retainer protruding through the bottom of the channel support. Secure the large washer by inserting an $\frac{1}{8}$" (3 mm.) diameter rod through the hole in the stem of the spring retainer.

Unscrew the eight bolts securing the channel support to the body. Remove the bolts, washers, stiffening plates and packing blocks. Remove the two bolts, nuts and washers securing the rear engine mounting plate to the gearbox.

Dismantling

Turn the assembly upside down and place on a press.

Compress the spring and remove the rod from the hole in the stem of the spring retainer.

Slowly release the pressure on the spring and remove the large washer.

Withdraw the spring retainer, spring, rubber spring seat and centre bush from the support channel bracket.

Re-assembling

When re-assembling the unit it is important that the correct type of spring is fitted in relation to the type of transmission.

Press the centre bush into the hole in the bottom of the channel support.

Apply adhesive to the recess in the channel support and press the rubber spring seat into position.

Place the spring in the rubber spring seat.

Place the spring retainer into position so that the stem protrudes through the centre of the spring and the washers welded to the two large lugs on the spring retainer are facing the side of the channel support, which has an extra cut-away portion in the flange.

Compress the spring until the stem of the spring retainer protrudes far enough through the bottom of the channel support to allow a large washer to be placed over the stem and an $\frac{1}{8}$" (3 mm.) rod to be inserted through the hole in the stem.

Refitting

Offer up the rear engine mounting assembly to the rear of the two lugs on the gearbox casing. There is an extra cut-away in one of the channel support flanges and this should be facing towards the front of the car.

Secure the spring retainer to the gearbox with two bolts, nuts and washers.

Jack up the gearbox until the channel support bracket holes line up with the holes in the body.

Insert the packing blocks between the body floor and the channel support bracket and secure with the four stiffening plates, eight set bolts and washers.

Remove the rod and the washer from the stem of the retainer.

Release the jack.

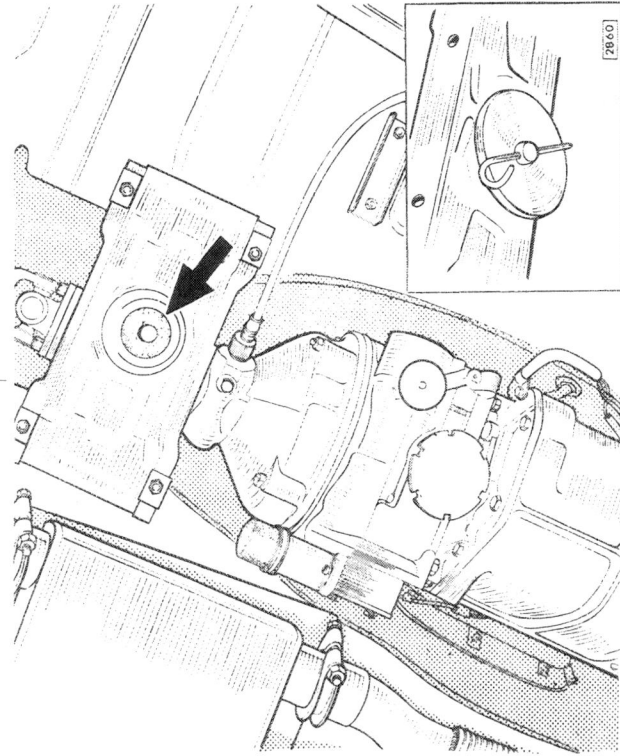

Fig. 75. *The hole at the bottom of the spring retainer stem is for assembly purposes only. The inset shows the large washer and rods in position prior to removal of the rear mounting.*

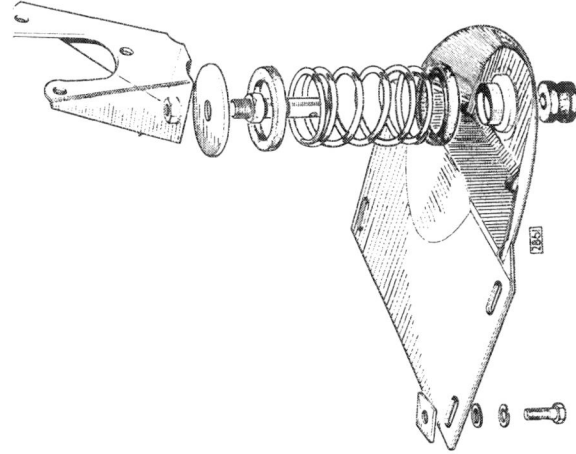

REAR ENGINE MOUNTING (Automatic Transmission Models)

Removal

Unscrew the six set bolts securing the ventilated cover plate to the bottom of the torque converter housing and remove cover plate. Place a piece of wood under the torque converter housing, taking care that it does not foul the torque converter.

Jack up under the piece of wood until the jack takes the weight of the engine.

Unscrew the self-locking nut A Fig. 78 on the engine stabiliser and remove the upper flanged washer B.

Mark the positions of the rear engine mounting bracket relative to the body floor so that the bracket can be refitted in its original position.

Remove the six bolts and packing washers from the rear engine support bracket, care being taken to note the number and positions of the various packing washers fitted between the bracket and the body floor.

Remove the two nuts and shakeproof washers attaching the rear engine support bracket to the two mounting rubbers.

Lower the jack slightly to facilitate the removal of the nuts and shakeproof washers securing the two mounting rubbers to the bracket attached to the transmission unit.

Refitting

After refitting the rubbers and rear mounting plate adjust the engine stabiliser as follows:

1. Screw the lower flanged washer up the stabilizer pin until the flange contacts the bottom of the stabiliser mounting. The washer is slotted on its upper face and can be screwed up the pin by engaging a thin bladed screwdriver in the slot through the centre hole of the rubber mounting.

2. Fit the upper flanged washer and tighten down with the self-locking nut.

Failure to observe the above procedure may cause engine vibration and or fouling of the transmission unit in its cowl due to the engine being pulled up on its mounting.

Rear Mounting Coil Spring (Automatic Transmission)

Free length 4$\frac{11}{16}$" (119 mm.).

Fig. 77. *Rear engine mounting (automatic transmission models).*

THE ENGINE STABILISERS

FRONT ENGINE STABILISERS (Fig. 77)

Description

There are two engine stabilisers mounted on the underframe which are attached to the front engine mounting brackets. Each stabiliser consists of a rubber bushed link pin (A), threaded at one end, on which are two nuts, B.

The link pin passes through a steel and rubber mounting consisting of flat steel washers (C), rubber bushes (E) and dished steel washers (F) fitted either side of the bracket on the underframe. The complete assembly is held in place by means of a self-locking nut. A distance tube (D) prevents the over-tightening of the assembly.

Fitting

In order to avoid pulling the engine down on its mountings it is IMPORTANT that the front stabilisers are assembled in the following manner:

With nuts (B) at the top of the threaded portion of the link pin, fit washer (C), distance tube (D), rubber bush (E) and dished washer (F).

Pass the link pin, with the above components mounted on it, through the hole in the underframe bracket and bolt up to the engine mounting bracket.

Fit the dished washer, rubber bush and flat washer to the portion of the link pin projecting through the underside of the underframe mounting.

Fit the self-locking nut but do not tighten.

Tighten nuts (B) until locked.

The self-locking nut can now be tightened.

REAR ENGINE STABILISER

The rear engine stabiliser consists of a rubber/steel mounting, attached to the body, which is connected to brackets on the clutch housing via a rubber bushed link pin. The link pin is threaded at its upper end and is connected to the rubber mounting by means of flanged washers and a self-locking nut.

Adjustment

It is MOST IMPORTANT that the stabiliser is assembled in the following manner as failure to observe this procedure may cause engine vibration and/or fouling of the gearbox in its cowl due to the engine having been pulled up on its mountings.

1. Screw the lower flanged washer (D, Fig. 78) up the stabiliser pin until the flange contacts the bottom of the stabiliser rubber mounting (C). The washer is slotted on its upper face and can be screwed up the pin by engaging a thin bladed screwdriver in the slot through the centre hole of the rubber mounting.

2. Fit the upper flanged washer (B) and tighten down with the self-locking nut (A).

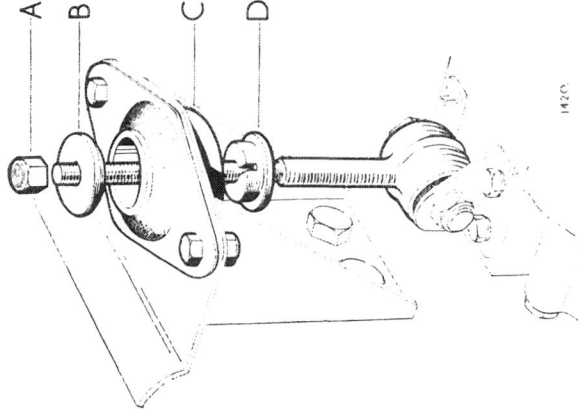

Fig. 78. *The rear engine stabiliser.*

Fig. 77. *Front engine stabilisers.*

SUPPLEMENTARY INFORMATION
TO
SECTION B "ENGINE"

AIR CLEANER

The air cleaner is of the paper element type and is fitted on top of the cylinder head.

No maintenance is necessary, but the element should be renewed every 12,000 miles (20,000 km.) or more frequently in dusty territories.

Roll back the sealing rubber between the carburetter elbow and the air cleaner. Slacken the two wing nuts securing the air cleaner to the

bracket on the cylinder head. Release the air cleaner by pulling it towards the left-hand wing valance. Release the two clips securing the end cover to the air cleaner. Withdraw the end cover and element. Remove the wing nut, washer, end cap and rubber ring securing the element to the end cover. When refitting the element ensure that the two rubber sealing rings are in their correct positions.

Fig. 79. Exploded view of air cleaner.

1. Shell	11. Rubber seal
2. End plate and element bracket	12. Air intake
3. Paper element	13. Gasket
4. Gasket	14. Dowel bolt
5. Plate	15. Adaptor
6. Wing nut	16. Drain tube
7. Washer	17. Clip
8. Wing nut	18. Intake hose
9. Washer	19. Clip
10. Bracket	20. Rubber seal

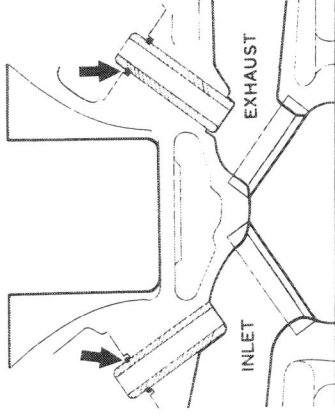

Fig. 80. Showing the fitted position of the valve guides in the cylinder head.

VALVE GUIDES

Commencing Engine Numbers

| 3.4 "S" | 7B 7090 |
| 3.8 "S" | 7B 60959 |

Commencing at the above engine numbers new valve guides are fitted. Circlips fitted to the valve guide ensure positive location of the guide in the cylinder head by engaging in counterbores machined in the valve guide bores of the head.

The outside diameter of the valve guide is reduced at the lower end to provide a "lead-in" when fitting the guide to the cylinder head.

REMOVAL

Heat the cylinder head by immersing in boiling water for 30 minutes. With a piloted drift, drive out the old valve guide from the combustion chamber end.

Ream the valve guide bore in the cylinder head to a diameter of $0.505'' + .0005''$ (.012 mm) $- .0002''$ (.005 mm).

If the bores are larger than these dimensions they should be reamed out to the following dimensions and the respective oversize valve guides fitted.

$0.510'' + .0005''$ (.012 mm)
$- .0002''$ (.005 mm)

REFITTING

Coat the valve guide with graphite grease and fit the circlip. Re-heat the cylinder head.

With a piloted drift, drive in the valve guide from the top until the circlip registers in the counterbore machined in the guide bore of the cylinder head. Visually check that the circlip has seated correctly.

REAR ENGINE MOUNTINGS

Commencing Chassis Numbers

	R.H. Drive	L.H. Drive
3.4 "S"	1B 2192	1B 25301
3.8 "S"	1B 52078	1B 76310

On cars with the above chassis numbers and onwards, a modified rear mounting for OVERDRIVE models is fitted. This is similar to that used only on cars with automatic transmission.

The parts introduced by this modification are not interchangeable with those previously fitted to Overdrive Transmission models.

SECTION C

CARBURETTERS AND FUEL SYSTEM

INDEX

CARBURETTERS

THE FUEL SYSTEM

CARBURETTERS

DESCRIPTION

The 3.4 'S' and 3.8 'S' models are fitted with twin S.U. H.D.6 type carburetters. The enrichment device for starting is in the form of an auxiliary carburetter attached to the front carburetter.

The H.D. type carburetter differs from the earlier type in that the jet glands are replaced by a flexible diaphragm, and idling mixture is conducted along a passage way, in which is located a metering screw, instead of being controlled by a throttle disc.

The jet (18)(Fig. 5) which is fed through its lower end, is attached to a synthetic rubber diaphragm (10) by means of the jet cup (9) and jet return spring cup (13), the centre of the diaphragm being compressed between these two parts ; at its outer edge it is held between the diaphragm casing (14) and the float chamber arm. The jet is controlled by the jet return spring (12) and the jet actuating lever (15), the latter having an external adjusting screw which limits the upward travel of the jet and thus controls the mixture adjustment ; screwing in (clockwise) enriches the mixture, and unscrewing it weakens the mixture

Throttle spindle glands

Provision is made for the use of throttle spindle glands consisting of the cork gland itself (25), a dished retaining washer (28), a spring (27) and a shroud (26). This assembly should not require servicing and can only be removed by dismantling the throttle spindle and disc.

Idling

The carburetter idles on the main jet and the mixture is conducted along the passage way (8) connecting the choke space to the other side of the throttle disc.

The quantity of mixture passing through the passage way and, therefore, the idling speed of the engine is controlled by the "slow-run" valve (5), the quality or relative richness of the mixture being determined by the jet adjusting screw. It follows that when idling, the throttle remains completely closed against the bore of the carburetter

DATA

Type		S.U. H.D.6 (twin)
Size		1¾" (4.45 cm.)
Jet needle type—		
3.4 litre	7 to 1 comp. ratio	TI.
	8 to 1 comp. ratio	TI.
	9 to 1 comp. ratio	TI.
3.8 litre	7 to 1 comp. ratio	TI.
	8 to 1 comp. ratio	TI.
	9 to 1 comp. ratio	TI.
Jet size		.10" (2.54 mm.)
Auxiliary starting carburetter needle type		425 8.

Note : The jet needle type is stamped on the side of the parallel portion of the needle. The auxiliary starting carburetter needle is stamped with the large number (e.g. 425) on the shoulder at the top of the needle and with the small number on the parallel portion of the needle.

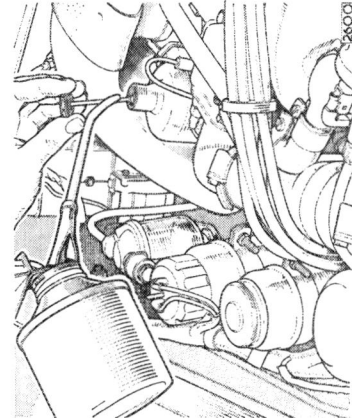

ROUTINE MAINTENANCE

Every 3,000 miles (5,000km.)

Lubricate Carburetter Piston Damper

Each carburetter is fitted with a hydraulic piston damper which, unless periodically replenished with oil, will cause poor acceleration and spitting back through the carburetter on rapid opening of the throttle.

To replenish with oil, unscrew the cap on top of suction chambers and lift out the damper valve which is attached to the cap. Fill the hollow piston spindle, which can be seen down inside the bore of the suction chamber, with S.A.E. 20 engine oil.

Checking Carburetter Slow Running

The idling speed of the engine should be 500 r.p.m. when the engine is at its normal working temperature.

On cars fitted with the all-syncromesh gearbox, the idling speed should be 700 r.p.m.

For cars equipped with automatic transmission the idling speed should be 600 r.p.m. with the selector in P or N.

If adjustment is required turn the two slow running volume screws (see Fig. 6) by **exactly equal amounts** until the idling speed observed on the revolution counter instrument is correct.

Every 6,000 miles (10,000km.)

Tune Carburetters

See instructions given on page C.7.

Cleaning Carburetter Filters

Removal of the bolt securing the petrol pipe banjo union to each float chamber will expose the filters. Remove the filters and clean in petrol; do not use a cloth as particles will stick to the gauze.

When refitting, insert the filter with the spring first and ensure that the fibre washers are replaced one to each side of the banjo union.

Fuel Feed Line Filter

The filter is attached to the right-hand wing valance and is of the glass bowl type with a flat filter gauze.

At the recommended intervals, or more frequently if the glass bowl shows the presence of water and or sediment, slacken the locking nut, swing the retaining clip to one side and remove the bowl, sealing washer and filter gauze.

Clean the filter gauze and bowl by washing in petrol. Examine the sealing washer and if necessary fit a new one.

Fig. 2. Topping up a carburetter hydraulic piston damper.

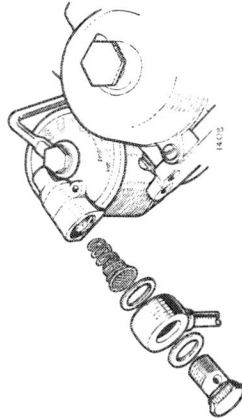

Fig. 3. Carburetter filter removal.

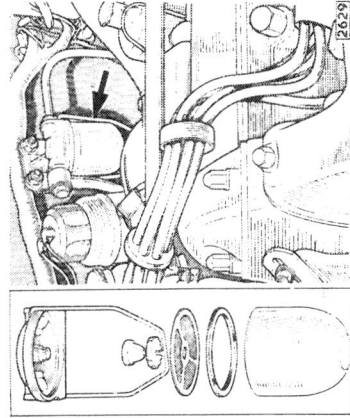

Fig. 4. Fuel feed line filter.

1. Front carburetter
2. Rear carburetter
3. Carburetter body
4. Ignition union adaptor
5. Gasket
6. Suction chamber and piston
7. Damper
8. Washer
9. Spring
10. Skid washer
11. Jet needle
12. Jet
13. Jet bearing
14. Nut—jet bearing
15. Spring
16. Jet unit housing
17. Float chamber
18. Float chamber lid
19. Float
20. Needle and seat
21. Float needle lever
22. Knurled pin
23. Gasket
24. Cap nut
25. Fibre serrated washer
26. Aluminium washer
27. Filter
28. Banjo bolt
29. Fibre washer
30. Thermostat body
31. Acceleration needle
32. Spring
33. Jet
34. Spring finger
35. Dust shield
36. Screw
37. Washer
38. Solenoid
39. Spring clip
40. Bracket
41. Connecting arm
42. Banjo bolt
43. Fibre Washer
44. Fibre Washer
45. Banjo Bolt
46. Fibre Washer
47. Aluminium washer
48. Slow running control valve
49. Spring
50. Neoprene washer
51. Brass washer
52. Connecting rod
53. Connecting rod coupling
54. Throttle return spring lever
55. Throttle operating lever
56. Anti-creep throttle switch
57. Insulator   lever
58. Gasket
59. Overflow pipe
60. Clip
61. Suction pipe
62. Neoprene tube
63. Elbow
64. Petrol feed pipe
65. Banjo bolt
66. Fibre washer
67. Petrol filter
68. Filter casting
69. Sealing washer
70. Filter gauze
71. Glass bowl
72. Retaining strap
73. Bracket
74. Bracket
75. Throttle return spring
76. Throttle return spring
77. Bracket
78. Throttle stop bracket
79. Dowel bolt
80. Intermediate throttle link
81. Trunnion
82. Throttle link rod
83. Intermediate throttle lever
84. Bracket

2775

Fig. 1. *Exploded view of the S.U. carburetter.*

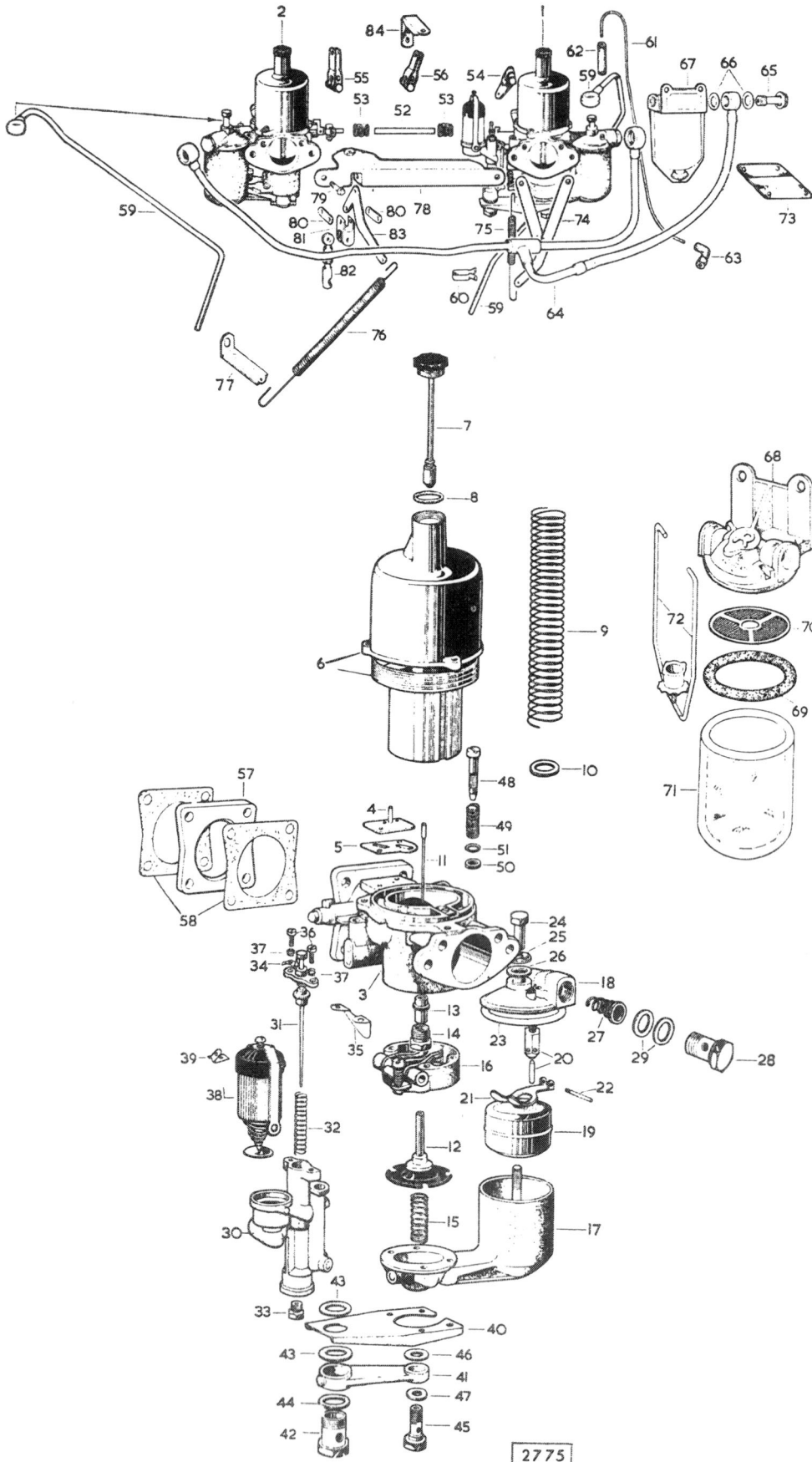

CARBURETTERS (Fig. 1)

Removal

Remove the air cleaner positioned across the cylinder head. Remove the air intake pipe by unscrewing the setscrews attaching the pipe to the carburetters. Remove both banjo bolts and the four fibre washers from the float chambers. Disconnect the two return springs (75 and 76) and the distributor vacuum pipes from the front carburetter. Remove the cover on the auxiliary starting carburetter solenoid on the side of the front carburetter (1) and disconnect the electrical cables. Remove the clip attaching the overflow pipes from the float chambers to the oil filter mounting screw and disconnect union connecting starter pipe to auxiliary starter carburetter.

Remove split pin, plain and spring washers from the connecting link pivot located on the manifold between front and rear carburetters and disconnect throttle link rod (82) from ball pin on bell crank lever.

Remove the four nuts and washers securing each carburetter to the inlet manifold. The carburetters can now be removed from the inlet manifold.

Refitting

Refitting is the reverse of the removal procedure.

CLEANING THE SUCTION CHAMBER AND PISTON

This should be done at approximate intervals of every twelve months or if the carburetter is dismantled for any reason. After detaching, clean the main inside bore of the suction chamber and the two outside diameters of the piston with a rag moistened in petrol or thinners and then reassemble in a dry and clean condition with a few spots of thin oil on the piston rod only. Do NOT use metal polish to clean the suction chamber and piston.

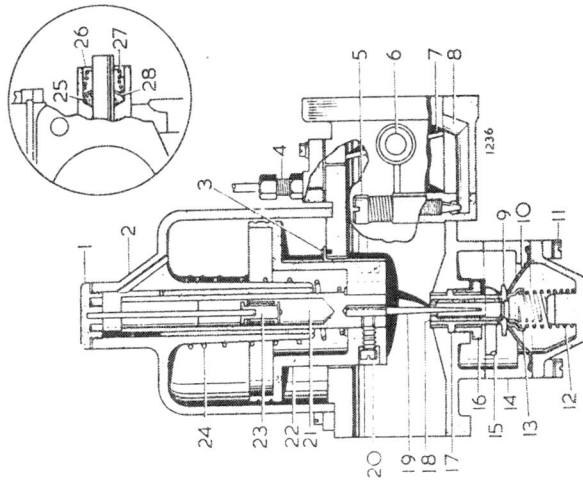

Fig. 5. *Sectioned view of the S.U. carburetter.*

1. Damper cap
2. Suction chamber
3. Piston guide
4. Union for vacuum advance/retard
5. Slow running volume screw
6. Throttle spindle
7. Throttle butterfly
8. Slow run passage
9. Jet cup
10. Diaphragm
11. Float chamber securing screw
12. Jet return spring
13. Return spring cup
14. Jet unit housing
15. Actuating lever
16. Nut—jet bearing
17. Jet bearing
18. Jet
19. Jet needle
20. Needle retaining screw
21. Oil reservoir
22. Piston
23. Damper
24. Piston return spring
25. Throttle spindle gland
26. Shroud for spring
27. Spring
28. Washer

Slacken one clamp bolt on the coupling between the throttle spindles. Check that both butterfly valves are fully closed by rotating both throttle spindles clockwise when viewed from the front. Tighten the coupling clamp bolt.

Screw in (rotate clockwise) the slow running volume screws until they are down fully on their seatings. Unscrew each screw 2 turns.

Remove the piston and suction chambers. Unscrew the mixture adjusting screws (B) until each jet is flush with the bridge of its carburetter. Replace the pistons and suction chambers and check (by means of the piston lifting pin) that each piston falls freely on to the bridge of its carburetter. Turn down the mixture adjusting screws 2½ turns.

Check that the hydraulic piston dampers are topped up with the recommended grade of engine oil.

Restart the engine and adjust to the desired idling speed of 500 r.p.m.* by moving each slow running volume screw an equal amount. By listening to the hiss in the intakes, adjust the slow running screws until the intensity of the hiss is similar on all intakes. This will synchronise the mixture flow of the carburetters.

* For cars equipped with automatic transmission the idling speed should be 600 r.p.m. in P or N.

CARBURETTER TUNING

Before tuning the carburetters, the sparking plug gaps, contact breaker gap, and static ignition timing should be checked and adjusted if necessary. The distributor centrifugal advance mechanism and vacuum advance operation should be checked and ignition timing set to the figure given under "General Data", with the centrifugal advance mechanisms in the static position. For final road test, adjustment of not more than six clicks of the micrometer adjustment at the distributor to either advance or retard is permitted.

The ignition setting is important since if retarded or advanced too far the setting of the carburetters will be affected. As the needle size is determined during engine development, tuning of the carburetters is confined to the correct idling setting.

If after tuning the carburetters, the idling setting and engine performance is not satisfactory, it will be necessary to check the cylinder compressions and the valve clearances.

Tuning

The air intake should be removed and the engine run until it has attained its normal operating temperature.

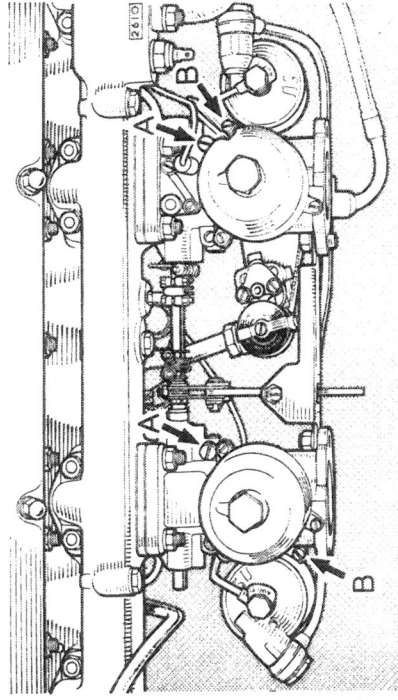

Fig. 6. *Carburetter tuning. A Slow running volume screw. B Mixture adjusting screw.*

When this is satisfactory the mixture should be adjusted by screwing both the mixture adjusting screws up (weaker) or down (richer) by the same amount until the fastest idling speed is obtained consistent with even firing.

As the mixture is adjusted, the engine will probably run faster and it may therefore be necessary to screw down the slow running volume screws in order to reduce the speed.

Now check the mixture strength by lifting the piston of the front carburetter by approximately $\frac{1}{32}''$ (.8 mm.) when, if :

(a) the engine speed increases and **continues to run faster,** this indicates that the mixture is too rich.

(b) the engine speed immediately decreases, this indicates that the mixture is too weak.

(c) the engine speed **momentarily** increases very slightly, this indicates that the mixture is correct.

Repeat the operation at the remaining carburetter and after adjustment recheck the front carburetter since the carburetters are interdependent.

When the mixture is correct, the exhaust note should be regular and even. If it is irregular with a splashy type of misfire and colourless exhaust, the mixture is too weak. If there is a regular or rhythmical type of misfire in the exhaust beat together with a blackish exhaust, then the mixture is too rich.

Float Chamber Fuel Level

When the fuel level setting is correct a $\frac{7}{16}''$ (11.1 mm.) test bar will just slide between the lid face and the inside curve of the float lever fork when the needle valve is in the "shut-off" position (Fig. 9).

If the float lever fails to conform with this check figure, it must be carefully bent at the start of the fork section to correct.

Ensure that, both forks are level and that the straight portion of the lever is perfectly flat.

When setting the fuel level ensure that the spring loaded plunger (A) in the "Delrin" needle is not compressed.

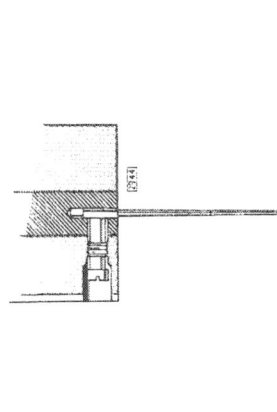

Fig. 7. *The jet needle must be positioned with the shoulder flush with the bottom face of the piston.*

Fig. 8. *The carburetter piston lifting pin : the first part of the movement is spring loaded free travel.*

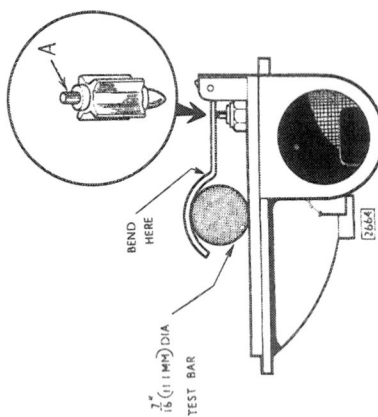

$\frac{7}{16}''$ (11·1 MM.) DIA
TEST BAR

BEND HERE

A

Fig. 9. *Checking the float lever setting, which controls the fuel level in the float chamber.*

CENTRING THE JET

Remove the carburetter from the engine as described on page C.6.

Remove the four setscrews securing the float chamber to the carburetter body. Remove the float chamber, jet housing and jet. Remove the hydraulic damper.

With a ring spanner slacken the jet locking nut approximately half a turn. Replace the jet and diaphragm assembly.

The jet is correctly centred when the piston falls, freely and hits the jet "bridge" with a metallic click. To centre the jet, push the jet and diaphragm assembly as high as possible with the hand and with a pencil or rod gently press the piston down on to the jet bridge ; centralisation will be facilitated if the piston is allowed to fall sharply onto the jet bridge several times. Tighten the jet locking nut.

The actual centring must be carried out with the setscrews holes in the jet diaphragm and carburetter in alignment. After tightening the jet locking nut the jet diaphragm must be kept in the same position relative to the carburetter body : the simplest way to do this is to mark one of the corresponding jet diaphragm and carburetter body setscrew holes with a soft pencil. Failure to do this may cause the centralisation to be upset.

Check that the centralisation is correct by noting if there is any difference in the sound of the piston hitting the jet bridge with the jet in its highest and lowest positions. If there is any difference in the sound, the procedure for centralising the jet will have to be repeated.

If difficulty in centring the jet is encountered after carrying out the above procedure, the jet needle can be lowered slightly in the piston to make the centralising effect more positive. The needle must, however, be restored to the normal position when checking the centralisation.

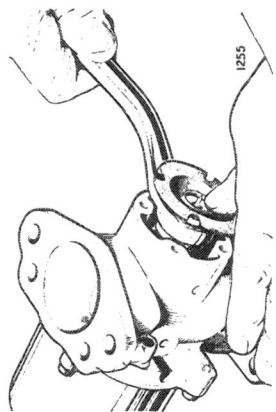

Fig. 10. *Centring the jet.*

1255

Warning: Take care not to bend the carburetter needle when carrying out this operation.

THE
AUXILIARY STARTING CARBURETTER

Description (Fig. 11)

The enrichment apparatus for starting is, in effect, an auxiliary carburetting system. The main body casting (1) containing a solenoid-operated valve and fuel metering system is illustrated as a separate unit attached by means of a ducted mounting arm to the base of the carburetter fuel inlet.

The auxiliary carburetter forms, therefore, a separate unit additional to the normal float chamber retained by the hollow cross-drilled bolt.

Fuel is supplied to the base of the jet (9), which is obstructed to a greater or lesser degree by the tapered slidable needle (10).

When the device is in action air is drawn from atmosphere through the air intake (7) and thence through the passage (8), being carburetted with fuel as it passes the jet (9). The mixture is thence carried upwards past the shank of the needle (10) through the passage (14) and so past the aperture provided between the valve (3) and its seating (2). From here it passes directly to the inlet manifold through an external feed pipe.

The device is brought into action by energizing the winding of the solenoid (5) from the terminals (6). The centrally located iron core (4) is thus raised magnetically, carrying with it the ball-jointed disc valve (3) against the load of the small conical spring and thus uncovering the aperture provided by the seating (2).

Considering the function of the slidable needle (10), it will be seen that this is loaded upwards in its open position by means of the light compression spring (11) which abuts against a disc (12) attached to the shank of the needle. The needle continues upwards through the vertically adjustable stop (13) in which it is slidable mounted and it finally terminates in an enlarged head.

Depression within the space surrounding the spring (11) is directly derived from that prevailing in the induction tract, and this exerts a downward force upon the disc (12), which is

provided with an adequate clearance with its surrounding bore. This tends to overcome the load of the spring (11) and to move the needle downwards, thus increasing the obstruction afforded by the tapered section which enters the jet (9).

The purpose of this device is to provide two widely different degrees of enrichment, the one corresponding to idling or light cruising conditions and the other to conditions of open throttle or full-power operation. In effect, under the former conditions the high induction depression prevailing will cause the disc (12) to be drawn downwards, drawing the tapered needle into the jet (9), while under the latter, the lower depression existing in the induction tract will permit the collar to maintain its upward position with the needle withdrawn from the jet.

The tuning elements concerned in this device are the size and degree of taper of the lower end of the needle (10), the diameter of the disc (12), the load provided by the spring (11) and the degree of movement permitted to the needle assembly, as determined by the adjustment of the stop (13).

The solenoid (5) is energized by means of a thermostatically operated switch housed in the inlet manifold water jacket. This is arranged to bring the apparatus into action at temperatures below about 30—35°C. (86—95°F.).

Adjustment

The engine must be at its normal running temperature and the carburetters correctly adjusted before any attempt is made to tune the auxiliary enrichment device.

As it can generally be assumed that the tapered form of the needle (10), the strength of the spring (11), and the diameter of the disc (12) have already been appropriately chosen, tuning is generally confined to the adjustment of the stop screw (13). It will be appreciated that the main purpose of this adjustment is to limit the downward movement of the needle, the head of which abuts against the upper surface of the stop screw at the lower extremity of its travel. The final downward movement of this needle determines, as has been described, the degree of enrichment provided under idling conditions

with the auxiliary carburetter in operation. An appropriate guide to its correct adjustment in this respect is provided by energizing the solenoid when the engine has already attained its normal temperature. The stop screw (13) should then be so adjusted that the mixture is distinctly although not excessively rich, that is to say, until the exhaust gases are seen to be discernibly black in colour, but just short of the point where the engine commences to run with noticeable irregularity.

Anti-clockwise rotation of the stop screw will, of course, raise the needle under these conditions and increase the mixture strength, while rotation in the opposite direction will have the opposite effect. In order to energize the solenoid under conditions when the thermostatic switch will normally have broken the circuit, it is merely necessary to short-circuit the terminal of the thermostatic switch directly to earth with a screwdriver and flick open the throttles when the starting device will be heard to come into operation with a pronounced hissing noise.

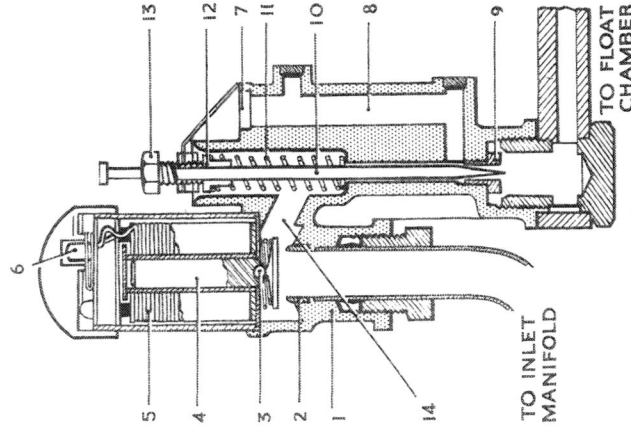

TO INLET MANIFOLD

TO FLOAT CHAMBER

1242

Fig. 11. *Sectioned view of the auxiliary starting carburetter.*

THE FUEL SYSTEM

THE PETROL PUMP (Fig. 14)

Description

The pump consists of three main assemblies, the main body casting (A); the diaphragm armature and magnet assembly (M) contained within the housing ; and the contact breaker assembly housed within the end cap (T2). A non-return valve assembly (C) is affixed to the end cover through the contact breaker chamber.

The main fuel inlet (B) provides access to an inlet air bottle (I) while access to the main pumping chamber (N) is provided by an inlet valve assembly. This assembly consists of a Melinex valve disc (F) permanently assembled within a pressed-steel cage, held in position by a valve cover (F1).

The outlet from the pumping chamber is provided by an identical valve assembly which operates in the reverse direction. Both inlet and outlet valve assemblies together with the filters are held in position by a clamp plate (H). The valve assemblies may be removed by detaching the clamp plate (H) after removing the self-tapping screws. A filter (E) is provided on the delivery side of the inlet valve assembly.' The delivery chamber (O) is bounded by a flexible plastic spring loaded diaphragm (L) contained by the vented cover (P). Sealing of the diaphragm (L) is provided by the rubber sealing ring (L2).

The magnetic unit consists of an iron coil housing, an iron core (Q), an iron armature (A1) provided with a central spindle (P1) which is permanently united with the diaphragm assembly (L1), a magnet coil (R) and a contact breaker assembly consisting of parts (P2). (U1), (U), (T1), (V). Between the coil housing and the armature are located 11 spherically edged rollers (S). These rollers locate the armature (A1) centrally within the coil housing and permit freedom of movement in a longitudinal direction.

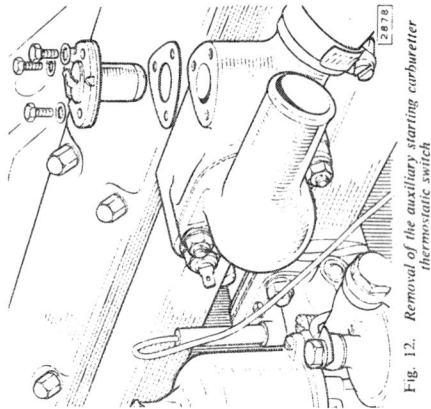

Fig. 14. *The Petrol Pump.*

WARNING : If at any time, it becomes necessary to blow through the fuel feed pipes the outlet pipes must be disconnected from the pumps. Failure to observe this procedure will cause the Melinex valves to be displaced or damaged.

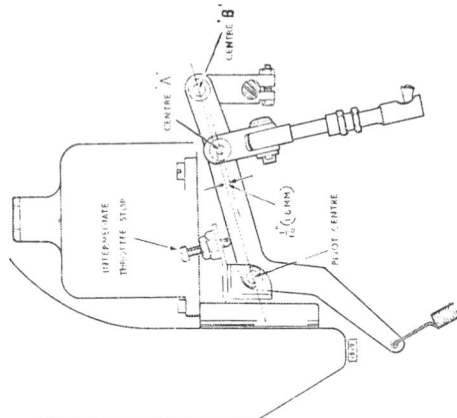

Fig. 12. *Removal of the auxiliary starting carburetter thermostatic switch*

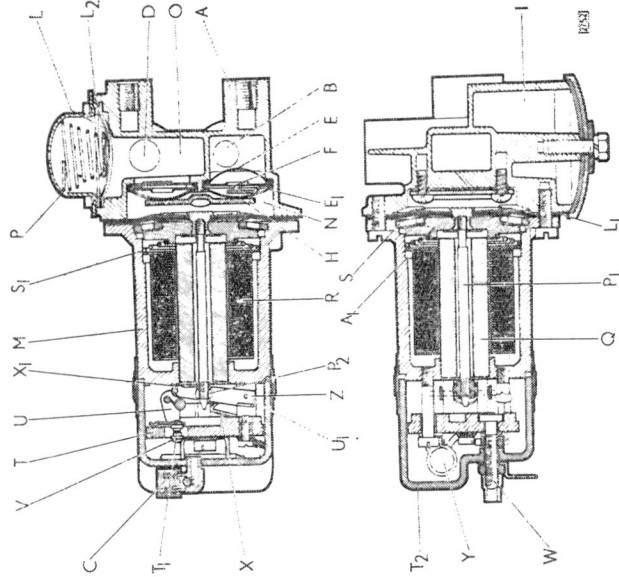

Fig. 13. *Throttle control linkage setting.*

Thermostatic Switch—Removal

The thermostatic switch which controls the operation of the auxiliary starting carburetter is situated at the front end of the inlet manifold water jacket.

Remove the electrical cable from the switch by removing the chrome plated domed nut.

If the radiator filler cap is securely tightened no appreciable amount of water will escape when the auxiliary starting carburetter switch is removed. Alternatively, a small amount of water can be drained from the radiator.

Remove the three securing setscrews and washers and withdraw the switch and the cork gasket.

Refitting

Refitting is the reverse of the removal procedure. A new cork gasket must be fitted when the switch is replaced. If any water has been drained from the radiator or has escaped during the removal of the switch, the radiator should be topped up to the correct level.

THROTTLE CONTROL LINKAGE SETTING

If carburetters have been removed or throttle linkage has been disturbed particular attention must be paid to the setting adjustment of the control linkage.

To adjust proceed as follows :—

(i) Disconnect front carburetter coupling and rear carburetter throttle lever by releasing clamp bolts. Check that both butterflies are fully closed and that the rear carburetter coupling bolt is clearing the manifold nut. With both carburetters fully closed retighten front coupling.

(ii) Unscrew intermediate throttle stop and push down on bell crank lever until centre "A" is 1/16" (1.6 mm.) below a line from centre "B" to pivot centre (Fig. 13). When in this position screw down stop on to intermediate throttle lever and lock in position. Lock lever to carburetter spindle.

(iii) Ensure that when throttle is closed the intermediate lever does not foul petrol connection pipe. Open throttle fully and check that both carburetters are in the fully open position.

The contact breaker consists of a bakelite pedestal moulding (T) carrying two rockers (U) and (U1) which are both hinged to the moulding at one end by the rocker spindle (Z). These rockers are interconnected at their top ends by means of two small springs arranged to give a throw-over action. A trunnion (P2) is carried by the inner rocker and the armature spindle (P1) is screwed into this trunnion. The outer rocker (U) is fitted with two tungsten points which contact with corresponding tungsten points which form part of the spring blade (V) connected to one end of the coil. The other end of the coil is connected to a terminal (W) while a short length of flexible wire (X), connecting the outer rocker to one of the screws holding the pedestal moulding onto the coil housing, provides an earth return to the body of the pump. It is important that the body of the pump be effectively earthed to the body of the vehicle by means of the earthing terminal provided on the flange of the coil housing.

Operation

When the pump is at rest the outer rocker (U) lies in the outer position and the tungsten points are in contact. Current passes from the Lucar connector (W) through the coil and back to the blade (V), through the points and to earth, thus energising the coil and attracting the armature (A1). The armature, together with the diaphragm assembly, then retracts thereby sucking petrol through the inlet valve into the pumping chamber (N). When the armature has travelled nearly to the end of its stroke, the throw-over mechanism operates and the outer rocker moves rapidly backwards, thus separating the points and breaking the circuit.

The spring (S1) then re-asserts itself forcing the armature and diaphragm away from the coil housing. This action forces petrol through the delivery valve at a rate determined by the re-quirements of the engine. As the armature nears the end of its stroke the throw-over mechanism again operates, the tungsten points remake contact and the cycle of operations is repeated.

The spring blade (V) rests against the small projection moulding (T) and it should be set so that, when the points are in contact, it is deflected away from the moulding. The gap at the points should be approximately .030" (.75 mm.) when the rocker (U) is manually deflected until it contacts the end face of the coil housing.

Removal (Fig. 20)

Remove both inlet and outlet pipes (39), (42), from the side of the pump by withdrawing the banjo bolt and washers. Disconnect the electrical feed cable to the pump by unscrewing the knurled knob on the end of the wire. Disconnect the earth cable from the side of the pump. Remove the two self-locking nuts attaching the pump to the bracket and withdraw the two washers from each stud. The pump can now be withdrawn from the bracket leaving the two rubber grommets in position. The rubber grommets in the bracket should be examined for deterioration and replaced if necessary, otherwise excessive petrol pump noise may result.

Refitting

Refitting is the reverse of the removal procedure.

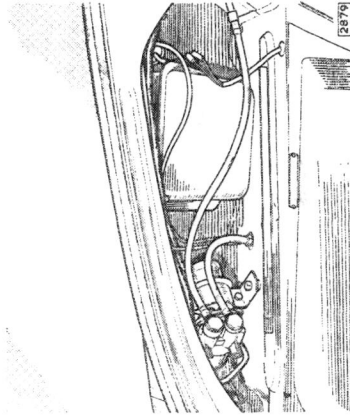

Fig. 15. *Location of the petrol pump.*

Replacing the Delivery Chamber Diaphragm

If the spring loaded diaphragm bounding the delivery chamber is ever removed, the following procedure should be adopted when replacing it.

Replace the perforated diaphragm disc first, depression towards the chamber. The rubber washer, transparent diaphragm shield, rubber diaphragm and 'O' ring are then replaced in that order. When replacing the cover, it is necessary to compress the spring so that the diaphragm is not stressed until firmly retained by the cover and 'O' ring. This may be done by passing a length of welding wire, suitably flattened at one end, through the hole in the top of the cover, down through the centre of the spring, and through the diamond shaped hole in the spring seat. The flat on the end of the wire should be such that it just passes through the widest part of the hole in the spring seat and when rotated 90° will retain the seat so that it can be fitted to compress the spring. With the spring compressed by holding the wire up with pliers, the cover should be tightened down evenly onto the 'O' ring already in position. When tightened fully the wire should be gently rotated back through the right angle to release the spring seat onto the diaphragm.

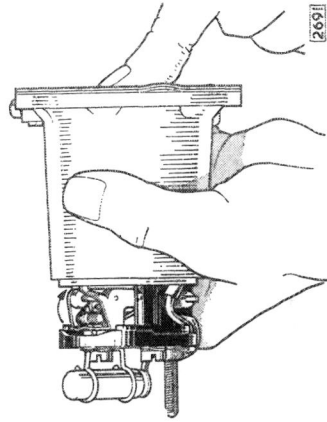

Fig. 16. *Checking the "throw-over" of the toggle mechanism —depress the diaphragm slowly and firmly, not jerkily.*

Resetting the Armature and Diaphragm Assembly

If the diaphragm assembly has been unscrewed it will be necessary to reset the armature spindle in the trunnion. The screw which secures the spring blade (V) should be slackened and the blade swung to one side so that the tungsten points no longer contact. The armature should then be screwed inwards until, as the diaphragm assembly is pushed in and out, the throw-over ceases.

The assembly should then be unscrewed gradually, a sixth of a turn at a time and pressed in and out continuously until when pressed in firmly and slowly, the throw-over mechanism just operates. It should then be unscrewed by a further amount equal to approximately $\frac{2}{3}$ of a turn (4–5 holes). The six screws which hold the coil housing to the main body may then be replaced.

It is now necessary to check that the top of the inner rocker (U1) has made contact with the end face of the coil housing as indicated at X1. If there is a visible or measurable gap there, the six screws should then be slackened off and then retightened until this condition of contact at (X1) is achieved. Finally, the spring blade (previously sprung to one side) should then be replaced in its normal position. The slot in this is provided with an adequate clearance round its attachment screw thus allowing it to be repositioned so that when the outer rocker operates to make and break contact between the tungsten points, one pair of points wipes over the centre-line of the other pair of points in a symmetrical manner.

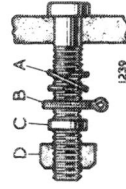

Fig. 17. *The terminal arrangement.*
A. *Double coil spring washer.*
B. *Cable tag.*
C. *Lead washer.*
D. *Countersunk nut.*

2833

Fig. 20. *The fuel system.*

1.	R.H. petrol tank	17.	Rubber gasket	
2.	Petrol filter	18.	Petrol gauge element	
3.	Washer		(R.H.)	
4.	Front vent pipe	19.	Petrol gauge element	
5.	Sleeve		(L.H.)	
6.	Rear vent pipe	20.	Rubber seal	
7.	Clip	21.	Locking ring	
8.	L.H. petrol tank	22.	Cover plate	
9.	Petrol filter	23.	Sealing pad	
10.	Washer	24.	Spacer	
11.	Front vent pipe	25.	Nyloc cap nut	
12.	Sleeve	26.	Fibre washer	
13.	Rear vent pipe	27.	Grommet	
14.	Clip	28.	Supporting cradle	
15.	Connecting tube	29.	Supporting cradle	
16.	Filler cap	30.	Supporting cradle	

31.	Supporting cradle	47.	T-piece	
32.	Locating pad	48.	Petrol pipe.	
33.	Distance pad	49.	Clip	
34.	Sealing rubber	50.	Grommet	
35.	Petrol pump	51.	Petrol filter	
36.	Clip	52.	Filter casting	
37.	Rubber packing	53.	Seating washer	
38.	Rubber mounting	54.	Filter gauze	
39.	Petrol pipe	55.	Glass bowl	
40.	Banjo bolt	56.	Retaining strap	
41.	Washer	57.	Banjo bolt	
42.	Petrol pipe	58.	Fibre washer	
43.	Petrol pipe	59.	Bracket	
44.	Banjo bolt	60.	Feed pipe	
45.	Fibre washer	61.	Banjo bolt	
46.	Clip	62.	Fibre washer	

·030 in. (·75 mm) MINIMUM

2690

Fig. 18. *If the contact blade "A" is held against the projection "B" there should be ·030' (·75 mm.) gap between the white rollers and the body of the pump. If necessary, set the tip of the blade to obtain the correct clearance.*

·070" (1·75MM) +·005"(·125MM)

·035" (·88MM) +·005"(·125MM)

3093

Fig. 19. *Rocker and contact clearances for modified type of outer rocker.*

The slightest stiffness in the rocker mechanism will seriously affect the working of the pump.

The rocker assembly, must, therefore, be fitted to the pedestal in such a way that it is a free-fit without side play. It is essential that the rockers are perfectly free to swing on the pivot pin and that the arms are not binding on the legs of the pedestal. If stiffness exists this can be remedied by using a pair of thin nosed pliers.

The contact blade should be fitted next to the bakelite pedestal, that is, underneath the tag. It should rest against the ridge on the pedestal when the points are apart and it must not be so stiff as to prevent the outer rocker from coming right forward when the points are in contact. The points should just make contact when the rocker is in its midway position.

To check that the assembly is correct, hold the blade against the pedestal ridge, being careful not to press on the overhanging position, and check that the gap between the white fibre rollers and the coil housing is ·030" (·75 mm).

On pumps having a modified type of outer rocker with two stop fingers for setting the various rocker and contact clearances, it is necessary to check and adjust the fingers before setting the rocker throwover. Check and adjust the fingers as follows:—

(a) Check the lift on the contact blade above the top of the pedestal and bend the stop finger if necessary.

(b) Check the gap between the rocker finger and coil housing. If necessary bend the rocker finger to obtain the prescribed clearance.

The assembly should then be unscrewed gradually, a sixth of a turn at a time. Press, in and out continuously until, when pressed in firmly and slowly, the throw-out mechanism just operates. It should then be unscrewed by a further amount equal to approximately ²/₃ of a turn (4-5 Holes). The six screws which hold the coil housing to the main body may then be replaced.

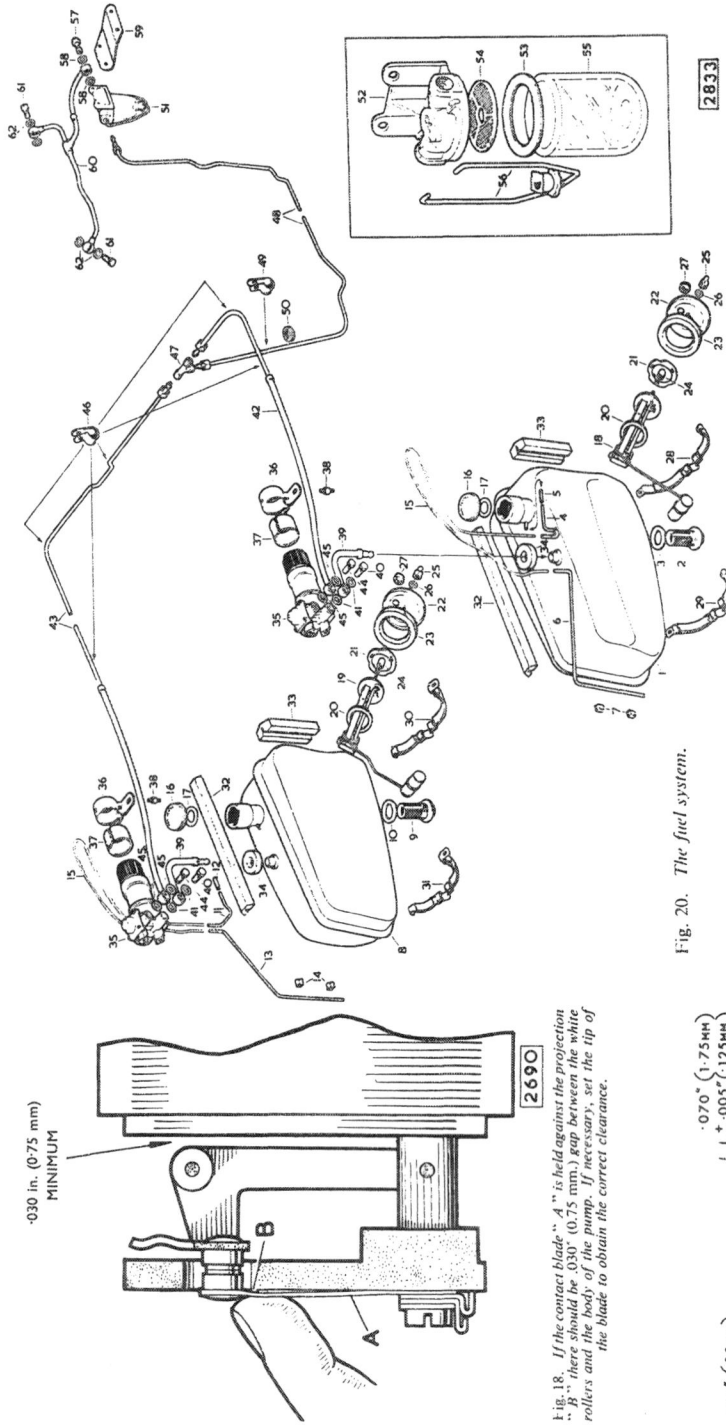

PETROL TANKS (Fig. 20)

Removal

It is not essential to drain the petrol tanks as they can be lowered vertically from their mountings. The car should be raised on a hoist to allow work to be carried out underneath.

Remove the three bolts attaching the exhaust silencers to the rubber mountings and the two bolts securing the exhaust pipes to the rear body coupling. The exhaust pipes can now be lowered.

Open the petrol filler doors and remove the filler cap (16).

Disconnect the flexible petrol pipes from the tanks by unscrewing the unions located behind the trim panels on either side of the luggage compartment. Break the snap connections to the petrol gauge units on either side of the luggage compartment.

Remove the setscrews securing the tank support cradles (28, 29, 30, 31) to the side members, unhook the cradles and remove the tanks.

Refitting

Refitting is the reverse of the removal procedure. Ensure that the electrical cables for the petrol tank units are drawn up through the cover plate aperture before the tanks are offered up to their mountings and that the grommets are replaced.

PETROL GAUGE TANK UNITS (Fig. 20)

Removal

Raise the car on a hoist and drain the petrol tanks.

Disconnect the electrical cables and remove the cover plate (22) together with the sealing pad (23). Turn the locking ring (21) anti-clockwise and remove. The petrol gauge element unit can now be withdrawn. When extracting or replacing the petrol gauge element ensure that the float arm is not damaged.

Refitting

Refitting is the reverse of the removal procedure.

Commencing Engine Numbers

3.4 "S"   7B 3392
3.8 "S"   7B 55353

On cars with the above engine numbers and onwards, Viton tipped needles are fitted to the carburetter float chambers. These are easily recognised as they have a black rubber tip.

If used as replacements for the previous needle and seat, they must be fitted together as an assembly.

COOLING SYSTEM

COOLING SYSTEM

The water circulation is assisted by an impeller type pump mounted on the front cover of the engine, the system being pressurised and thermostatically controlled. Water is circulated from the base of the radiator block via the water pump through the cylinder block and cylinder head water passages and returned to the radiator header tank via the inlet manifold water jacket. The radiator holds approximately 6¼ pints (3.85 litres) of water. A fan, mounted on the spindle of the water pump draws in air through the radiator block.

DATA

Total capacity—including heater	..	..	..	..		22 pints (12.5 litres)
Water pump—type	..	..	..	..		Centrifugal
—drive	..	..	..	..		Fan belt
Fan belt—angle of ' V '	..	..	..	..		40°
Fan—No. of blades	..	..	..	..		12
Fan to engine speed ratio	..	..	..			0.9 : 1
Cooling system control	..	..	..			Thermostat
Thermostat data	..	..	..	..		See page D.10

Radiator cap :

Make and type	..	..	..	..		A.C.—relief valve
Release pressure	..	..	..	..		4 lb. per sq. in. (0.28 kg./cm.²)
Release depression	..	..	..	..		¼ lb. (0.23 kg.)

Radiator flow figures : Water at 62° F. (17° C.)

lb. per sq. in.	galls./min.	litres/min.
1	15.5	70.5
1.5	19.0	86.5
2	22	100.0
2.5	24.5	111.5
3	26.5	120.0
3.5	29.0	131.8
4	31.0	140.9

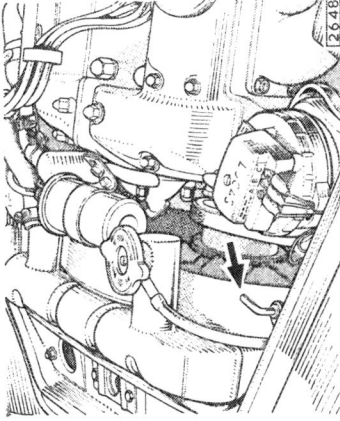

Fig. 1. Radiator drain tap remote control.

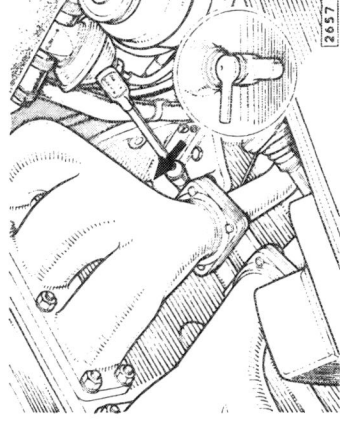

Fig. 2. Cylinder block drain tap.

ROUTINE MAINTENANCE

DAILY

Checking Radiator Water Level

Every day, check the level of the water in the radiator and, if necessary, top up to the bottom of the filler neck.

Use water that is as soft as is procurable; hard water produces scale which in time will affect the cooling efficiency of the system.

Periodically

Care of the Cooling System

The entire cooling system should occasionally be flushed out to remove sediment. To do this, open the radiator block and cylinder block drain taps and insert a water hose into the radiator filler neck. Allow the water to flow through the system, with the engine running at a fast idle speed (1,000 r.p.m.) to cause circulation, until the water runs clear.

Since deposits in the water will in time cause fouling of the surfaces of the cooling system with consequent impaired efficiency it is desirable to retard this tendency as much as possible by using water as nearly neutral (soft) as is available. One of the approved brands of water inhibitor may be used with advantage to obviate the formation of deposits in the system.

Check the radiator water level after running the engine and top up if necessary.

Important

When refilling the cooling system following complete drainage, depress the " HEAT " button on the heater controls to allow the heater system to be filled with coolant. Re-check the level after running the engine for a short period.

FROST PRECAUTIONS

Anti-freeze—Important

During the winter months it is strongly recommended that an anti-freeze compound with an inhibited Ethylene Glycol base be used in the proportions laid down by the anti-freeze manufacturers.

It should be remembered that if anti-freeze is not used or is not of sufficient strength it is possible owing to the action of the thermostat for the radiator to "freeze-up" whilst the car is being driven even though the water in the radiator was not frozen when the engine was started.

Before adding anti-freeze solution the cooling system should be cleaned by flushing. To do this, open the radiator block and cylinder block drain taps and insert a water hose into the radiator filler neck. Allow the water to flow through the system, with the engine running at 1,000 r.p.m. to cause circulation, until the water runs clear. The cylinder head gasket must be in good condition and the cylinder head nuts pulled down correctly, since if the solution leaks into the crankcase a mixture will be formed with the engine oil which is likely to cause blockage of the oil-ways with consequent damage to working parts. Check the tightness of all water hose connections, water pump and manifold joints. To ensure satisfactory mixing, measure the recommended proportion of water and anti-freeze solution in a separate container and fill the system from this container, rather than add the solution direct to the cooling system. Check the radiator water level after running the engine and top up if necessary. Topping up must be carried out using anti-freeze solution or the degree of protection may be lost. Topping up with water will dilute the mixture possibly to an extent where damage by frost will occur.

Important

When refilling the cooling system following complete drainage, depress the "HEAT" button on the heater controls to allow the heater system to be filled with coolant. Re-check the level after running the engine for a short period.

Engine Heater

Provision is made on the right-hand side of the cylinder block for the fitment of an American standard engine heater element No. 7 manufactured by "James B. Carter Ltd., Electrical Heating and Manufacturing Division, Winnipeg, Manitoba, Canada" or "George Bray & Co. Ltd., Leicester Place, Blackman Lane, Leeds 2, England".

Warning

The fitting of an engine heater does not obviate the use of "anti-freeze" during the winter months.

RADIATOR (Fig. 3)

The radiator is pressurised by means of the radiator filler cap. This incorporates a pressure relief valve which is designed to hold a pressure of up to 4 pounds per square inch (0.28 kg./cm²) above atmospheric pressure inside the system. When the pressure rises above four pounds the spring loaded valve lifts off its seat and the excess pressure escapes via the overflow pipe. As the water cools down again a small valve, incorporated in the centre of the pressure valve unit, opens and restores atmospheric pressure should a depression be caused by the cooling of the water.

By raising the pressure inside the cooling system, the boiling point of the coolant is raised by approximately six degrees thus reducing the risk of coolant loss from boiling.

Removal

Drain the radiator by operating the remote control at the top rear of the radiator header tank, and disconnect top and bottom radiator hoses.

Unscrew the four nuts securing the cowl and allow it to rest on the water pump housing behind the fan. Remove the split pin securing the control rod to the drain tap and unscrew the drain tap from the radiator block. Remove the two setscrews securing the sides of the radiator to the body. Remove the two securing nuts at the bottom of the radiator. Carefully lift out the radiator taking care not to damage the matrix on the fan blades. Remove the fan cowl.

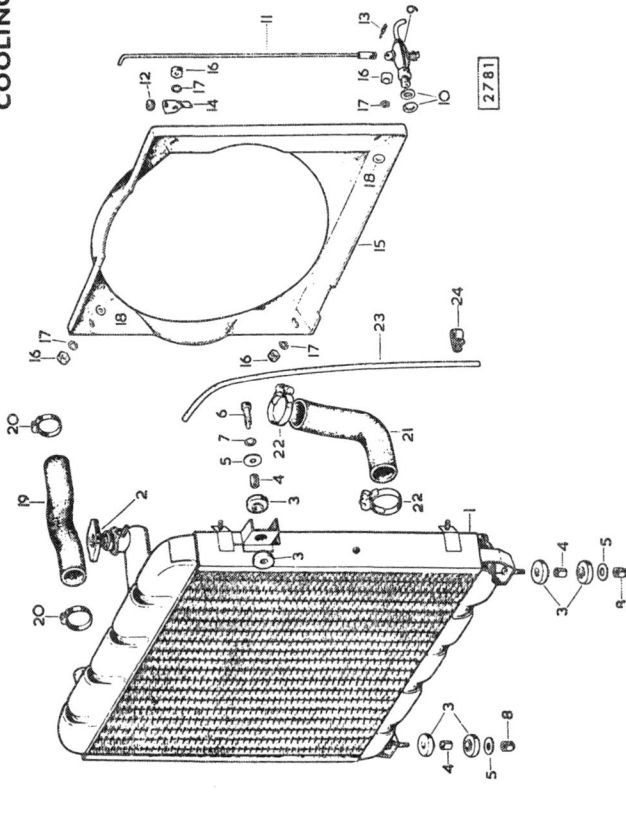

Fig. 3. *Radiator.*

1. Radiator block	14. Drain tap control rod bracket
2. Filler cap	15. Cowl
3. Rubber pad	16. Nut
4. Distance tube	17. Shakeproof washer
5. Special washer	18. Washer
6. Bolt	19. Hose
7. Shakeproof washer	20. Clip
8. Self-locking nut	21. Hose
9. Drain tap	22. Clip
10. Fibre washer	23. Overflow pipe
11. Control rod	24. Clip
12. Grommet	
13. Split pin	

Important : Always keep the radiator block in an upright position so that there is no danger of sediment which may have accumulated on the bottom of the tank passing into the narrow core passages and causing a blockage.

Cleansing the System

Periodically the cooling system should be flushed out and cleansed with a suitable cleansing compound. The procedure is as follows :

Drain the system by removing the two drain taps and when the engine has cooled flush the system with cold water. When this has drained away, replace the taps and fill the system to the normal level with a solution of suitable cleansing compound. Beware of splashing compound on to the paintwork as this would cause serious damage. The engine must now be run for the period prescribed by the makers of the compound after which the system must be drained, thoroughly flushed and refilled with soft water.

Refitting

Refitting is the reverse of the removal procedure.

AUTOMATIC FAN BELT TENSIONER

A spring-loaded jockey pulley is fitted on the right-hand side of the engine. This pulley maintains the correct fan belt tension without any periodic adjustment being required. If it should become necessary to replace the fan belt, the following instructions should be carried out.

FAN BELT

Removal

Slacken the two bolts securing the dynamo to the mounting bracket. Remove the nut and unscrew the bolt securing the top dynamo link to the dynamo. Slacken the bolt securing the dynamo link to the engine and press the dynamo in towards the cylinder block.

Remove the fan belt by pressing the jockey pulley inwards.

Refitting

Place the new fan belt in position on the water pump, jockey and crankshaft pulleys.

Press the jockey pulley inwards and pass the fan belt over the dynamo pulley.

Insert the top bolt through the dynamo adjusting link and screw into the lug on the dynamo.

Pull the dynamo away from the cylinder block as far as possible and tighten the top securing bolt.

Replace the locknut on the top bolt.

Tighten the bolt securing the link to the engine and also the two bottom dynamo mounting bolts.

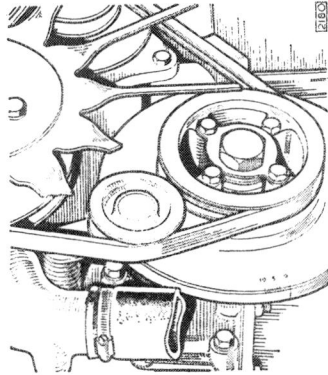

Fig. 6. *The automatic fan belt tensioner.*

FAN

Removal

Remove the radiator as described on page D.6.

Slacken the dynamo adjusting link bolt and the two dynamo bolts and nuts underneath the dynamo. Press the jockey pulley inwards and remove the fan belt. Before removing the setscrews securing the fan to the hub, mark the positions of the semi-circular balance piece(s) relative to the fan and fan hub. (On initial assembly the ends of the balance piece(s) and fan are marked with a centre punch and a small hole is drilled through the balance piece(s), hub and fan to assist re-assembly). Remove the fan from the hub by unscrewing the four setscrews fitted with shakeproof washers.

Refitting

Refitting is the reverse of the removal procedure, but attention should be paid to the removal notes in order to preserve the balance of the assembly.

If it becomes necessary to replace any part of the assembly then it should be rebalanced as shown in Fig. 4. Static balancing is effected by varying the position of the semi-circular balance piece(s) which are retained by setscrews securing the fan to the hub. These should be arranged so that the fan remains at rest in any position. After re-balancing, the ends of the balance piece(s) and the fan should be marked with a centre punch, a small hole drilled through the balance piece(s), hub and fan and the old hole filled in with solder.

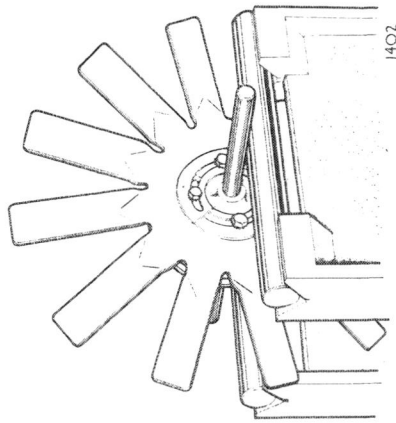

Fig. 4. *Balancing the fan assembly.*

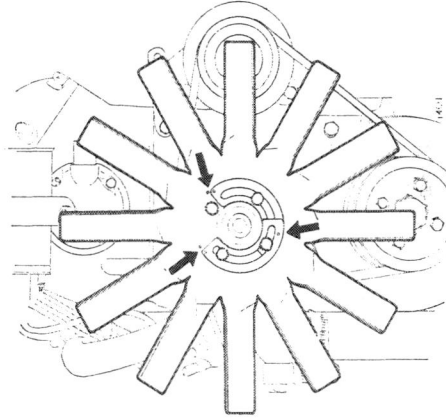

Fig. 5. *Fan, showing balance pieces and location hole.*

THERMOSTAT

This is a valve incorporated in the cooling system which restricts the flow of coolant through the radiator until the engine has reached its operating temperature. When the engine temperature rises to a pre-determined figure (see "Thermostat Data") the thermostat valve commences to open and allows the water to circulate round the radiator. The flow of water increases as the temperature rises until the valve is fully open. Included in the system is a water by-pass utilizing a slot in the thermostat housing integral with the water outlet pipe. This allows the coolant to by-pass the radiator until the thermostat opening temperature is attained, thus providing a rapid warming up of the engine and in cold weather an early supply of warm air to the interior of the car via the heater.

Removal

Drain sufficient water from the system to allow the level to fall below the thermostat by operating the remote control of the drain tap situated at the top left-hand side of the radiator block. Slacken the clip and remove the top water hose from the elbow pipe on the thermostat housing. Remove the two nuts and spring washers securing the water outlet elbow and remove elbow. Lift out the thermostat, noting the gasket between the elbow pipe and thermostat housing.

Checking

Thoroughly clean the thermostat and check that the small hole in the valve is clear. Check the thermostat for correct operation by immersing in a container of cold water together with a thermometer and stirrer. Heat the water, keeping it well stirred and observe if the characteristics of the thermostat are in agreement with data given under "Thermostat Temperatures".

Refitting

Refitting is the reverse of the removal procedure.

Always fit a new gasket between the elbow pipe and the thermostat housing. Ensure that the recess in the thermostat housing and all machined faces are clean.

DATA

Thermostat Identification

Jaguar Part No.	Initial opening Temp. °C.	Fully open Temp. °C.	Remarks
C.20766	70 - 75	85	
C.20766/1	80 - 84	93	High setting for extreme winter conditions.

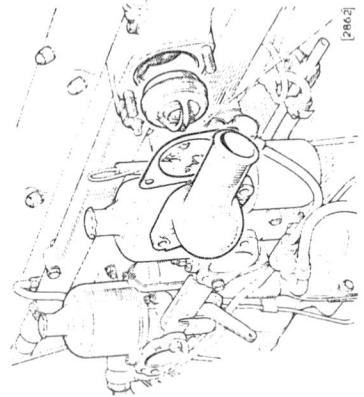

Fig. 7. Removal of the thermostat.

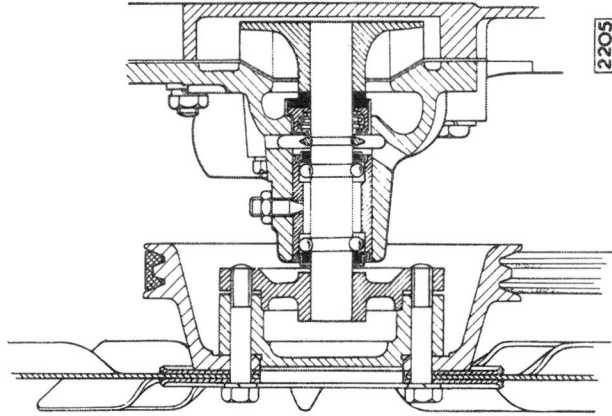

WATER PUMP

The water pump (Fig. 8) is of the centrifugal vane impeller type, the impeller being mounted on a steel spindle which in turn runs in a double row of ball bearings. These are sealed at their ends to exclude all dirt and to retain the lubricant. The main seal on the pump spindle is located in the pump housing by a metal cover and the carbon face maintains a constant pressure on the impeller by means of a thrust spring inside the seal. A hole drilled in the top of the casting acts as an air vent and leads into an annular groove in the casting into which stray water is directed by means of a rubber thrower on the pump spindle. A drain hole at the bottom of the groove leads away any water and prevents seepage into the bearing.

Removal

Remove the radiator as described on page D.6.

Slacken the two bolts securing the dynamo to the mounting bracket. Remove the nut and unscrew the bolt securing the top dynamo link to the dynamo. Slacken the bolt securing the dynamo link to the engine and press the dynamo in towards the cylinder block.

Remove the fan belt by pressing the jockey pulley inwards.

Remove the fan and fan pulley as described on page D.8. Detach hose connections from the water pump.

Unscrew the set bolts and nuts and remove the water pump from the timing cover. Note the gasket between the pump and the timing cover.

Fig. 8. Sectioned view of water pump.

Refitting

Refitting is the reverse of the removal procedure. Care should be taken to renew the pump to timing cover gasket, lightly smearing with grease before fitting.

WATER TEMPERATURE GAUGE

The indicator head is attached to the instrument panel and operates on a thermal principle using a bi-metal strip surrounded by a heater winding. The transmitter unit is mounted in the inlet manifold water jacket adjacent to the thermostat.

For the full description and fault analysis of this instrument refer to Section P "Electrical and Instruments."

Dismantling

Remove the fan hub by means of a suitable extractor as shown in Fig. 9. Slacken locknut and remove Allen head locating screw

Withdraw the spindle and impeller assembly from the pump casting. This assembly must not be pushed out by means of the shaft or the bearing will be damaged. A tube measuring $1\frac{3}{32}"$ (27.77 mm.) outside diameter and $\frac{31}{32}"$ (24.61 mm.) inside diameter must be used to push out the assembly from the front of the pump.

Press out the spindle from the impeller as shown in Fig. 10 and remove seal and rubber thrower. The spindle and bearing assembly cannot be dismantled any further.

Checking

Thoroughly clean all parts of the pump except the spindle and bearing assembly in a suitable cleaning solvent.

Note : The bearing is a permanently sealed and lubricated assembly and therefore must not be washed in the solvent.

Inspect the bearing for excessive end play and remove any burrs, rust or scale from the shaft with fine emery paper, taking the precaution of covering the bearing with a cloth, to prevent emery dust from entering the bearing. If there are any signs of wear or corrosion in the bearing bore or on the face in front of the impeller the housing should be renewed.

Re-assembly

Install shaft and bearing assembly into the pump body from the rear and line up the location hole in the bearing with the tapped hole in the body. Fit locating screw and locknut. Place the rubber thrower in its groove on the spindle in front of the seal. Coat the outside of the brass seal housing with a suitable water resistant jointing compound and fit into the recess in the pump casting. Push the seal into its housing with the carbon face towards the rear of the pump. Ensure that the seal is seated properly.

Press on impeller as shown in Fig. 11 until the rear face of the impeller is flush with the end of the spindle. In a similar manner press the fan hub on to the spindle until it is flush with the end."

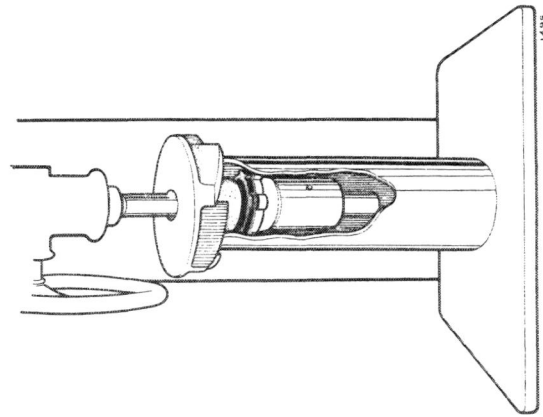

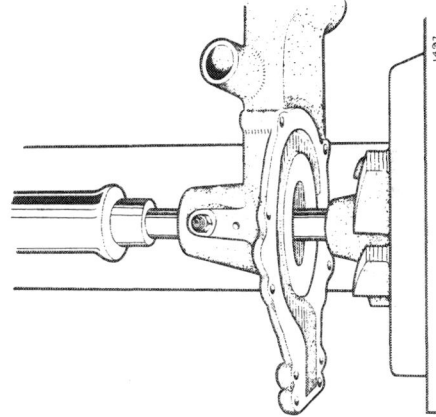

Fig. 10. *Removing water pump impeller from pump spindle.*

Fig. 11. *Fitting impeller.*

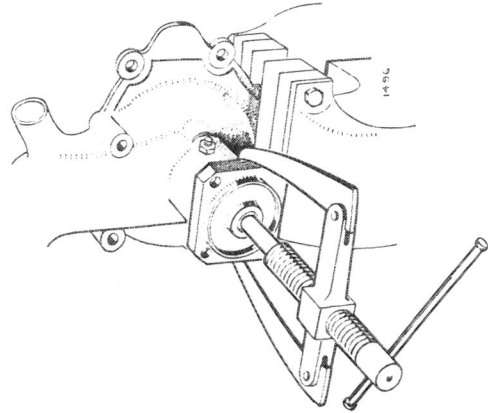

Fig. 9. *Withdrawing fan hub from spindle.*

SECTION E

CLUTCH

ADDENDA

The following corrections should be noted when using SECTION E.

Page	Correction
E.6.	"Dunlop Disc Brake Fluid" should read "Castrol/Girling Crimson Clutch/Brake Fluid".

COOLING SYSTEM

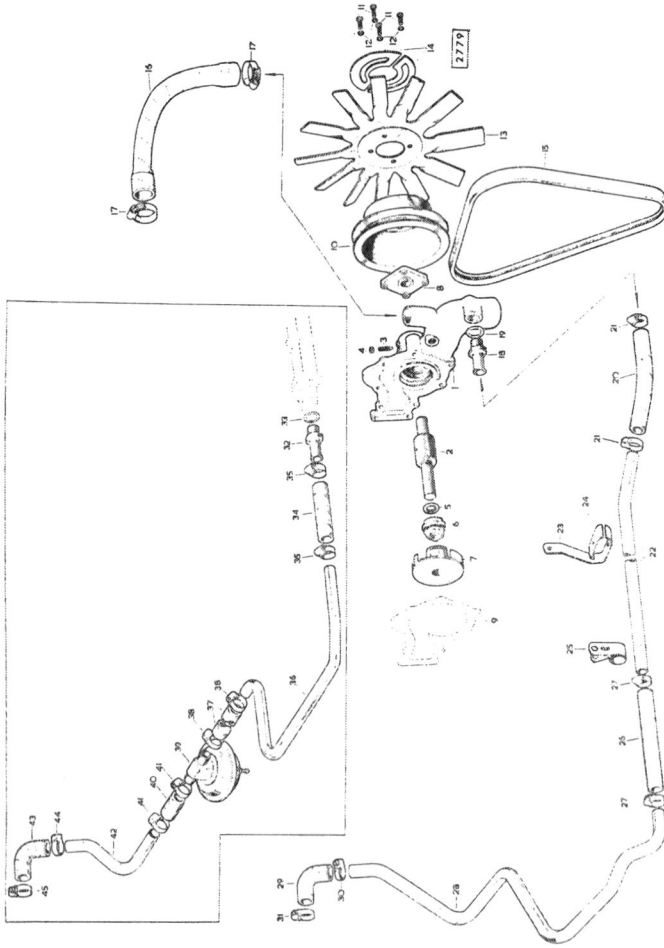

Fig. 12. *Exploded view of water pump and heater connections.*

1. Water Pump Body
2. Spindle
3. Allen screw
4. Nut
5. Thrower
6. Seal
7. Impeller
8. Carrier
9. Gasket
10, Fan Pulley
11. Set Screw
12. Shakeproof Washer
13. Fan
14. Balance Piece
15. Belt
16. By-Pass Water Hose
17. Clip
18. Adaptor for Heater Return
19. Copper Washer
20. Hose
21. Clip
22. Front Return Pipe
23. Hanger Bracket
24. Clamp
25. Clip
26. Hose
27. Clip
28. Rear Return Pipe
29. Elbow Hose
30. Clip
31. Clip
32. Adaptor
33. Copper Washer
34. Hose
35. Clip
36. Feed pipe
37. Hose
38. Clip
39. Valve (Vacuum operated)
40. Hose
41. Clip
42. Heater Pipe
43. Elbow Hose
44. Clip
45. Clip

INDEX

INDEX (continued)

CLUTCH

DESCRIPTION

The clutch is of the single dry plate type and consists of a spring loaded driven plate assembly, a cover assembly and a graphite release bearing.

The operating mechanism consists of a pendant-type foot pedal, coupled by a push rod to an independent master cylinder, integral with which is a fluid reservoir. This is connected by piping and a flexible hose to a slave cylinder mounted on the clutch housing. Depressing the clutch pedal moves the piston in the master cylinder and imparts thrust to the slave cylinder piston which in turn, operates the graphite release bearing by means of a push rod and operating fork. The bearing is forced against the clutch release lever plate which causes the release levers to withdraw the pressure plate and thus release the clutch driven plate.

DATA

	3.4 litre	3.8 litre
Make	Borg and Beck	Borg and Beck
Model	10 A6—G	10 A6—G
Outside diameter	9·84″—9·87″ (249 mm.—250 mm.)	9·84″—9·87″ (249 mm.—250 mm.)
Inside diameter	6·75″—6·76″ (171 mm.—172 mm.)	6·75″—6·76″ (171 mm.—172 mm.)
Type	Single dry plate	Single dry plate
Clutch release bearing	Graphite	Graphite
Operation	Hydraulic	Hydraulic
Clutch thrust springs—number	12	12
—colour	Yellow/Light green	Black
—free length	2·68″ (68 mm.)	2·68″ (68 mm.)
Driven plate—type	Borglite	Borglite
Driven plate damper springs—number	6	6
—colour	Red/Cream	Brown/Cream

ROUTINE MAINTENANCE

(Standard Transmission)

Every 3,000 miles (5,000 km.)

Check Fluid Level

The clutch is operated hydraulically from a master cylinder situated at the rear of the engine compartment on the driver's side of the car. The hydraulic fluid is stored in a reservoir also situated on the driver's side of the car and it is important that the level does not fall below the line marked "Fluid Level".

Clutch Free Travel

There should be ¹⁄₁₆″ (1·5 mm.) free travel measured on the operating rod between the slave cylinder and clutch withdrawal lever.

This free travel is most easily felt, after removal of pedal return spring, by moving the operating rod towards the slave cylinder and then returning towards the withdrawal lever to the fullest extent. Adjustment is effected by slackening the locknut, and turning the operating rod. Screwing the rod into the knuckle joint will increase the free travel, screwing the rod out will decrease the free travel. Always replace return spring after adjustment.

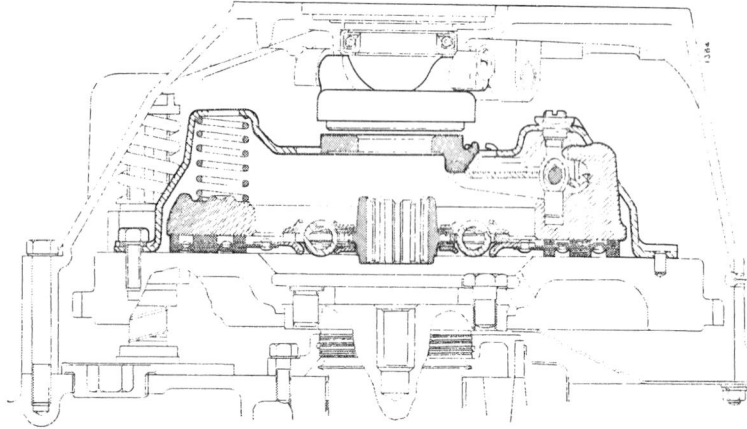

Fig. 1. Sectional view of clutch.

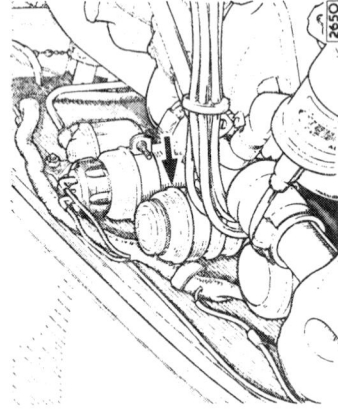

Fig. 2. Clutch fluid reservoir—right hand drive.

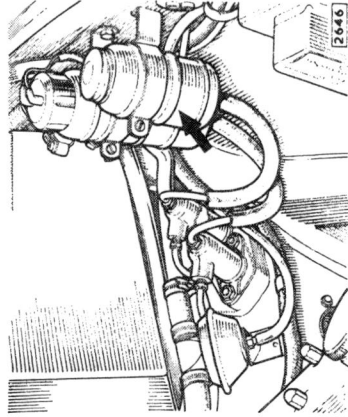

Fig. 3. Clutch fluid reservoir—left hand drive.

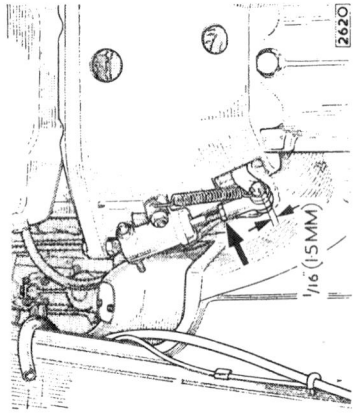

Fig. 4. Adjustment of the clutch free travel is effected on the rod between the clutch slave cylinder and withdrawal lever.

FLUSHING THE SYSTEM

Should the fluid in the system become thick or "gummy" after many years in service, or after a vehicle has been laid up for some considerable time, the system should be drained, flushed and re-filled. It is recommended that this should be carried out once every five years.

Pump all fluid out of the hydraulic system through the bleeder screw of the clutch slave cylinder. To the bleeder screw on the slave cylinder connect one end of a rubber tube, and allow the other end to fall into a container, slacken the screw one complete turn and pump the clutch pedal by depressing it quickly and allowing it to return without assistance; repeat, with a pause in between each operation, until no more fluid is expelled. Discard the fluid extracted.

Fill the supply tank with industrial methylated spirit and flush the system as described above. Keep the supply tank replenished until at least a quart of spirit has passed through the bleeder screw.

Remove the master cylinder and pour off any remaining spirit. Refit the master cylinder, re-fill with clean brake fluid and "bleed" the system.

Note : If the system has been contaminated by the use of mineral oil, etc., the above process will not prove effective. It is recommended that the various units, including the pipe lines, be dismantled and thoroughly cleaned and that all rubber parts, including flexible hoses should be renewed. The contaminated fluid should be destroyed immediately.

REMOVAL AND REFITTING A FLEXIBLE HOSE

In some cases, the cause of a faulty clutch mat be traced to a choked flexible hose. Do not attempt to clear the obstruction by any means except air pressure, otherwise the hose may be damaged. If the obstruction cannot be cleared the hose must be replaced by a new one.

Removal

To renew a flexible hose, adopt the following procedure :

Unscrew the tube nut from the hose union, then unscrew the locknut and withdraw the hose from the bracket. Disconnect the hose at the other end.

BLEEDING THE SYSTEM

"Bleeding" the clutch hydraulic system (expelling air) is not a routine maintenance operation and should only be necessary when a portion of the hydraulic system has been disconnected or if the level of the fluid in the reservoir has been allowed to fall. The presence of air in the hydraulic system may result in difficulty in engaging gear owing to the clutch not disengaging fully.

The procedure is as follows :

Fill up the master cylinder reservoir with brake fluid exercising great care to prevent the entry of dirt. Attach a rubber bleed tube to the nipple on the slave cylinder on the right-hand side of the clutch housing and allow the tube to hang in a clean glass jar partly filled with brake fluid. Unscrew the nipple one complete turn. Depress the clutch pedal slowly, **tighten the bleeder nipple before the pedal reaches the end of its travel** and allow the pedal to return unassisted.

Repeat the above procedure, closing the bleed nipple at each stroke, until the fluid issuing from the tube is entirely free of air, care being taken that the reservoir is replenished **frequently** during this operation, for should the level be allowed to drop appreciably air will enter the system.

On completion, top up the master cylinder reservoir to the line marked "Fluid Level."

Do not on any account use the fluid which has been bled through the system to replenish the reservoir as it will have become aerated. Always use fresh fluid straight from the tin.

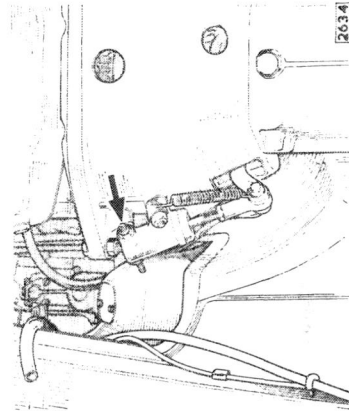

Fig. 6. *Position of clutch bleed nipple.*

All internal parts should be dipped in clean brake fluid and assembled wet, as the fluid acts as a lubricant. When assembling the rubber parts use the fingers only.

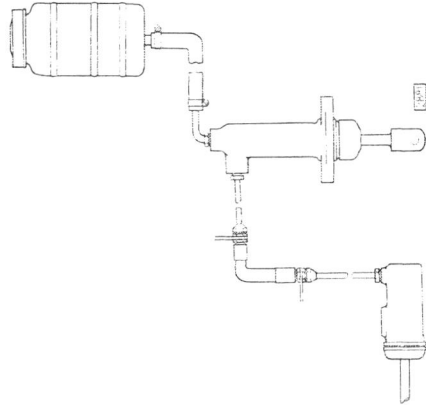

Recommended Hydraulic Fluid

Dunlop Disc Brake Fluid is recommended.

This conforms to S.A.E. 70 R3 specification.

Where this is not available, only fluid guaranteed to conform to S.A.E. 70 R3 specification, which is fully miscible with Dunlop Disc Brake Fluid, may be used as an alternative.

In the event of deterioration of the rubber seals and hoses due to the use of an incorrect fluid, all the seals and hoses must be replaced and the system thoroughly flushed and refilled with one of the above fluids. (See "Flushing the System").

HYDRAULIC SYSTEM—GENERAL INSTRUCTIONS

Should it be found necessary to dismantle any part of the clutch system (that is, master cylinder or slave cylinder), the operation must be carried out under conditions of scrupulous cleanliness. Clean the mud and grease off the unit before removal from the vehicle and dismantle on a bench covered with a sheet of clean paper. Do not swill a complete unit, after removal from the vehicle, in paraffin, petrol or trichlorethylene (trike) as this would ruin the rubber parts and, on dismantling, give a misleading impression of their original condition. Do not handle the internal parts, particularly rubbers, with dirty hands. Place all metal parts in a tray of clean brake fluid to soak; afterwards dry off with a clean lint-free cloth, and lay out in order on a sheet of clean paper. Rubber parts should be carefully examined and if there is any sign of swelling or perishing they should be renewed: in any case it is usually good policy to renew **all** rubbers. The main castings may be swilled in any of the normal cleaning fluids, but all traces of the cleaner must be dried out before assembly. In the case of the master cylinder make sure that the by-pass port is clear by probing with a bent piece of wire not exceeding .018" (0.46 mm.) diameter.

If the by-pass port is clogged, rapid wear of the release bearing or clutch slip will result due to pressure being built up in the system.

Fig. 5. *Clutch hydraulic system.*

Refitting

When refitting a hose, first ensure that it is not twisted or "kinked" (this is MOST IMPORTANT) then pass the hose union through the bracket and, whilst holding the union with a spanner to prevent the hose from turning, fit the locknut and the shakeproof washer; connect up the pipe by screwing on the tube-nut.

THE MASTER CYLINDER (Fig. 7)

Principle of Operation

The master cylinder is mechanically linked to the clutch pedal and provides the hydraulic pressure necessary to operate the clutch. The components of the master cylinder are contained within the bore of a body which at its closed end has two 90° opposed integral pipe connection bosses. Integrally formed around the opposite end of the cylinder is a flange provided with two holes for the attachment bolts. In the unloaded condition a spring loaded piston (4), carrying two seals (5 and 13) is held against the underside of a circlip (14) retained dished washer (3) at the head of the cylinder. A hemispherically ended push-rod (1) seats in a similarly formed recess at the head of the piston. A fork end on the outer end of the push-rod provides for attachment to the pedal. A rubber dust excluder (2), the lip of which seats in a groove, shrouds the head of the master cylinder to prevent the intrusion of foreign matter.

A cylindrical spring support locates around the inner end of the piston and a small drilling in the end of the support is engaged by the stem of a valve (11). The larger diameter head of the support locates in a central blind bore in the piston. The valve passes through the bore of the vented spring support and interposed between the spring support and an integral flange formed on the valve is a small coiled spring (9). A lipped rubber seal (10) registers in a groove around the end of the valve. This assembly forms a recuperation valve which controls fluid flow to and from the reservoir.

When the foot pedal is in the OFF position the master cylinder is fully extended and the valve is held clear of the base of the cylinder by the action of the main spring. In this condition the master cylinder is in fluid communication with the reservoir, thus permitting recuperation of any fluid loss sustained, particularly during the bleeding operation.

1. Push rod
2. Dust excluder
3. Dished washer
4. Piston
5. Cup seal
6. Body
7. Return spring
8. Spring support
9. Valve spring
10. Seal
11. Valve
12. Spring support
13. Sealing ring
14. Circlip

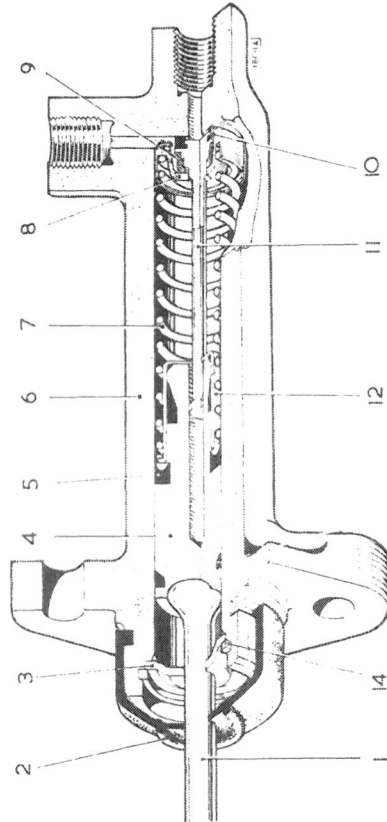

Fig. 7. Sectional view of the master cylinder.

Detach the master cylinder push-rod from the clutch pedal from inside the car by removing the split pin and withdrawing the clevis pin. Remove the clutch master cylinder from the housing situated inside the engine compartment by removing two nuts.

Reseating the Master Cylinder Seals

Ease the dust excluder clear of the head of the master cylinder.

With suitable pliers remove the circlip; this will release the push rod complete with dished washer.

Withdraw the piston and remove both seals.

Withdraw the valve assembly complete with springs and supports. Remove the seal from the end of the valve.

Lubricate the new seals and the bore of the cylinder with brake fluid, fit the seal to the end of the valve ensuring that the lip registers in the groove. Fit the seals in their grooves around the piston.

Insert the piston into the spring support, ensuring that the head of the valve engages the piston bore.

Lubricate the piston with Castrol Rubber Grease H.95/59 and slide the complete assembly into the cylinder body taking particular care not to damage or twist the seals. The use of a fitting sleeve is advised.

Position the push-rod and depress the piston sufficiently to allow the dished washer to seat on the shoulder at the head of the cylinder. Fit the circlip and check that it fully engages the groove.

Fill the dust excluder with clean Castrol H.95/59 Rubber Grease.

Master Cylinder Push-rod—Free Travel

To ensure that this piston returns to the fully extended position clearance is provided between the enlarged head of the push-rod, the piston and dished washer. As this washer also forms the return stop for the clutch pedal, no means of adjustment is necessary.

Refitting

Secure the master cylinder to the vehicle by fitting the fixing nuts at the flange. Connect the pipes to the inlet and outlet connections, the push rod to the pedal, and bleed the system. Check for leaks by depressing the clutch pedal once or twice and examining all hydraulic connections.

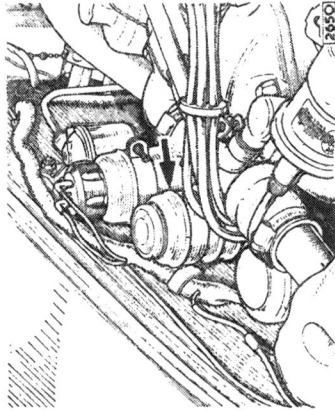

Fig. 8. Clutch fluid reservoir—right hand drive.

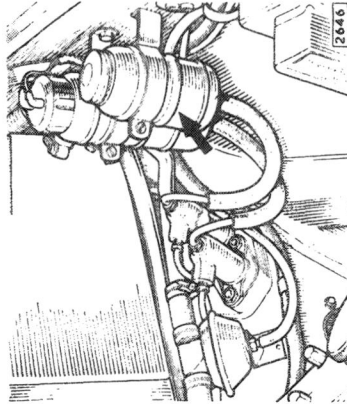

Fig. 9. Clutch fluid reservoir—left hand drive.

When a load is applied to the foot pedal the piston moves down the cylinder against the compression of the main spring. Immediately this movement is in excess of the valve clearance the valve closes under the influence of its spring and isolates the reservoir. Further loading of the pedal results in the discharge of fluid under pressure from the outlet connection, via the pipe lines to the clutch slave cylinder.

Removal of the load from the pedal reverses the sequence, the action of the main spring returns the master cylinder to the extended position.

Removal

Drain the clutch reservoir and detach the inlet and outlet pipes from the clutch master cylinder, by unscrewing the two union nuts.

THE SLAVE CYLINDER (Fig. 10)

The clutch slave cylinder consists of a body (4) which incorporates two threaded connections and is bored to accommodate a piston (5) against the inner face of which a rubber cup (3) is loaded by a cup filler (2) and a spring (1); the travel of the piston is limited by a circlip (6) fitted in a groove at the end of the bore. A rubber boot (7) through which a push-rod passes, is fitted on to the body to prevent the intrusion of dirt or moisture.

One of the connections in the body receives a pipe from the clutch master cylinder, whilst the other is fitted with a bleeder screw : the connection for the pipe is parallel to the mounting flange on the body.

Removal

To remove from the vehicle, disconnect the pipe, detach the rubber boot from the body and remove the fixing screws; leave the push-rod attached to the vehicle. If the boot is not being renewed it may be left on the push-rod.

Dismantling

Remove the circlip (6) from the end of the bore and apply a **low** air pressure to the open connection to expel the piston (5) and the other parts; remove the bleeder screw.

Assembling

Prior to assembly, smear all internal parts and the bore of the body with Rubberlube.

Fit the spring (1) in the cup filler (2) and insert these parts, spring uppermost, into the bore of the body (4). Follow up with the cup (3), lip leading, taking care not to turn back or buckle the lip; then insert the piston (5), flat face innermost, and fit the circlip (6) into the groove at the end of the bore.

Refitting

Fit the rubber boot (7) on the push-rod, if removed previously, and offer up the slave cylinder to the vehicle, with the push-rod entering the bore. Secure the cylinder with the fixing screws and stretch the large end of the boot into the groove on the body. Fit into their respective connections the bleeder screw and the pipe from the clutch master cylinder.

"Bleed" the clutch as described on page E.7.

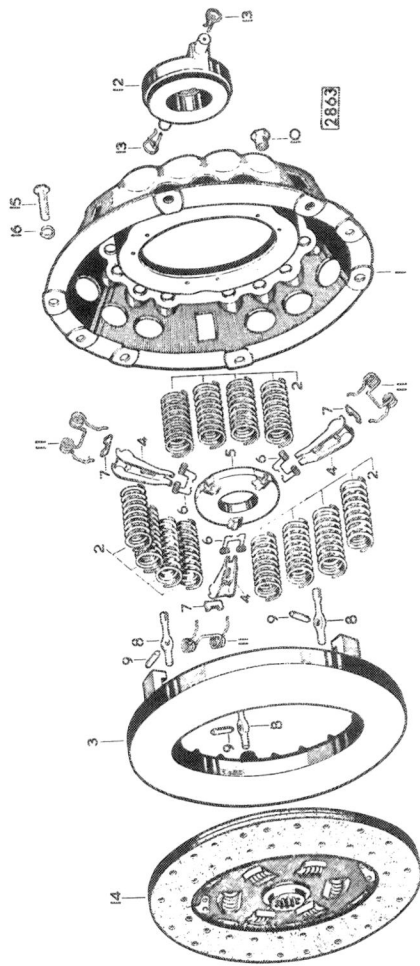

Fig. 11. *Exploded view of clutch assembly.*

1. Cover
2. Thrust spring
3. Pressure plate
4. Release lever
5. Release lever plate
6. Release lever retainer
7. Release lever strut
8. Release lever eyebolt
9. Eyebolt pin
10. Adjustment nut
11. Anti-rattle spring
12. Release bearing and cup assembly
13. Release bearing retainer
14. Driven plate assembly
15. Securing bolt
16. Spring washer

Fig. 10. *Exploded view of clutch slave cylinder.*

THE CLUTCH UNIT

(Fig. 11)

The driven plate assembly (14) is of the flexible centre type, in which a splined hub is indirectly attached to a disc and transmits the power and overrun through a number of coil springs held in position by shrouds.

The cover assembly consists of a pressed steel cover (1) and a cast iron pressure plate (3) loaded by thrust springs (2) the number of which vary with the model. Mounted on the pressure plate are release levers (4), which pivot on floating pins (9), retained by eye bolts (8). Adjustment nuts (10) are screwed on to the eye bolts and secured by staking. Struts (7) are interposed between lugs on the pressure plate and the outer end of the release levers. Anti-rattle springs (11) restrain the release levers and retainer springs (6) connect the release lever plate (5) to the levers.

The graphite release bearing (12) is shrunk into a bearing cup which is mounted on the throw-out forks and held by the release bearing retainer springs (13).

GENERAL INSTRUCTIONS

When overhauling the clutch the following instructions should be noted and carried out:

Clutch Cover Assembly

Before dismantling the clutch, suitably mark the following parts so that they can be re-assembled in the same relative positions to each other to preserve the balance and adjustment; clutch cover, lugs on the pressure plate and the release levers.

When re-assembling make sure that the markings coincide and, if new parts have been fitted which would affect the adjustment, carefully set the release levers (see page E.16).

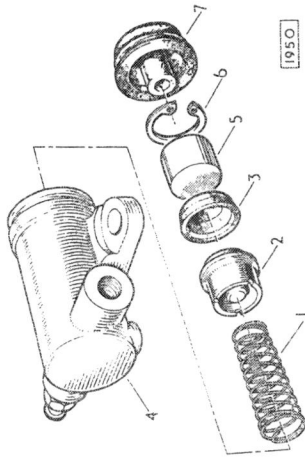

If a new pressure plate has been fitted, it is essential that the complete cover assembly, should be re-balanced, for which reason it is not a practical proposition where special equipment is not available.

Before assembly, thoroughly clean all parts and renew those which show appreciable wear. A very slight smear of grease such as Lockheed Expander Lubricant or Duckham's Keenol K.O.12 should be applied to the release lever pins, contact faces of the struts, eyebolts seats in the clutch cover, drive lug sides on the pressure plate and the plain end of the eyebolts.

Release Bearing

If the graphite release bearing is badly worn it should be replaced by a complete bearing assembly.

CONDITION OF CLUTCH FACINGS

The possibility of further use of the friction facings of the clutch is sometimes raised, because they have a polished appearance after considerable service. It is natural to assume that a rough surface will give a higher frictional value against slipping, but this is not correct.

Since the introduction of non-metallic facings of the moulded asbestos type, in service, a polished surface is a common experience, but it must not be confused with a glazed surface which is sometimes encountered due to the conditions discussed below.

The ideal smooth or polished condition will provide a normal contact, but a glazed surface may be due to a film or a condition introduced, which entirely alters the frictional value of the facings. These two conditions might be simply illustrated by the comparison between a polished wood, and a varnished surface. In the former the contact is still made by the original material, whereas in the latter instance, a film of dried varnish is interposed between the contact surfaces.

The following notes are issued with a view to giving useful information on this subject:

(a) After the clutch has been in use for some little time, under perfect conditions (that is, with the clutch facings working on true and polished or ground surfaces of correct material, without the presence of oil, and with only that amount of slip which the clutch provides for under normal conditions) then the surface of the facings assumes, a high polish, through which the grain of the material can be clearly seen. This polished facing is of mid-brown colour and is then in a perfect condition.

(b) Should oil in small quantities gain access to the clutch in such a manner as to come in contact with the facings it will burn off, due to the heat generated by slip which occurs under normal starting conditions. The burning off of this small amount of lubricant, has the effect of gradually darkening the facings, but, provided the polish on the facings remains such that the grain of the material can be clearly distinguished, it has very little effect on clutch performance.

(c) Should increased quantities of oil or grease obtain access to the facings, one or two conditions, or a combination of the two, may arise, depending upon the nature of oil, etc.

1. The oil may burn off and leave on the surface facings a carbon deposit which assumes a high glaze and causes slip. This is a very definite, though very thin deposit, and in general it hides the grain of the material.

2. The oil may partially burn and leave a resinous deposit on the facings, which frequently produces a fierce clutch, and may also cause a "spinning" clutch due to a tendency of the facings to adhere to the flywheel or pressure plate face.

3. There may be a combination of (1) and (2) conditions, which is likely to produce a judder during clutch engagement.

(d) Still greater quantities of oil produce a black soaked appearance of the facings, and the effect may be slip, fierceness, or judder in engagement, etc., according to the conditions. If the conditions under (c) or (d) are experienced, the clutch driven plate should be replaced by one fitted with new facings, the cause of the presence of the oil removed and the clutch and flywheel face thoroughly cleaned.

ALIGNMENT

Faulty alignment will cause excessive wear of the splines in the hub of the driven plate, and eventually fracture the steel disc around the hub centre as a result of "swash action" produced by axial movement of the splined shaft.

PEDAL ADJUSTMENT

This adjustment is most important and the instructions given should be carefully followed; faulty adjustment falls under two headings :—

(a) Insufficient free (or unloaded) pedal travel may cause a partly slipping clutch condition which becomes aggravated as additional wear takes place on the facings, and this can result in a slipping clutch leading to burning out unless corrected. Over-travel of effective pedal movement only imposes undue internal strain and causes excessive bearing wear.

(b) Too much free pedal movement results in inadequate release movement of the bearing and may produce a spinning plate condition that is, dragging clutch rendering clean changes impossible.

REMOVAL

To remove the clutch, the engine and gearbox must first be removed (refer to Section B "Engine").

Slacken the clutch mounting screws a turn at a time by diagonal selection until the thrust spring pressure is released. Remove the setscrews and withdraw the complete clutch assembly from the flywheel. Remove the driven plate assembly and take care to maintain the driven plate faces in a clean condition. Observe that the clutch and flywheel are balanced as an assembly. This location is indicated by balance marks "B" stamped on the clutch and flywheel (Fig. 23).

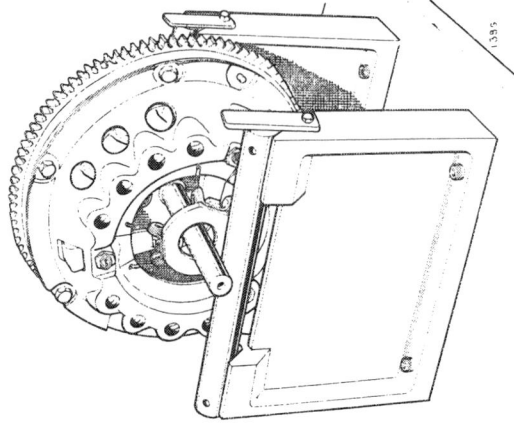

Fig. 12. *Clutch and flywheel balance.*

DISMANTLING

Before dismantling, mark all the major components.

To dismantle the clutch, either bolt the assembly to the baseplate of the Churchill fixture, to a spare flywheel, or place the clutch on the bed of a press with blocks under the pressure plate in such a manner that the cover is free to move downwards when pressure is applied.

Having compressed the clutch in one of these various ways, unscrew the nuts (Fig. 13), (considerable torque is initially necessary in order to break off the squeezed-in portion of each nut), and slowly release the clamping pressure. Lift the cover and the thrust springs off the pressure plate and remove the release lever mechanism. Fig. 14 shows the method whereby the strut is disengaged from the lever, after which the threaded end of the eye-bolt and the inner end of the lever are held as close together as possible to enable the shank of the eyebolt to clear the hole in the pressure plate.

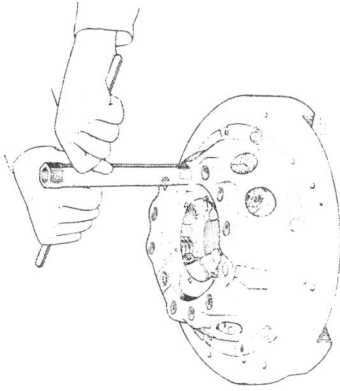

Fig. 13.   Removal of the adjustment nuts.

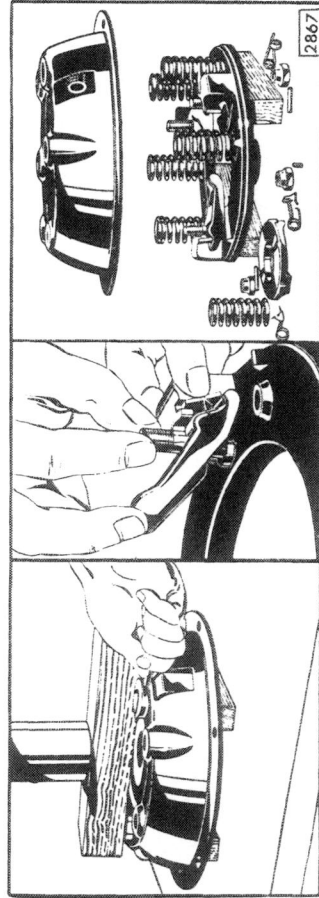

Fig. 14.   Dismantling the clutch assembly using a ram press.

ASSEMBLING (Fig. 11)

It is essential that all major components be returned to their original positions if the balance of the assembly is to be preserved.

Fit a pin (9) into an eyebolt (8) and locate the parts within a release lever (4). Hold the threaded end of the eyebolt and the inner end of the lever as close together as possible and, with the other hand, engage the strut (7) within the slots in a lug on the pressure plate, with the other end of the strut push outwards towards the periphery of the plate. Offer up the lever assembly, first engaging the eyebolt shank within the hole in the plate, then locate the strut within the groove in the lever. Fit the remaining levers in the same way, not forgetting to lubricate all contact faces.

Place the pressure plate on the baseplate of the Churchill fixture, on a spare flywheel, or on blocks on the bed of a press and position the thrust springs (2) on the bosses of the plate. Having arranged all the springs, and ensuring that the anti-rattle springs (11) are fixed within the cover, rest the cover on the springs, carefully aligning the pressure plate lugs with the cover slots. If the Churchill fixture or a spare flywheel is being used, move the clutch to align the holes in the cover flange with the tapped holes in the flywheel or baseplate and then clamp the cover down with the fixing screws, turning them a little at a time to avoid distortion. If a press is being used, arrange a block across the cover and compress the assembly. Then screw the adjustment nuts (10) into an approximately correct position.

The release levers must now be set to the correct height, adopting any of three methods elsewhere described after which the adjusting nuts should be locked by punching them into the eyebolt slots. After setting the levers, fit the release lever plate.

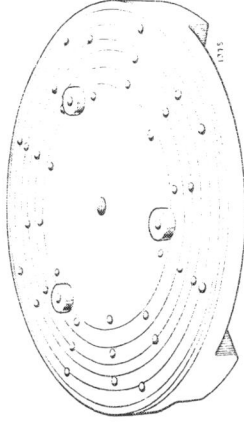

Fig. 15.   The special base plate for clutch adjustment.

ADJUSTING THE RELEASE LEVERS

To ensure satisfactory operation, correct adjustment of the release levers is essential. In service, the original adjustment made by the makers never needs attention and re-adjustment is only necessary if the clutch has been dismantled.

To facilitate adjustment of the release levers the gauge plates once produced by the clutch manufacturer can be utilized. As numerous Traders still possess these plates details as to their identification are given on page E.20.

An alternative method of lever adjustment is to use the universal fixture known as the No. 99 manufactured by V. L. Churchill & Co. Ltd., which caters for the 6¼"—11" clutch.

Finally, where neither a gauge plate nor Churchill tool is available the levers may be set using the actual driven plate as a gauge and these three methods are described as follows:—

(1) Using a Borg & Beck gauge plate (Fig. 16)

(a) Mount the clutch on the actual or a spare flywheel (1), or alternatively clamp it down to a flat surface, with the gauge plate (4) occupying the position normally taken by the driven plate. The ground lands of the gauge plate should each be located under a release lever (5).

(b) Adjust the levers by turning the eyebolt nuts (6) until the levers are just in contact with a short straight edge resting upon the boss of the gauge plate.

(c) Having made a preliminary setting some attempt must be made to operate the clutch several times in order to settle the mechanism. Normally, this operation can be carried out in a drilling machine or light press having a suitable adaptor, arranged to bear upon the lever tips.

(d) Carry out a further check and re-adjust if necessary.

(2) Using the Churchill Fixture

This tool, which is illustrated in Fig. 18, provides for the accurate adjustment of the levers; additionally, it affords a convenient fixture upon which to dismantle and assemble the unit. A device is included to operate the clutch and thereby to settle the working parts after assembly. To use the tool, adopt the following procedure, which also indicates the additional operations when dismantling and assembling the clutch.

Remove from the box the gauge finger (6), the pillar (4) and the actuator (2) and consult the code card to determine the reference of the adaptor (5) and the spacers appropriate to the clutch which is being serviced.

Rest the base plate (1) on a flat surface, wipe it clean and place the spacers (3) upon it in the positions quoted on the code card. Place the clutch on the spacers, aligning it with the appropriate tapped holes in the base, arranging it so that the release levers are as close to the spacers as possible.

Screw the actuator into the centre hole in the base plate and press the handle down to clamp the clutch. Then screw the set bolts provided firmly into the tapped holes in the baseplate using a speed brace: remove the actuator.

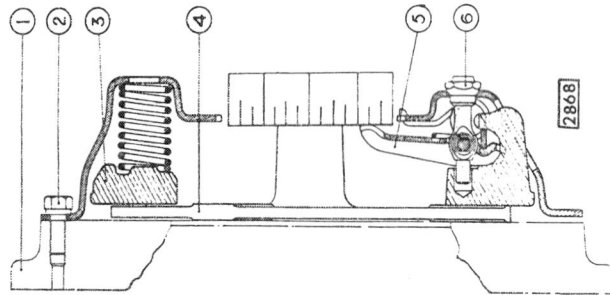

Fig. 16. Release lever adjustment.

Fig. 17. The gauge plate.

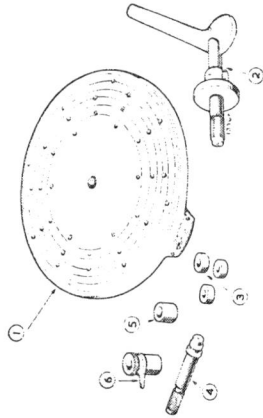

Fig. 18. The base plate and accessories.

(3) Using the Actual Driven Plate

This method of setting the levers is not highly accurate and should only be resorted to when neither a gauge plate nor Churchill Fixture is available. The drawback to this method lies in the fact that although the driven plate is produced to close limits, it is difficult to ensure absolute parallelism. Although the error in the plate is small, it is magnified some five-fold at the lever tip due to the lever ratio.

The method to be adopted is as follows:

(a) Mount the clutch on the flywheel with the driven plate in its normal position or clamp the assembly to any flat surface having a hole within it to accommodate the boss of the driven plate.

(b) Consult the chart on Page E.20 to ascertain the height of the lever tip from the flywheel and adjust the levers until this dimension is achieved.

(c) Having made a preliminary setting slacken the clamping pressure, turn the driven plate through a right angle, reclamp the cover and check the levers against any lack of truth in the driven plate.

REFITTING

Place the driven plate on the flywheel taking care that the larger part of the splined hub faces the gearbox. Centralise the plate on the flywheel by means of the dummy shaft (a constant pinion shaft may be used for this purpose. Fig. 22). Secure the cover assembly with six setscrews and spring washers, tightening the screws a turn at a time by diagonal selection. Ensure that the "B" stamped adjacent to one of the dowel holes coincides with the "B" stamped on the periphery of the flywheel (Fig. 23). Do not remove the dummy shaft until all the setscrews are securely tightened, otherwise the driven plate will come off centre and difficulty will be met in engaging the constant pinion shaft into the bush in the rear end of the crankshaft.

Fig. 22. Centralising driven plate.

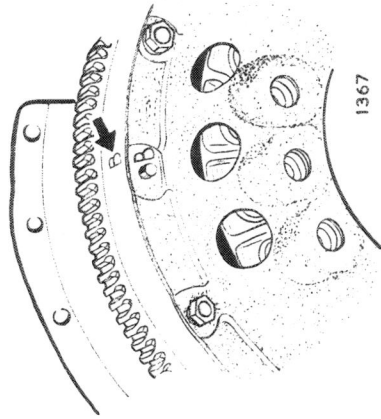

Fig. 23. Balance marks on the clutch and flywheel.

Remove the adjusting nuts (Fig. 13) and gradually unscrew the set bolts to relieve the load of the thrust springs (Fig. 19). Lift the cover off the clutch and carry out whatever additional dismantling may be desired.

After carrying out the necessary servicing of the clutch components, re-assemble the parts on the clutch pressure plate, place the cover upon it and transfer the assembly to the base plate, resting on the spacers and aligned correctly.

Carefully bolt the cover to the base plate and screw the adjusting nuts on to the eyebolts until flush with the tops of the latter. Screw the actuator into the base (Fig. 20) and pump the handle a dozen times to settle the clutch mechanism. Remove the actuator. Screw the pillar firmly into the base and place upon it the appropriate adaptor, recessed face downwards, and the gauge finger.

Turn the adjusting nuts until the finger just touches the release levers, pressing downwards on the finger assembly to ensure that it is bearing squarely on the adaptor (Fig. 21).

Remove the finger adaptor and pillar, replace the actuator and operate the clutch a further dozen times. Replace the pillar and check the lever setting, making any final correction.

Finally, lock the adjusting nuts. The cylindrical portion of the nut must be peened into the slot in the eyebolt, using a blunt chisel and hammer.

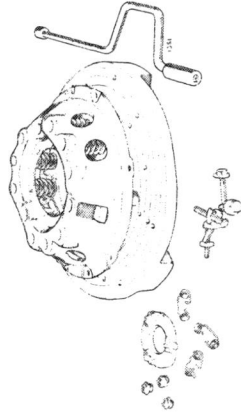

Fig. 19. Removing clutch cover assembly.

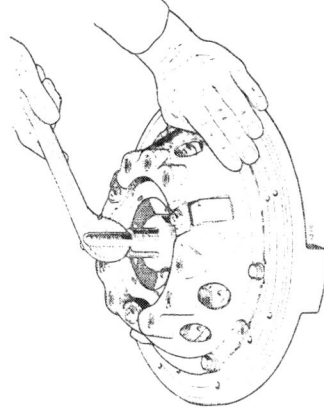

Fig. 20. Screwing actuator into base plate.

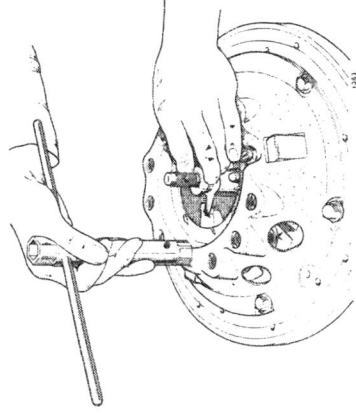

Fig. 21. Using finger assembly to adjust release levers.

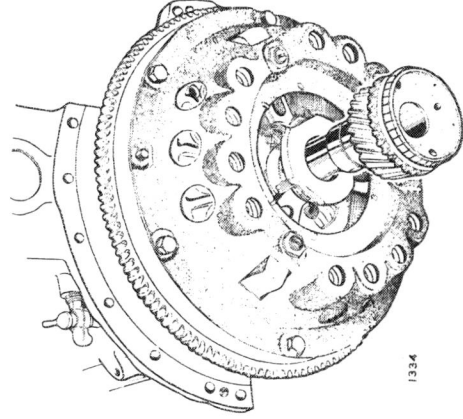

FAULT-FINDING

SYMPTOM	CAUSE	REMEDY
Drag or Spin	(a) Oil or grease on the driven plate facings	Fit new facings
	(b) Misalignment between the engine and splined clutch shaft	Check over and correct the alignment
	(c) Air in clutch system	"Bleed" system
	(d) Bad external leak between the clutch master cylinder and the slave cylinder	Renew pipe and unions
	(e) Excessive clearance between the release bearing and the release lever plate	Adjust to $\frac{1}{16}$" (1.58 mm.) clearance (non-hydrostatic clutch)
	(f) Warped or damaged pressure plate or clutch cover	Renew defective part
	(g) Driven plate hub binding on splined shaft	Clean up splines and lubricate with small quantity of high melting point grease such as Duckham's Keenol
	(h) Distorted driven plate due to the weight of the gearbox being allowed to hang on clutch plate during assembly	Fit new driven plate assembly using a jack to take overhanging weight of the gearbox
	(i) Broken facings of driven plate	Fit new facings, or replace plate
	(j) Dirt or foreign matter in the clutch	Dismantle clutch from flywheel and clean the unit, see that all working parts are free
		CAUTION: Never use petrol or paraffin for cleaning out clutch
Fierceness or Snatch	(a) Oil or grease on driven plate facings	Fit new facings and ensure isolation of clutch from possible ingress of oil or grease
	(b) Misalignment	Check over and correct alignment
	(c) Worn out driven plate facings	New facings required
Slip	(a) Oil or grease on driven plate facings	Fit new facings and eliminate cause of foreign presence
	(b) Failure to adjust at clutch slave cylinder to compensate for loss of release bearing clearance consequent upon wear of the driven plate facings ($\frac{1}{16}$" (1.58 mm.) clearance is necessary between the release bearing and the release lever plate)	Adjust push rod as necessary (non-hydrostatic clutch).
	(c) Seized piston in clutch slave cylinder	Renew parts as necessary

DATA FOR CLUTCH LEVER TIP SETTING

Clutch Model	Driven Plate	Gauge Plate Part No.	Lever tip height from flywheel face Dimension "A"	Gauge Plate Land Thickness Dimension "C"	Gauge Plate Dia.	Remarks
10″	Borglite	CG14322	1.955″ (49.65 mm.)	0.330″ (8.381 mm.)	8.375″ (212.7 mm.)	Dimension "A" 2.45″ (62.23 mm.) if taken with Release Lever Plate in position.

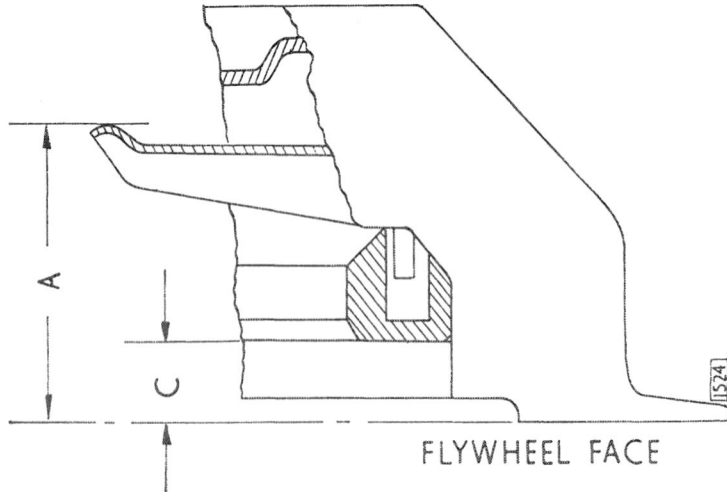

FLYWHEEL FACE

Fig. 24. *Dimensions for clutch lever tip setting.*

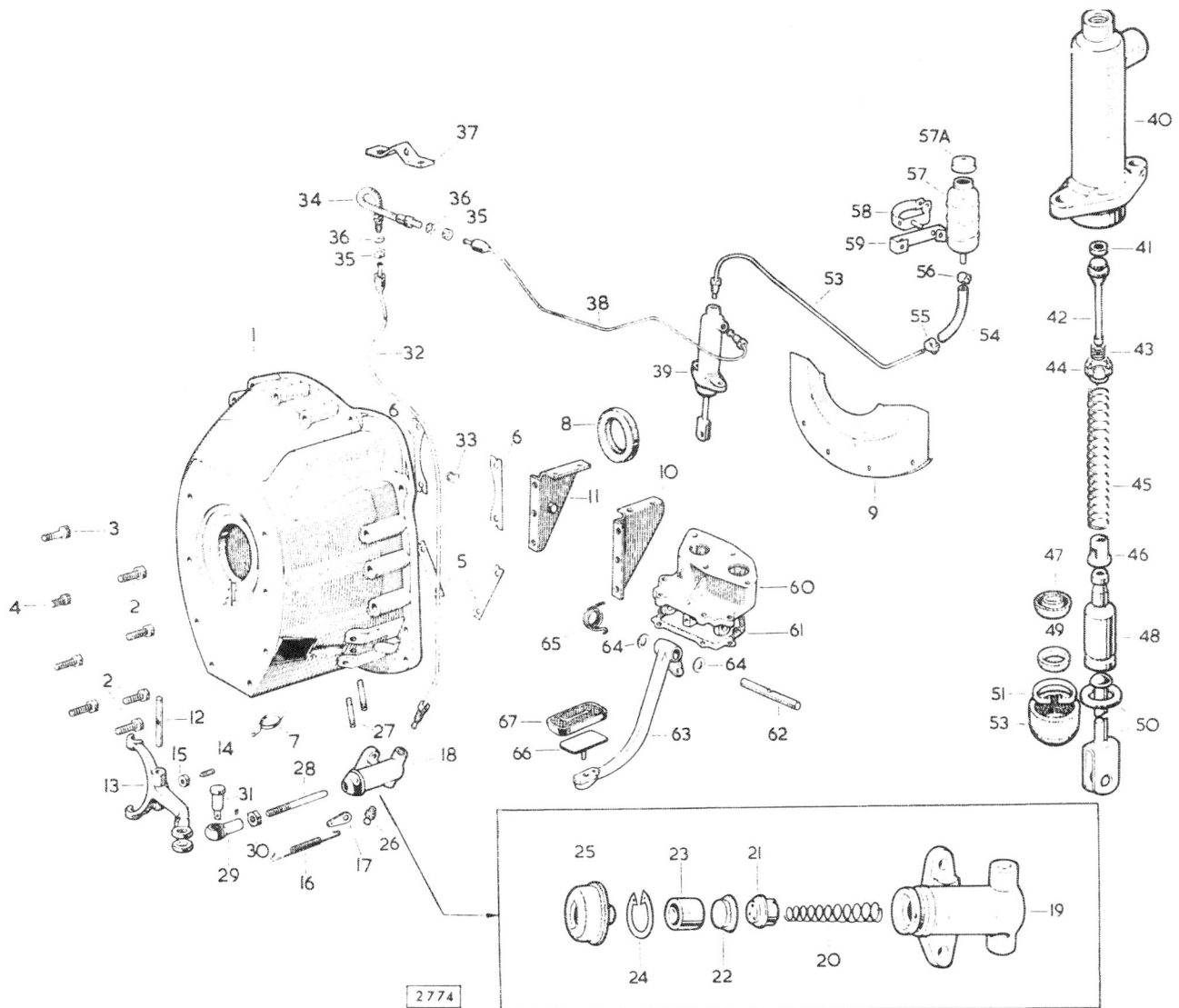

Fig. 25. *Clutch operating system.*

1. Clutch housing	26. Bleeder screw
2. Bolt	27. Stud
3. Bolt	28. Operating rod
4. Bolt	29. Adjuster assembly
5. Plate	30. Nut
6. Plate	31. Pivot pin
7. Cover	32. Pipe assembly
8. Oil seal	33. Clip
9. Cover plate	34. Flexible hose
10. Support bracket	35. Nut
11. Support bracket	36. Washer
12. Shaft	37. Bracket
13. Fork	38. Pipe assembly
14. Screw	39. Clutch master cylinder
15. Nut	40. Body
16. Return spring	41. Recuperation seal
17. Plate	42. Valve
18. Slave cylinder	43. Spring
19. Body	44. Spring support
20. Spring	45. Main spring
21. Cup filler	46. Spring support
22. Seal	47. Cup seal
23. Piston	48. Piston
24. Circlip	49. Static seal
25. Rubber boot	50. Push rod

51. Circlip
52. Dust excluder
53. Pipe assembly
54. Low pressure hose
55. Clip
56. Clip
57. Clutch fluid container
58. Clip
59. Clutch mounting bracket
60. Clutch pedal housing
61. Gasket
62. Shaft
63. Clutch pedal
64. Washer
65. Return spring
66. Steel pad
67. Rubber pad

SUPPLEMENTARY INFORMATION
TO
SECTION E "CLUTCH"

FAULT-FINDING (continued)

SYMPTOM	CAUSE	REMEDY
Judder	(a) Oil, grease or foreign matter on driven plate facings	Fit new facings or driven plate
	(b) Misalignment	Check over and correct alignment
	(c) Pressure plate out of parallel with flywheel face in excess of the permissible tolerance	Re-adjust levers in plane, and, if necessary, fit new eyebolts
	(d) Contact area of friction facings not evenly distributed. Note that friction facing surface will not show 100% contact until the clutch has been in use for some time, but the contact actually showing should be evenly distributed round the friction facings	This may be due to distortion, if so fit new driven plate assembly
	(e) Bent splined shaft or buckled driven plate	Fit new shaft or driven plate assembly
Rattle	(a) Damaged driven plate, broken springs, etc.	
	(b) Worn parts in release mechanism	
	(c) Excessive back lash in transmission	Fit new parts as necessary
	(d) Wear in transmission bearings	
	(e) Bent or worn splined shaft	
	(f) Graphite release bearing loose on throw out fork	
Tick or Knock	Hub splines worn due to misalignment	Check and correct alignment then fit new driven plate
Fracture of Driven Plate	(a) Misalignment distorts the plate and causes it to break or tear round the hub or at segment necks	Check and correct alignment and introduce new driven plate
	(b) If the gearbox during assembly be allowed to hang with the shaft in the hub, the driven plate may be distorted, leading to drag, metal fatigue and breakage	Fit new driven plate assembly and ensure satisfactory re-assembly
Abnormal Facing Wear	Usually produced by overloading and by the excessive slip starting associated with overloading	In the hands of the operator

HYDROSTATIC CLUTCH SLAVE CYLINDER

Commencing Engine Numbers

3.4 "S" 7B 5213
3.8 "S" 7B 58367

Commencing at the above engine numbers an hydrostatic clutch operating slave cylinder is fitted. Normal clutch wear is automatically compensated for by this slave cylinder and no clearance adjustment is necessary.

When replacing the new operating slave cylinder it is IMPORTANT that the operating rod adjustment dimension as shown in Fig. 26 is adhered to. To obtain this dimension proceed as follows:—

1. Extract the clevis pin securing the operating rod to the clutch lever.

2. Release the fork end nut.

3. Push the clutch operating lever away from the slave cylinder until resistance is felt and retain in this position.

4. Push the operating rod to the limit of its travel into the slave cylinder and adjust the fork end to a dimension of .75" (19 mm) between the centre of the fork end and the centre of the clutch operating lever. Tighten the locknut.

5. Release the operating rod and connect the fork end to the lever. Refit the clevis pin.

6. Bleed the clutch slave cylinder in the normal manner.

INTRODUCTION OF DIAPHRAGM SPRING CLUTCH

Commencing Engine Numbers

3.4 "S" 7B 6572
3.8 "S" 7B 60391

Commencing at the above engine numbers a Borg and Beck diaphragm spring clutch is fitted to all "S" type cars.

The Borg and Beck diaphragm spring clutch is serviced ONLY in the U.K. by fitting an exchange unit which is available from Jaguar Cars Ltd. Spares Division, Coventry.

Individual parts are available from the same source for the repair of this clutch in Overseas Markets where exchange units may not be readily available. IT IS ESSENTIAL, when overhauling the diaphragm spring clutch, to rigidly observe the service instructions detailed below and particular attention is drawn to the necessary special tools required.

The diaphragm spring is riveted inside the cover pressing with two fulcrum rings interposed between the shoulders of the rivets and the cover pressing. The diaphragm spring also pivots on these two fulcrum rings. Depressing the clutch pedal actuates the release bearing causing a corresponding deflection of the diaphragm spring thus pulling the pressure plate from the driven plate and freeing the clutch.

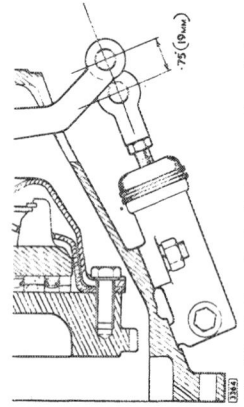

Fig. 26. *Hydrostatic clutch slave operating cylinder.— setting dimension.*

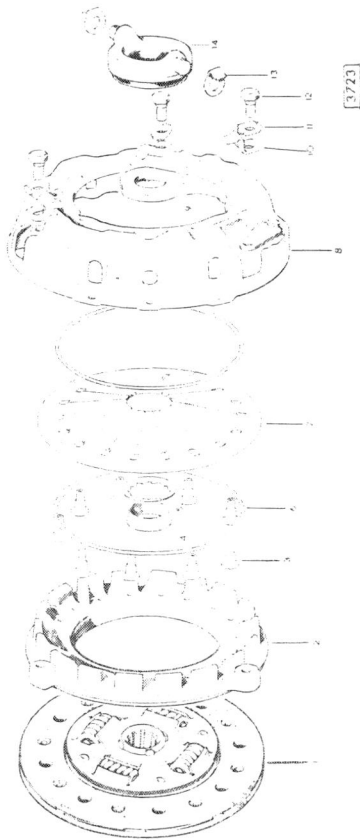

Fig. 27. Exploded view of the diaphragm spring clutch.

1. Driven plate
2. Pressure plate
3. Rivet
4. Centre sleeve
5. Belleville washer
6. Fulcrum ring
7. Diaphragm spring
8. Cover pressing
9. Release plate
10. Retainer
11. Tab washer
12. Setscrew
13. Retainer
14. Release bearing

DISMANTLING

Removing the Release Plate

The centrally mounted release plate is held in position by a small centre sleeve which passes through the diaphragm spring and belleville washer into the release plate.

To free the plate, collapse the centre sleeve with a hammer and chisel. To avoid any possible damage whilst carrying out this operation, support the release plate in the locating boss of the special tool which should be held firmly in a vice.

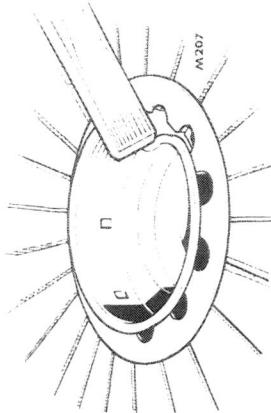

Fig. 28. Collapsing the centre sleeve with a hammer and chisel.

Separating the Pressure Plate from Cover Pressing

Knock back the locking tabs and remove the three setscrews securing the pressure plate to the straps riveted to the cover pressing.

These straps within the cover pressing must NOT be detached as this is an assembly reduced to its minimum as a spare part.

Dismantling the Cover Assembly

Remove the rivets securing the diaphragm spring and fulcrum rings by machining the shank of the rivets using a spot face cutter.

IT IS ESSENTIAL that the thickness of the cover is not reduced in excess of .005" (.127 mm) at any point.

The remaining portion of the rivets may be removed with a standard pin punch.

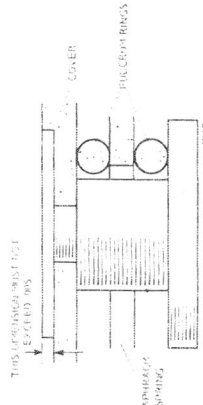

Fig. 29. Do not reduce the thickness of the cover pressing in excess of .005" (.127 mm).

---

exceeds this figure, the cover pressing must be replaced.

To achieve a satisfactory result when riveting the diaphragm spring into the cover pressing, a special tool must be fabricated to the specifications given in Fig. 31.

All parts except the spring can be made from mild steel.

Ref: Qty. Description
A 6 ⅜" flat washer
B 3 ⅜" nut
C 3 1¼" diameter setscrew
D 1 Spring (Minimum load of 100 lbs. fitted length)
E 1 Washer ½" I.D. × 1½" O.D. × ¼" thick
F 1 Tube ½" I.D. × 3¼" long
G 2 Washer ⅞" I.D. × 1½" O.D. × ⅛" thick
H 1 Bolt 1" Whit. × 6" long

REBUILDING

The Cover Assembly

Prior to rebuilding, check the cover pressing for distortion. Bolt the cover firmly to a flat surface plate and check that the measurement taken from the cover flange to the machined land inside the cover pressing does not vary by more than .007" (.2 mm). If the measurement

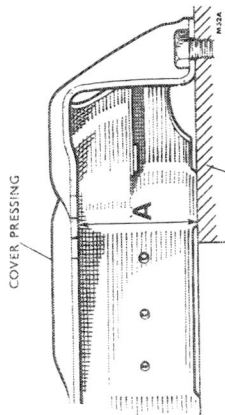

Fig. 30. The measurement "A" must not vary by more than .007" (.2 mm).

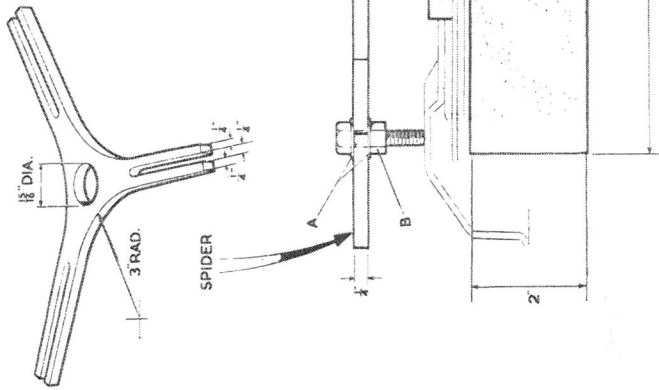

Fig. 31. Dimensions of the special tool for compressing the diaphragm spring when riveting the spring to cover pressing.

Position the fulcrum ring inside the cover pressing so that the location notches in the fulcrum ring engage a depression between two of the larger diameter holes in the cover pressing.

Place the diaphragm spring on the fulcrum ring inside the cover and line up the long slots in the spring with the small holes in the cover pressing. Locate a further fulcrum ring on the diaphragm spring so that the location notches are diametrically opposite the location notches in the first ring. Fit new shouldered rivets ensuring that the shouldered portion of each seats on the machined land inside the cover.

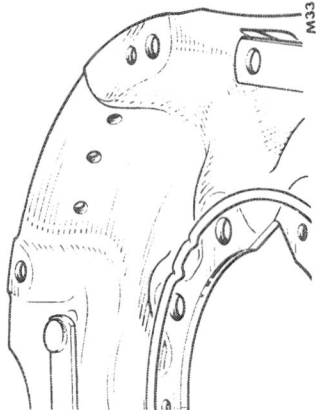

Fit the collar to the large bolt and fit the large bolt complete with spring, spider and collar into the tapped hole in the base. Position the three setscrews on the spider so that they contact the cover pressing. Tighten down the centre bolt until the diaphragm spring becomes flat and the cover pressing is held firmly by the set-screws.

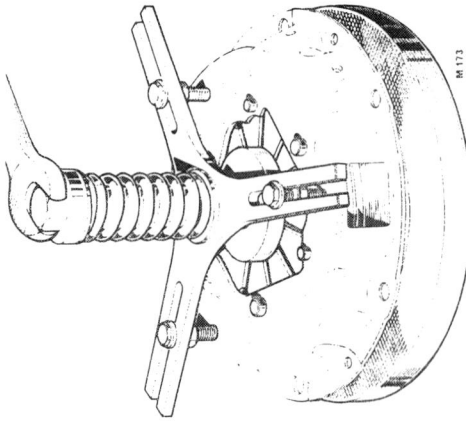

Fig. 32. *Assembly of cover pressing and fulcrum ring.*

Place the base of the special tool on to the rivet heads. Invert the clutch and base plate.

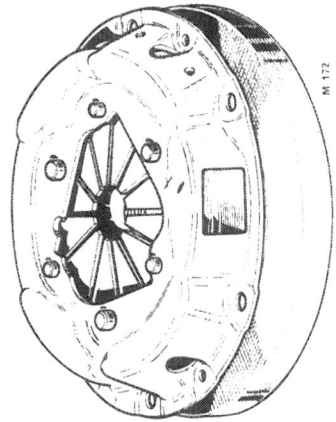

Fig. 33. *Clutch and base plate inverted.*

Assembling the Pressure Plate to Cover Pressing

Before assembling the pressure plate to the cover pressing, examine the plate for any signs of wear. Should it have been damaged or have excessive scoring, it is strongly recommended that a new pressure plate is fitted. If, however, renewal of the pressure plate is not possible grinding of the original unit may be undertaken by a competent machinist bearing in mind that incorrect grinding of the plate may seriously affect the operation of the clutch. IN NO CIRCUMSTANCES MAY THE PRESSURE PLATE BE GROUND TO A THICKNESS OF LESS THAN 1.070" (27.178 mm).

Position the pressure plate inside the cover assembly so that the lugs on the plate engage the slots in the cover pressing. Insert the three setscrews through the straps which are riveted to the cover pressing and lock with the tab-washers.

Fitting a New Release Plate

A special tool (Part No. SSC.805) is available from Automotive Products Ltd. Service &

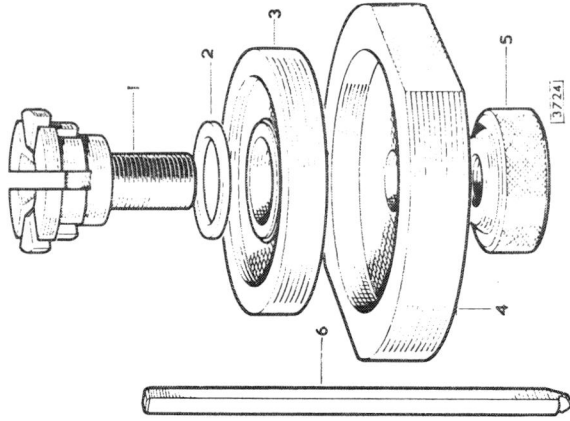

Fig. 34. *Tighten down the large nut until the diaphragm spring is compressed flat.*

Rivet securely with a hand punch.

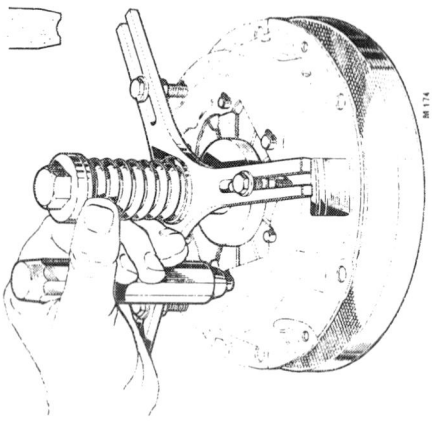

Fig. 35. *Riveting with a hand punch.*

Fig. 36. *Special tool (SSC. 805)*

1. Staking guide
2. Washer
3. Locating boss
4. Base plate
5. Knurled nut
6. Punch

[3724]

Spares Division, Banbury, England, for completion of this operation. Ensure that all parts of the clutch and the special tool are clean.

Grip the base of the tool in a vice and place the locating boss into the counterbore of the base plate. Place the release plate, face downwards, into the counterbore of the locating boss.

Apply a little high melting point grease to the tips of the diaphragm spring fingers and position the clutch, pressure plate friction face upwards, on to the release plate. Ensure that the diaphragm spring fingers locate between the small raised pips on the release plate.

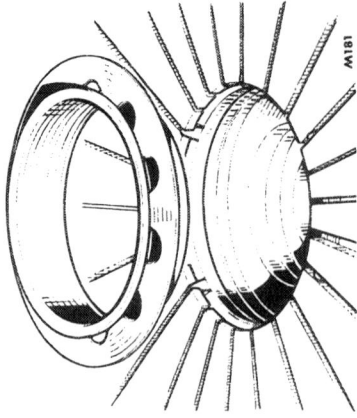

Fig. 37. *Fitting the sleeve and belleville washer*

Place the belleville washer, concave surface towards the spring, on to the centre of the diaphragm spring and then push the centre sleeve through the spring into the release plate.

Drop the special washer into the sleeve and insert the staking guide into the centre of the assembly. Fit the knurled nut to the thread of the staking guide and tighten down until the whole assembly is solid. Using the special punch, stake the centre sleeve in six places into the groove in the release plate.

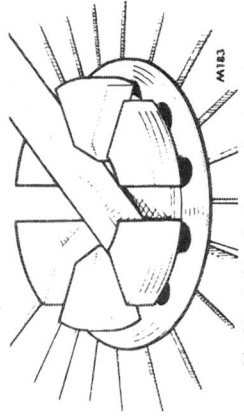

Fig. 38. *Staking the sleeve to the release plate.*

GEARBOX AND OVERDRIVE

INDEX

GEARBOX

	Page
Gearbox Ratio Data	F.4
Data	F.4
Routine Maintenance	F.5
Recommended Lubricants	F.5
Gearbox—To remove and refit	F.8
Gearbox—To dismantle	F.8
Dismantling the mainshaft	F.9
Dismantling the constant pinion shaft	F.10
Gearbox—To re-assemble	F.11
Checking layshaft end float	F.11
Assembling the mainshaft	F.12
Assembling the second gear synchro assembly	F.13
Fitting the second gear assembly to the mainshaft	F.13
Assembling the third/top synchro assembly	F.14
Fitting the third/top synchro assembly to the mainshaft	F.15
Assembling the constant pinion shaft	F.15
Assembling the gears to the casing	F.16
Fitting the top cover	F.16
Fitting the extension	F.16
Fitting the clutch housing	F.16

OVERDRIVE

Description	F.17
Method of operation	F.17

INDEX (continued)

	Page
Routine Maintenance	F.19
Recommended Lubricants	F.19
Dismantling and re-assembling :	
Removing the overdrive from the gearbox	F.20
Dismantling the overdrive	F.20
Inspection	F.20
Re-assembling the overdrive	F.21
Refitting the overdrive to the gearbox	F.22
Components :	
The operating valve	F.23
The hydraulic system	F.24
The pump valve	F.24
The Pump :	
Dismantling	F.25
Assembly	F.25
Hydraulic Pressure	F.26
The Accumulator Piston and Spring :	
Removal	F.26
The Control System :	
Wiring Circuit	F.27
Operating solenoid	F.27
Data	F.28
Operating Instructions	F.29
Fault Finding	F.30
Service Tools	F.30

GEARBOX AND OVERDRIVE

The gearbox is of the four-speed type with synchromesh on the second, third and top gears; these gears are of single helical form and are in constant mesh. The first and reverse gears have spur teeth which slide into mesh.

The overdrive (fitted as an optional extra) is of the Laycock de Normanville type and is dealt with separately at the end of this section.

GEARBOX

	Gearbox Ratios	Overall Ratios (Standard model) 3·4 and 3·8 litre	Overall Ratios (Overdrive model) 3·4 and 3·8 litre
Gearbox prefix*	GB	GB	GB
Gearbox suffix	JS	JS	JS
First and reverse	3·377 : 1	11·954 : 1	12·731 : 1
Second	1·86 : 1	6·584 : 1	7·012 : 1
Third	1·283 : 1	4·541 : 1	4·836 : 1
Top	1 : 1	3·54 : 1	3·77 : 1
Overdrive	·778 : 1		2·933 : 1
Axle ratio		3·54 : 1	3·77 : 1

* The letter "N" at the end of the prefix letters "GB" indicates that a gearbox mainshaft suitable for the attachment of an overdrive is fitted.

Ordering Spare Parts

It is essential when ordering spare parts for an individual gearbox, to quote the prefix and suffix letters in addition to the gearbox number.

The gearbox number is stamped on a lug situated at the left-hand rear corner of the gearbox casing and on the top cover.

DATA

Second gear end-float on mainshaft—·002″ to ·004″ (·05 to ·10 mm.)
Third gear end-float on mainshaft—·002″ to ·004″ (·05 to ·10 mm.)
Layshaft end-float on countershaft—·002″ to ·004″ (·05 to ·10 mm.)

ROUTINE MAINTENANCE

Every 3,000 miles (5,000 km.)

Gearbox Oil Level

Check the level of the oil in the gearbox with the car standing on level ground.

A combined level and filler plug is fitted on the left-hand side of the gearbox. Clean off any dirt from around the plug before removing it.

The level of the oil should be to the bottom of the filler and level plug hole.

Overdrive Oil Level—Important

The oil for the lubrication and operation of the overdrive unit is fed from the gearbox casing and therefore checking the gearbox oil level will also check the level of oil in the overdrive unit, but as this unit is hydraulically controlled extra attention should be paid to exercising absolute cleanliness when replenishing with oil. It is also important that the oil level is not allowed to fall appreciably otherwise the operation of the overdrive may be affected.

Every 12,000 miles (20,000 km.)

Changing the Gearbox Oil

The draining of the gearbox should be carried out at the end of a run when the oil is hot and therefore will flow more freely. The drain plug is situated at the front end of the gearbox casing.

After all the oil has drained replace the drain plug and refill the gearbox with the recommended grade of oil through the combined filler and level plug hole situated on the left-hand side of the gearbox casing; the level should be to the bottom of the hole.

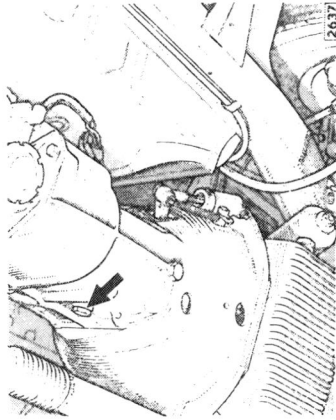

Fig. 1. The gearbox filler and level plug.

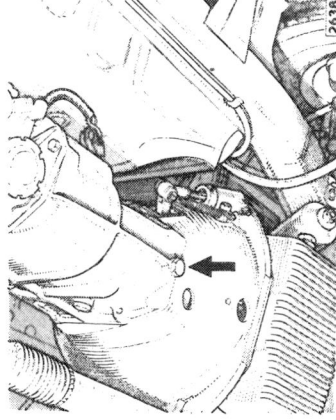

Fig. 2. Gearbox drain plug standard transmission

Recommended Lubricants

Mobil	Castrol	Shell	Esso	B.P.	Duckham	Regent Caltex/Texaco
Mobiloil A	Castrol XL	Shell X-100 30	Esso Extra Motor Oil 20W/30	Energol SAE 30	NOL 30	Havoline 30

Capacities

	Imperial	U.S.	Litres
Gearbox (without Overdrive) … … …	2½ Pints	3 Pints	1·5

1. Gearbox case
2. Drain plug
3. Fibre washer
4. Oil filler plug
5. Fibre washer
6. Locking plate
7. Setscrew
8. Spring washer
9. Ball bearing
10. Circlip
11. Ball bearing
12. Collar
13. Circlip
14. Fibre washer
15. Gasket
16. Gearbox extension
17. Gasket
18. Oil seal
19. Speedometer drive gear
20. "O" ring
21. Dowel screw
22. Striking rod, first/second gears
23. Striking rod, third/top gears
24. Striking rod, reverse gear
25. Stop
26. Change speed fork, first/second gears
27. Change speed fork, third/top gears
28. Change speed fork, reverse gear

29. Selector, third/top gears
30. Plunger
31. Spring
32. Locking ball
33. Spring
34. Dowel screw
35. Ball
37. Top cover
38. Switch
39. Gasket
40. Gasket
41. Bolt
42. Bolt
43. Bolt
44. Spring washer
45. Dowel
46. Ball
47. Plunger
48. Spring
49. Breather
50. Fibre washer
51. Stud
52. Welch washer
53. Welch washer
54. Plug
55. Copper washer

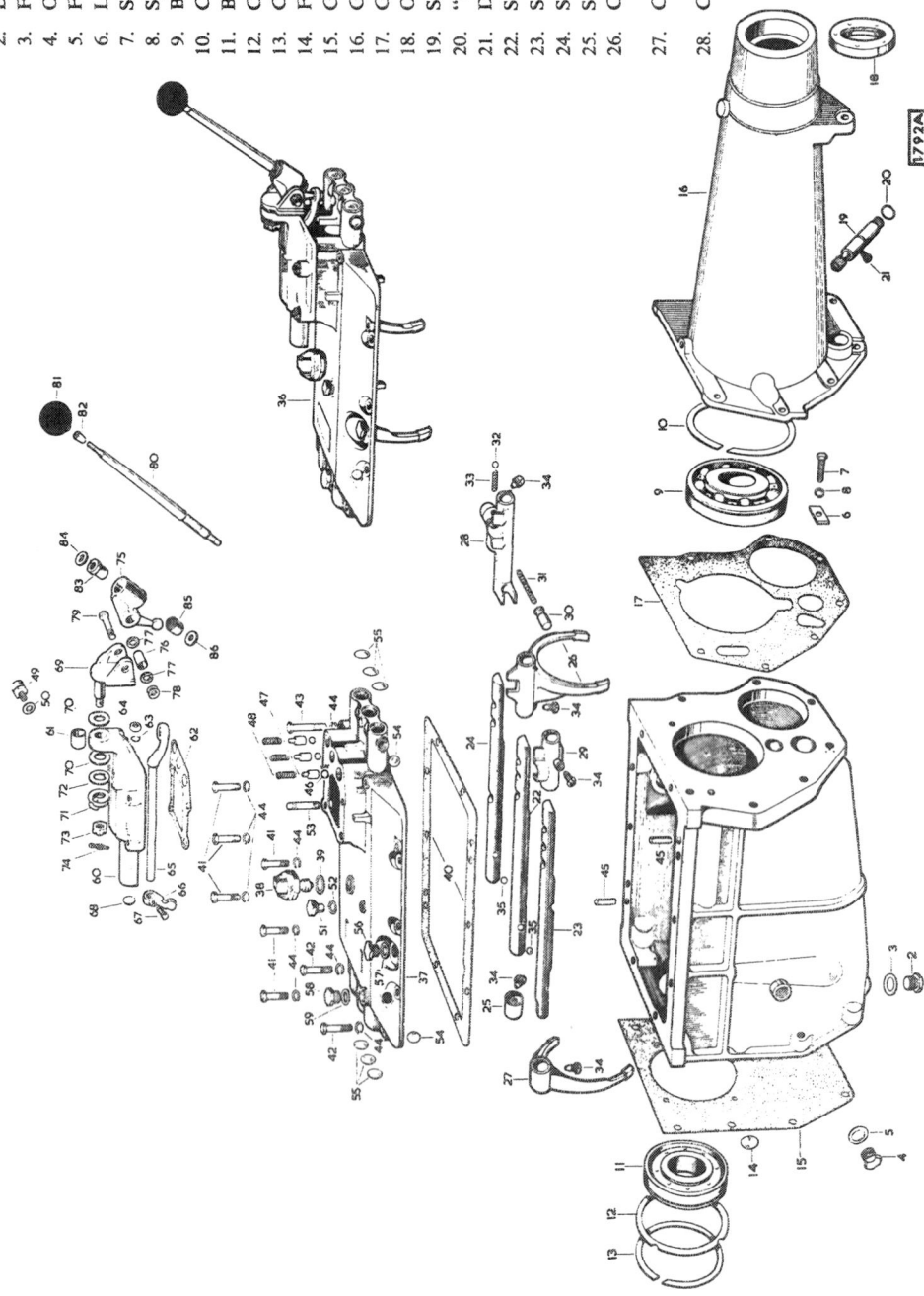

Fig. 4. Exploded view of gearbox casing and top cover.

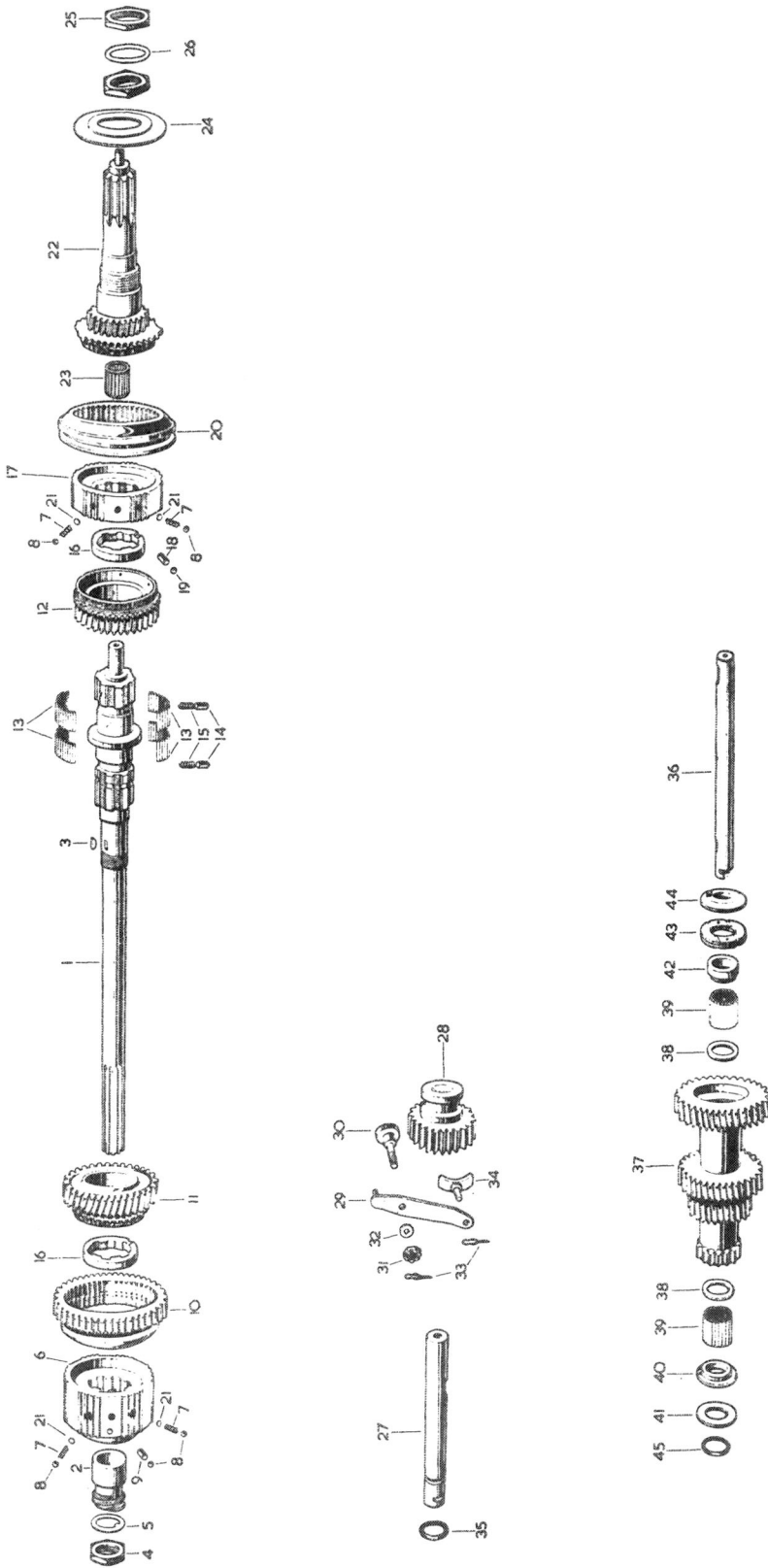

1. Mainshaft
2. Speedometer driving gear
3. Key
4. Nut
5. Tab washer
6. Synchronising sleeve second gear
7. Spring
8. Ball
9. Plunger
10. Frist speed mainshaft gear
11. Second speed mainshaft gear
12. Third speed mainshaft gear
13. Needle roller
14. Plunger
15. Spring
16. Thrust washer
17. Synchronising sleeve
18. Plunger
19. Ball
20. Operating sleeve
21. Shim
22. Constant pinion shaft
23. Roller bearing
24. Oil thrower
25. Locknut
26. Tab washer
27. Reverse spindle
28. Reverse gear
29. Lever
30. Fulcrum pin
31. Slotted nut
32. Plain washer
33. Split pin
34. Reverse slipper
35. Sealing ring
36. Countershaft
37. Gear unit on countershaft
38. Retaining ring
39. Needle roller
40. Thrust washer
41. Thrust washer
42. Retaining ring
43. Thrust washer
44. Thrust washer
45. Sealing ring

Fig. 5. *Exploded view of gears.*

1628

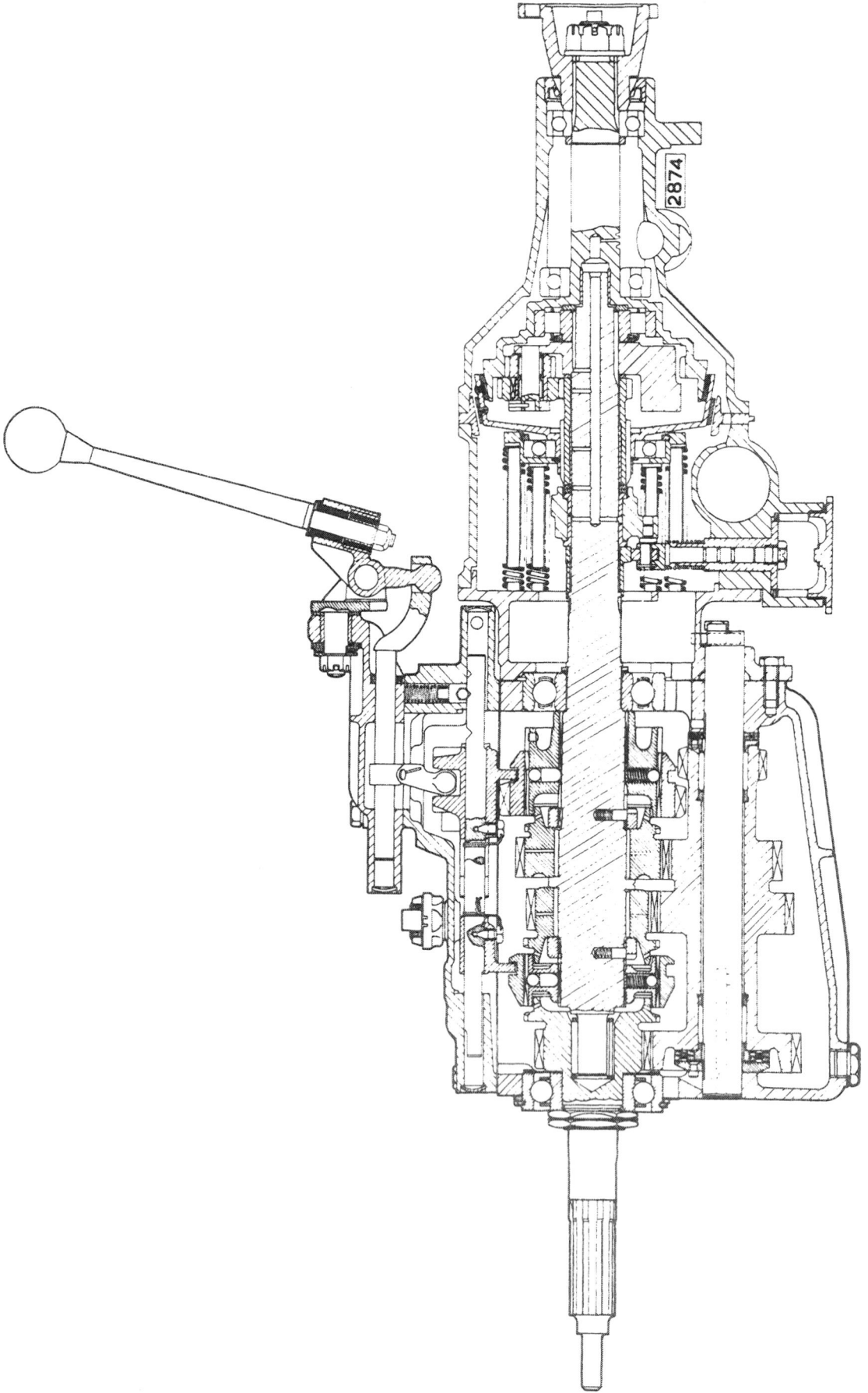

Fig. 3. *Sectional drawing of the gearbox.*

2874

GEARBOX—TO REMOVE AND REFIT

In order to remove the gearbox (and overdrive if fitted) it is necessary to remove the engine and gearbox as an assembly. For particulars of removal refer to Section B of this manual.

GEARBOX—TO DISMANTLE

Drain the gearbox by removing plug and fibre washer situated at base of the casing. Place gearbox in neutral and remove the ten setscrews with spring washers securing the top cover. Lift off top cover noting that this is located by two dowels fitted in the gearbox case. Remove and scrap the gasket.

Remove the clutch slave cylinder from the clutch housing. Detach the spring clips and remove the clutch release bearing. Release the locknut and remove the Allen headed screw securing the clutch fork to shaft. Withdraw shaft downwards and remove clutch fork. From inside the clutch housing remove the locking wire from the two bolts and tap back the tabs on the locking washers. Unscrew the eight bolts and remove washers. Unscrew the eight bolts and remove the clutch housing

Remove the locking screw retaining the speedometer driven gear bush in the extension. Withdraw the driven gear and bearing.

Remove the fibre blank from the front end of the layshaft.

On non-overdrive gearboxes remove the seven set-screws securing the rear extension to the gearbox casing. (Do not disturb the layshaft/reverse idler locking plate.) Withdraw the extension complete with shafts at the same time inserting a dummy countershaft into the countershaft bore at the front of the gearbox casing (see Fig. 7). The dummy shaft and countershaft must be kept in contact until the countershaft is clear of the casing.

Fig. 6. The gearbox top cover removed showing the layout of the mainshaft gears.

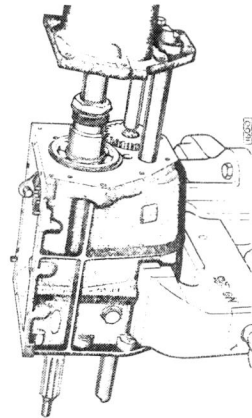

Fig. 7. Showing the removal of the extension: note the dummy countershaft inverted at the front end of the casing.

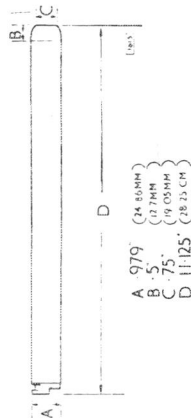

A .979 (24.86MM)
B .5" (12.7MM)
C .75" (19.05MM)
D 11.125" (28.25 CM)

Fig. 8. Dummy countershaft dimensions.

Engage top and first gears. On non-overdrive gearboxes tap back the tab washer securing the locknut at the rear of the mainshaft and unscrew the locknut. Withdraw the speedo drive gear. Remove the woodruff key from the mainshaft. Withdraw the dummy countershaft allowing the layshaft gear unit to drop into the bottom of the casing.

On gearboxes equipped with an overdrive, remove the circlip, plain washer and shims from behind the gearbox rear bearing.

Rotate the constant pinion shaft until the two cut-away portions of the driving gear are facing the top and bottom of the casing. Tap the mainshaft to the front to knock the constant pinion shaft with ball-bearing forward out of the case (see Fig. 9). Remove the constant pinion shaft and withdraw the roller bearing from the shaft spigot. Continue to tap mainshaft forward until free of the rear bearing. Tap the bearing rearward out of casing.

Push the reverse gear forward out of engagement to clear the mainshaft first speed gear. Lift the front end of the mainshaft upwards and remove complete with all mainshaft gears forward out of the casing leaving the layshaft in the bottom of the casing (see Fig. 10).

Draw reverse wheel rearwards as far as it will go to clear layshaft first speed gear. Lift out layshaft gear unit observing inner and outer thrust washers fitted at each end of the gears. Take care not to lose any needles which are located at each end of the gear unit.

Push reverse gear back into the case and remove through top. Note bush which is a press fit in reverse gear.

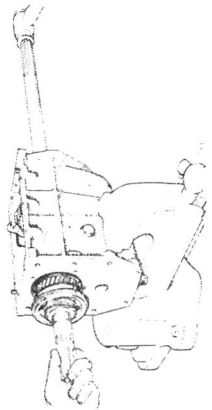

Fig. 9. The constant pinion shaft is removed by tapping the mainshaft forward.

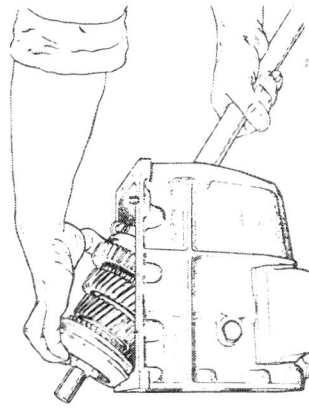

Fig. 10. Removing the mainshaft from the gearbox casing.

DISMANTLING THE MAINSHAFT

Withdraw the top third gear operating and synchronising sleeves forward off the shaft. Press the operating sleeve off the synchronising sleeve and remove the six synchronising balls and springs. Remove the interlock plungers and balls from the synchro sleeve.

Withdraw the second gear synchronising sleeve complete with first speed gear rearwards off the shaft. Press the first speed gear off the synchronising sleeve and remove the six synchronising sleeve and plunger from the synchro sleeve.

Press in the plunger locking the third speed gear thrust washer (see Fig. 11) and rotate washer until splines line up, when washer can be withdrawn. Remove the washer forward off shaft followed by third speed gear, taking care not to lose any needles which will emerge as the gear is removed. Remove the spring and plunger.

Press in the plunger locking the second speed gear thrust washer (see Fig. 12) and rotate washer until splines line up, when washer can be withdrawn. Remove the washer rearwards off shaft followed by second speed gear, taking care not to lose any needles which will emerge as the gear is removed. Remove the spring and plunger.

DISMANTLING THE CONSTANT PINION SHAFT

Knock back tab washer securing locknuts and remove locknuts (right-hand thread). Withdraw the bearing from the shaft and remove the oil thrower.

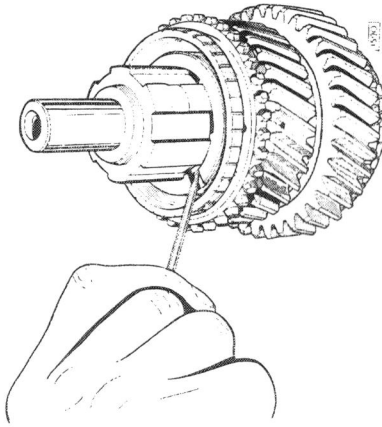

Fig. 11. *Depressing the third speed thrust washer locking plunger.*

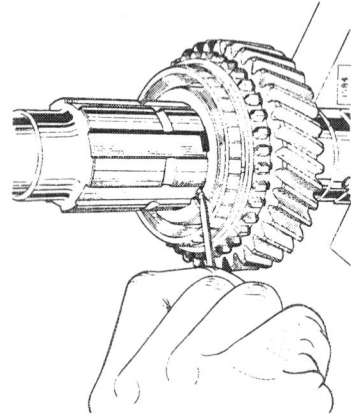

Fig. 12. *Depressing the second speed thrust washer plunger.*

GEARBOX—TO RE-ASSEMBLE

CHECKING LAYSHAFT END-FLOAT

Check the clearance between bronze thrust washer and the casing at rear of layshaft (see Fig. 13). The end-float should be ·002" to ·004" (·05 to ·10 mm.). Thrust washers are available in thicknesses of ·152", ·156", ·159", ·162" and ·164" (3·86, 3·96, 4·04, 4·11 and 4·17 mm.) to provide a means of adjusting the end-float.

Note: The gearbox must not be gripped in a vice when checking the end-float otherwise a false reading will be obtained.

Remove dummy countershaft and insert a thin rod in its place.

Place bushed reverse gear in slipper and draw gear rearwards as far as possible to give clearance for fitting layshaft gear unit.

ASSEMBLING THE MAINSHAFT

Fit the needle rollers (41 off) behind the shoulder on the mainshaft and slide the second speed gear, synchronising cone to rear, on to rollers. Apply grease to the needle rollers to facilitate assembly. Fit the second speed thrust washer spring and plunger into plunger hole. Slide thrust washer up shaft and over splines. Align large hole in synchro cone and with a steel pin compress plunger and rotate thrust washer into locked position with cut-away in line with plunger. Check the end-float of the second speed gear on the mainshaft by inserting a feeler gauge between the thrust washer and the shoulder on the mainshaft. The clearance should be ·002" to ·004" (·05 to ·10 mm.). Thrust washers are available in the following thicknesses to enable the end-float to be adjusted:

·471"–·472"—(11·96/11·99 mm.)

·473"–·474"—(12·01/12·03 mm.)

·475"/·476"—(12·06/12·09 mm.)

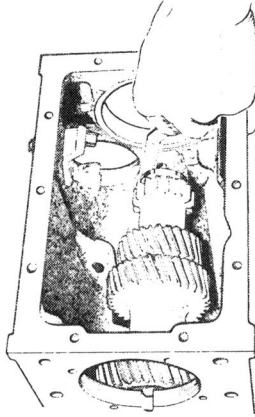

Fig. 13. *Checking layshaft end-float.*

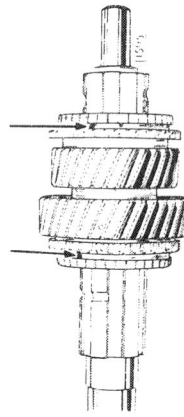

Fig. 14. *Showing the holes through which the thrust washer locking plungers are depressed.*

Fit the needle rollers (41 off) in front of the shoulder on the mainshaft and slide the third speed gear, synchronising cone to front, on to rollers. Apply grease to the needle rollers to facilitate assembly. Fit the third speed thrust washer spring and plunger into plunger hole. Slide thrust washer down shaft and over splines. Align large hole in synchro cone and with a steel pin compress plunger and rotate thrust washer into locked position with cut-away in line with plunger. Check the end-float of the third gear on the mainshaft by inserting a feeler gauge between the thrust washer and the shoulder on the mainshaft. The clearance should be ·002" to ·004" (·05 to ·10 mm.). Thrust washers are available in the following thicknesses to enable the end-float to be adjusted:

·471"/·472"—(11·96/11·99 mm.)
·473"/·474"—(12·01/12·03 mm.)
·475"/·476"—(12·06/12·09 mm.)

ASSEMBLING THE SECOND GEAR SYNCHRO ASSEMBLY

Fit the springs and balls (and shims if fitted) to the six blind holes in the synchro sleeve. Fit the first speed gear to the second speed synchronising sleeve with the relieved tooth of the internal splines in the gear in line with the stop pin in the sleeve (see Fig. 15). Compress the springs by means of a hose clip or by inserting the assembly endwise in a vice and slowly closing the jaws. Slide the operating sleeve over the synchronising sleeve until the balls can be heard and felt to engage the neutral position groove.

It should require 62 to 68 lb. (28 to 31 kg.) pressure to disengage the synchronising sleeve from the neutral position in the operating sleeve. In the absence of the necessary equipment to check this pressure, grip the operating sleeve in the palms of the hands and press the synchronising sleeve with the fingers until it disengages from the neutral position; it should require firm finger pressure before disengaging. Shims can be fitted underneath the springs to adjust the pressure of the balls against the operating sleeve.

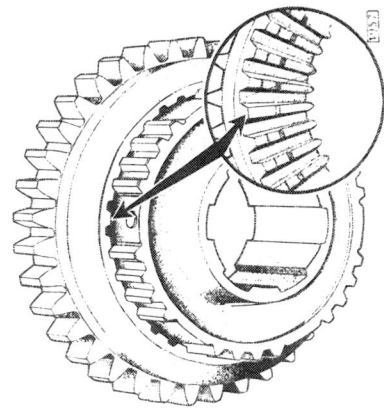

Fig. 15. When fitting the first speed gear to the second speed synchro sleeve the relieved tooth on the internal splines must be in line with the stop pin in the sleeve.

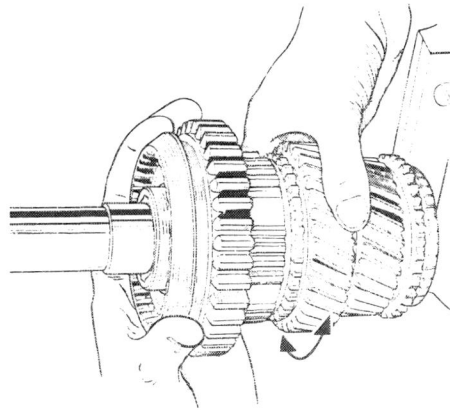

Fig. 16. With first gear engaged and slight downward pressure on the synchro assembly the second speed gear should be free to rotate.

FITTING THE SECOND GEAR ASSEMBLY TO THE MAINSHAFT

Fit the first speed gear second speed synchro assembly, to the mainshaft (any spline) and check that the synchro sleeve slides freely on the mainshaft, when the ball and plunger is not fitted. If it does not, try the sleeve on different splines on the mainshaft and check for burrs at the end of the splines.

Remove the synchro assembly from the mainshaft, fit the ball and plunger and refit to the same spline on the mainshaft.

Check the interlock plunger as follows:
Slide the outer operating sleeve into the first gear position as shown in Fig. 16.

With slight downward pressure on the synchro assembly the second speed gear should rotate freely without any tendency for the synchro cones to rub.

If the synchro cones are felt to rub, a longer plunger should be fitted to the synchro sleeve. Plungers are available in the following lengths: ·490", ·495" and ·500" (12·4, 12·52 and 12·65 mm.).

ASSEMBLING THE THIRD/TOP SYNCHRO ASSEMBLY

Fit the springs and balls (and shims if fitted) to the six blind holes in the inner synchronising sleeve. Fit the wide chamfer end of the operating sleeve to the large boss end of inner synchronising sleeve (see Fig. 17) with the two relieved teeth in operating sleeve in line with the two ball and plunger holes in the synchronising sleeve (see Fig. 18). Compress the springs by means of a hose clip or by inserting the assembly endwise in a vice and slowly closing the jaws. Slide the operating sleeve over the synchronising sleeve until the balls can be heard and felt to engage the neutral position groove.

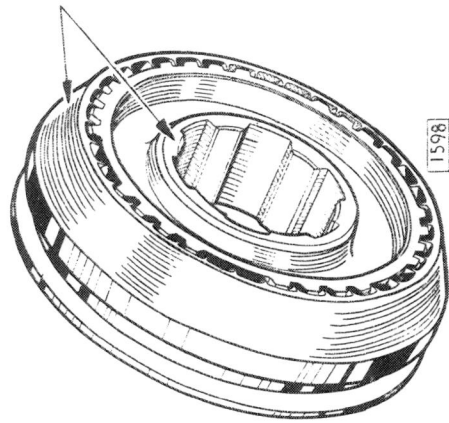

Fig. 17. The wide chamfer end of the operating sleeve must be fitted to the same side as the large boss of the inner synchro sleeve.

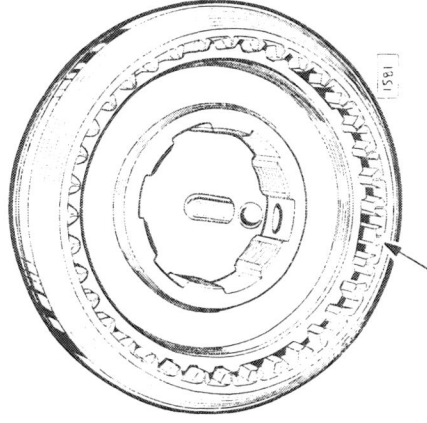

Fig. 18. The relieved teeth in the operating sleeve must be in line with the interlock plunger holes in the synchro sleeve.

ASSEMBLING THE CONSTANT PINION SHAFT

Fit the oil thrower followed by ball-bearing on to shaft with circlip and collar fitted to outer track of bearing. Screw on nut (right-hand thread) and fit tab washer and locknut. Fit the roller race into the shaft spigot bore.

ASSEMBLING THE GEARS TO THE CASING

Enter the mainshaft through the top of the casing and pass to the rear through bearing hole in case. Fit a new gasket to the front face of casing. Offer up the constant pinion shaft at the front of the case with cut-away portions of toothed driving member facing the top and bottom of the casing. Tap the constant pinion shaft to the rear until the collar and circlip on the bearing butt against the casing. Holding the constant pinion shaft in position tap in the rear bearing complete with circlip.

Lift the layshaft cluster into mesh with the thin rod and insert a dummy countershaft through the countershaft bore in front face of the casing (see Fig. 22).

Engage top and first gears. On non-overdrive gearboxes fit the woodruff key and speedo drive gear to the mainshaft. Fit the tab washer and locknut and secure. Place gearbox in neutral.

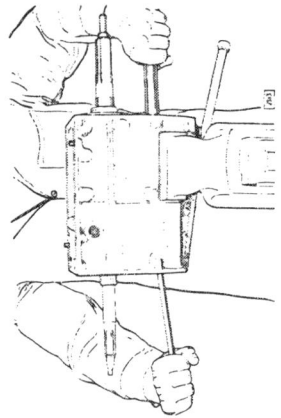

Fig. 22. Lifting the layshaft into mesh and inserting dummy countershaft.

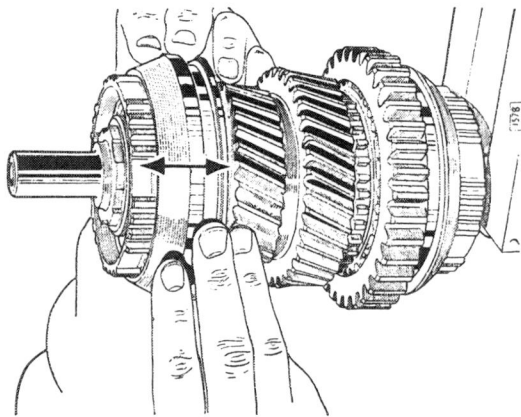

Fig. 20. Checking third speed interlock plunger. With third speed engaged there should be approximately $\frac{3}{32}$" (2.5 mm.) axial movement without drag.

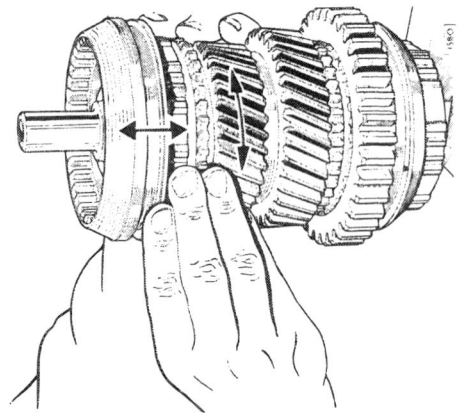

Fig. 21. Checking the fourth (top) gear interlock plunger. With the top gear engaged there should be approximately $\frac{3}{16}$" (4.76 mm.) axial movement without any drag. With top gear still engaged and with slight downward pressure exerted on synchro assembly the third speed gear should be free to rotate.

Plungers are available in the following lengths:

.490", .495" and .500" (12.4, 12.52 and 12.65 mm.).

Next slide the operating sleeve into the top gear position as shown in Fig. 21.

Lift and lower the synchro assembly; it should be possible to move the assembly approximately $\frac{3}{32}$" (2.5 mm.) without any drag being felt. Also with slight downward pressure exerted on the synchro assembly the third speed gear should be free to rotate without any tendency for the synchro cones to rub.

If it is found that the synchro assembly does not move freely a shorter top gear plunger should be fitted. If the third gear synchro cones are felt to rub, a longer top gear plunger should be fitted; looking at the wide chamfer end of the outer operating sleeve, the top gear plunger is one in line with the relieved tooth in the operating sleeve.

Plungers are available in the following lengths:

.490", .495" and .500" (12.4, 12.52 and 12.65 mm.).

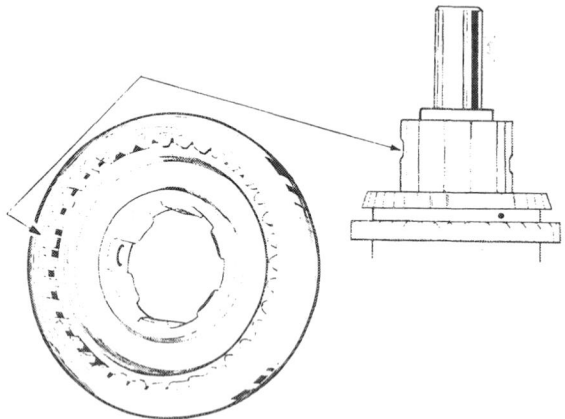

Fig. 19. The relieved tooth at the wide chamfer end of the outer operating sleeve must be in line with the foremost groove in the mainshaft.

It should require 52 to 58 lb. (24 to 26 kg.) pressure to disengage the synchronising sleeve from the neutral position in the operating sleeve. In the absence of the necessary equipment to check this pressure, grip the operating sleeve in the palms of the hands and press the synchronising sleeve with the fingers until it disengages from the neutral position; it should require firm finger pressure before disengaging. Shims can be fitted underneath the springs and balls to adjust the pressure of the balls against the operating sleeve.

FITTING THE THIRD TOP SYNCHRO ASSEMBLY TO THE MAINSHAFT

Fit the interlock balls and plungers, balls first, to the holes in the synchronising sleeve.

When fitting the third speed top gear synchro assembly to the mainshaft note the following points:

(a) There are two transverse grooves on the mainshaft splines which take the third top synchro assembly, and the relieved tooth at the wide chamfer end of the outer operating sleeve must be in line with the **foremost** groove in the mainshaft (Fig. 19). Failure to observe this procedure will result in the locking plungers engaging the wrong grooves thereby preventing full engagement of top and third gears.

(b) The wide chamfer end of the outer operating sleeve must be facing forward, that is, towards the constant pinion shaft end of the gearbox.

The inner sleeve must slide freely on the mainshaft, when the balls and plungers are not fitted. If it does not, check for burrs at the ends of the splines.

Fit the two balls and plungers to the holes in the inner synchro sleeve and refit the synchro assembly to the mainshaft observing points "a" and "b" above.

Check the interlock plungers as follows:

Slide the third top operating sleeve over the third speed gear dogs as shown in Fig. 20. With the third gear engaged lift and lower the synchro assembly; it should be possible to move the assembly approximately $\frac{3}{32}$" (2.5 mm.) without any drag being felt. If it is found that the synchro assembly does not move freely a shorter third speed plunger should be fitted; looking at the wide chamfer end of the outer operating sleeve this is the plunger that is not opposite the relieved tooth in the operating sleeve.

On gearboxes equipped with an overdrive, fit the shim(s), plain washer and circlip behind the rear bearing. Fit as many shims as are necessary to eliminate all end-float from the mainshaft (Fig. 23).

Fit the clutch operating fork and insert shaft. Fit the locking screw and locknut. Fit the release bearing and spring clips. Engage slave cylinder with operating rod and slide on to studs. Fit the spring anchor plate to lower stud and secure with the nuts. Fit the return spring.

FITTING THE TOP COVER

Fit a new gasket on to top face of case. Offer up the top cover, noting that this is located by two dowels and secure in position with ten setscrews and spring washers. (Two long screws at rear and two short screws at front.) Fit the gearbox drain plug and fibre washer.

FITTING THE EXTENSION

Fit a new gasket to the rear face of the gearbox casing. Offer up the extension complete with counter and reverse shafts and tap into position, driving the dummy countershaft forward out of the casing. Secure the extension with seven setscrews and spring washers. Fit a new fibre washer at the front end of the countershaft. Fit the speedo driven gear and bearing to the extension.

FITTING THE CLUTCH HOUSING

Fit a new oil seal into the clutch housing, lip of oil seal facing the gearbox.

Fit the clutch housing and secure with the eight bolts and three tab washers and locking wire.

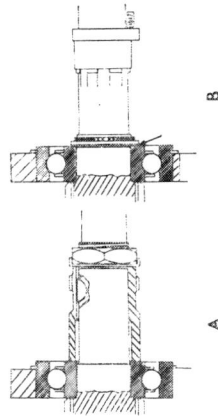

Fig. 23. *Showing the rear bearing retaining arrangement.*

"A" Standard Gearbox
"B" Overdrive Gearbox
The arrow indicates the shims

OVERDRIVE

DESCRIPTION

The Laycock de Normanville overdrive unit consists of a hydraulically controlled epicyclic gear housed in a casing at the rear of the gearbox.

When engaged, the overdrive reduces the engine speed in relation to the road speed thus permitting high road speeds with low engine revolutions. Consequently, the use of the overdrive results in fuel economy and reduced engine wear.

METHOD OF OPERATION (Fig. 24)

Power Input

The power input enters the overdrive unit through the extension to the gearbox driven shaft (D) and by means of cam (C) operates the plunger-type hydraulic pump (G). This, in turn, builds up pressure against the spring loaded piston (F) in the accumulator cylinder placed across the bottom of the main casing.

The Sun-wheel

The sun-wheel (O) is integral with the splined sleeve (M) which is free to rotate on the input shaft. Immediately behind the sun-wheel and splined to shaft (D) is the planet carrier (N) in which are mounted the three planet wheels (T).

The Uni-direction Clutch

This operates from the input shaft on to which is splined the inner member (S). The other components of this clutch are the rollers (P) and the outer member (W) which is attached to the combined annulus (U) and output shaft (Q). The drive is transmitted from the input shaft through the clutch inner member and the rollers which are forced up the inner member's inclined faces (R) wedging the whole clutch

solid. The clutch then drives the annulus output shaft.

The Cone Clutch

The cone clutch (K) is mounted on the splined sleeve (M) on which it is free to slide. The cone clutch springs (V), which hold the inner lining (L) in contact with the corresponding cone of the annulus (U), maintain the clutch in the direct drive position. This prevents a free-wheel condition when the car tries to over-run the engine. Engine braking is, therefore, always available.

Power is transmitted by way of the cone clutch, inner lining and the annulus when the reverse gear is engaged as the uni-directional clutch is in-operative.

Hydraulic Operation

Overdrive is brought into operation by rotating the operating shaft (E) thus lifting the operating valve (A). This action allows the stored hydraulic pressure in the accumulator to be applied to the two pistons (B). The pistons move the clutch (K) forward away from the annulus (U) overcoming the springs (V). During the forward movement of the clutch (K) the drive from the engine to the wheels is maintained by the roller clutch.

The hydraulic operation causes the outer lining (J) of the cone clutch (K) to contact the brake ring (H), bringing the sun-wheel (O) and sleeve (M) to rest.

This action is effected without shock as the clutch (K,J,H) is oil immersed. The input drive now passes from shaft (D) to the planet carrier (N) and the rotation of the planet wheels (T) around the stationary sun-wheel causes both the annulus and the output shaft (U,Q) to be driven faster than the input shaft (D). In this condition the outer member (W) of the roller clutch over-runs the inner member (S).

Because the sun-wheel can move neither backwards nor forwards, there is always engine braking available in overdrive gear.

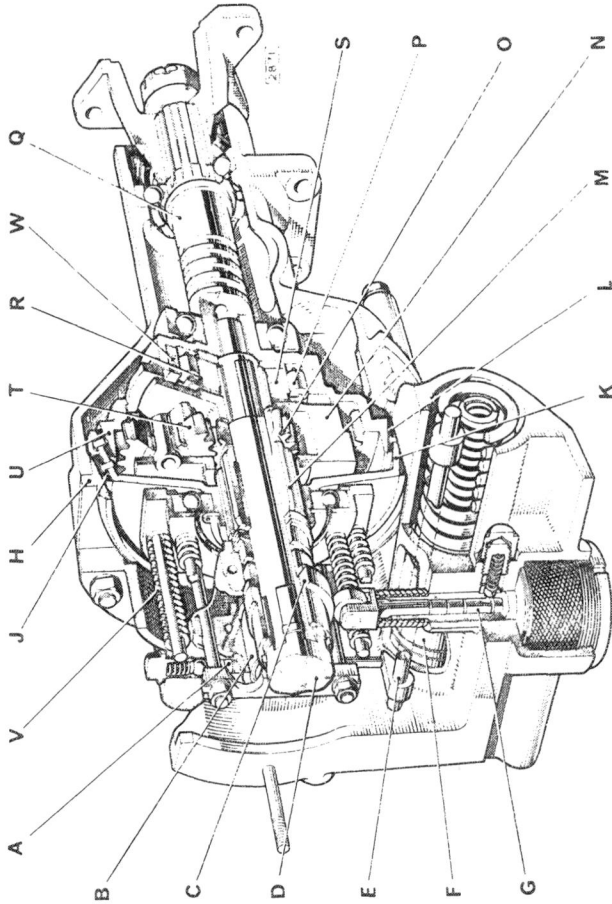

Fig. 24. Cut-away view of overdrive unit.

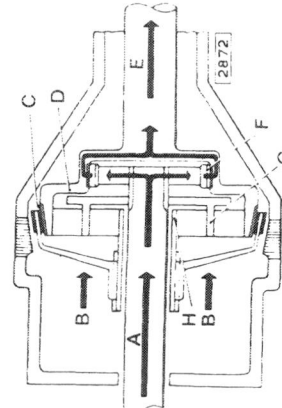

Fig. 25. In direct drive.

A From Gearbox
B Spring Pressure
C Annulus and Sun-wheel locked

D Annulus
E To Propeller Shaft
F Uni-directional clutch

G Planet wheel and carrier
H Sun-wheel

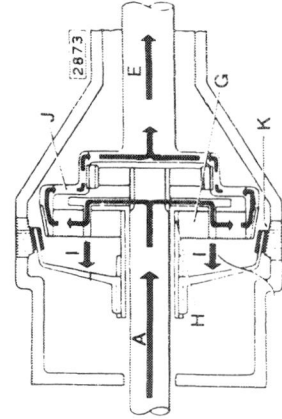

Fig. 26. In overdrive.

G Planet wheel and carrier
H Sun-wheel
I Hydraulic pressure

J Annulus overdriven by planet wheels.
K Locked cone clutch holds Sun-wheel.

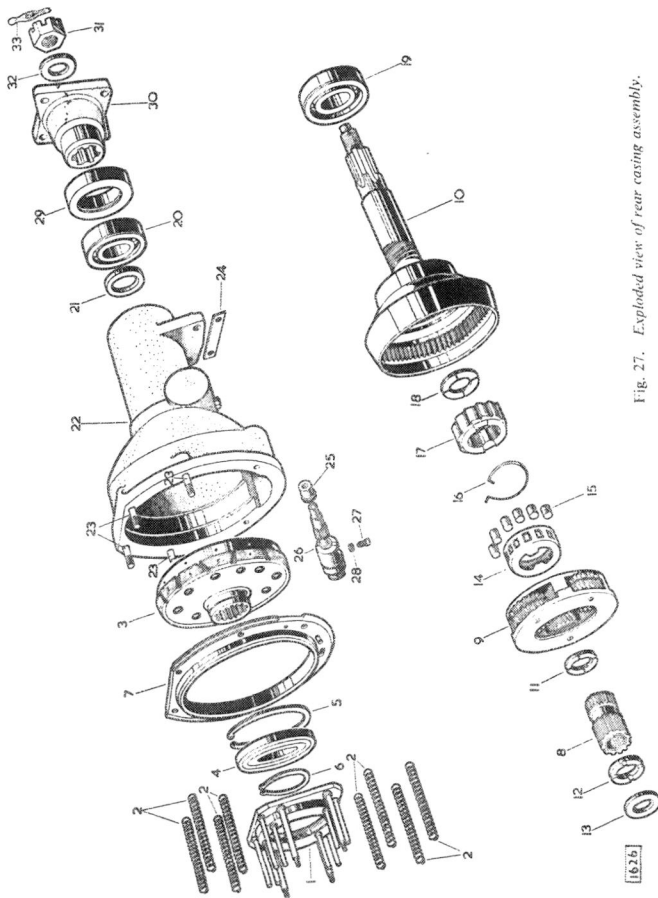

1. Extension for attachment to gearbox
2. Gasket
3. Bearing spacing washer
4. Circlip
5. Front casing
6. Welch washer plug
7. Operating valve plug
8. Operating valve shaft
9. Operating valve cam lever
10. "O" ring seals
11. Operating and restrictor valve
12. Operating valve ball
13. Ball plunger
14. Plunger spring
15. Valve and pressure take-off plug
17. Copper washer
17. Pump guide peg
18. Valve setting lever
19. Setting lever pin
20. Stud
21. Stud
22. Long gearbox stud
23. Rear casing stud
24. Operating valve shaft collar

25. Breather plug
26. Pump plunger
27. Pump body
28. Pump plunger spring
29. Pump plug
30. Pump securing screw
31. Spring washer
32. Non return valve ball
33. Ball plunger
34. Plunger spring
35. Screwed plug
36. Sealing washer
37. Gauze filter
38. Drain plug
39. Sealing washer
40. Pump operating eccentric
41. Cone clutch operating piston
42. Piston bridge piece
43. Nut
44. Tab washer
45. Accumulator piston assembly
46. Set of six piston rings, four narrow, two wide

47. Piston housing

48. 3/8 litre spacer tube
48a. 3/4 litre spacer tube
49. 3/8 litre piston spring (large)
50. 3/8 litre piston spring (small)
50a. 3/4 litre piston spring
51. Solenoid mounting bracket assembly
52. Rubber buffer for solenoid plunger
53. Gasket
54. Nut
55. Spring washer
56. Bolt, holding accumulator spring in tension
57. Solenoid lever
58. Pinch bolt
59. Nut
60. Spring washer
61. Cover plate
62. Gasket
63. Operating solenoid
64. Bolt
65. Spring washer

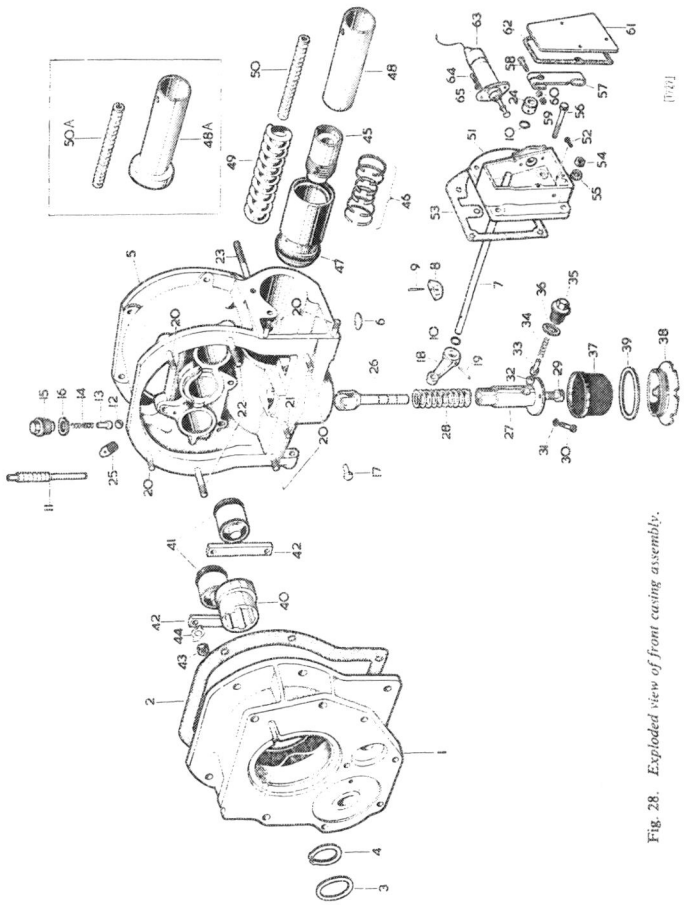

Fig. 28. *Exploded view of front casing assembly.*

Fig. 27. *Exploded view of rear casing assembly.*

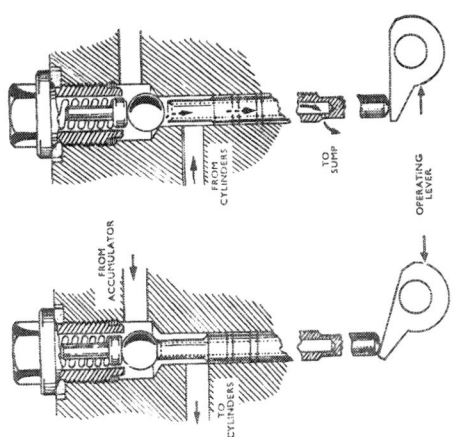

Fig. 26. *Principle of the operating valve.* (See Fig. 28).

*In the case of the 3·8 litre model there are twelve springs (4 inner and 8 outer).

1. Clutch thrust ring assembly
2. Springs *
3. Sliding member
4. Ball bearing
5. Circlip
6. Brake ring
7. Sun wheel assembly
8. Planetary carrier assembly
9. Annulus assembly
10. Thrust washer (phosphor bronze)
11. Thrust washer (phosphor bronze)
12. Thrust washer (phosphor bronze)
13. Thrust washer (steel)
14. Cage for uni-directional clutch
15. Rollers
16. Spring
17. Inner member for uni-directional clutch
18. Thrust washer
19. Ball bearing
20. Ball bearing
21. Spacing washer
22. Rear casing assembly
23. Stud
24. Packing
25. Pilot bush
26. Speedometer drive assembly
27. Locking screw
28. Spring washer
29. Oil seal
30. Flange
31. Nut
32. Plain washer
33. Split pin

Page F.18

ROUTINE MAINTENANCE

Oil Level—Important

The oil for lubrication and operation of the overdrive unit is fed from the gearbox casing and therefore checking the gearbox oil level will also check the level of oil in the overdrive unit, but as this unit is hydraulically controlled extra attention should be paid to exercising absolute cleanliness when replenishing with oil. It is also important that the oil level is not allowed to fall appreciably otherwise the operation of the overdrive may be affected.

Oil Changing

The oil for the overdrive unit is common with that in the gearbox but draining the gearbox casing will not drain oil from the overdrive unit. When draining the gearbox the large brass drain plug in the base of the overdrive unit should be removed using Churchill Tool No. J.3. Clean the magnet in the drain plug and remove the overdrive oil pump filter. This oil filter is accessible through the drain plug aperture and can be withdrawn by hooking the end of a piece of malleable wire in the centre hole. Avoid damaging the gauze.

Thoroughly wash the filter gauze and allow to dry; when refitting engage the flange of the oil pump inside the overdrive unit with the top edge of the filter and engage the small hole in its base with the small button on the inside face of the drain plug. Fully tighten the drain plug utilizing the Churchill Tool No. J.3.

Refill the gearbox and overdrive with oil through the gearbox filler and level plug hole. **Recheck the level after the car has been run** as a certain amount of oil will be retained in the hydraulic system of the overdrive.

Particular attention should be paid to maintaining absolute cleanliness when filling the gearbox and overdrive with oil as any foreign matter that enters may seriously affect the operation of the overdrive.

Every 3,000 miles (5,000 km.)

Check the oil level of the gearbox and overdrive and top up as necessary through the filler and level plug hole in the side of the gearbox casing.

Every 12,000 miles (20,000 km.)

Drain and refill the gearbox units. Two drain plugs, one for the gearbox and one for the overdrive unit, are situated at the base of their respective castings.

Both must be removed to drain the transmission completely. A common filler and level hole is situated on the left-hand side of the gearbox.

After draining the oil, remove the overdrive oil pump filter and clean the filter gauze by washing in petrol. The filter is accessible through the drain plug aperture being situated between the base of the oil pump and drain plug.

After refilling the gearbox and overdrive with oil, recheck the level after the car has been run, since a certain amount of oil will be retained in the hydraulic system of the overdrive unit.

Fig. 30. *Overdrive drain plug.*

Recommended Lubricants

Mobil	Castrol	Shell	Esso	B.P.	Duckham	Regent Caltex/Texaco
Mobiloil A	Castrol XL	X-100 30	Esso Extra Motor Oil 20W/30	Energol SAE 30	NOL 30	Havoline 30

Gearbox and Overdrive Oil Capacity

Imperial Pints	U.S. Pints	Litres
4	4¾	2¼

DISMANTLING AND RE-ASSEMBLING

If trouble should arise necessitating dismantling the unit further than is described in the previous section it will be necessary to remove the overdrive unit from the car. The engine, gearbox and overdrive unit are removed together. Remove the gearbox and clutch housing from the engine. Detach the clutch housing from the gearbox casing.

BEFORE COMMENCING ANY DISMANTLING OPERATIONS IT IS IMPORTANT THAT THE HYDRAULIC PRESSURE IS RELEASED FROM THE SYSTEM. DO THIS BY OPERATING THE OVERDRIVE 10-12 TIMES.

REMOVING THE OVERDRIVE FROM THE GEARBOX

1. The overdrive unit is separated from the gearbox at the joint between the gearbox rear extension and the overdrive front casing which are attached by seven studs **two of which are extra long.**

2. Remove the five nuts on the short studs before those on the longer studs are touched.

3. Slacken the two nuts on the long studs by equal amounts to release the compression of the clutch springs.

4. Remove the two nuts when the overdrive unit can be withdrawn off the mainshaft.

DISMANTLING THE OVERDRIVE

1. Remove the clutch springs from their pins, noting that the four inner springs **are shorter than the four outer springs.**

   In the instance of the overdrive fitted to 3-8 litre models the clutch springs number twelve; four short, situated in the centre and the remaining eight around the outside.

2. The two bridge pieces against which the operating pistons bear can now be removed. Each is secured by two ¼" nuts locked by tab washers. Withdraw the two operating pistons.

3. The pump valve can be dismantled without removing the solenoid bracket from the housing and there is no need to disturb the latter unless it is necessary to remove the accumulator piston and spring. (2 springs, 3-8 litre model.)

4. Remove the six nuts securing the two halves of the housing and separate them, removing the brake ring which is spigotted into the clutch carrier two pieces. Lift out the clutch carrier assembly. Remove the clutch sliding member complete with the thrust ring and bearing, the sun-wheel and thrust washers. Take out the inner member of the uni-directional clutch, the rollers, cage, etc.

5. If it is necessary to remove the planet gears from the carrier, the three Mills pins securing the planet bearing shafts must be extracted before the latter can be knocked out.

6. To remove the annulus, first take off the coupling flange at the rear of the unit, remove the speedometer gear, and drive out the annulus from the back. The front bearing will come away on the shaft leaving the rear bearing in the housing.

INSPECTION

Each part should be thoroughly inspected after the unit is dismantled and cleaned to ensure what parts should be replaced. It is important to appreciate the difference between parts which are worn sufficiently to affect the operation of the unit and those which are merely "worn-in".

1. Inspect the front casing for cracks, damage, etc. Examine the bores of the operating cylinders and accumulator for scores and wear. Check for leaks from the plugged ends of the oil passages. Ensure that the blanking plug beneath the accumulator bore is tight and not leaking. Inspect the support bushes in the centre bore for leaks.

2. Examine the clutch sliding member assembly. Ensure that the clutch linings are not burned or worn. Inspect the pins for clutch springs and bridge pieces and see that they are tight in the thrust ring and not disturbed. Ensure that the ball-bearing is in good condition and rotates freely. See that the sliding member slides easily on the splines of the sun-wheel.

3. Inspect the clutch springs for distortion.

4. Inspect the teeth of the gear train for damage. If the sun-wheel or planet bushes are worn the sun-wheel or planet bushes are worn the gears will have to be replaced since it is not possible to fit new bushes in service because they have to be bored to the pitch line of the teeth.

5. Inspect steel and bronze thrust washers.

6. Inspect the uni-directional clutch. See that the rollers are not chipped and that the inner and outer members of the clutch are free from damage. Make sure that the outer member is tight in the annulus. Ensure that the spring is free from distortion.

7. Inspect the ball bearings on the output shaft and see that there is no roughness when they are rotated slowly.

8. Inspect the mainshaft splines for nicks and burrs. See that the oil holes are open and clean.

9. Inspect the oil pump for wear on the pump plunger and roller pin. Ensure that the plunger spring is not distorted. Inspect the valve seat and ball and make sure that they are free from nicks and scratches.

10. Inspect the operating valve for distortion and damage and see that it slides easily in its bore in the front casing.

RE-ASSEMBLING THE OVERDRIVE

The unit can be re-assembled after all the parts have been thoroughly cleaned and checked to ensure that none are damaged or worn.

1. Assemble the annulus into the rear casing, not forgetting the spacing washer which fits between a shoulder on the shaft and the rear ball-bearing. This washer is available in different thicknesses for selective assembly and should allow no end float of the annulus (output shaft) and no pre-loading of the bearings.
   Selective washers are furnished in the following sizes:

Part No.	Size
C.5981	.146" ± .0005" (3.70 mm. ± 0.013 mm.)
C.5694	.151" ± .0005" (3.83 mm. ± 0.013 mm.)
C.5695	.156" ± .0005" (3.95 mm. ± 0.013 mm.)
C.5696	.161" ± .0005" (4.07 mm. ± 0.013 mm.)
C.5697	.166" ± .0005" (4.20 mm. ± 0.013 mm.)

2. Replace the thrust washer and uni-directional clutch inner member with its rollers and cage. A fixture (Fig. 31) is needed for retaining the rollers in position when assembling the clutch. Ensure that the spring is fitted correctly, so that the cage urges the rollers up the ramps on the inner member.

3. Fit the pump cam on to gearbox mainshaft, offer up the front housing to the gearbox rear extension and secure temporarily with two nuts. In order to determine the amount of end-float of the sun-wheel which should be .008" to .014" (.20 to .35 mm.), an extra thrust washer of known thickness should be assembled with the two normally used in front of the sun-wheel.

4. Fit the planet carrier, with its planet gears, over the sun-wheel with the marked teeth of the planets radially outwards as shown in Fig. 32 and with the assembly in this position offer it up to the annulus.

5. Assemble the brake ring to the front casing, then offer up the front and rear assemblies, leaving out the clutch sliding member with its springs, etc. The gap between the flanges of the brake ring and rear casing should be measured. This gap will be less than the thickness of the extra thrust washer by the

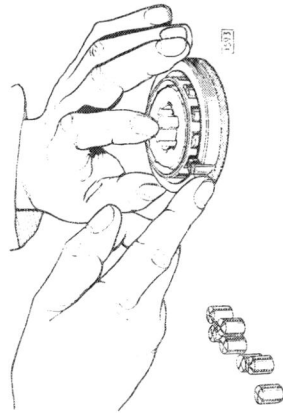

Fig. 31. *Assembly of the uni-directional clutch (Tool No. L.178).*

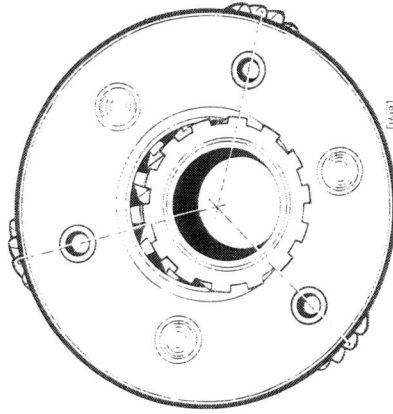

Fig. 32. *Assembly of the planet gears; note the positions of the marked teeth.*

amount of end-float of the sun-wheel. If this is between the limits specified the unit may be stripped down again and re-assembled without the extra thrust washer. The clutch sliding member, bridge peices, etc., must then be replaced.

6. If the indicated end-float is more or less than that required it must be adjusted by replacing the steel thrust washer at the front of the sun-wheel by one of less or greater thickness, as required. Washers of varying thicknesses are stocked for this purpose. Seven sizes are available, as follows :

Jaguar Part No.	Size
C.5943	.113"—.118" (2.87 mm.—2.99 mm.)
C.5944	.107"—.104" (2.71 mm.—2.64 mm.)
C.5945	.101"—.102" (2.55 mm.—2.58 mm.)
C.5946	.095"—.096" (2.41 mm.—2.44 mm.)
C.5947	.089"—.090" (2.25 mm.—2.28 mm.)
C.5948	.083"—.084" (2.10 mm.—2.13 mm.)
C.5949	.077"—.078" (1.95 mm.—1.98 mm.)

7. Care must be taken to ensure that the thrust washers at the front and rear of the sun-wheel are replaced in their correct positions. At the front of the sun-wheel the steel washers fit next to the head of the support bush in the housing and the bronze washer between the steel one and the sun-wheel. At the rear the steel washer is sandwiched between the two bronze washers. The latter are similar and their positions interchangeable. It is essential when assembling the gear train to ensure that the planets are turned to their correct relative positions as shown in Fig. 32.

REFITTING THE OVERDRIVE TO THE GEARBOX

1. Place the overdrive unit upside down in a vice.

2. Fit the oil pump operating cam on the gearbox mainshaft with the long plain end facing the gearbox, and with the back of the cam towards the bottom of the casing.

3. Ensure that the splines in the uni-directional clutch and planet carrier are in alignment. These splines are visible at the bottom of the overdrive unit. If alignment is necessary, use the dummy mainshaft as shown in Fig. 33.

4. Turn the bottom set of splines anti-clockwise into alignment using a long screwdriver. Check the alignment using a dummy mainshaft. (No. L.185).

5. Ensure that clutch springs are over their respective bosses on the gearbox rear extension. Press the gearbox down to test the cushioning of springs.

6. Fit two nuts to the long studs and tighten up until there is approximately ¾" (19.05 mm.) gap between the overdrive casing and the gearbox rear extension, meanwhile ensuring that the oil pump cam does not drop down off the splines on the mainshaft.

7. Enter two screwdrivers into the gap between the overdrive casing and the gearbox rear extension, with one, compress the oil pump plunger, and with the other, lever the cam down into alignment with the plunger roller (Fig. 34).

8. Tighten the two nuts on the long studs by equal amounts until the remaining five nuts can be started. Fully tighten the seven nuts by turning by equal amounts.

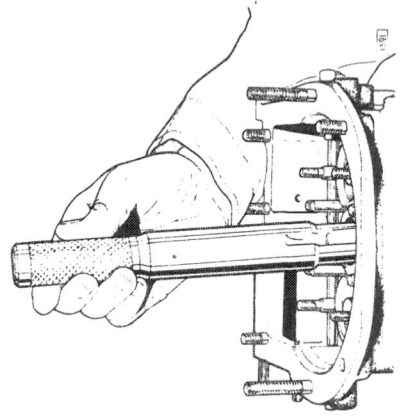

Fig. 33. Aligning the splines (Tool No. L.185).

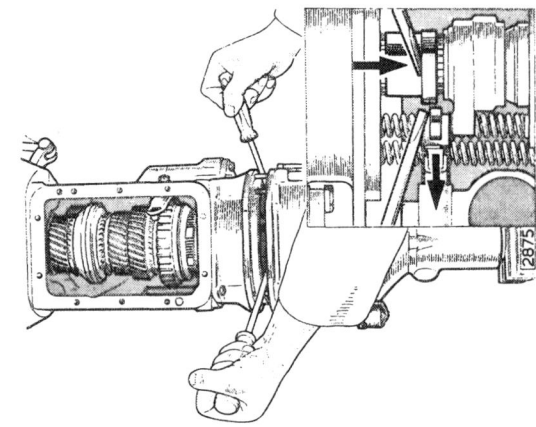

Fig. 34. When fitting the overdrive to the gearbox, compress the oil pump plunger and align the operating cam with the plunger roller.

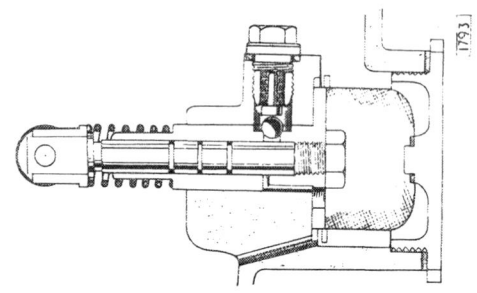

Fig. 35. The oil pump.

COMPONENTS

BEFORE COMMENCING ANY DIS-MANTLING OPERATIONS IT IS IMPORT-ANT THAT THE HYDRAULIC PRESSURE IS RELEASED FROM THE SYSTEM. DO THIS BY OPERATING THE OVERDRIVE 10-12 TIMES.

With the ignition switched on and the gear lever in top gear position operate the overdrive switch when the solenoid will be heard to energise.

THE OPERATING VALVE

Having gained access to the unit by removing the gearbox cowling unscrew the valve plug and remove the spring and plunger. The ball valve will then be seen inside the valve chamber (Fig. 29).

The ball should be lifted $\frac{1}{32}$" (0.79 mm.) off its seat when the overdrive control is operated. If the ball does not lift by this amount the fault lies in the control mechanism. Located on the right-hand side of the unit, and pivoting on the valve operating cross shaft which passes right through the housing is a valve setting lever. In its outer end is a $\frac{3}{16}$" (4.76 mm.) diameter hole which corresponds with a similar hole in the housing when the unit is in overdrive (i.e. when the ball is lifted $\frac{1}{32}$" (0.79 mm.) off the valve seat). If the two holes do not line up it will be necessary to adjust the control mechanism.

To adjust, remove the solenoid bracket cover plate on the opposite side of the overdrive front casing. Slacken the clamp bolt securing the valve operating lever to the valve operating cross shaft. Rotate the shaft until the $\frac{3}{16}$" (4.76 mm.) diameter rod can be inserted through the valve setting lever into the corresponding hole in the casing (Fig. 36) and tighten the clamp bolt. Check lift of ball after adjustment.

A small magnet will be found useful for removing the ball from the valve chamber. The valve can be withdrawn by inserting the point of a pencil into the top, but care must be taken not to damage the ball seating at the end of the valve.

Near the bottom of the valve will be seen a small hole breaking through to the centre drilling. This is the jet for restricting the exhaust of oil from the operating cylinders. Ensure that this jet is not choked.

If the ball valve is not seating correctly the ball should be tapped sharply on to its seat using a copper drift for the purpose.

THE HYDRAULIC SYSTEM

If the unit fails to operate and the ball valve is found to be seating and lifting correctly check that the pump is functioning.

Jack up both rear wheels of the car, then with the engine ticking over and the valve plug removed, engage top gear. Watch for oil being pumped into the valve chamber. If none appears, then the pump is not functioning.

The pump (Fig. 35) is of the plunger type and delivers oil via a non-return valve to the accumulator. Possible sources of trouble are (1) failure of the non-return valve due to foreign matter on the seat or to a broken valve spring and (2) breakage of the spring holding the pump plunger in contact with the cam.

The pump is self-priming but failure to deliver oil after the system has been drained and refilled indicates that the air bleed is choked, causing air to be trapped inside the pump.

In the unlikely event of this happening it will be necessary to remove the pump and clean the flat on the pump body, and the bore of the casting into which it fits.

THE PUMP VALVE

Access to the pump is gained through a cover on the left-hand side of the unit. Proceed as follows :

1. Remove drain plug and drain off oil.
2. Remove cover from solenoid bracket.
3. Remove solenoid body.
4. Slacken off clamping bolt in operating lever and remove lever complete with solenoid plunger.
5. Remove distance collar from valve operating shaft.
6. The solenoid bracket is secured by two $\frac{5}{16}$" (7.94 mm.) diameter studs and two $\frac{5}{16}$" (7.94 mm.) diameter bolts, the heads of which are painted red (see Fig. 38). REMOVE THE NUTS FROM THE STUDS BEFORE TOUCHING THE BOLTS. THIS IS IMPORTANT. The two bolts should now be slackened off together, releasing the tension on the accumulator spring.
7. Remove the solenoid bracket.

8. Unscrew the valve cap and take out the spring, plunger and ball.

Re-assembly is the reverse of the above operations. Ensure that the soft copper washer between the valve cap and pump housing is nipped up tightly to prevent oil leakage.

Reset the valve operating lever, proceeding as follows :

Before clamping up the valve operating lever and replacing the solenoid bracket cover, rotate the valve operating shaft until a $\frac{3}{16}$" (4.76 mm.) diameter pin inserted in the setting lever registers in a corresponding hole in the casing. Allow approximately .008" (.2mm) end float on the shaft then push the solenoid plunger fully home at the same time holding the lever fork lightly against the large collar on the plunger.

Tighten the clamp bolt to secure the lever to the shaft. With the plunger still hard home in the solenoid, set the adjustable stop by means of a $\frac{1}{4}$" A F Allen key until there is a gap of .150"/.155" (3.75/3.87 mm.) between the end of the plunger and the stop. When this gap has been obtained tighten the locknut against the solenoid bracket to secure the adjustable stop. Remove the $\frac{3}{16}$" (4.76 mm) diameter pin from the setting lever, energise the solenoid, then check the alignment of the setting holes together with the .150/.155" (3.75/3.87 mm) gap

THE PUMP

Dismantling

1. Remove drain plug and drain off oil.
2. Remove pump valve as described above.
3. Remove the filter after unscrewing the securing bolt.
4. Take out the two cheese head screws securing the pump body flange and extract the pump body. A special extractor tool (Fig. 40) is available for this purpose. This screws into the bottom of the pump body in place of the screwed plug.

Assembling

The pump body should be tapped into the maincase without the blanking plug in position and should be aligned and driven home using guide pegs and replacer tool (No. L.184).

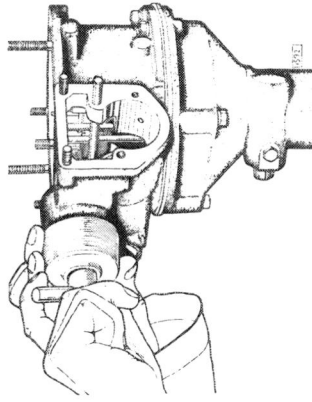

Fig. 39. *Extracting the oil pump (Tool No. L.184).*

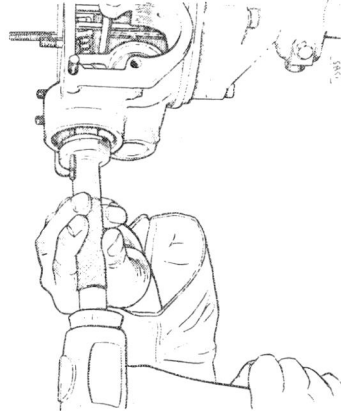

Fig. 40. *Replacing the oil pump (Tool No. L.184).*

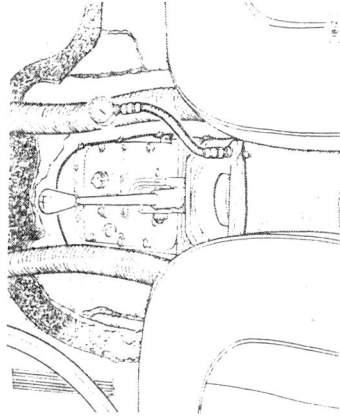

Fig. 41. *Checking the overdrive hydraulic pressure utilizing the Churchill Tool No. L.188.*

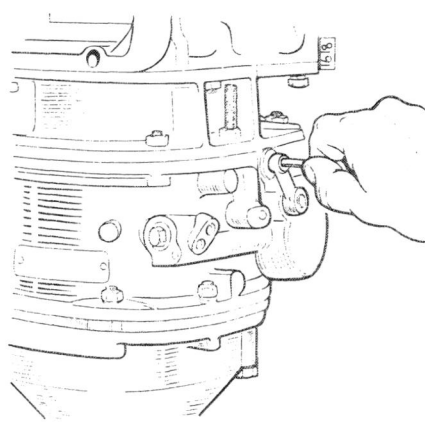

Fig. 36. *Aligning the hole in the valve setting lever with the corresponding hole in the housing.*

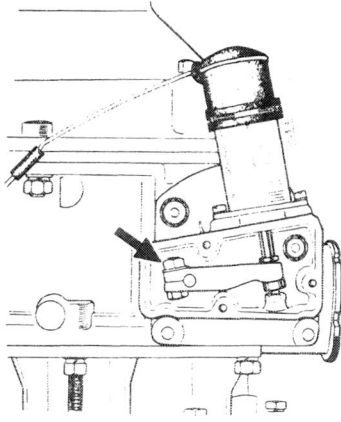

Fig. 37. *Valve operating lever clamp bolt.*

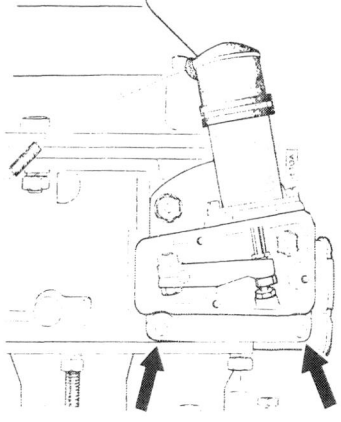

Fig. 38. *Removing the solenoid bracket—remove the nuts from the studs before unscrewing the two setscrews.*

HYDRAULIC PRESSURE

The required hydraulic pressures are as follows:

3.4 litre .. .. 420—440 lb. per sq. in. (29.53—30.93 kg./cm².).

3.8 litre .. .. 540—560 lb. per sq. in. (37.97—39.37 kg./cm².).

Hydraulic test equipment is available from V. L. Churchill & Co. Ltd. (see page F.30). The pipe union should be screwed into the overdrive casing after removing the operating valve plug.

Low pressure often indicates leakage or a broken accumulator spring.

THE ACCUMULATOR PISTON AND SPRING

Removal

1. Proceed as described in operations 1 to 7 in the paragraph headed "The Pump Valve".

2. The accumulator spring or springs in the instance of the 3.8 litre model, followed by the piston housing assembly and piston can now be withdrawn (see Fig. 42).

3. Withdraw the piston from the piston housing. It is important to appreciate that correct fitting of the piston rings is of vital importance to the efficient working of the unit. Check that the rings are not gummed up due to use of an unsuitable lubricant or have excessive clearance in the grooves. Check also that the rubber "O" rings on the piston housing are not damaged. It is advisable to renew all "O" rings each time the unit is stripped down.

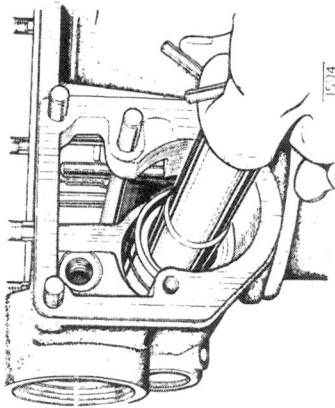

Fig. 42. *Extracting the accumulator piston housing (Tool No. L.182 or L.216).*

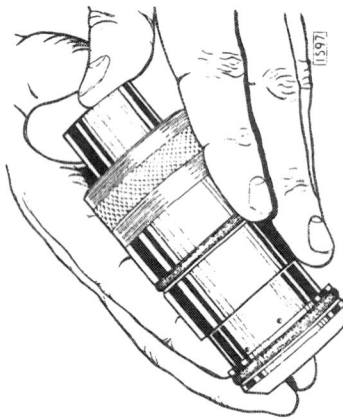

Fig. 43. *Fitting the accumulator piston (Tool No. L.179).*

Fig. 44. *Fitting an "O" ring to the accumulator piston (Tool No. L.181 or L.217).*

---

mally closed contacts. These contacts are connected in series with the closing coil.

When the contacts of the relay close or when both the manual switch and the top gear switch are closed, both coils in the solenoid become energised and actuate the soft iron plunger. Movement of the plunger opens the solenoid internal switch and cuts the low resistance closing coil, the **magnetism due to the high resistance coil alone being sufficient to keep the plunger in** the overdrive position.

Note : The normal current consumption of the system should be approximately 1.0—1.5 amperes. A defective solenoid will be accompanied by a current of approximately 18—20 amperes.

The solenoid should never be bench tested as, when the solenoid plunger is not loaded the impact of the plunger will tend to bend the contact arm thus preventing the points from opening.

THE CONTROL SYSTEM

The solenoid, which actuates the overdrive, is controlled by two switches; a manual switch, mounted in the facia panel and a top-gear switch mounted on the gearbox cover which will only close when top gear is selected.

To enable a change into overdrive to be made:

1. The car must be in top gear.
2. The driver must operate the manual switch.

Wiring Circuit

This is a simple circuit where the control switch is connected in series between the top gear switch and the solenoid. The control switch does not incorporate a warning light and there is no relay in the circuit.

Operation of the solenoid is effected when the switches are closed.

Operating Solenoid

The solenoid is made up of a soft iron plunger, a holding coil of high resistance, a closing coil of low resistance and a pair of nor-

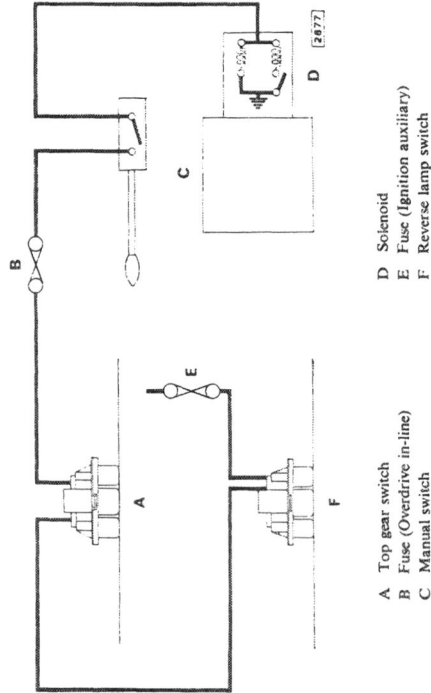

Fig. 45. *Circuit diagram.*

A Top gear switch
B Fuse (Overdrive in-line)
C Manual switch

D Solenoid
E Fuse (Ignition auxiliary)
F Reverse lamp switch

DATA

	Dimensions	Clearances
Pump		
Plunger diameter	.3742"/.3746"	
Bore for plunger in pump body ..	.3748"/.3758"	.0002"/.0016"
Plunger spring fitted load at top of stroke ..	9.493 lb. at 1.137"	
Valve spring load	2⅝ lb. at 1¹⁹/₃₂" long	
Pin for roller	.2497"/.2502" dia.	
Bore for pin in roller	.2510"/.2520"	.0008"/.0023"
Gearbox Mainshaft		
Diameter at steady bushes	1.1544"/1.1553"	
Steady bush internal diameter ..	1.1582"/1.1592"	.0029"/.0048"
Shaft diameter at sun-wheel bush ..	1.1544"/1.1553"	
Sun-wheel bush internal diameter ..	1.1582"/1.1592"	.0029"/.0048"
Shaft diameter at rear steady bush ..	.6235"/.6242"	
Rear steady bush internal diameter ..	.6250"/.6260"	.0008"/.0025"
Gear Train		
End float of sun-wheel		.008"/.014"
Piston Bores		
Accumulator bore : 3.4 litre ..	1.1245"/1.1255"	
3.8 litre ..	1.4995"/1.5005"	
Operating piston bores	1.3745"/1.3755"	
Miscellaneous		
Clutch movement from direct to overdrive ..	.080"/.120"	
Hydraulic Pressure		
3.4 litre	420/440 lb. per sq. in. (29.53/30.93 kg./cm².)	
3.8 litre	540/560 lb. per sq. in. (37.97/39.37 kg./cm².)	

OPERATING INSTRUCTIONS

When brought into operation, the overdrive reduces the engine speed in relation to the road speed. This permits high road speeds with low engine revolutions resulting in considerable fuel economy and reduced engine wear.

Operation

The overdrive operates in top gear only and is brought into operation by means of a lever behind the steering wheel on the right hand side of the column.

Driving

Use of the clutch pedal when changing into or out of overdrive is unnecessary but to ensure maximum smoothness or operation, particularly when changing down from overdrive to top gear, the accelerator pedal should be slightly depressed.

For driving in towns, heavy traffic or hilly country when the maximum flexibility and low speed performance is required the overdrive manual switch should be placed in the "OUT" position which will bring the drive into the normal top gear ratio.

For normal driving in open country the overdrive should be brought into operation when the required cruising speed has been obtained.

The following table gives the relationship between engine revolutions per minute to road speed in miles and kilometres per hour for top gear and overdrive:

Road Speed		Engine Revolutions per minute	
		3.4 and 3.8 litre models Axle Ratio 3.77 : 1	
Kilometres per hour	Miles per hour	Top Gear 3.77 : 1	Overdrive Top 2.93 : 1
16	10	495	385
32	20	990	770
48	30	1485	1154
64	40	1980	1540
80	50	2475	1925
96	60	2970	2310
112	70	3465	2693
128	80	3960	3078
144	90	4455	3463
160	100	4950	3848
176	110	5445	4232
192	120	—	4617

Note : The figures in the above table are theoretical and make no allowance for changes in tyre radius due to the effect of centrifugal force.

SUPPLEMENTARY INFORMATION
TO
SECTION F "GEARBOX AND OVERDRIVE"

INTRODUCTION OF 4-SPEED ALL SYNCHROMESH GEARBOX

	Commencing Chassis Numbers	
	R.H. Drive	*L.H. Drive*
3.4 "S"	1B 2192	1B 25301
3.8 "S"	1B 52078	1B 76310
	plus 1B 52034	
	and 1B 52036	

Commencing at the above chassis numbers, a new gearbox with synchromesh on all four forward gears is fitted. This gearbox cannot be used as a replacement for the earlier unit.

Description

The gearbox is of the four speed type with baulk ring synchromesh on all four forward gears. With the exception of reverse, the detents for the gears are incorporated in the synchro assemblies, the three synchro balls engaging with grooves in the operating sleeve. The detent for reverse is a spring loaded ball which engages with a groove in the selector rod.

Two interlock balls and a pin located at the front of the selector rods, prevent the engagement of two gears at the same time.

The gears are pressure fed at approximately 5 lb./sq. in. from a pump driven from the rear of the mainshaft on standard transmission cars and pressure fed by the overdrive oil pump on overdrive transmission cars.

DATA

Ratios:—

1st. Gear	...	3.04 : 1
2nd. Gear	...	1.973 : 1
3rd. Gear	...	1.328 : 1
4th. Gear	...	1.00 : 1
Reverse gear	...	3.49 : 1
1st. Gear — end float on mainshaft	...	.005" – .007" (.13 – .18 mm)
2nd. Gear – end float on mainshaft	...	.005" – .008" (.13 – .20 mm)
3rd. Gear — end float on mainshaft	...	.005" – .008" (.13 – .20 mm)
Countershaft gear unit end float	...	.004 – .006 (.10 – .15 mm)

Clutch slip in reverse or freewheel condition on overrun

1. Solenoid operating lever out of adjustment.
2. Partially blocked restrictor jet in operating valve.
3. Solenoid stop screw incorrectly adjusted.

SERVICE TOOLS

The following tools for servicing the overdrive unit are obtainable from V. L. Churchill & Co. Ltd., P.O. Box No. 3, London Rd., Daventry, Northants.

Part No.	Description
L.176	Drive Shaft Oil Seal Remover.
L.177A	Drive Shaft Oil Seal Replacer and cone clutch and spring thrust housing dismantling tool.
L.178	Freewheel Assembly Ring
L.179	Piston and Ring Fitting Tool, 1⅛" (28.57 mm.) diameter (3.4)
L.180	Piston and Ring Fitting Tool, 1⅜" (34.9 mm.) diameter (3.8)
L.181	Accumulator "O" Ring Fitting Tool (3.4 litre).
L.217	Accumulator "O" Ring Fitting Tool (3.8 litre).
L.182	Accumulator Piston Housing Remover (3.4 litre).
L.216	Accumulator Piston Housing Remover (3.8 litre).
L.183	Pump Barrel Remover.
L.184	Pump Barrel Replacer.
L.185A	Dummy Drive Shaft.
L.186	Mainshaft Bearing Replacer.
L.188	Hydraulic Test Equipment.
J.3	Drain Plug Remover.

GEARBOX AND OVERDRIVE

FAULT FINDING

When an overdrive unit does not operate properly it is advisable first to check the level of oil and, if below the low level mark, top up with fresh oil and test the unit again before making any further investigations.

Faulty units should be checked for defects in the order listed below.

Should the electrical control not operate the electrical circuits should be checked from the diagram.

Overdrive does not engage

1. Insufficient oil in the gearbox.
2. Solenoid not operating due to fault in electrical system.
3. Solenoid operating lever out of adjustment.
4. Insufficient hydraulic pressure due to pump non-return valve incorrectly seating (probably dirt on seat).
5. Insufficient hydraulic pressure due to worn accumulator.
6. Leaking pump non-return valve due to foreign matter on ball seat or broken valve spring.
7. Insufficient hydraulic pressure due to leaks or broken accumulator springs.
8. Damaged gears, bearings or shifting parts within the unit requiring removal and inspection of the assembly.

Overdrive does not release

IMPORTANT : If the overdrive does not release, do NOT reverse the car, otherwise extensive damage may be caused.

1. Fault in electrical control system.
2. Blocked restrictor jet in operating valve.
3. Solenoid operating lever adjustment.
4. Sticking clutch.
5. Solenoid stop screw incorrectly adjusted.

Clutch slip in overdrive

1. Insufficient oil in unit.
2. Worn clutch linings.
3. Solenoid operating lever out of adjustment.
4. Insufficient hydraulic pressure due to pump non-return valve incorrectly seating (probably dirt on seat).
5. Insufficient hydraulic pressure due to worn accumulator.

ROUTINE MAINTENANCE

Every 3,000 Miles (5,000km)

Gearbox Oil Level

Every 3,000 miles (5,000 km) check the level of oil in the gearbox with the car standing on level ground. A combined level and filler plug is fitted on the left hand side of the gearbox. Clean off any dirt from around the plug before removal.

The level of oil should be to the bottom of the level and filler plug hole.

Overdrive Oil Level — Important

The oil for the lubrication and operation of the overdrive is fed from the gearbox and therefore checking the gearbox oil level will automatically check the level of oil in the overdrive unit, but, as this unit is hydraulically controlled, extra attention should be paid to exercising absolute cleanliness when replenishing with oil. It is also important that the oil level is not allowed to fall appreciably, otherwise the operation of the overdrive will be affected.

Every 12,000 Miles (20,000km).

Changing the Gearbox Oil

The draining of the gearbox oil should be carried out at the end of a long run when the oil is hot and will, therefore, run more freely. The drain plug is situated at the front of the gearbox casing.

After all the oil has been drained, replace the drain plug and fill the gearbox with the recommended grade of oil through the combined level and filler plug hole.

Overdrive Oil Changing

The oil for the overdrive unit is common to that of the gearbox but draining the gearbox oil will not drain the overdrive unit.

When draining the gearbox, remove the overdrive filter plug, (situated in the side of the overdrive unit) filter and magnetic washers. Thoroughly wash the filter and magnetic washers.

When dry, refit the filter, magnetic washers and filter plug. Fully tighten the filter plug and refill the gearbox and overdrive units.

Recheck the level after the car has been run as a certain amount of oil will remain in the hydraulic system of the overdrive.

Particular attention should be paid to maintaining absolute cleanliness when filling the gearbox and overdrive with fresh oil.

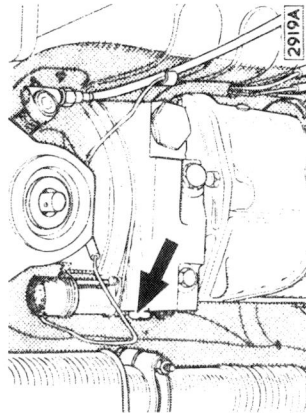

Fig. 46. *Overdrive filter plug.*

RECOMMENDED LUBRICANTS

Mobil	Castrol	Shell	Esso	B.P.	Duckham	Regent/Caltex/Texaco
Mobilube GX 90	Castrol Hypoy	Spirex E.P. 90	Esso Gear Oil G.P. 90/140	Gear Oil S.A.E. 90 E.P.	Hypoid 90	Multigear Lubricant E.P. 90

1.	Gearbox casing
2.	Oil drain plug
3.	Oil filter plug
4.	Fibre washer
5.	Circlip
6.	Ball bearings
7.	Ball bearings
8.	Circlip
9.	Collar
10.	Fibre washer
11.	Gasket
12.	Gasket
13.	Remote control
14.	Striking rod 1st/2nd gears
15.	Striking rod 3rd/top gears
16.	Striking rod, reverse gear
17.	"O" ring
18.	Stop
19.	Stop
20.	Change speed fork
21.	Change speed fork
22.	Locating arm
23.	Plunger
24.	Spring
25.	Ball
26.	Spring
27.	Setscrew
28.	Nut
29.	Dowel screw
30.	Roller
31.	Ball
32.	Top cover
33.	Switch
34.	Gasket
35.	Gasket
36.	Dowel
37.	Ball
38.	Plunger
39.	Spring
40.	Spring
41.	Welch washer
42.	Breather elbow
43.	Nut
44.	Gearbox breather assembly
45.	Distance piece
46.	Hose
47.	Clip
48.	Pivot jaw
49.	Bush
50.	Fibre washer
51.	Self-locking nut
52.	Spring washer
53.	"D" washer
54.	Selector lever
55.	Bush
56.	Fibre washer
57.	Spring washer
58.	Pivot pin
59.	Self-locking nut
60.	Change speed lever
61.	Knob
62.	Cone
63.	Upper bush
64.	Washer
65.	Lower bush
66.	Self-locking nut.

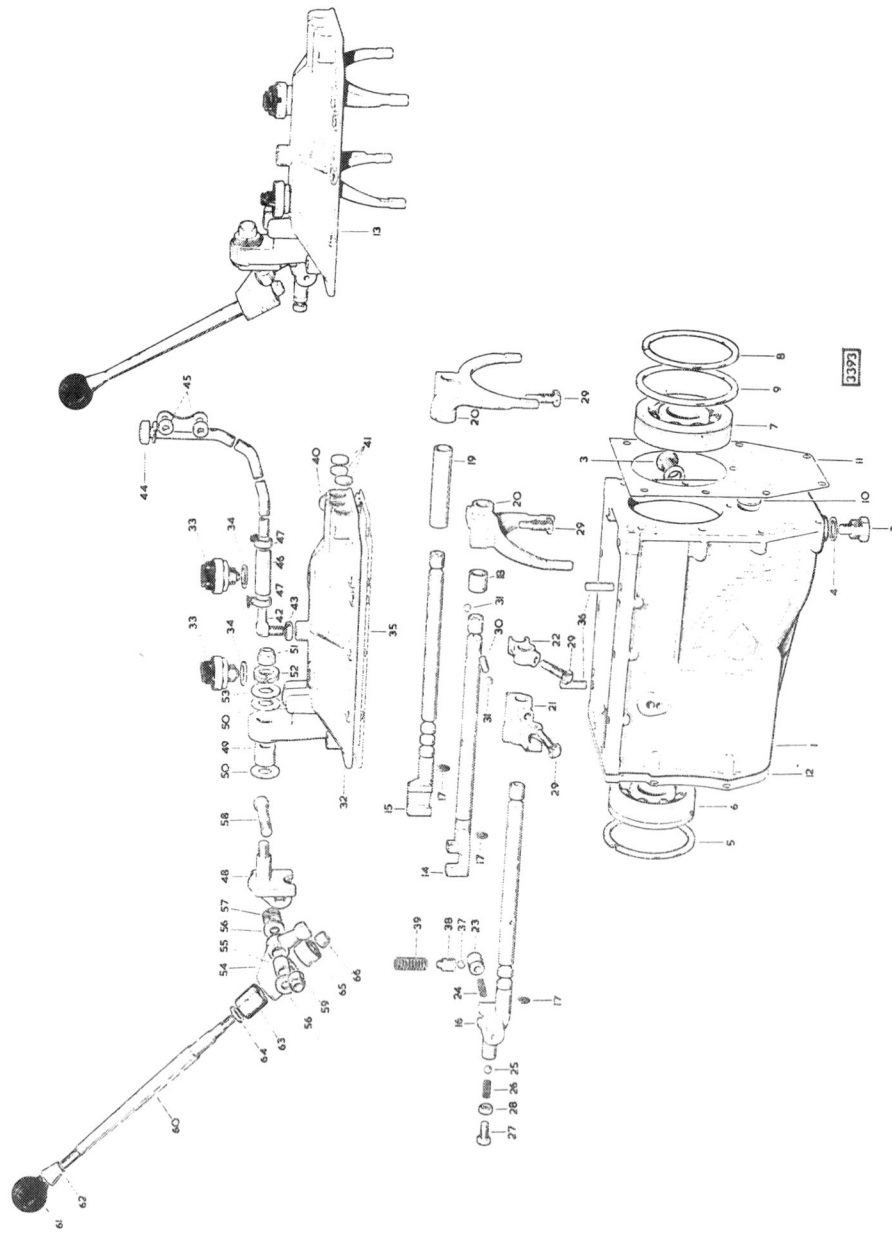

Fig. 47. Exploded view of gearbox and top cover.

1. Mainshaft
2. Nut
3. Tab washer
4. Reverse gear
5. 1st gear
6. Bearing sleeve
7. Needle roller
8. Spacer
9. Synchro hub
10. Operating sleeve
11. Thrust member
12. Plunger
13. Detent ball
14. Spring
15. Synchro ring
16. 2nd gear
17. 3rd gear
18. Needle roller
19. Spacer
20. Spacer
21. Synchro hub
22. Operating sleeve
23. Thrust member
24. Plunger
25. Detent ball
26. Spring
27. Synchro ring

28. Nut
29. Tab washer
30. Plug
31. Constant pinion shaft
32. Roller bearing
33. Spacer
34. Oil thrower
35. Nut
36. Tab washer
37. Reverse spindle
38. Key
39. Reverse idler gear
40. Lever assembly
41. Setscrew
42. Fibre washer
43. Tab washer
44. Reverse slipper
45. Split pin
46. Countershaft
47. Key
48. Cluster gear unit
49. Needle roller
50. Retaining ring
51. Rear thrust washer
52. Front thrust washer
53. Outer thrust washer.

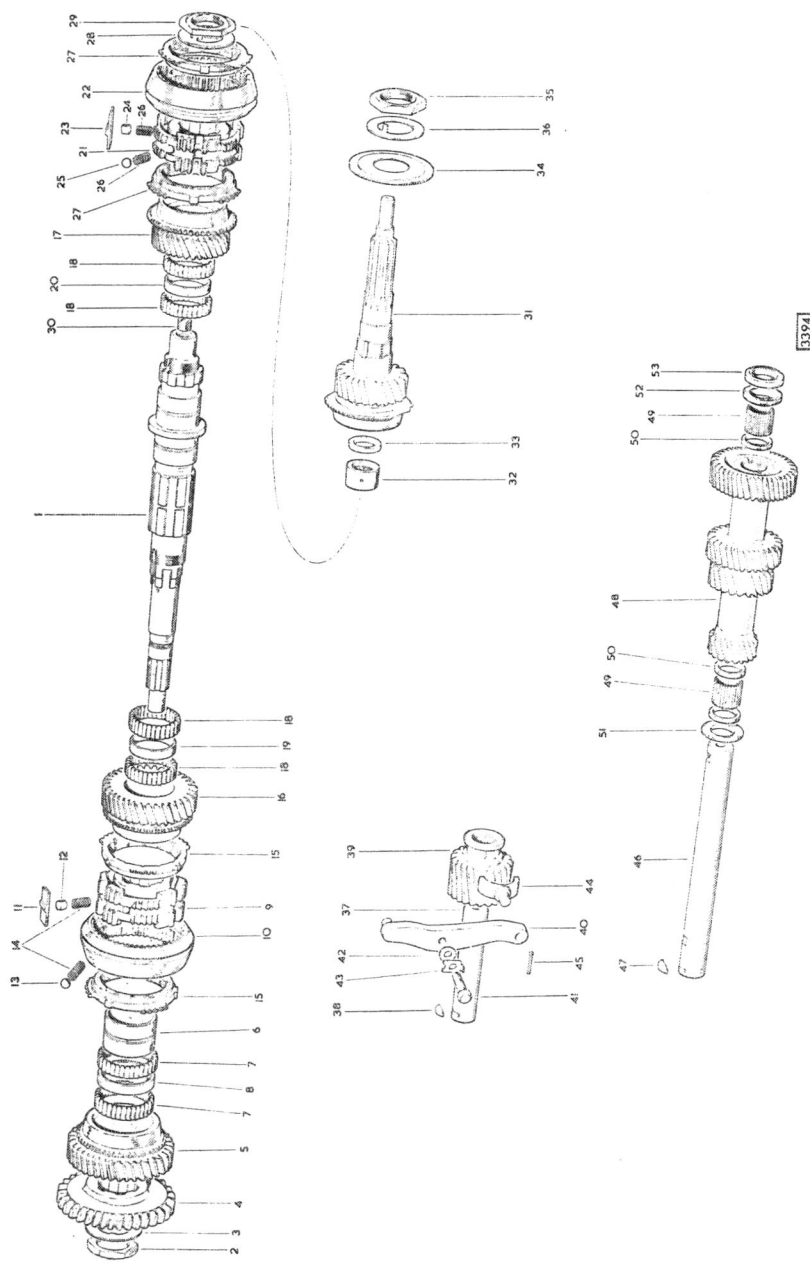

Fig. 48. *Exploded view of gears.*

GEARBOX

Removal

The 4-speed all synchromesh gearbox unit is removed in the same manner as the 3 speed synchromesh gearbox as detailed on page F.8.

DISMANTLING

Remove Clutch Housing

Detach the springs and remove the carbon thrust bearing.

Unscrew the two nuts and detach the clutch slave cylinder.

Unscrew the Allen screw, push out the fulcrum pin and detach the clutch fork.

Tab back the locking tabs, break the locking wire, remove the eight setscrews and detach the clutch housing.

Remove Top Cover

Place the gear lever in neutral.

Remove the eight setscrews and two nuts and lift off the top cover.

Remove Rear Extension

Engage first and reverse gears to lock the unit. Tap back the tab washer and remove the flange nut. Withdraw the flange.

Remove the speedometer pinion and bush assembly after unscrewing the retaining screw.

Remove the six setscrews and withdraw the extension. Collect the distance piece, oil pump driving pin and oil filter.

Remove Oil Pump

From the inside face of the rear extension break the staking and withdraw the three countersunk setscrews securing the oil pump to the gear housing. Withdraw the housing by entering two of the setscrews into the tapped holes in the housing; screw in the two screws evenly until the housing is free.

Mark the gears with marking ink so they can be replaced the same way up in the housing.

Remove Countershaft

Remove the fibre plug from in front of the countershaft. Drive out the countershaft from the front of the casing.

Important: Ensure that the rear washer, pegged to the casing, drops down in a clockwise direction looking from the rear to avoid trapping the washer with the reverse gear when driving the mainshaft forward. This can be most easily affected by pushing the washer down with a piece of stiff wire bent at right angles.

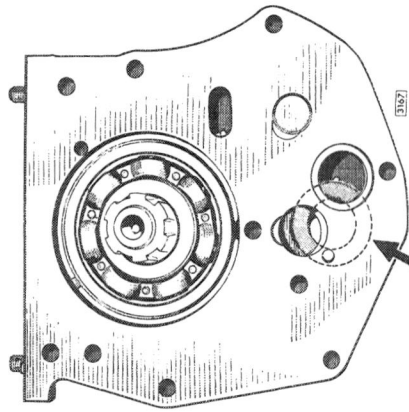

Fig. 49. Ensure that the washer (indicated by arrow) drops down in a clockwise direction.

Remove Constant Pinion Shaft

Rotate the constant pinion shaft until the cutaway portions of the driving gear are facing the top and bottom of the casing otherwise the gear will foul the gear of the countershaft.

With the aid of two levers, ease the constant pinion shaft and front bearing assembly forward from the casing.

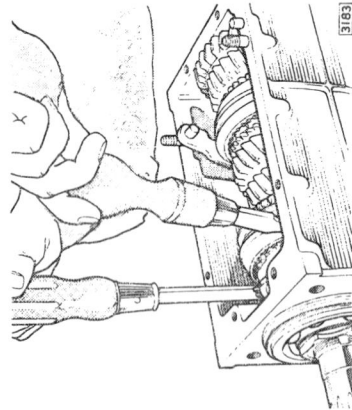

Fig. 50. With the aid of two levers, ease the constant pinion shaft forward.

Remove Mainshaft

Rotate the mainshaft until one of the cutaway portions of the 3rd top synchro hub is in line with the countershaft otherwise the hub will foul the constant gear of the countershaft.

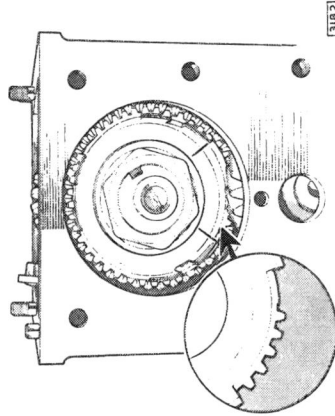

Fig. 51. Rotate the mainshaft until one of the cutaway portions of the 3rd/top synchro hub is in line with the countershaft.

Tap or press the mainshaft through the rear bearing ensuring that the reverse gear is kept tight against the first gear.

Remove the rear bearing from the casing and fit a hose clamp to the end of the mainshaft to prevent the reverse gear from sliding off the shaft.

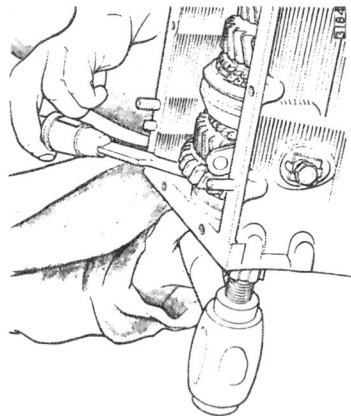

Fig. 52. Tapping the mainshaft through the rear bearing.

Slacken the reverse lever bolt to allow the lever to be moved freely back and forth.

Lift out the cluster gear and collect the needle roller bearings and retaining rings.

Withdraw the reverse idler shaft and lift out the gear. Note the locking key on the gear.

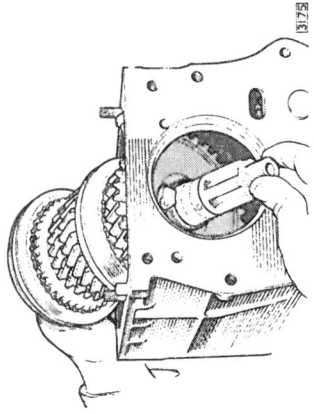

Fig. 53. Removal of mainshaft. Note the hose clamp fitted to retain reverse gear.

Dismantling the Constant Pinion Shaft Assembly

Remove the roller bearing and spacer from inside the constant pinion shaft.

Tap back the tab washer and remove the large nut, tab washer and oil thrower. Tap the shaft sharply against a metal plate to dislodge the bearing.

Dismantling the Mainshaft

Note: The mainshaft needle roller bearings are graded in diameter and must be kept in sets for their respective positions.

Remove the hose clip and withdraw the reverse gear.

Withdraw first gear and collect the 120 needle roller bearings, spacer and sleeve.

Withdraw the 1st/2nd synchro assemblies and collect the two loose synchro rings.

Remove 2nd gear and collect the 106 needle rollers. The spacer remains on the mainshaft.

Tap back the tab washer and remove the nut retaining the third/top synchro assembly on the mainshaft. Withdraw the assembly and collect the two loose synchro rings.

Withdraw the 3rd gear and collect the 106 needle rollers.

Dismantling the Synchro Assembly

Completely surround the synchro assembly with a cloth and push out the synchro hub from the operating sleeve. Collect the synchro balls and springs, thrust members balls and springs.

Dismantling Top Cover

Unscrew the self-locking nut and remove the double coil spring, washer, flat washer and fibre washer securing the gear lever to the top cover. Withdraw the lever and collect the remaining fibre washer.

Remove the locking wire and unscrew the selector rod retaining screws.

Withdraw the 3rd top selector rods and collect the selector, spacing tube and interlock ball. Note the loose interlock pin at the front of the 1st 2nd selector rod.

Withdraw the reverse selector rod and collect the reverse fork, stop spring and detent plunger.

Withdraw the 1st 2nd selector rod and collect the fork and short spacer tube.

GEARBOX ASSEMBLY

GEARBOX—ASSEMBLY

Building the Synchro Assemblies

The assembly for the 1st 2nd and 3rd top synchro assemblies is the same. Although the 3rd top and 1st 2nd synchro hubs are similar in appearance they are not identical and, to distinguish them, a groove is machined on the edge of the 3rd top synchro hub. (Fig. 54).

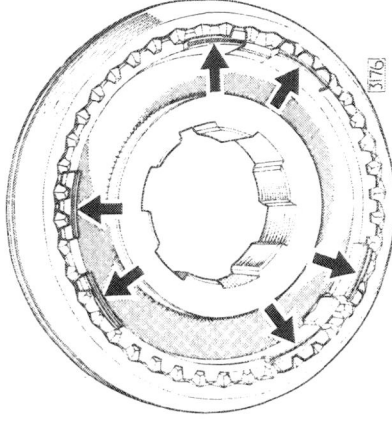

Fig. 54. *Identification grooves—3rd top synchro assembly.*

Assemble the synchro hub to the operating sleeve with:—

1. The wide boss of the hub on the opposite side to the wide chamfer end of the sleeve (Fig. 55).

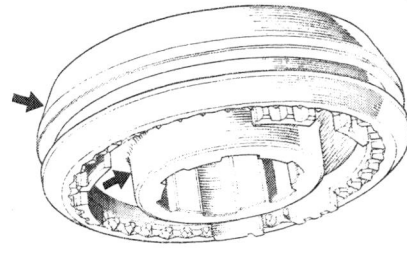

Fig. 55. *Assembly of synchro hub.*

2. The three balls and springs in line with the teeth having three detent grooves (Fig. 56).

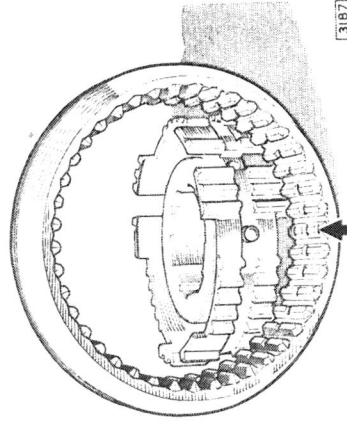

Fig. 56. *Fitting the synchro hub in the sleeve.*

Pack up the synchro hub so that the holes for the ball and springs are exactly level with the top of the operating sleeve. (Fig. 57).

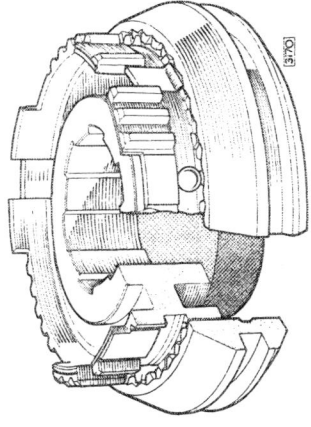

Fig. 57. *Fitting the springs, plungers and thrust members.*

Fit the three springs, plungers and thrust members to their correct positions with grease; press down the thrust members as far as possible. Fit the three springs and balls to the remaining holes with grease.

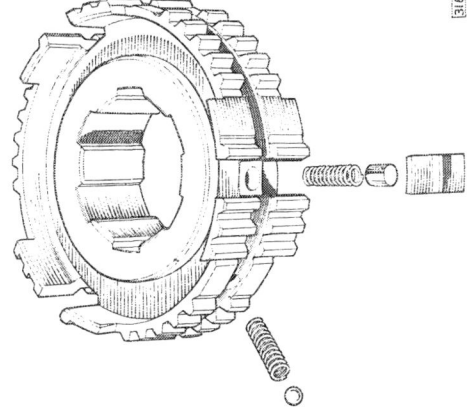

Fig. 58. *Showing the relative positions of the detent ball, plunger and thrust members.*

Compress the springs with a large hose clamp or piston ring clamp as shown in Fig. 59 and carefully lift off the synchro assembly from the packing piece.

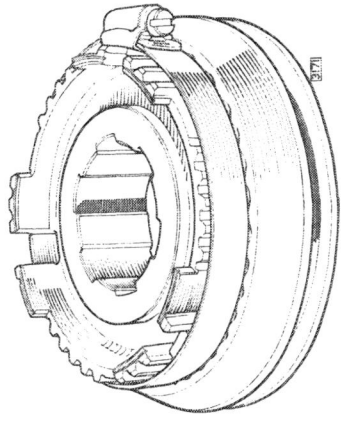

Fig. 59. *Compressing the springs.*

Depress the hub slightly and push down the thrust members with a screwdriver until they engage the neutral groove in the operating sleeve. (Fig. 60).

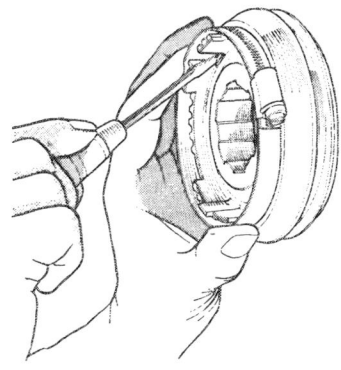

Fig. 60. *Pushing down the thrust members.*

Finally, tap the hub down until the balls can be heard and felt to engage the neutral groove. (Fig. 61).

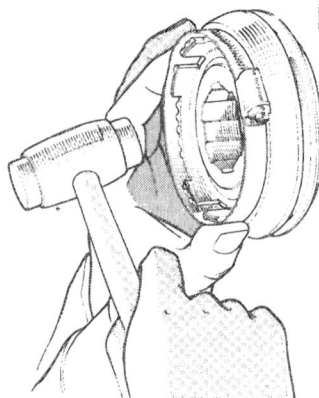

Fig. 61. Tapping the hub into position.

Assembling the Cluster Gear

Fit one retaining ring in the front end of the cluster gear. Locate the 29 needle roller bearings in position with grease and fit the inner thrust washer ensuring that the peg on the washer locates in a groove machined on the front face of the cluster gear.

Fit a retaining ring, 29 needle rollers and a second retaining ring to the rear of the cluster gear.

Check the Cluster Gear End-Float

Fit the reverse idler gear, lever and idler shaft.

Fit the pegged rear thrust washer to its boss on the casing with grease.

Locate the outer thrust washer to the front of the cluster gear with grease: lower the cluster gear carefully into position. Insert a dummy shaft and check the clearance between the rear thrust washer and the cluster gear. The correct clearance is .004–.006" (.10–.15 mm) and is adjusted by means of the outer front thrust washer. This washer is available in the following selective thicknesses.

Part No.	Thickness
C1862.3	.152" (3.86 mm)
C1862	.156" (3.96 mm)
C1862.1	.159" (4.04 mm)
C1862.2	.162" (4.11 mm)
C1862.4	.164" (4.17 mm)

Assembling the Constant Pinion Shaft

Assembling is the reverse of the dismantling procedure but care must be taken to ensure that the bearing is seated squarely on the shaft.

Assembling the Mainshaft

The assembling of the mainshaft is the reverse of the dismantling procedure but the following instructions should be noted.

1. The end float of the gears on the mainshaft is given in "Data" at the beginning of this section and, if found to be excessive, end float can only be restored by fitting new gears.

2. The needle roller bearings which support the gears on the mainshaft are graded on diameter and rollers of one grade only must be used for an individual gear. The grades are identified by a .1, .2 and .3 after the part number.

3. Fit a hose clip to prevent reverse gear sliding off the shaft when assembling the mainshaft to the casing.

Assembling the Gears to the Casing

Withdraw the dummy shaft from the cluster gear and, at the same time, insert the thin rod keeping both the dummy shaft and the rod in contact until the dummy is clear of the casing. The thin rod allows the cluster gear to be lowered sufficiently in the casing for insertion of the mainshaft.

Fit a new paper gasket to the front face of the casing.

Enter the mainshaft through the top of the casing and pass the rear of the shaft through the rear bearing hole.

Enter the constant pinion shaft and front bearing assembly through the bearing hole in the front of the casing with the cutaway portion of the driving gear at the top and bottom.

Tap the assembly into position entering the front end of the mainshaft into the spigot bearing of the constant pinion shaft. Clamp the constant pinion shaft in position and, with a hollow drift, tap the rear bearing into position.

Withdraw the thin rod from the front bore of the cluster gear approximately half way and lever the cluster gear upwards, rotating the mainshaft and constant pinion shaft gently until the cluster gear meshes. Carefully insert the countershaft from the rear and withdraw the thin rod. Fit the key locating the countershaft in the casing.

Refitting the Rear Extension

Coat the gears and the housing with oil and refit the oil pump gears the same way round as removed. Secure the pump housing with three setscrews and retain by staking.

Fit a paper gasket to the rear face of the gearbox casing.

Fit the distance piece and driving pin to the oil pump in the rear extension. Offer up the rear extension and secure with six setscrews.

Fit the speedometer driving gear to the mainshaft. Fit the speedometer driven gear and bush with the hole in the bush in line with the hole in the casing and secure with the retaining screw.

Fit a new oil seal to the rear extension with the lip facing forward; fit a new gasket to the rear cover face.

Fit the rear cover to the extension noting that the setscrew holes are offset.

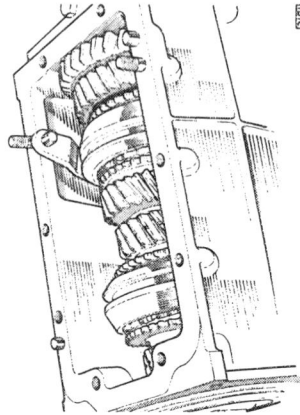

Fig. 62. Assembled gearbox prior to fitting the top cover.

Assembling the Top Cover

Assembling the top cover is the reverse of the dismantling procedure. When assembling the selector rods do not omit the interlock balls and pin.

Renew the "O" rings on the selector rods.

Fit the reverse plunger and spring. Fit the ball and spring and start the screw and locknut: press the plunger inwards as far as possible and tighten the screw to lock the plunger. Slowly, slacken the screw until the plunger is released and the ball engages with the circular groove in the plunger. Hold the screw and tighten the locknut.

Fitting the Top Cover

Fit a new paper gasket.

Ensure that the gearbox and the top cover are in the neutral position. Ensure that the reverse gear is out of mesh with the reverse idler gear on the mainshaft by pushing the lever rearwards.

Engage the selector forks with the grooves in the synchro assemblies. Secure the top cover with setscrews noting the different lengths.

Refitting the Clutch Housing

Refitting the clutch housing is the reverse of the removal procedure. Always fit a new oil seal to the clutch housing; this seal has a metal flange and should be pressed in fully.

Note: After refitting the gearbox, run the car in top gear as soon as possible to prime the oil pump.

OVERDRIVE "A" TYPE COMPACT UNIT

The "A" type compact overdrive unit is fitted to all "S" type cars having the all synchromesh gearbox.

The operation, construction, routine maintenance etc. are similar to those detailed for the earlier units but dismantling and re-assembly procedures for the new unit are listed.

Removal

Remove the four short studs which pass through the front flange of the casing and the two long studs at the bottom of the unit which pass right through the main and rear casings. There is no spring tension to release and, after removal of the units from these studs, the complete assembly can be withdrawn off the mainshaft leaving the adaptor plate in place.

The overdrive can be divided into four main assemblies.

1. Front casing and brake ring.
2. Clutch sliding member.
3. Planet carrier and gear train.
4. Rear casing and annulus.

DISMANTLING

IMPORTANT: Scrupulous cleanliness must be maintained throughout all service operations, even minute particles of dust, dirt or lint from cleaning cloths may cause damage or, at best, interfere with correct operation.

Prepare a clean area in which to lay out the dismantled unit and some clean containers to receive the small parts.

Hold the overdrive with the front casing uppermost in a vice fitted with suitable soft jaws. Release the tab washers locking the four nuts retaining the operating piston bridge pieces; remove the nuts, washers and bridge pieces.

Loosen the solenoid by the two screws to allow the front casing to be removed.

Remove the four nuts which secure the front and rear casings; separate the casings. The brake ring is spigoted into each half and may remain attached to the front half. If not, a few taps with a mallet around its flange will remove the brake ring from rear casing.

Lift out the clutch sliding member with the thrust ring, bearing and sunwheel.

Lift out the planet carrier and gear train.

The overdrive is now separated into its four main assemblies.

Front Casing and Brake Ring

Remove the operating valve plug and lift out the spring, plunger and ball. Insert a piece of stiff wire in the central bore of the operating valve and draw it down taking care to avoid damaging the seating at the bottom of the valve. Ensure that the small hole which breaks through into the central bore near the top of the valve is not choked. This hole provides the exhaust of oil from the operating cylinders.

Remove the operating pistons by gripping the centre boss with a pair of pliers and rotate gently whilst pulling.

Solenoid — Removal

Remove the rectangular plate by withdrawing the four retaining setscrews. Remove the two screws securing the solenoid and pull off the solenoid. Ease the plunger from the yoke of the valve operating lever.

Accumulator — Removal

Remove the large nut from the bottom of the unit. The length of thread on the plug is sufficient to allow all compression to be released from the spring before the plug is completely unscrewed. The spring, support pin and washer will come away with the plug.

The accumulator piston has a groove inside the bore and a piece of stiff wire can be hooked into this to enable the piston to be withdrawn.

The Pump Non-Return Valve—Removal

Remove the centre plug in the bottom of the unit. Unscrew the valve body using Churchill Tool No. L.213 and remove the spring, support pin and $\frac{7}{32}$" ball.

Pump—Removal

Remove the pump locating screw.

Extract the pump using Churchill Tool No. L.183A and adaptor, L.183A-2. The plunger and spring will also come out when the body is withdrawn.

Oil Filter—Removal

Unscrew the plug situated immediately below the solenoid cover plate. Withdraw the cylindrical gauze filter and magnetic washers.

Clutch Sliding Member and Sunwheel

Remove the circlip from the sunwheel and slide off the corrugated washer and sunwheel.

Planet Carrier Assembly

At this stage, inspect all teeth for any signs of damage or chipping and assess the assembled bearings for any excessive clearance.

Note: Planet gears are not available separately for servicing and a complete planet carrier assembly should be fitted if damage or wear necessitates replacement of any part.

Rear Casing and Annulus

Remove the circlip and oil thrower. Remove the uni-directional clutch by placing the Special Assembly Ring, Churchill Tool No. L.178, centrally over the front face of the annulus and lifting the inner member of the uni-directional clutch up into it. This will ensure that the rollers will not fall out of the retaining cage.

If further dismantling is desirable, remove the assembly ring and allow the rollers to come out. The hub will readily come from the cage exposing the spring. Remove the bronze thrust washer fitted between the hub of the uni-directional clutch and the annulus.

Annulus—Removal

Remove the speedometer dowel screw; withdraw the drive bush and pinion. Remove the coupling flange and oil seal.

Press the annulus forward out of the rear casing. The front and rear bearings will remain in the rear casing with the speedometer drive gear sandwiched between them.

Remove the circlip and drive out the speedometer drive gear and rear bearing. Drive out the front bearing.

INSPECTION

INSPECTION

Front Casing and Brake Ring

Inspect the front casing for damage or cracks. Examine the bores of the operating cylinders and accumulator for scores or wear.

Check for signs of leaks from the plugged ends of the oil passages. Ensure that the sealing disc in the front face of the casing is tight and not leaking.

Inspect the centre bush for wear or damage.

Check the operating pistons for signs of scoring and replace the sealing ring if there is any sign of damage or distortion.

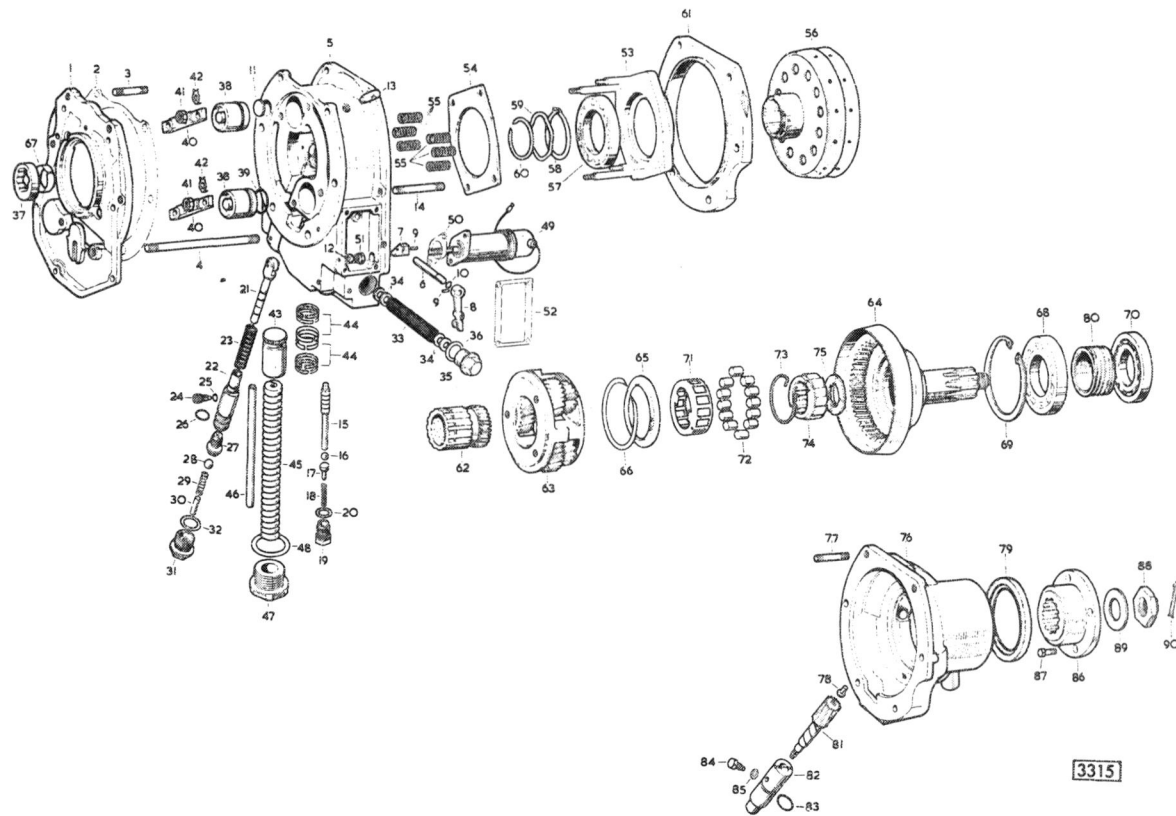

Fig. 63. *Exploded view of overdrive unit.*

1. Adaptor plate	31. Plug	61. Brake ring
2. Gasket	32. Copper washer	62. Sunwheel
3. Stud	33. Filter	63. Planetary carrier
4. Stud	34. Magnetic ring	64. Annulus
5. Front casing	35. Plug	65. Oil thrower
6. Main operating valve shaft	36. Washer	66. Spring ring
7. Cam	37. Oil pump operating cam	67. Spring clip
8. Lever	38. Operating piston	68. Ball bearing
9. Roll pin	39. "O" Ring	69. Circlip
10. "O" Ring	40. Bridge piece	70. Ball bearing
11. Welch washer	41. Nut	71. Cage for uni-directional clutch
12. Rubber stop	42. Tab washer	72. Roller
13. Breather	43. Accumulator piston	73. Cage spring
14. Stud	44. Piston ring	74. Inner member
15. Main operating valve	45. Spring	75. Thrust washer
16. Ball $\frac{5}{16}$" dia.	46. Support rod	76. Rear casing
17. Plunger	47. Plug	77. Stud
18. Spring	48. Washer	78. Thrust button
19. Plug	49. Solenoid	79. Oil seal
20. Copper washer	50. Gasket	80. Speedometer driving gear
21. Oil pump plunger	51. Nut	81. Speedometer driven gear
22. Body	52. Gasket	82. Bearing assembly
23. Spring	53. Thrust ring	83. "O" Ring
24. Screw	54. Retaining plate	84. Screw
25. Fibre washer	55. Springs	85. Copper washer
26. "O" Ring	56. Sliding member	86. Flange
27. Non-return valve body	57. Ball bearing	87. Bolt
28. Ball $\frac{7}{32}$" dia.	58. Circlip	88. Slotted nut
29. Spring	59. Corrugated washer	89. Washer
30. Support rod	60. Snap ring	90. Split pin.

Check the pump roller and bush for any undue wear. Check the pump plunger for wear and scores. Check the pump body for wear and scores. Check the valve seat and ball to ensure they are free from nicks and scratches. Check the pump spring for distortion.

Check the accumulator piston for signs of wear and scores. Check that there are no broken piston rings. Check the accumulator spring for distortion or collapse.

Inspect the operating valve for distortion or damage. See that it slides easily in the bore of the front casing. Check that the ball seat is clean and free from scratches. Check that the restriction jet is clear. Check the ball and spring for distortion.

Clean the filter thoroughly in petrol. Remove all metallic particles from the magnetic washers.

Check the brake rings for signs of wear, scoring or cracks.

Clutch Sliding Member

Inspect the clutch linings on the sliding member for signs of excessive wear or charring. If this is evident, the complete sliding member must be replaced.

Inspect the pins for the bridge pieces on the thrust ring and see that they are not distorted but are a tight fit.

Inspect the ball race and check for noisy rotation. This can be a source of noise when running in direct gear.

Check the clutch springs for any signs of distortion or collapse.

Planet Carrier and Gear Train

The gears and bearings should have been inspected previously as detailed under "Dismantling Planet Carrier Assembly".

Inspect the teeth on the sunwheel for signs of damage or chipping. If the bush is worn, a complete new gear must be fitted.

Rear Casing and Annulus

Ensure that the rollers of the uni-directional clutch are not worn or chipped and that the inner and outer members are not worn or damaged. Check that the cage, particularly the two ears, is not damaged. Check that the spring is not distorted or broken.

Inspect the gear teeth of the annulus for damage. Inspect the conical surface for signs of wear. Check the bronze spigot bearing fitted in the annulus under the uni-directional clutch. If damaged, a new annulus and bearing must be fitted as a complete assembly.

Check the ball races for smooth running. Inspect the rear oil seal. If removed, a new seal must always be fitted. Check the teeth of the speedometer pinion for wear.

RE-ASSEMBLY

Front Casing and Brake Ring

Insert the pump plunger, spring and body into the central hole in the bottom of the casing taking care to locate the flat of the plunger against the thrust button which is situated below the centre bush. Tap the pump body home until the annular groove lines up with the locating screw hole in the casing. Insert the screw ensuring that the dowel locates in the groove.

Re-seat the non-return valve by lightly tapping with a copper drift and screw in the non-return valve body using Churchill Tool. No. L.213. Fit the ball, spring, support pin, copper washer and plug. Tighten the plug ensuring that the spring is located in the plug recess.

Accumulator

Carefully insert the piston into the casing using Churchill Tool No. L.304. Insert the spring, support pin, metal washers and fit the fibre washer and plug.

Operating Pistons

When inserting the pistons, carefully ease the rubber sealing rings into the cylinder bores. The centre bosses of the pistons face towards the front of the unit.

Operating Valve

Insert the operating valve into the casing ensuring that the hemispherical end engages on the flat of the small cam on the operating shaft. Drop in the $\frac{1}{8}$" ball, plunger and spring. Screw in and tighten the operating valve plug.

Oil Filter

Fit the two magnetic washers, filter, further two magnetic washers and filter plug.

Rear Casing and Annulus

Press the front bearing into the rear casing using Churchill Tool No. L.303. Fit the retaining circlip. Support the inner race of the bearing using Churchill Tool No. L.303 and then press the annulus until the bearing abuts on the locating shoulder. Fit the speedometer driving gear. Using the same tool, press the rear bearing on to the tail shaft and into the casing simultaneously. Press in the rear oil seal using Churchill Tool No. L.305 until it is flush with the end of the rear casing. Fit the bolts and press on the coupling flange. Fit the washer and slotted nut. Tighten the nut to a torque figure of 1200—1560 lb. in. and fit the split pin.

Ensure that the "O" ring is serviceable and insert the speedometer pinion and bush. Turn the annulus to engage the gear and align the holes in the casing and bush. Fit the locating screw and copper washer.

Assembling and Fitting the Uni-Directional Clutch

Assemble the spring into the cage of the uni-directional clutch. Fit the inner member into the cage and engage it into the other end of the spring. Engage the slots of the inner member with the tongues of the roller cage and check that the spring rotates the cage to urge the rollers, when fitted, up the inclined faces of the inner member. The cage is spring loaded anti-clockwise when viewed from the front.

Place this assembly, front face downwards, into the special assembly ring, Churchill Tool No. L.178, and fit the rollers through the slots in the tool turning the clutch clockwise until all the rollers are in place.

Fit the thrust washer and replace the uni-directional clutch using the special tool to enter the rollers into the outer member in the annulus. Fit the oil thrower and retaining circlip.

Planet Carrier and Gear Train

Special care must be taken when assembling the planet carrier assembly to the annulus and sunwheel. Turn each gear respectively until a dot marked on one tooth of the large gear is positioned radially outwards. Insert the sunwheel to mesh with the planet gears keeping the dots in the same position and insert this assembly into mesh with the internal gear of the annulus. Insert the dummy mainshaft, Churchill Tool No.

L.185A engaging the splines of the planet carrier and the uni-directional clutch.

Clutch Sliding Member

Press the thrust bearing into the thrust ring and then press this assembly on to the hub of the clutch sliding member taking care not to damage the linings. Secure the assembly in position by fitting the circlip on the hub of the sliding member. Slide this assembly on to the sunwheel splines until the inner lining is in contact with the annulus and then fit the corrugated washer and circlip.

Final Assembly

Fit the retaining plate over the bolts of the thrust ring bearing assembly. Smear a good quality jointing compound on to both faces of the brake ring flange and tap this home into the front casing. Insert the clutch return springs into the pockets of the front casing and then attach the brake ring and front casing to the rear casing. The clutch spring pressure will be felt as the two casings go together and the four nuts should be progressively tightened until the two faces meet. Fit the two bridge pieces, nuts and new tab washers.

Fit the solenoid plunger in the fork of the operating lever and, after fitting a new gasket to the solenoid flange, fasten the solenoid to the flange with two setscrews.

Adjust the solenoid operating lever until a $\frac{3}{16}$" diameter pin pushed through the hole in the lever registers with the hole in the casing. Then screw the nut on to the plunger when, with the plunger pushed right home in the solenoid, the nut just contacts the fork in the lever. Remove the $\frac{3}{16}$" diameter pin.

Re-check by energising the solenoid and noting the alignment of the holes. When the solenoid is energised, its consumption should be about 1 amp. If consumption is 15.—20 amps. it is an indication that the solenoid plunger is not moving far enough to switch from the operating to the holding coil and the lever must be adjusted.

Secure the solenoid cover plate with four setscrews and lockwashers.

Refitting the Overdrive to the Gearbox

Check that the cam is not unduly worn and that the flat spring ring on the gearbox mainshaft is not distorted and does not protrude above the crown of the splines.

Rotate the shaft to position the cam with its highest point uppermost. The lowest point will now coincide with the overdrive pump roller. The mainshaft should not be turned again until the overdrive has been fitted and it is advisable to engage bottom gear.

Remove the dummy mainshaft from the overdrive. The splines will now be correctly lined up and it is most important that the coupling flange is not turned until the unit has been fitted to the gearbox.

Fit a new paper joint to the front face of the overdrive unit. Fit the overdrive to the gearbox carefully ensuring that the pump roller "rides" on the cam which is chamfered for this purpose and that the overdrive pushes right up to the adaptor plate by hand pressure only. If it will not, the splines have been misaligned. In this case, remove the overdrive, re-align the splines by rotating the inner member of the uni-directional clutch anti-clockwise. This can be done with a long screwdriver.

When the overdrive has been fitted, tighten the four nuts on the front casing flange also the two nuts on the long studs which go through the rear casing.

DATA

		Dimensions New		Clearance No.
Pump:				
Plunger diameter	..	.3742″/.3746″	}	.0002″/.0016″
Pump body bore	..	.3746″/.2758″		
Pump Roller Bush:				
Outside diameter of bush	..	.3736″/.3745″	}	.0005″/.0023″
Inside diameter of roller	..	.3750″/.3759″		
Inside diameter of bush	..	.2510″/.2518″	}	.0007″/.0020″
Outside diameter of pin	..	.2497″/.2502″		
Accumulator:				
Piston diameter	..	1.1232″/1.1241″	}	.0004″/.0023″
Bore diameter	..	1.1245″/1.1255″		
Operating Pistons:				
Piston diameter	..	1.3732″/1.3741″	}	.0004″/.0023″
Bore diameter	..	1.3745″/1.3755″		
Operating Valve:				
Valve diameter	..	.2494″/.2497″	}	.0003″/.0012″
Bore diameter	..	.2500″/.2506″		
Overdrive Mainshaft:				
Diameter at oil transfer bush	..	1.1544″/1.1533″	}	.0029″/.0048″
Inside diameter of bush	..	1.1582″/1.1592″		
Diameter at sunwheel	..	1.1544″/1.1533″	}	.0029″/.0048″
Inside diameter at sunwheel bush	..	1.1582″/1.1592″		
Diameter at spigot bearing	..	.6235″/.6242″	}	.0008″/.0025″
Inside diameter of spigot bearing	..	1.6250″/.6260″		

Clutch Movement from Direct to Overdrive .090″/.100″

SECTION FF

AUTOMATIC TRANSMISSION

AUTOMATIC TRANSMISSION

DESCRIPTION

The automatic transmission assembly consists of a three-element hydraulic torque converter followed by two planetary gear sets which permit the elimination of the clutch pedal and normal gear lever. The planetary gear sets incorporate freewheels and are controlled by hydraulically operated bands and disc clutches.

This section deals principally with the setting of the linkages and controls for automatic transmission models.

The dismantling, testing and servicing of the automatic transmission unit is dealt with in a separate publication — "Service Manual for the Jaguar Automatic Transmission".

Anti-Creep System (Fig. 10)

The anti-creep system is a special braking feature which prevents the car from creeping forward when stopped on level ground or slight grades, if the ignition is switched ON. Apply the footbrake to stop the car and then remove the foot from the brake pedal. The car will not creep either backwards or forwards. Either movement of the accelerator pedal or turning off the ignition key will free the anti-creep action.

The system consists of a solenoid valve which holds brake pressure on the rear wheel brakes whenever the anti-creep circuit is closed.

A pressure control switch operated by the transmission rear pump pressure is used to open the anti-creep circuit when the car is moving forward and to close the circuit when the car is stationary or moving in reverse. In addition a switch is incorporated in the throttle linkage which opens and closes the anti-creep circuit as the accelerator is depressed or released.

With the ignition switched ON, the accelerator released (anti-creep throttle switch closed) and the car stationary, pressure control switch closed, the anti-creep circuit is completed and the solenoid valve is energised. When the brakes are applied under these conditions the anti-creep solenoid valve will retain hydraulic pressure at the rear wheel brakes to prevent creeping.

INDEX

DATA

Maximum torque ratio of converter	2·15 : 1
Low gear reduction	2·308 : 1
Intermediate gear reduction	1·435 : 1
Direct drive — no converter	1 : 1
Reverse gear reduction	2·009 : 1
Rear axle ratio	3·54 : 1

AUTOMATIC GEAR CHANGES:

Upshifts	m.p.h.	k.p.h.
Low to intermediate — light throttle ..	11	18
Low to intermediate — full throttle ..	40	64
Intermediate to direct — light throttle ..	23	37
Intermediate to direct — full throttle ..	64	103
Intermediate to direct — after "kick-down" ..	78	126

Downshifts		
Direct to intermediate — closed throttle ..	16	26
Intermediate to low — closed throttle ..	4	6
Direct to intermediate — "kick-down" ..	Up to 68	109
Parking pawl permitted to engage ..	Below 3 to 5	5 to 8
Reverse gear permitted to engage	Below 10	16
Manual change from drive to low to be avoided ..	Above 45	72

ROUTINE MAINTENANCE

The fluid necessary for the operation of the torque converter is common with that used in the transmission. The total capacity of the transmission assembly is approximately 15 Imperial pints (18 U.S. pints — 8·5 litres), but, when draining the transmission, a small quantity of fluid will remain in the unit and the amount required to refill it will be that needed to bring the fluid level to the FULL mark on the dipstick as described in "Drain and Refill Transmission".

Every 3,000 miles (5,000 Km.)

Check Transmission Fluid Level

1. Raise the bonnet. The dipstick is situated forward of the carburetter adjacent to the radiator top water hose.

2. Set the handbrake firmly with the car on a level floor. Select the P (Park) position and start the engine. Apply the footbrake and move the selector lever to L (Low). Run the engine at 800 r.p.m. for 2 or 3 minutes to raise the transmission fluid temperature.

3. Clean the end of the filler tube. Remove the dipstick and wipe it dry. With the foot still on the brake and the selector lever at L (Low), run the engine at its normal idling speed and check the fluid level. Add sufficient fluid to bring the level up to the FULL mark on the dipstick. DO NOT OVER-FILL. The space between the FULL and LOW marks on the dipstick represents approximately one pint.

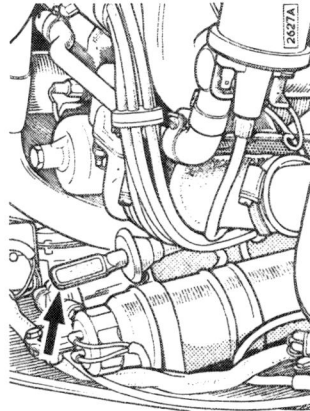

Fig. 1. *Automatic transmission dipstick.*

Every 12,000 miles (20,000 Km.)

Drain and Refill Transmission (Fig. 2)

1. Raise the bonnet. The dipstick is situated forward of the front carburetter adjacent to the radiator top water hose.

2. Apply the handbrake firmly with the car on a level floor. Set the selector lever in the P (Park) position and start the engine. Apply the footbrake, move the selector lever to L (Low) and raise the transmission fluid temperature by running the engine at 800 r.p.m. for 2 or 3 minutes.

3. Stop the engine. Clean the end of the filler tube.

4. Remove the transmission oil pan drain plug (A).

5. Remove the converter housing cover plate and rotate the converter until the drain plug is in position for draining. Remove the converter drain plug (B).

6. To facilitate draining, remove the square-headed converter pressure take-off plug from the bottom of the housing attached to the left-hand side of the transmission casing (C).

7. After fluid has drained, refit and tighten the drain plugs in the transmission oil pan and converter. Refit the converter housing cover plate. Refit and tighten the converter pressure take-off plug.

8. Pour 10 Imperial pints (12 U.S. pints — 5·7 litres) of the recommended grade of fluid into the transmission through the filler tube.

9. Set the selector lever in the P (Park) position and start the engine. Apply the footbrake and move the selector lever to L. Run the engine at 800 r.p.m. for 2 or 3 minutes to transfer fluid from the transmission case to the converter.

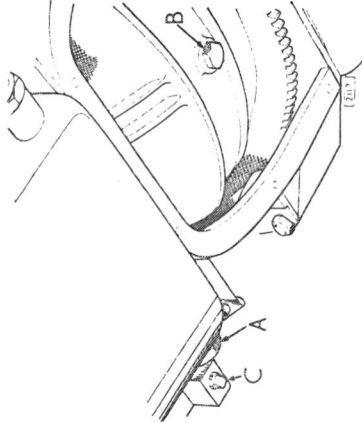

Fig. 2. *Automatic transmission drain plug (the converter housing cover plate has been removed).*

10. With the foot brake applied and the selector lever at L (Low) run the engine at its normal idling speed and add fluid (approximately 5 Imperial pints; 6 U.S. pints — 2·8 litres) to bring the level up to the FULL mark on the dipstick. DO NOT OVERFILL.

Mobil	Castrol	Shell	Esso	B.P.	Duckham	Regent Caltex/Texaco
ATF 210	Castrol T.Q.F.	Shell Donax T7	Esso Glide	Autran B	Q-matic	Texamatic Type F

OPERATION

The transmission assembly consists of a three-element hydraulic torque converter followed by two planetary gear sets which permit the elimination of the clutch pedal and normal gear-shift lever. The planetary gear sets incorporate free-wheels and are controlled by hydraulically-operated band disc clutches.

The manual control lever allows selection of the following conditions:

P (Park). A pawl is mechanically engaged with teeth on the main shaft. A hydraulic interlock prevents engagement at speeds above 3 to 5 m.p.h. (5 to 8 k.p.h.).

N (Neutral). All clutches are disengaged and there is no drive beyond the torque converter.

D (Drive). Automatic changes between the low gear and intermediate gear and direct drive.

Warning: The P (Park) position must not be selected whilst the car is moving in Reverse. Always bring the car to a stop and apply the handbrake firmly before selecting P (Park).

Changes from low to intermediate gear and intermediate to direct drive depend upon the combination of road speed and throttle position; the larger the throttle opening the higher the speed at which the change occurs. This is achieved by mechanically combining the motions of a mechanical centrifugal governor and the throttle linkage. The resultant motion operates a hydraulic valve.

Depression of the accelerator pedal beyond normal travel causes a "kick-down" change from direct to intermediate gear. Below 52 m.p.h. (84 k.p.h.) a downshift from direct to intermediate gear can be obtained by depressing the accelerator to the full throttle position short of "kick-down". No "kick-down" downshift is possible for intermediate to low gear.

The torque converter and a gear reduction are operative in the low intermediate gears. Direct drive is obtained by coupling the engine directly to the main shaft by a disc clutch. The relevant road speeds are given on page FF.4.

Manual L (Low). A low gear train and the torque converter are operative and no automatic change can occur. Manual changes between L (Low) and D (Drive) may be made while the car is in motion but changes into L (Low) should be avoided at speeds above 45 m.p.h. (72 k.p.h.).

R (Reverse). A reverse-gear train and the torque converter are operative. A hydraulic interlock prevents engagement of the reverse clutch at forward speeds above 5 m.p.h. (8 k.p.h.).

Electrical connection to the starter is made only when N (Neutral) and P (Park) are selected. An anti-creep device traps brake fluid pressure when the car is stationary after the brakes have been applied. Opening the throttle releases the fluid.

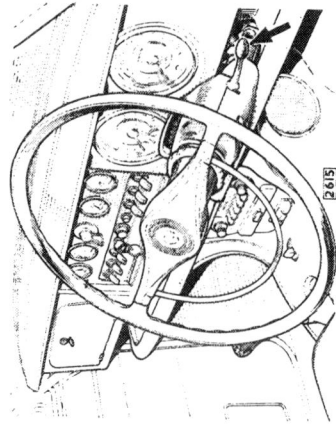

Fig. 3. *Automatic transmission selector lever.*

Selector

The operation of the automatic transmission is controlled by the position of the selector lever which is indicated by the quadrant pointer. The quadrant is situated in front of the steering wheel and is marked P, N, D, L and R. The lever must be raised when selecting P, L or R and when moving from P to any other position.

When the ignition is switched on, the letters P, N, D, L, R, in the quadrant behind the steering wheel become illuminated; when the side-lights are switched on the illumination is automatically dimmed.

To start the engine the selector lever must be in the P or N position.

P (Park) provides a safe, positive lock on the rear wheels when the car is stopped. Movement of the selector lever to the P (Park) position actuates a mechanical locking device in the transmission which prevents the rear wheels from turning in either direction. For this reason, should the car be pushed from front or rear with sufficient force, the car will skid on the rear tyres. This condition is quite similar to that encountered when a car with conventional transmission is parked in gear or with the handbrake applied firmly. The fact that the engine may be started with the selector in P (Park) position is convenient when parked on an incline.

The P (Park) position must not be selected whilst the car is moving in Reverse. Always bring the car to a stop and apply the handbrake firmly before selecting P.

When the car is stopped on a hill and the P (Park) position is selected, the parking mechanism may become very firmly engaged due to the load on the pawl. To disengage the parking pawl under these conditions the following procedure should be adopted:

To release transmission from P (Park) when facing UP HILL.

1. Start the engine.
2. Release the handbrake.
3. Select D (Drive) and hold lever in this position (irrespective of the direction in which it is desired to move off).
4. Depress accelerator slowly until the car moves forward, indicating the release of the parking pawl.
5. The car is now "free" and can be driven away in the desired direction.

To release transmission from P (Park) when facing DOWN HILL.

1. Start the engine.
2. Release the handbrake.
3. Select R (Reverse) and hold lever in this position (irrespective of the direction in which it is desired to move off).
4. Depress the accelerator slowly until the car moves backwards, indicating the release of the parking pawl.
5. The car is now "free" and can be driven away in the desired direction.

N (Neutral) position permits idling of the engine without the possibility of setting the car into motion by pressure on the accelerator and may be used when starting the engine. It is inadvisable to engage neutral for coasting.

D (Drive) provides the normal forward driving range and includes automatic shifting between the low, intermediate and direct drive ranges. Virtually all forward driving, accelerat-ing and stopping can be done with the lever in the D (Drive) position. Once the engine is started and the lever is moved to D (Drive) it can be left in this position for all normal driving. When accelerating, the transmission shifts automatically from low to intermediate between 11 and 40 m.p.h. (18 and 64 k.p.h.) and from intermediate to direct between 23 and 64 m.p.h. (37 and 103 k.p.h.) depending on the position of the accelerator pedal. On deceleration, it will shift automatically from direct drive to intermediate at approximately 16 m.p.h. (26 k.p.h.) and from intermediate to low at approximately 4 m.p.h. (6 k.p.h.).

L (Low) is an emergency engine power range for use on unusually long and steep grades or for braking on descents, for extra heavy pulling and for rocking the car out of mud, sand or snow.

R (Reverse) position of the selector lever provides reverse driving range.

Intermediate Speed Hold. A switch mounted on the facia provides a means for the driver to obtain a downshift from direct to intermediate without depressing the accelerator pedal (as advised under the heading "Additional Power and Acceleration") and to retain the drive in the intermediate range. This will be found convenient for overtaking or when hill climbing.

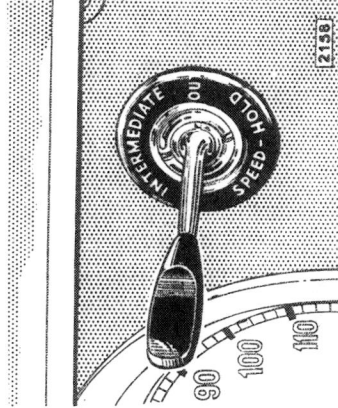

Fig. 4. Intermediate speed hold switch.

With the switch in the IN position no upshift will take place between intermediate and direct drive; placing the switch lever in the OUT position will cause the transmission to shift to direct drive, provided the normal upshift speed has been obtained.

Warning: DO NOT allow the maximum permitted engine revolutions to be exceeded through allowing the Intermediate Speed Hold to remain in operation longer than necessary, or by switching in the Hold at speeds in excess of 75 m.p.h. (121 k.p.h.).

Additional Power and Acceleration in D (Drive) range can be obtained as follows:

(a) Below 52 m.p.h. (84 k.p.h.) depress the accelerator pedal to the full throttle position to effect a change into the intermediate range; the drive will continue in the intermediate range until the release of the accelerator or approximately 64 m.p.h. (103 k.p.h.) is reached.

(b) Between 52 m.p.h. and 68 m.p.h. (84 k.p.h. and 109 k.p.h.) depress the accelerator pedal all the way to the floorboard to effect a kickdown change into intermediate range; the drive will continue in the intermediate range until release of the accelerator or approximately 78 m.p.h. (126 k.p.h.) is reached.

Hard pulling, such as is encountered in deep snow, mud or other adverse driving conditions, is best accomplished in the L (Low) range.

Rocking out of mud, sand or snow is accomplished with the accelerator pedal slightly depressed and held steady while making quick alternate selections of L (Low) and R (Reverse) ranges.

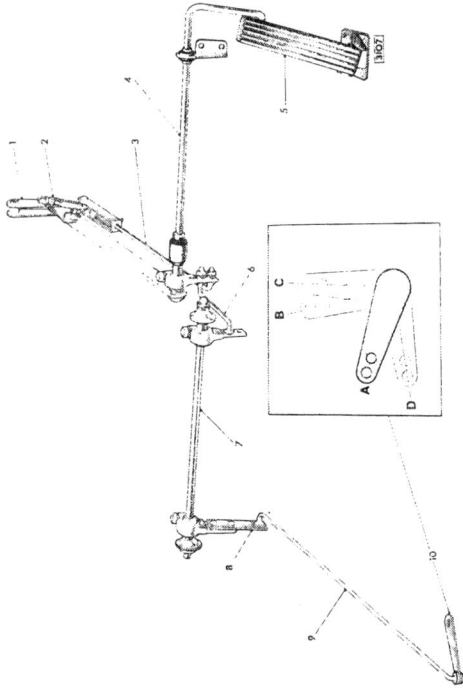

Fig. 5.   Adjustment of the governor control lever, R.H. Drive.

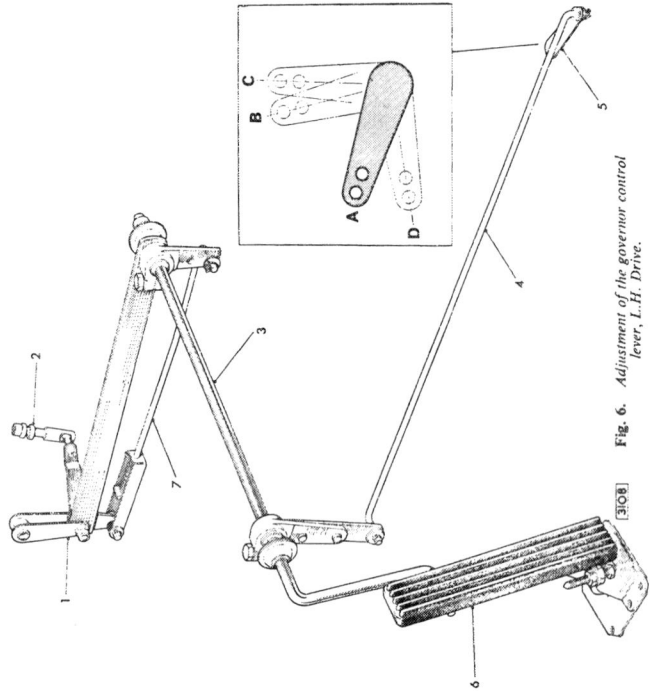

Fig. 6.   Adjustment of the governor control lever, L.H. Drive.

AUTOMATIC TRANSMISSION

Anti-creep is a special braking feature which prevents the car from creeping forward when stopped on level ground or slight grades, as long as the ignition key is turned on. Apply the footbrake to stop the car and then remove the foot from the brake pedal. The car will not creep forward or backward. Any movement of the accelerator pedal, or turning of the ignition key, releases the anti-creep action.

Push starting may sometimes be necessary, as in the case of a flat battery. Turn the ignition key ON, place the selector lever in the N (Neutral) position. The car may now be pushed and when it has reached 15 to 20 m.p.h. (24 to 32 k.p.h.) move the selector lever to D (Drive) or L (Low) position. Do not tow the car to start the engine — it may overtake the tow car.

Engine braking, for descending long mountainous grades, is easily secured by bringing the car speed below 45 m.p.h. (72 k.p.h.) and momentarily depressing the accelerator while placing the selector lever in the L (Low) position

Prolonged idling is sometimes unavoidable. In such cases, as a safety precaution, move the selector lever to the P (Park) or N (Neutral) position.

Towing of the car should be done with the selector in the N (Neutral) position. When towing the car for an appreciable distance the propeller shaft must be disconnected.

SERVICE ADJUSTMENTS

Accelerator to Governor Lever Adjustment

Before adjusting the linkage from the accelerator pedal to the governor lever, the carburetter idling speed should be checked and, if necessary, adjusted.

In order to check the idling speed, disconnect the control rod from the governor control lever and proceed as follows:

(a) Slacken the locknut retaining the anti-creep throttle switch and screw the switch downwards out of operation.

(b) Adjust the idling speed of the engine to 500 r.p.m. by rotating the three throttle adjusting screws by exactly equal amounts.

(c) Reset the anti-creep throttle switch as described on page FF.13.

Adjusting the Governor Lever

Depress the accelerator pedal to the full throttle position at the carburetters. DO NOT depress the pedal sufficiently hard to overcome the "kick-down" overtravel spring in the vertical link situated on the right-hand side of the bulkhead for right-hand drive cars and left-hand side for left-hand drive cars.

Turn the governor lever to the full throttle position, that is, the position where solid resistance is felt before overcoming the cam detent.

With the accelerator pedal and governor lever in the above positions, adjust the length of the governor lever control rod at the large knurled nut. The ball pin should slip easily into the inner hole of the governor lever.

Check the kickdown operation on road test.

Starter cut-out and Reverse Light Switch

The combined starter cut-out and reverse light switch is attached to a bracket situated at the bottom of the manual selector control rod.

The switch is operated by a centrally pivoted arm which is attached to the selector rod.

The starter cut-out serves to close the starter motor operating circuit only when the manual selector lever is in the P (Park) or N (Neutral) position. This ensures that the engine cannot be started when the transmission is in any of the driving ranges.

The reverse light switch serves to close the reversing light circuit when the manual selector lever is in the R (Reverse) position and the ignition switch is ON.

Adjustment

To ensure correct operation of the starter cut-out and reverse light inhibitor, it is most important that the following instructions be adhered to:

Slacken the starter/reverse light switch securing bolt. Move the gear selector lever until the gear indicator is halfway between the L (Low) and R (Reverse) position. Rotate the starter/reverse light switch until the hole in the lever is in line with the hole in the switch base plate. Place a piece of wire through the two holes and tighten the nut securing the switch to the upper steering column (see Fig. 7).

Remove the wire.

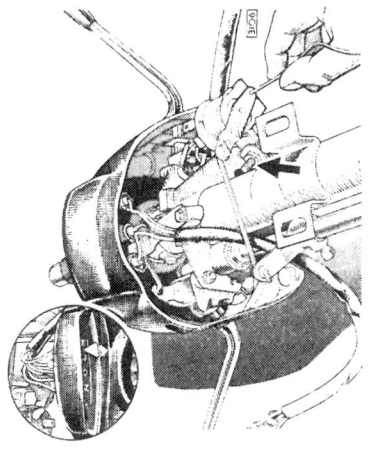

Fig. 7. Setting the starter/reverse inhibitor.

Manual Selector Cable Adjustment

Adjustment of the cable should only be necessary if the cable has been removed from the car. The exhaust system will have to be removed to provide access to the selector valve lever.

1. Place the selector valve lever in the D (Drive) position, that is, the centre of the five positions that can be obtained (see Fig. 8).

2. Pass the outer cable through the transmission mounted-locating bracket. Secure the cable to the selector valve lever. It is most important that the cable runs in a straight line to the selector valve lever.

3. Measure the distance between the inner and outer cables at the top end. The dimension on right-hand drive cars should be 4¾ in. (12·06 cm.) and left-hand drive cars 3 11/16 in. (9·36 cm.).

4. Carry out any necessary adjustment on the lower end of the cable. Lock up the collet on the lower ball joint.

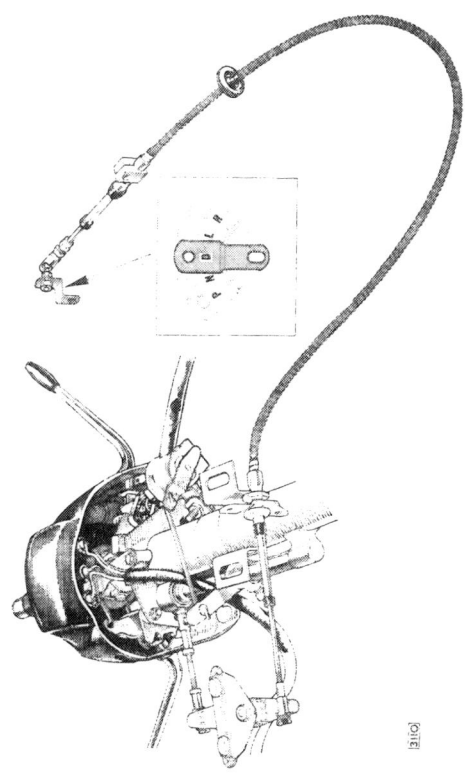

Fig. 8. Adjustment of the manual selector cable.

5. Place the gear selector lever in the D (Drive) position. The upper ball joint should now be in line with the hole in the selector lever control rod.

6. Connect the cable ball joint to the control rod and check the operation of the transmission in the five positions of the manual selector lever quadrant.

Anti-Creep Throttle Switch Adjustment

The anti-creep throttle switch is attached to a bracket situated forward of the front carburetter. For a full description of the system see page FF.3.

To adjust the switch, carry out the following procedure:

Slacken the locknut securing the anti-creep throttle switch to its bracket and screw the switch downwards so that the plunger in the centre of the switch is not in contact with the operating lever.

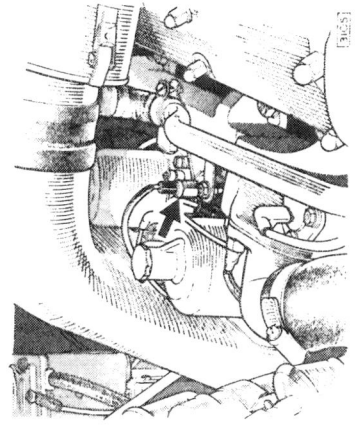

Fig. 9. Anti-creep throttle switch.

AUTOMATIC TRANSMISSION

Ensure that the throttle adjusting screws are on their stops and check the idling speed of the engine which should be 500 r.p.m. If it is necessary to adjust the idling speed, rotate the three throttle adjusting screws by exactly equal amounts.

Adjust the position of the switch in the bracket so that the plunger in the centre of the switch is FULLY DEPRESSED by the operating lever with the carburetter throttles in the normal idling position. Tighten the locknut.

To test the anti-creep system, carry out the following procedure:

1. Drive the car at approximately 10 m.p.h. (16 k.p.h.) with the hand control selector lever in D (Drive) position and quickly apply and then immediately release the brakes.

Do not touch the accelerator pedal. The anti-creep pressure switch should open the anti-creep circuit so that the car rolls freely with no evidence of brake drag after the brake pedal is released.

2. Whilst driving the car on level ground, release the accelerator pedal and apply the brakes, bringing the car to a complete stop. The car should not creep forward until the accelerator is depressed.

3. When the car is standing still on level ground with the engine idling and the selector lever is in one of the driving positions, the anti-creep system prevents the car from creeping. To release the anti-creep system, touch the accelerator pedal lightly and release it. The car will now creep slowly from a standstill.

PROPELLER SHAFT

SECTION G

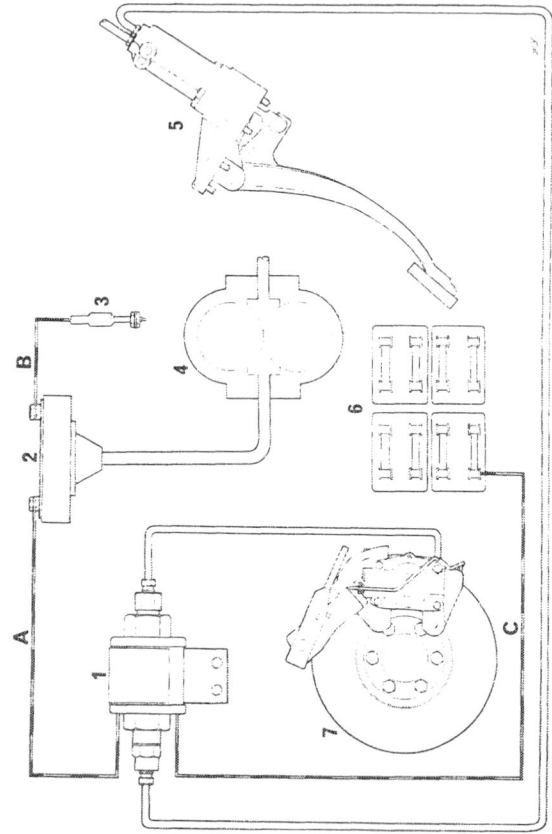

Fig. 10. *Anti-creep system.*

PROPELLER SHAFT

DESCRIPTION

Standard transmission cars are fitted with a fixed length propeller shaft with a universal joint at each end and a sliding spline at the front end.

Overdrive and automatic transmission models are provided with a propeller shaft which has a universal joint at each end and a sliding spline encased in a rubber gaiter.

ROUTINE MAINTENANCE

The propeller shaft universal joints and sliding spline are of the "sealed for life" type which do not require periodic lubrication.

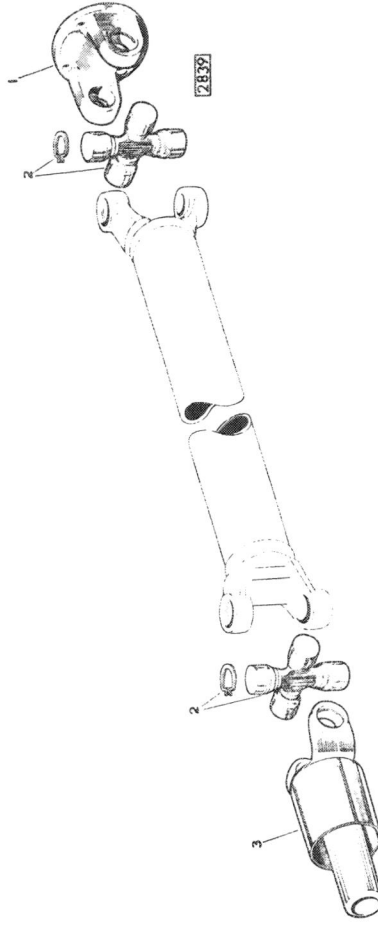

Fig. 1. *Standard transmission propeller shafts.*

1. Flange yoke
2. Journal assembly
3. Sleeve yoke assembly

INDEX

PROPELLER SHAFT

PROPELLER SHAFT (Standard Transmission models)

Removal

Jack up one of the rear wheels clear of the ground to enable the propeller shaft to be rotated. Place blocks at each side of the other rear wheel.

Remove the nuts and washers from the four bolts attaching the propeller shaft to the rear axle flange.

Separate the two flanges and withdraw the propeller shaft from the splines at the rear of the gearbox mainshaft.

Refitting

Refitting is the reverse of the removal procedure.

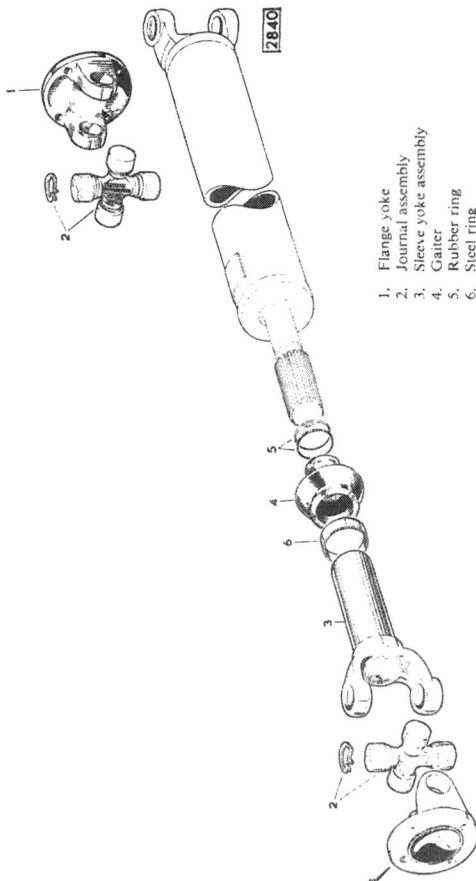

1. Flange yoke
2. Journal assembly
3. Sleeve yoke assembly
4. Gaiter
5. Rubber ring
6. Steel ring

Fig. 2. *Overdrive and automatic transmission propeller shaft.*

PROPELLER SHAFT (Overdrive and Automatic Transmission Models)

Removal

Jack up one of the rear wheels clear of the ground to enable the propeller shaft to be rotated. Place blocks at each side of the other rear wheel.

Remove the nuts and washers from the flange attaching the propeller shaft to the front flange. Remove the nuts and washers from the flange attaching the propeller shaft to the rear axle flange. Compress the sliding joint when the propeller shaft can be removed.

Refitting

Refitting is the reverse of the removal procedure.

THE UNIVERSAL JOINTS
Examine and Check for Wear

The parts most likely to show signs of wear after long usage are the bearing races and spider journals. Should looseness in the fit of these parts, load markings or distortion be observed they should be renewed as a unit as worn needle bearings used with a new spider journal or new needle bearings with a worn spider journal will wear more rapidly, making another replacement necessary in a short time.

It is essential that the bearing races are a light drive fit in the yoke trunnion.

In the rare event of wear having taken place in the yoke cross holes, the holes will have become oval and the yokes must be removed.

In the case of wear of the cross holes in a fixed yoke, which is part of the tubular shaft, only in cases of emergency should these be replaced. They should normally be replaced by a complete assembly.

The other parts likely to show signs of wear are the splined sleeve yoke and splined shaft. A total of ·004″ (·1 mm.) circumferential movement, measured on the outside diameter of the spline, should not be exceeded. If wear has taken place above this limit the complete propeller shaft should be replaced.

Dismantling

Remove the sliding joint from the splined shaft, after releasing the retaining ring from the rubber gaiter.

Clean the paint and dirt from the rings and top of bearing races. Remove all the snap rings by pinching together with a pair of pliers and prising out with a screwdriver. If a ring does not snap out of its groove readily, lightly tap end of bearing brace to relieve the pressure against the ring.

Hold the joint in the hand and with a soft nosed hammer tap the yoke lug as shown in Fig. 3.

The top bearing will gradually emerge and can finally be removed with the fingers (see Fig. 4).

If necessary, tap the bearing race from inside with a small diameter bar, taking care not to damage the bearing race (see Fig. 5).

Repeat this operation for the opposite bearing. The splined sleeve yoke or flange yoke can now be removed. Rest the two exposed

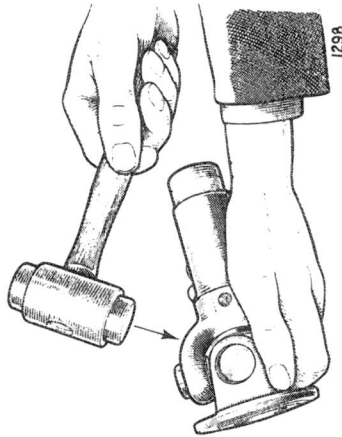

Fig. 3. *Tapping the yoke to remove the bearing.*

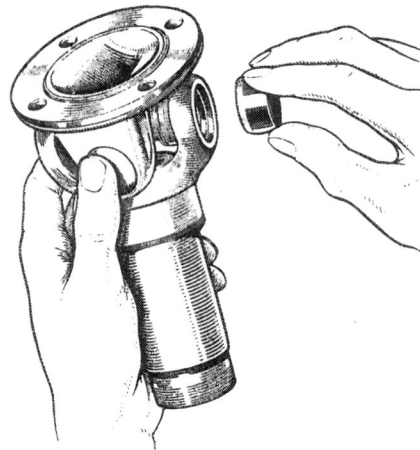

Fig. 4. *Withdrawing the bearing from the universal joint.*

trunnions on wood or lead blocks, then tap yoke with a soft nosed hammer to remove the two remaining bearing races. Wash all parts in petrol.

Assembling

Insert the journal cross into the flange yoke, tilting it to engage in the yoke bores.

Fit one of the bearing races in the yoke bore, and using a soft, round drift about $\frac{5}{16}''$ (8 mm.) smaller in diameter than bearing diameter, tap it into the yoke bore. Fit a circlip and ensure it is correctly located in the groove.

Assemble the opposite bearing into the yoke bore introducing the bearing from the bottom. Repeat this operation for the other two bearings, and fit new circlips.

Wipe off any superfluous grease from the last unassembled peg prior to fitting the last race.

Proceed to fit the unit package assembly at the other end in a similar manner.

Ensure that all the circlips are sitting correctly in their respective grooves. Check the races are bearing against the circlip. Check for free movement of the journals.

Important

When replacing the sliding joint it must be refitted with its fixed yoke in line with the fixed yoke at the end of the propeller shaft tube. Arrows are stamped on the two parts to facilitate alignment. (See Fig. 7).

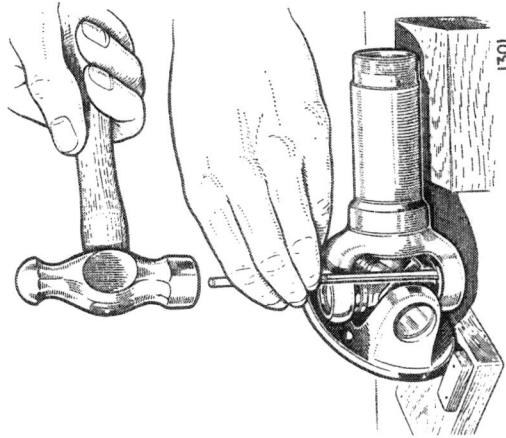

Fig. 5. *Tapping out a bearing with a small diameter rod.*

Fig. 6. *Separating the universal joint yokes.*

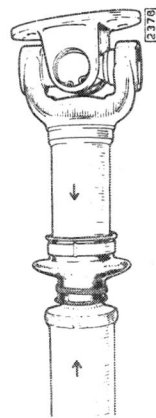

Fig. 7. *Showing the arrows on the sliding joints.*

SECTION H

REAR AXLE

ADDENDA

The following corrections should be noted when using SECTION H.

Page	Correction
H.18	"Fig. 24 Exploded view of output shaft assembly. (Early cars)."
H.19	Fig. 26 caption "with adaptor SL14-2" should read "with adaptor SL14-1".
H.20	Fig 27 caption "with adaptor SL14-5" should read "with adaptor SL14-3."
H.21	Fig. 29 caption "with adaptor SL550-5" should read "with adaptor SL550-4."

INDEX

INDEX (continued)

REAR AXLE

DESCRIPTION

The rear axle unit (Fig. 1) is of the Salisbury 4.HU type and is mounted independently from the hubs and road wheels. Short drive shafts with universal joints at each end are coupled to the axle output shafts. These output shafts also provide a mounting for the discs of the inboard mounted disc brakes.

A Thornton "Powr-Lok" limited slip differential is fitted as standard to the 3·8 litre model and is available as an optional extra on the 3·4 litre model.

The axle gear ratio is stamped on a tag attached to the assembly by one of the rear cover screws.

The axle serial number is stamped on the gear carrier housing.

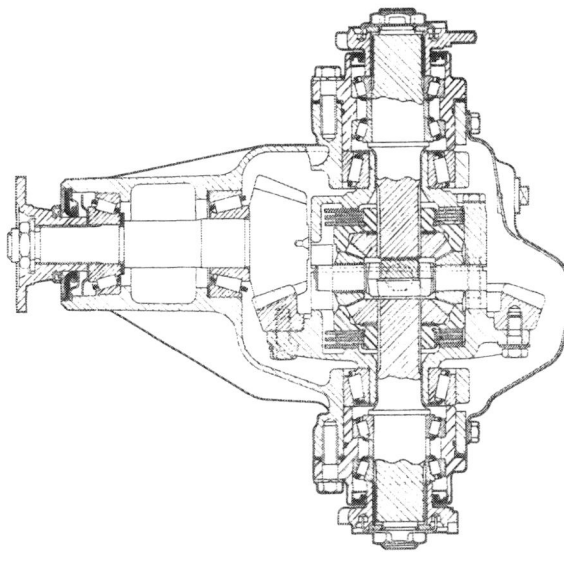

Fig. 1. *Sectional view of the rear axle unit. (Powr-Lok differential illustrated)*

DATA

Output shaft end float	-001″ to -003″ (-02 to -07 mm.)
Differential bearing pre-load	-006″ to -010″ (-15 to -25 mm.) total shim allowance
Pinion bearing pre-load	8 to 12 lb. in. (-09 to -14 kg. m.)
Backlash	As etched on drive gear—minimum -004″ (-10 mm.)

Tightening torque :

Drive gear bolts	70 to 80 lb. ft. (9·7 to 11·1 kg. m.)
Differential bearing cap bolts ..	60 to 65 lb. ft. (8·3 to 9·0 kg. m.)
Pinion nut	120 to 130 lb. ft. (16·6 to 18·0 kg. m.)
Thornton "Powr-Lok" differential bolts ..	40 to 45 lb. ft. (5·5 to 6·2 kg. m.)

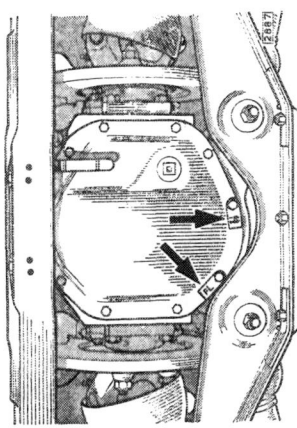

Fig. 2. *Showing the location of the axle ratio and "Powr-Lok" tags.*

AXLE RATIOS

Standard model	.. ..	3·54 : 1 (46 × 13)
Automatic transmission model		3·54 : 1 (46 × 13)
Overdrive model	.. ..	3·77 : 1 (49 × 13)

Reconditioning Scheme (Great Britain only)

Although full servicing instructions for the rear axle are given in this section it is recommended that, wherever possible, advantage is taken of the factory reconditioning scheme particularly in view of the intricate adjustments and the number of special tools required.

Reconditioned axles are supplied on an exchange basis and comprise an axle complete less half shafts, hubs and brake details; rear axles for overhaul should therefore be returned in this condition.

RECOMMENDED LUBRICANTS

Component	Mobil	Castrol	Shell	Esso	B.P.	Duckham	Regent Caltex/ Texaco
Rear Axle	Mobilube GX 90	Castrol Hypoy	Spirax 90 EP	Esso Gear Oil GP 90/140	Gear Oil SAE EP 90	Hypoid 90	Multigear Lubricant EP 90
Rear wheel bearings	Mobilgrease MP	Castrolease LM	Retinax A	Esso Multipurpose Grease H	Energrease L2	LB10	Marfak All-purpose

CAPACITIES

Imperial Pints	U.S. Pints	Litres
2¾	3¼	1.5

1.	Gear carrier	19.	Roller bearing	36. Screw
2.	Screw	20.	Oil slinger	37. Washer
3.	Lockwasher	21.	Oil seal	38. Oil seal
4.	Cover	22.	Gasket	39. Bolt
5.	Plug	23.	Companion flange	40. Nut
6.	Gasket	24.	Nut	41. Breather
7.	Elbow	25.	Washer	42. Differential case
9.	Setscrew	26.	Drive shaft and flange	43. Friction plate
10.	Lockwasher	27.	Roller bearing	44. Friction plate
11.	Roller bearing	28.	Spacing collar	45. Friction disc
12.	Crown wheel and pinion	29.	Shim	46. Ring
13.	Setscrew	30.	Drive shaft bearing housing	47. Side gear
14.	Lock strap	31.	'O' ring	48. Pinion mate gear
15.	Roller bearing	32.	Nut	49. Shaft
16.	Shim	33.	Lockwasher	50. Bolt
17.	Distance washer	34.	Shim	
18.	Shim	35.	Screw	

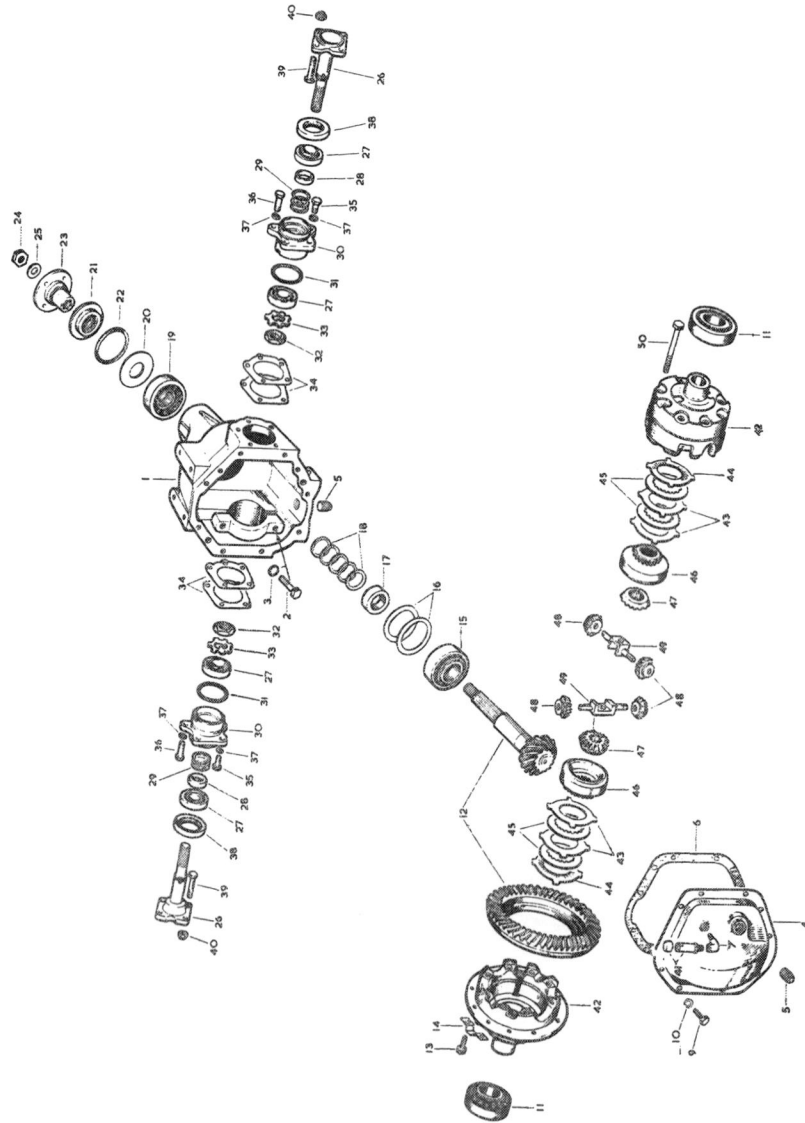

JSP125

Fig. 3.   *Exploded view of rear axle unit (latest condition illustrated).*

ROUTINE MAINTENANCE

The rear axle half shaft universal joints are of the "sealed for life" type and require no periodic lubrication.

Every 3,000 miles (5,000 km.)

Checking Rear Axle Oil Level

Check the level of the oil in the rear axle with the car standing on level ground.

A combined filler and level plug is fitted in the rear of the axle casing accessible from underneath the car. Clean off any dirt from around the plug before removing it.

The level of the oil should be to the bottom of the filler and level plug hole; use only HYPOID oil of the correct grade and since different brands may not mix satisfactorily, draining and refilling is preferable to replenishing if the brand of oil in the axle is unknown.

Every 12,000 miles (20,000 km.)

Changing the Rear Axle Oil

The draining of the rear axle should be carried out at the end of a run when the oil is hot and will therefore flow more freely. The drain plug is situated in the base of the differential casing.

After the oil has drained, replace the drain plug and refill the rear axle with the recommended grade of oil after removal of the combined filler and level plug situated in rear cover.

The level of the oil should be to the bottom of the filler and level plug hole when the car is standing on level ground.

Use only HYPOID oil of the correct grade.

Note : The rear axles of all new cars are filled with a special oil. This oil must not be drained off at the 1,000 mile (1600 km.) Free Service.

After 6,000 miles (10,000 km.) this special oil may be drained off and the rear axle refilled with one of the normal Hypoid 90 lubricants.

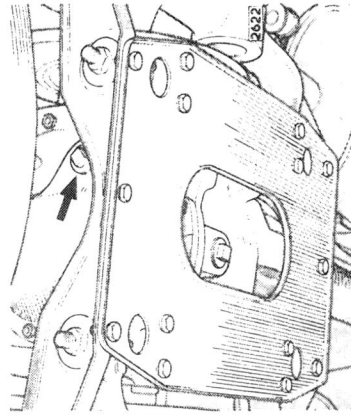

Fig. 4. *Rear axle filler and level plug.*

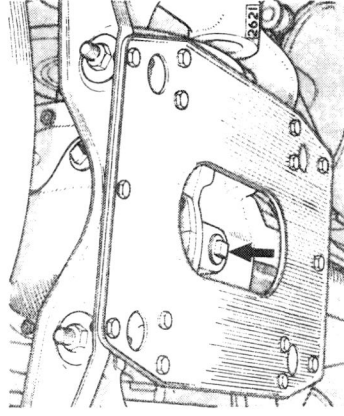

Fig. 5. *Rear axle drain plug.*

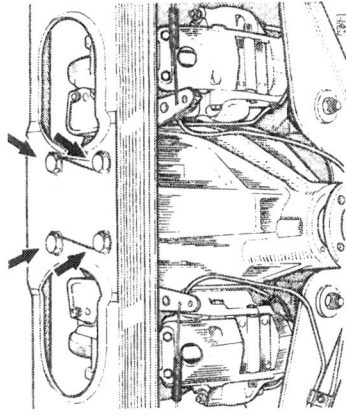

Fig. 6. *Showing the top axle casing mounting bolts.*

THE AXLE UNIT

REMOVAL

The following removal and refitting operations are described assuming the rear suspension is removed from the car. If it is possible for the operations to be carried out with the rear suspension in position on the car the fact will be noted in the text.

Remove the rear suspension assembly from the car (as described in section K "Rear Suspension"). Invert the suspension assembly on a bench and remove the 14 bolts securing the tie plate. Remove the tie plate and disconnect the four hydraulic damper and spring units. Remove the four self-locking nuts securing the halfshaft universal joint to the brake disc and axle output shaft flange. Owing to heat dissipation from the brake disc, it is most important that the locknuts fitted on the output shaft flange studs are of the metal and not nylon self-locking type. Withdraw the half shaft from the bolts noting the number of camber shims. Remove one self-locking nut from the inner wishbone fulcrum shaft and drift out the shaft. Remove the hub, halfshaft, wish-bone and radius arm assembly and repeat the procedure at the other side. Disconnect the hydraulic feed pipes at the brake calipers. Turn the suspension assembly over and remove the locking wire from the four differential carrier mounting bolts. Unscrew the mounting bolts and remove the cross beam from the differential carrier by tilting forward over the nose of the pinion.

REFITTING

Refitting is the reverse of the removal procedure, it should be noted however, that the inner wishbone fulcrum shaft self-locking nut should be tightened to a torque of 55 lb. ft. (7·6 kg. m.). The four differential carrier mounting bolts on the top of the cross beam should be tightened to a torque of 75 lb. ft. (10·4 kg. m.).

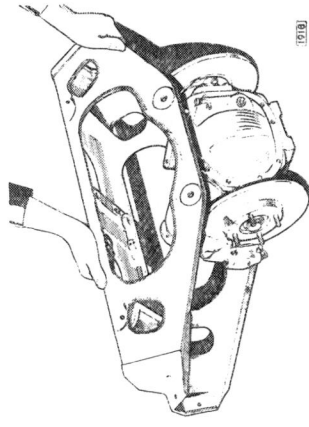

Fig. 7. *Removing the cross beam from the axle unit.*

THE REAR HUBS

REMOVAL

It is not necessary to remove the rear suspension unit from the car to carry out this operation.

Jack up and support the rear end of the car and remove the appropriate road wheel. Withdraw the split pin and remove the castellated nut and washer from the halfshaft. Using the extractor, Special Tool No. J.1.C. as shown in Fig. 8, withdraw the hub and hub carrier assembly from the splined end of the halfshaft retaining the inner oil seal track and end float spacer. Remove the lower wishbone outer fulcrum shaft (as described in Section K "Rear Suspension") and remove the hub and hub carrier assembly.

Note : Since it is necessary to press the hub assembly on to the half shaft before refitting to the rear suspension assembly, remove the halfshaft as follows. Remove the front hydraulic damper and spring unit (as described in Section K "Rear Suspension"). Remove the four steel type self-locking nuts securing the halfshaft inner universal joint to the axle shaft output flange and brake disc. Withdraw the halfshaft from the bolts noting the number of camber shims fitted.

DISMANTLING

Invert the hub carrier so that the inner hub bearing is at the top and press out the hub (Fig. 9) with the outer bearing inner race and the outer oil seal track in place, discarding the outer oil seal. Remove the three setscrews and withdraw the water deflector. Prise out the inner oil seal and remove the inner bearing inner race. Drift out the outer races of the inner and outer bearings if necessary. Withdraw the outer bearing inner race with a suitable extractor.

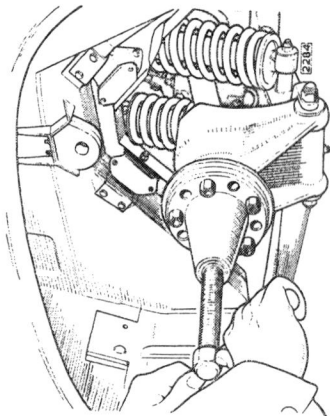

Fig. 8. Removing the rear hub with the aid of the extractor (Churchill Tool No. J.1.C.)

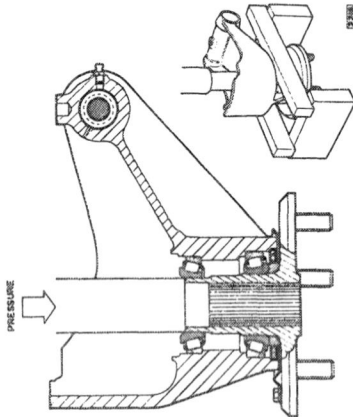

Fig. 10. Removing the inner race from the hub using Churchill Tool No. SL.14 with adaptor J.16.A.

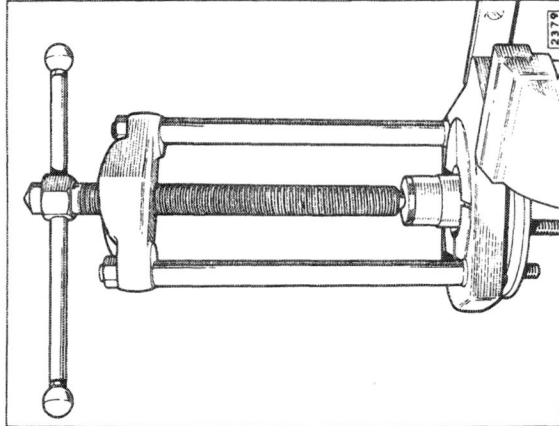

Fig. 9. Pressing the hub from the hub carrier.

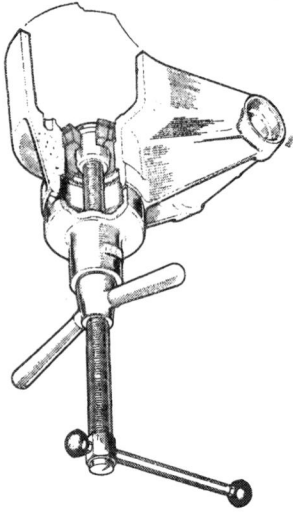

Fig. 11. Replacing the hub bearing outer races using Churchill Tool No. J.20A with adaptors J.20A—1.

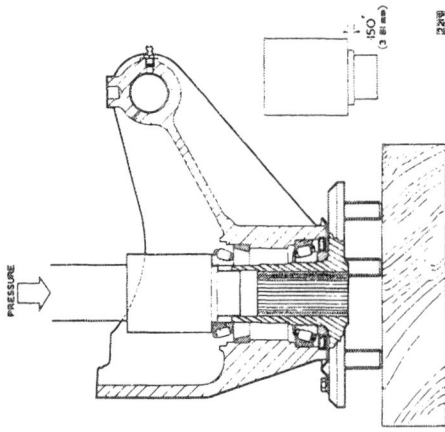

Fig. 12. Pressing in the hub inner bearing inner race using the master spacer (Churchill Tool No. J.15).

ASSEMBLING

If new bearings are to be fitted, press new inner and outer bearing outer races into the hub carrier ensuring that they seat correctly in their recesses.

With the hub carrier held so that the outer bearing will be at the top, place the outer bearing inner race in position and press the outer oil seal into its recess. Fit the water deflector and press the hub with the outer oil seal track in position into the outer bearing inner race until the hub is fully home.

Hub Bearing End-float

Hold the hub and hub carrier vertically in a hand press with the inner end of the hub upper-most. Place the inner bearing inner race on the hub, fit the masterspacer (Special Tool No. J.15) into the race and press the race onto the hub (Fig. 12) until the master spacer contacts the hub. This will ensure a certain amount of end-float. Remove the hub and hub carrier from the hand press and secure in a vice in order to measure the end-float. With the inner end of the hub upper-most, and the master spacer in position as before, fit a dial gauge (Special Tool No. J.13) to the hub as shown in Fig. 13. Tap the hub carrier down-wards, zero the dial gauge and using two screw-drivers or similar levers between the hub and hub carrier, move the hub carrier upwards to its full-lest extent. Note the reading on the dial gauge. Having determined the measured end-float, a spacer must be fitted in place of the special collar to give an end-float of ·002"—·006" (·051—·152 mm.). Spacers are supplied in thicknesses of ·109"—·151" (2·77—3·87 mm.) in steps of ·003" (·076 mm.) as shown in the following table:

Spacer Letter		Thickness inches	mm.
A	..	·109	2·77
B	..	·112	2·85
C	..	·115	2·92
D	..	·118	3·00
E	..	·121	3·07
F	..	·124	3·15
G	..	·127	3·23
H	..	·130	3·30
J	..	·133	3·38
K	..	·136	3·45
L	..	·139	3·53
M	..	·142	3·61
P	..	·145	3·68
Q	..	·148	3·75
R	..	·151	3·87

For example, assume the end-float measured to be ·025" (·64 mm.). Subtract the nominal end-float of ·004" (·10 mm.) from the measured end-float giving ·021" (·53 mm.). Since the Master Spacer is ·150" (3·81 mm.) thick, the thickness of the spacer to be fitted will be ·150"—·021" i.e. ·129" (3·28 mm.). The nearest spacer is ·130" (3·30 mm.) so a letter H spacer should be fitted.

When the hub assembly and half shaft have been refitted to the rear suspension the end-float should be checked using the dial indicator as shown in Fig. 15.

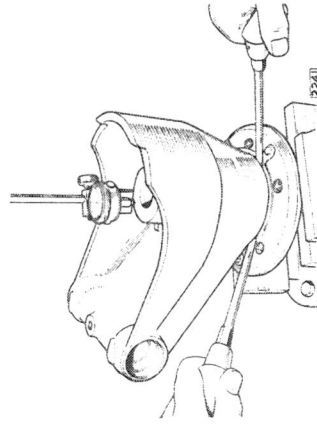

Fig. 13. *Checking the hub bearing end-float with a dial tester indicator (Churchill Tool No. J.13).*

Fitting the Hub Assembly to the Halfshaft

To fit the hub assembly to the halfshaft it will be necessary to use a hand-press (see Fig. 14). Ensure that both the splines of the halfshaft and the hub are free of grease by using a suitable solvent. Place the inner oil seal track and end float spacer on the halfshaft. Apply a drop or two of "Loctite" (available in 10 c.c. bottles, Part No. 9035) to the halfshaft splines for about an inch from the threaded end using a small paint brush to ensure even spreading. Only use "Loctite" sparingly as no additional benefit will be achieved by using large amounts. Introduce the halfshaft into the hub and engage the splines. Place the assembly on the hand-press and press the hub onto the halfshaft. Fit the washer and castellated nut, tighten to 140 lb. ft. (19·3 kg. m.) torque and fit the split pin.

Note: To obtain the best results from the "Loctite" sealant, the joint should be allowed to set for 24 hours, that is, this period should be allowed to elapse before the car is run.

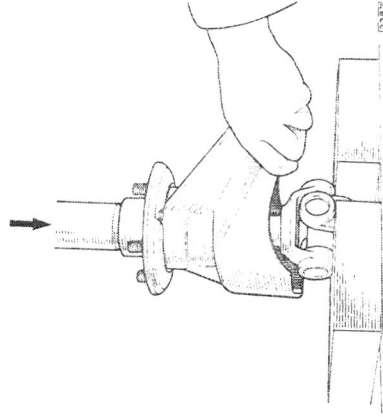

Fig. 14. *Fitting the hub to the halfshaft using a press.*

REFITTING

The hub assembly and halfshaft are refitted to the rear suspension as described under "Half shaft Refitting".

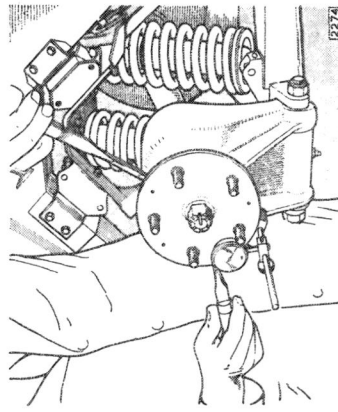

Fig. 15. *Checking the hub bearing end-float when in position (Churchill Tool No. J.13).*

THE HALFSHAFTS

REMOVAL

Proceed as described under Rear Hub Removal until the hub assembly can be withdrawn. Remove the front hydraulic damper and spring unit (as described in Section K "Rear Suspension"). Remove the four steel type self-locking nuts securing the halfshaft inner universal joint to the axle output shaft flange and brake disc. Withdraw the halfshaft from the bolts noting the number of camber shims fitted.

REFITTING

Refit the hub assembly to the halfshaft as described in the Rear Hub section and proceed as follows.

Replace the camber shims and place the halfshaft and hub into position with the halfshaft inner universal joint over the four bolts. Fit the four steel-type self-locking nuts and tighten up. Refit the front hydraulic damper and spring unit (as described in Section K "Rear Suspension"). Refit the lower wishbone outer fulcrum shaft (as described in Section K "Rear Suspension"). If the halfshaft has been renewed, it will be necessary to refer to Section K "Rear Suspension" for checking the wheel camber.

THE UNIVERSAL JOINTS
Examine and Check for Wear

The part most likely to show wear after long usage are the bearing races and spider journals. Should looseness in the fit of these parts, load markings or distortion be observed, they should be renewed as a unit, as worn needle bearings used with a new spider journal or new needle bearings with a worn spider journal will wear more rapidly, making another replacement necessary in a short time.

It is essential that the bearing races are a light drive fit in the yoke trunnion.

In the rare event of wear having taken place in the yoke cross holes, the holes will have become oval and the yokes must be removed.

In the case of wear of the cross holes in a fixed yoke, which is part of the shaft, only in a case of emergency should these be replaced. They should normally be replaced by a complete assembly.

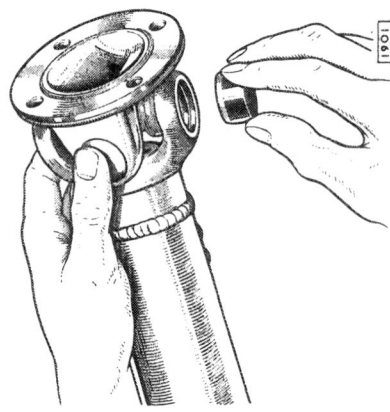

Fig. 16. Withdrawing the halfshaft.

Dismantling

Clean the paint and dirt from the rings and top of bearing races. Remove all the snap rings by pinching together with a pair of pliers and prising out with a screwdriver. If a ring does not snap out of its groove readily lightly tap the end of the bearing race to relieve the pressure against the ring.

Hold the joint in the hand and with a soft nosed hammer tap the yoke lug as shown in Fig. 17.

The bearing will gradually emerge and can finally be removed with the fingers (see Fig. 19).

If necessary tap the bearing race from inside with a small diameter bar taking care not to damage the bearing race (see Fig. 20).

Repeat the operation for the opposite bearing. The flange yoke can now be removed. Rest the two exposed trunnions on wood or lead blocks, then tap the yoke with a soft nosed hammer to remove the two remaining bearing races. Wash all parts in petrol.

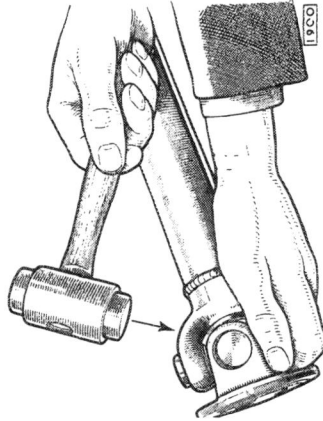

Fig. 17. Tapping the yoke to remove the bearing.

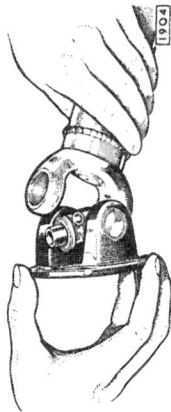

Fig. 18. Separating the universal joint yokes.

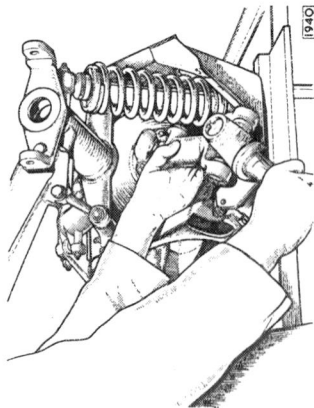

Fig. 19. Withdrawing the bearing from the universal joint.

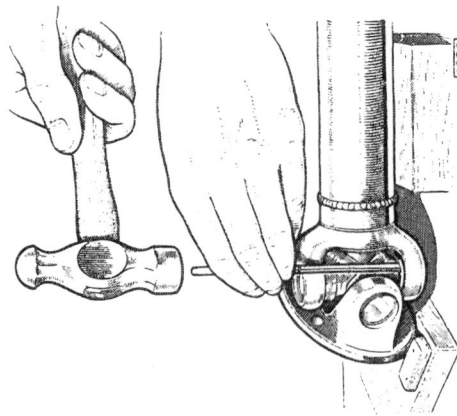

Fig. 20. Tapping out a bearing with a small diameter bar.

Assembling

Insert the journal in the yoke holes and using a soft round drift with a flat face $\frac{1}{32}$" (8 mm.) smaller in diameter than the hole in the yoke, tap the bearings into position. Repeat this operation for the other three bearings. Fit new snap rings and ensure that they are correctly located in their grooves. If the joint appears to bind, tap lightly with a wooden mallet to relieve any pressure of the bearings on the end of the journal.

Should any difficulty be encountered when assembling the needle rollers in the housing, smear the wall of the race with vaseline. It is advisable to install new cork gaskets and gasket retainers on the spider assembly using a tubular drift.

THE DIFFERENTIAL UNIT

The Thornton "Powr-Lok" limited slip differential is fitted as standard on 3·8 litre models and is optional on 3·4 litre cars.

Warning

When a car is equipped with a Thornton "Powr-Lok" differential the engine must NOT be run with the car in gear and one wheel off the ground otherwise, owing to the action of the differential, the car may drive itself off the jack or stand.

If it is desired to turn the transmission by running the engine with the car in gear **both** wheels must be jacked up clear of the ground.

THE POWR-LOK DIFFERENTIAL—DESCRIPTION

The limited slip differential has two pinion shafts with two mates to each other. The pinion shafts are mounted at right angles to each other but do not make contact at their intersection. Double ramps with flat surfaces at each end of the pinion shafts, mate with similar ramps in the differential case. Clearance in the differential case permits slight peripheral movement at the ends of the pinion shafts.

When a driving force is applied to the differential case, the pinion shafts, pinion mates and differential side gears splined to the axle shafts, rotate as a unit. Resistance to turning at the wheel forces the pinion shafts to slide up the differential case ramps, pushing the pinion shafts and side gears apart they apply load to the clutch plates thus restricting turning between the axle shafts and the differential case. Both axle shafts have now become clutched to the differential case to a varying degree dependent upon the amount of torque transmitted. This in effect locks the axle shafts to the differential case in the normal straight ahead driving position, which reduces spinning of either rear wheel should it leave the road or encounter poor traction such as ice, snow, sand, loose gravel or oil patches.

Due to the lateral movement of the pinion shafts in the differential case, a little more back-lash may be apparent in a limited slip rear axle. Slight chatter may also occur when one wheel is on a slippery surface, this is due to surge torque

PRINCIPLE OF OPERATION

The conventional differential divides the load equally between both driving wheels. In this connection, it should be remembered that the conventional differential will always drive the wheel which is easiest to turn. This is a definite disadvantage under adverse conditions of driving where the traction of one wheel is limited.

The main purpose of the limited slip differential is to overcome this limit-action. Many times the torque of the slipping wheel is provided to the driving wheel, thus permitting improved operation under all conditions of driving. The torque is transmitted from the differential case to the cross pins and differential pinions to the side gears in the same manner as torque is applied in the conventional differential.

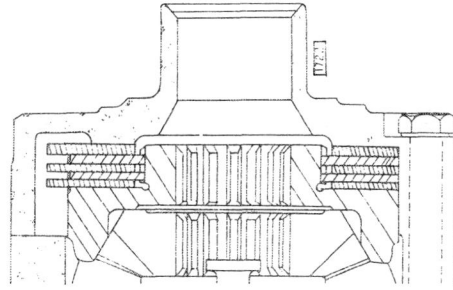

Fig. 21. *Sectional view showing the friction discs and plates.*

---

clutches E. This provides a torque ratio between the axle shafts which is based on the amount of friction in the differential and the amount of load that is being applied to the differential.

When turning a corner, this process is in effect partially reversed. The differential gears become a planetary set, with the gear on the inside of the curve becoming the fixed gear of the planetary. The outer gear of the planetary over-runs as the outside wheel on the curve has a further distance to travel. With the outer gear over-running and the inner gear fixed, the pinion mates A, Fig. 23, are caused to rotate, but inasmuch as they are restricted by the fixed gear, they first must move pinion mate shafts B back down the cam surface C relieving the thrust loads on the plate clutches E. Thus when turning the corner, the differential, for all practical purposes, is similar to a conventional differential and the wheels are free to rotate at different speeds.

On straight driving, the clutches are engaged and thus prevent momentary spinning of the wheels when leaving the road or when encountering poor traction. In turning a corner, the load is relieved from the clutch surface so that wear is reduced to a minimum.

Power Flow in Forward Driving

Under normal starting and operating conditions the torque or power flow in both the limited slip and conventional type differential is transmitted equally to each axle shaft and wheel. However, when sudden patches of ice, loose gravel or oil are encountered, the limited slip differential will not permit the wheel with the lesser traction to spin, gain momentum and swerve the car when a dry surface is regained.

Power Flow in Turns

In turning, the limited slip differential gives normal differential action and permits the outer wheel to turn faster than the inner wheel. At the same time the differential applies the major driving force to the inside rear wheel, improving stability and cornering.

Power Flow with Poor Traction

When traction conditions under the rear wheels are dissimilar, the driving force with an ordinary differential is limited by the wheel with the poorer traction. Typically, in this situation, the wheel with the poorer traction spins and the vehicle remains immobile. The limited slip differential enables the wheel with the better traction to apply the major driving force to the road.

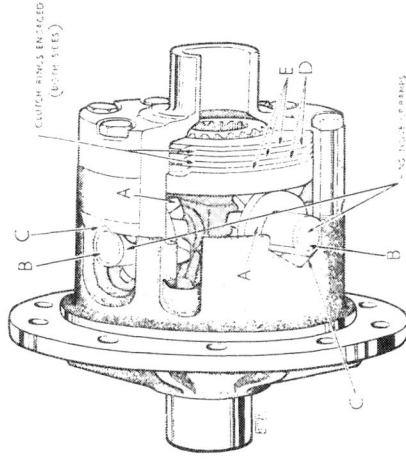

Fig. 22. *Operation of limited slip differential (straight driving).*

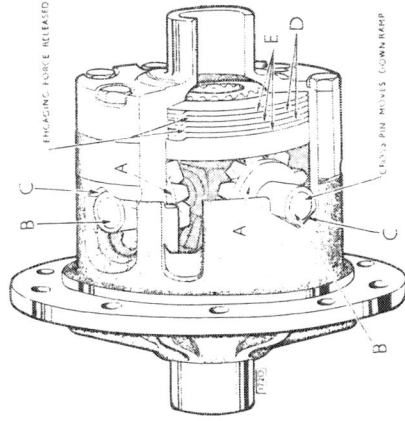

Fig. 23. *Operation of limited slip differential (cornering).*

The driving forces move the cross pins B, Fig. 22, up the ramp of the cam surfaces C, applying a load to the clutch rings D and restricts turning of the differential through the friction

Action on Rough Roads

Bumps do not adversely affect wheel action when wheels are controlled by the limited slip differential. The free wheel does not spin and gain momentum. There is no sudden wheel stoppage to cause car swerve or tyre scuffing and wheel hop is reduced.

THE OUTPUT SHAFTS

Removal

Remove the brake caliper and disc as described under "Removing the Differential Assembly from the Carrier."

Unscrew and remove the five bolts securing the output shaft bearing housings, bearings and adjustment shims, noting the number of pre-load shims.

Dismantling

Unlock the tab washer and remove the nut, tab washer and plain washer. Press the output shaft with the inner bearing inner race, spacing sleeve and end-float shims in position through the flange and bearing housing. If it is necessary to replace the bearings, remove the end-float shims and spacing collar, and using a suitable extractor withdraw the inner bearing inner race from the shaft. Drift out the inner bearing outer race and using a suitable sized tube on the outer race, press out the complete outer bearing and the oil seal. If it is necessary to reset the output shaft end-float withdraw the oil seal and the outer bearing inner race.

Assembling

Press in the new inner and outer bearing outer races ensuring that they are fully home in the recesses. The races must be fitted so that the bearings will be opposed. Press the inner bearing inner race on to the shaft ensuring that it is fully home against the shoulder and that the race is fitted the correct way round. Fit the spacing sleeve and the end-float shims. Fit the output shaft into the bearing housing and place the outer bearing inner race on the shaft from the opposite end. Do not fit the oil seal at this stage. Fit the output shaft flange with the plain washer and a new tab washer, fit the nut and tighten.

Check the end-float with a dial gauge, this should be ·001″—·003″ (·025—·076 mm.). Should adjustment be necessary remove the flange nut, tab and plain washers and withdraw the flange and outer bearing inner race. Add or remove shims to obtain the correct clearance. Adding shims increases the end-float and removing shims decreases it. When the correct end-float is obtained replace the outer bearing inner race and press a new oil seal into position, flush with the casing and with the lip inwards. Refit the flange and the plain tab washers ensuring that the two tags on the tab washer locate in the holes on the flange. Tighten the nut and turn one or more tabs up securing the nut. Ensure that these tabs lie as flat on the nut as possible.

Refitting

See "Drive Gear Mesh Adjustment and Differential Bearing Pre-load Setting", page H.25.

REMOVING THE DIFFERENTIAL ASSEMBLY FROM THE CARRIER

Remove the axle as described on page H.9.

Knock up the locking tabs and unscrew the brake caliper mounting bolts.

Remove the caliper noting the number of small round shims between the caliper and the shims and differential carrier. Remove the brake disc.

Drain the lubricant from the gear carrier and remove the gear carrier rear cover. Flush out the unit thoroughly so that the parts can be carefully inspected.

Unscrew the five bolts securing the output shaft bearing housing. Withdraw the output shaft, bearing housing, bearings and adjustment shims noting the number of pre-load shims.

Repeat for the other drive shaft. Remove the two bolts holding each differential bearing cap and withdraw the differential unit.

Pinion Removal

Remove the pinion nut and washer. Withdraw the universal joint companion flange with a suitable puller. PRESS the pinion out of the outer bearing. It is important that the pinion should be pressed out, not driven out, to prevent damage to the outer bearing. The pinion having been pressed from its outer bearing may now be removed from the differential casing.

Note : Keep all shims intact.

Remove the pinion oil seal together with the oil slinger and outer bearing cone. Examine the outer bearing for wear and if replacement is required extract the bearing outer race using Tool No. SL.550 and SL.550-5 shown in Fig. 25. If the correct tool is not available and the bearing cup is to be replaced it is possible to drive out the cup, the shoulder locating the bearing being recessed to facilitate this operation. Remove the pinion inner bearing outer race as shown in Fig. 25 if the bearing requires replacement or adjustment of the pinion setting is to be undertaken. Take care of the shims fitted between the bearing cup and the housing abutment face. If the inner bearing is to be replaced it may be driven out but the correct service tool should be used when the bearing is removed in order to carry out pinion setting adjustment.

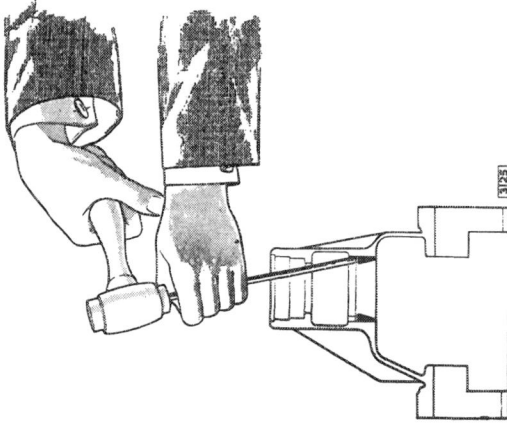
Fig. 25. Removing the pinion bearing outer race.

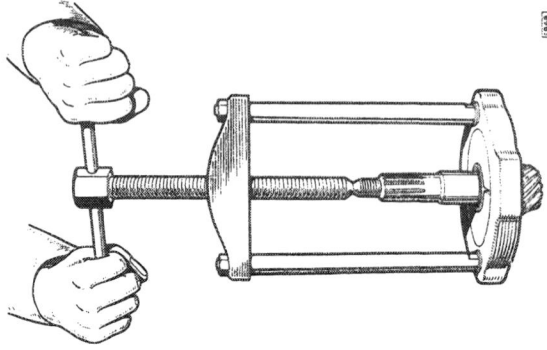
Fig. 26. Withdraw the pinion inner bearing race using Churchill Tool No SL.14 with adaptor SL.14-2.

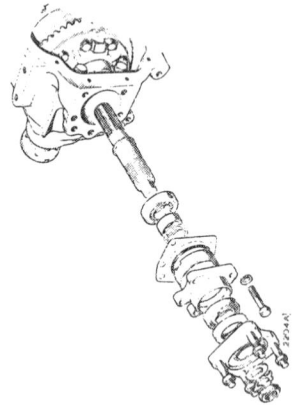
Fig. 24. Exploded view of output shaft assembly.

DISMANTLING THE DIFFERENTIAL UNIT (POWR-LOK TYPE)

Knock back the locking tabs from the drive gear securing setscrews. Remove the securing setscrews and tap the drive gear from the differential case with a rawhide mallet.

In the absence of any mating or aligning marks as shown in Fig. 30, scribe a line across the two half casings to facilitate assembly.

Remove the eight bolts (9, Fig. 31) securing the two halves of the differential casing.

Split the casing and remove the clutch discs (3) and plates (2 and 4) from one side.

Remove the differential side gear ring (5).

Remove the pinion side gear (6) and the pinion mate cross shafts (7) complete with the pinion mate gears.

Separate the cross shafts (10).

Remove the remaining side gear and the side gear ring.

Extract the remaining clutch discs and plates.

ASSEMBLING THE DIFFERENTIAL UNIT (POWR-LOK TYPE)

Refit the clutch plates and discs alternately into the flange of the casing.

Fit the two "Belleville" clutch plates (i.e. curved plates) so that the convex side is against the differential casing (see Fig. 31).

Fit the side gear ring so that the serrations on the gear mesh with the serrations in the two clutch discs.

Place one of the side gears into the recess of the side gear ring so that the splines in both align.

Fit the cross shafts together.

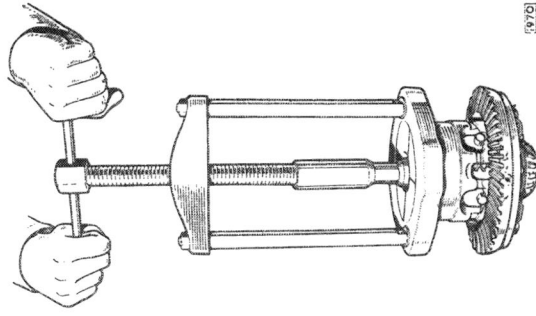

Fig. 27. Withdrawing a differential bearing using Churchill Tool No. SL.14 with adaptor SL.14-5.

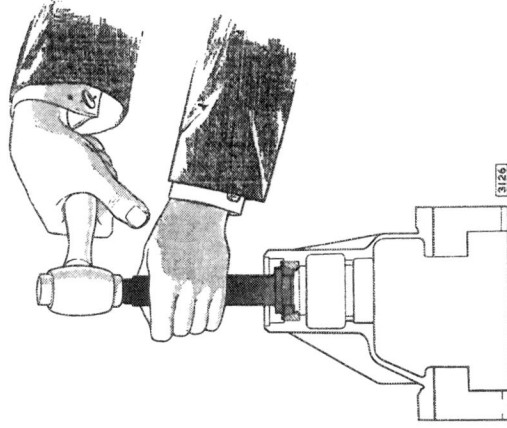

Fig. 28. Tightening the differential casing bolts with the output shaft in position.

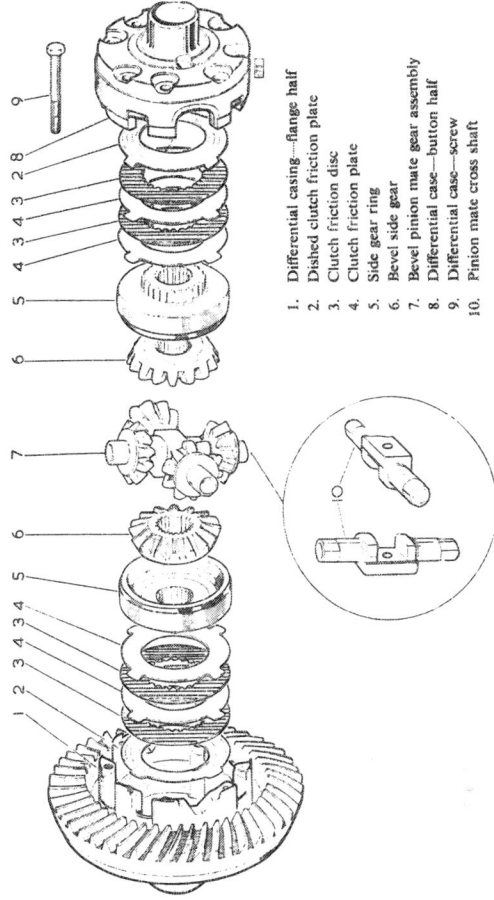

Fig. 29. Replacing a pinion outer bearing outer race using Churchill Tool No. 550 with adaptor SL.550-5.

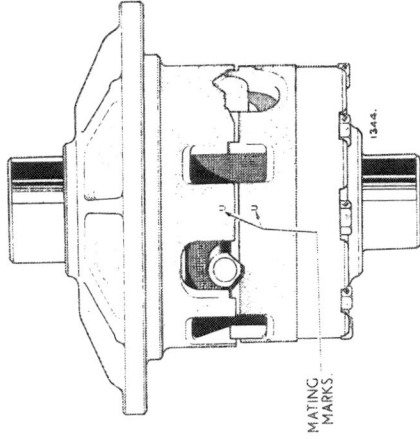

MATING MARKS.

Fig. 30. Alignment marks on the differential casing.

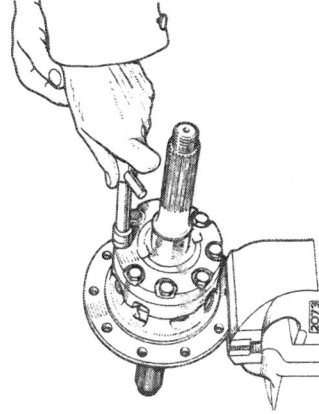

1. Differential casing—flange half
2. Dished clutch friction plate
3. Clutch friction disc
4. Clutch friction plate
5. Side gear ring
6. Bevel side gear
7. Bevel pinion mate gear assembly
8. Differential case—button half
9. Differential case—screw
10. Pinion mate cross shaft

Fig. 31. Exploded view of the Thornton "Powr-Lok" differential.

A. Pinion Drop .. 1·5" (38·1 mm.)
B. Zero Cone Setting 2·625" (66·67 mm.)
C. Mounting Distance 4·312" (108·52 mm.)
D. Centre Line to
Bearing Housing 5·495" (139·57 mm.)
to
5·505" (139·83 mm.)

Refit the pinion mate cross shafts complete with pinion mate gears ensuring that the ramps on the shafts coincide with the mating ramps in the differential case.

Assemble the remaining side gear and side gear ring so the splines in both align.

Refit the remaining clutch plates and discs to the side gear ring.

Offer up the button half of the differential case to the flange half in accordance with the identification marks and position the tongues of the clutch friction plates so they align with the grooves in the differential case. Assemble the button half to the flange half of the differential case with eight bolts but do not tighten at this juncture.

Check the alignment of the splines in the side gear rings and side gears by inserting two output shafts, then tighten the eight bolts to a torque of 40—45 lb. ft. (5·5 to 6·2 kg. m.) while the output shafts are in position. Failure to observe this instruction, particularly with the differential unit having the dished clutch friction plates, will render it difficult or impossible to enter the output shafts after the eight bolts have been tightened.

Refit the drive gear to the differential case, having first ensured that the locating faces are not damaged, by aligning the bolt holes on the gear and case and tapping the gear into position with a rawhide mallet. Fit the securing setscrews using NEW locking straps and tighten to a torque of 70 to 80 lb. ft. (9·7 to 11·1 kg. m.) knock up the tabs around the heads of the setscrews.

Checking for Wear

With one output shaft locked, the other output shaft must not turn radially more than ¾" (19 mm.) measured on a 6" (152 mm.) radius.

PINION ADJUSTMENT

Refit the pinion outer bearing outer race using Tool No. SL.550 with the adaptor SL.550 —4. Refit the pinion inner bearing outer race (as shown in Fig. 29) with the original shims in position between the outer race and its abutment shoulder.

Press the inner bearing inner race onto the pinion using a hand-press and a length of tube. Ensure that the tube contacts only the inner portion of the race and not the roller retainer. Place the pinion into position, turn the gear carrier over and support the pinion with a suitable block of wood. Fit the original outer bearing shims to the pinion shank so that they seat on the shoulder of the shank.

Fit the outer bearing inner race, companion flange, washer and nut only, omitting the oil slinger and oil seal assembly and tighten the nut. It will now be necessary to check the pinion cone setting as follows :—

Pinion Cone Setting

The correct pinion cone setting is marked on the ground end of the pinion as shown on the inset in Fig. 34. The serial number of the matched drive gear and pinion assembly is marked above the cone setting, it is most important that similarly marked drive gears and pinions are kept in their matched sets as each pair is lapped together at the factory. The letters on the left and right of the pinion should be disregarded.

Hold the gear carrier so that the ground end of the pinion is uppermost. Take the pinion cone setting gauge (Tool No. SL.3) and remove the magnetic keeper from the gauge post. Using the setting block on a surface plate as shown in Fig. 34, set the dial test gauge to zero on the 4

PRESSURE

Fig. 32. Replacing the differential bearing using Churchill Tool No. SL.550 with adaptor SL.550-1.

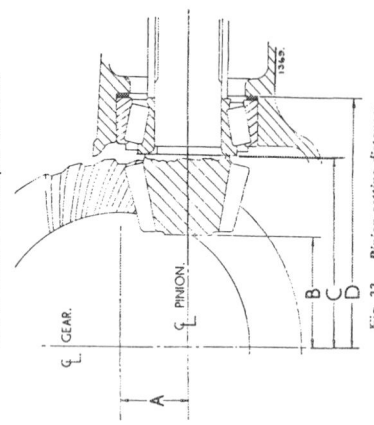

Fig. 33. Pinion setting distances.

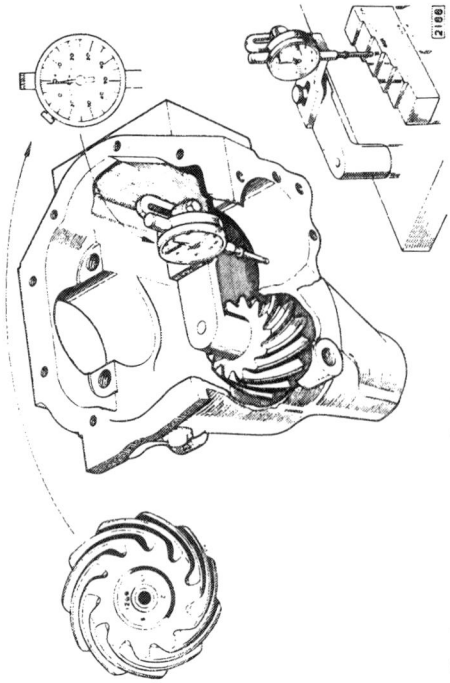

HA Setting. Place the dial gauge post on the end of the pinion as shown in Fig. 34, so that the plunger of the dial gauge registers in the differential bearing bore. Check the pinion cone setting by moving the gauge plunger in the differential bore; the actual reading being the minimum obtained. If the cone setting is correct, the reading on the dial gauge will be the same as the figure marked on the pinion end. For example, if the setting marked on the pinion is –2 then the reading on the dial gauge must also be –2.

If the pinion setting is incorrect, it will be necessary to remove the pinion assembly (as described on page H.19) and remove the pinion inner bearing outer race. Withdraw the shim pack and add or remove shims as necessary. Adding shims to the pack will decrease the gauge reading, that is, increase the number on the gauge if negative (—) and decrease the number if positive (+); removing shims will increase the reading: shims are available in ·003", ·005" and ·010" (·076, ·127 and ·254 mm.) thicknesses.

Example, assume the required pinion cone setting distance (marked on the pinion end) to be –2, if on checking with the dial gauge, the reading is –7 it will be necessary to remove a ·005" (·127 mm.) thick shim in order to reduce the gauge reading to –2.

Replace the inner bearing outer race, fit the pinion and check the cone setting as described before.

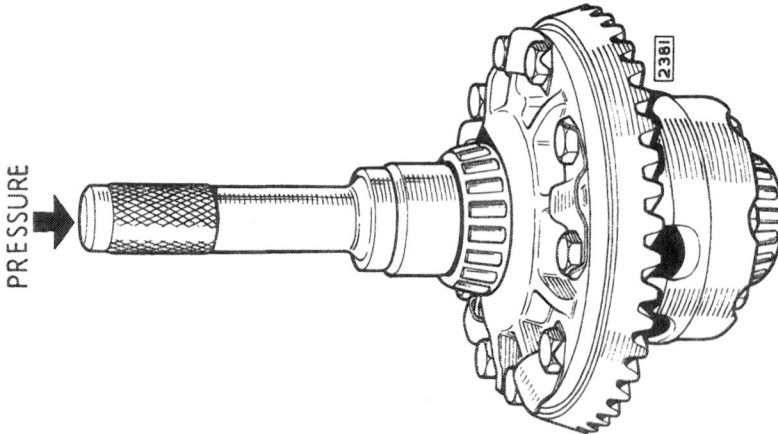

Fig. 34. Checking the pinion cone setting using Churchill Tool No. SL.3. Also illustrated are the pinion setting marks located on the end of the pinion.

Drive Gear Mesh Adjustment and Differential Bearing Pre-load Setting

Install the drive shafts without any shims between the drive shaft bearing housing and the differential carrier. Note the condition of the "O" ring on the bearing housing and renew if necessary. Fit three bolts evenly spaced around each bearing housing. Set up a dial indicator on the differential carrier with the plunger of the gauge against one of the drive gear teeth as nearly in line with the direction of tooth travel as possible (as shown in Fig. 37). Move the drive gear by hand to check the backlash; the correct backlash will be etched on the sloping face of the drive gear. If the backlash reading is incorrect, move the drive gear towards or away from the pinion as necessary until the correct backlash reading is obtained. To move the drive gear in the required direction, it will be necessary to tighten the bolts in the drive shaft housing on one side of the differential carrier and slacken the bolts on the other side.

When the correct backlash has been obtained, measure the gap between the drive shaft bearing housing and the differential carrier on each side using a set of feeler gauges. Note the gap, having first checked around the circumference of the housing to ensure that the gap is even. Make up a shim pack to fill the gap on each side but subtract ·003" (·076 mm.) from the pack to give the correct preload on the differential bearings. The shims are available in thicknesses of ·003", ·005", ·010" and ·030" (·076, ·127, ·254 and ·762 mm.).

For example: Assume that the backlash etched on the drive gear is ·007" (·178 mm.), when this figure has been obtained as described previously, the gap on one side is ·054" (1·37 mm.) and ·046" (1·17 mm.) on the other, then the amount of shims to be fitted will be ·054"—·003", that is, ·051" (1·30 mm.) and ·046"—·003", that is, ·043" (1·09 mm.) to the other side.

Finally, fit the output shafts with the shims in position to the differential carrier, fit the five bolts to each housing and tighten up. The drive gear mesh adjustment should now be checked as described in "Tooth Contact" on page H.27.

Fig. 37. *Checking the backlash and the drive gear location.*

Fig. 35. *Differential bearing cap markings.*

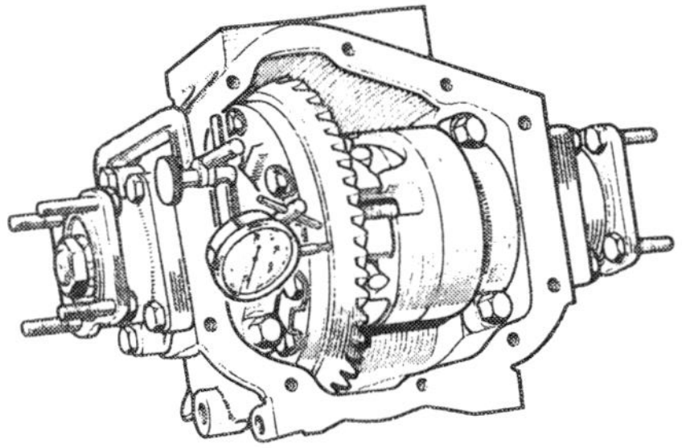

Fig. 36. *Checking the drive gear run-out.*

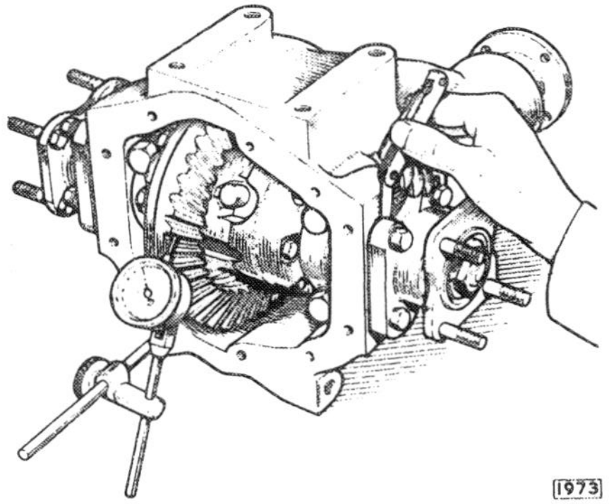

When the correct pinion setting has been obtained, check the pinion bearing pre-load. There should be no end play in the pinion and a slight resistance to turning should be felt. The correct pinion bearing pre-load is given as a torque figure under "Data" on page H.4. Inadequate pre-load will result in pinion deflection under load whilst excessive pre-load will lead to pitting and failure of the bearings.

To adjust the pre-load, add or remove shims at the shim pack between the outer bearing inner race and the shoulder on the pinion shank. Removing shims will increase the pre-load and adding shims will decrease the pre-load; shims are available in thicknesses of ·003", ·005", ·010" and ·030" (·076, ·127, ·254 and ·762 mm.). It is most important that the shims behind the inner bearing outer race which control the pinion cone setting are not disturbed when setting the pre-load.

DIFFERENTIAL BEARING PRE-LOAD AND DRIVE GEAR ADJUSTMENT

With the pinion (less the oil seal and oil slinger) installed in the differential carrier, fit the differential assembly. Fit the differential bearing caps noting that the numerals marked on the bearing caps and the end cover face correspond as shown in Fig. 35. Fit the cap bolts and tighten to a torque of 60 to 65 lb. ft. (8·3 to 9·0 kg. m.).

Drive Gear Runout

Mount a dial indicator on the gear carrier with the plunger of the gauge against the back face of the drive gear as shown in Fig. 36. Turn the drive gear by hand and check the run-out on the back face which should not exceed ·005" (·13 mm.). If the run-out exceeds this figure, the differential assembly should be removed, the drive gear withdrawn from the assembly and the locating surfaces on the drive gear and differential casing cleaned and burrs removed.

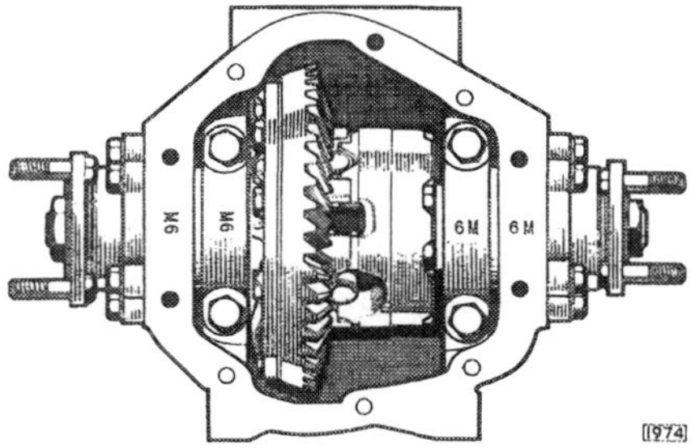

FINAL ASSEMBLY

Remove the pinion flange nut, washer and the companion flange, and fit the oil slinger. Place the oil seal gasket into position in the oil seal recess, then fit the oil seal so that the lip of the seal faces inwards and the dust excluder flange is uppermost. Fit the installation collar Tool No. SL.4 and tighten down the pinion nut and washer to drive the assembly home as shown in Fig. 38. Remove the installation collar, fit the companion flange, washer and pinion nut and tighten to a torque of 120 to 130 lb. ft. (16·6 to 18·0 kg. m.).

Fit the differential carrier rear cover gasket, renewing if necessary, fit the rear cover and secure with setscrews and spring washers. Do not omit to refit the "Powr-Lok" (P.L.) and axle ratio tags which are also secured by the cover setscrews for identification purposes. Check that the drain plug is tightened and fill the axle with one of the recommended lubricants specified on page H.5. Replace the filler plug, check the tightness of the cover setscrews and check the complete unit for oil leaks.

Refit the brake discs and calipers, centralising the calipers by means of the adjusting shims (as described in Section L "Brakes"). Fit new tab washers to the mounting bolts, tighten the bolts to a torque of 55 lb. ft. (7·6 kg. m.) and secure the bolt heads with the tab washers.

TOOTH CONTACT

After setting the backlash to the required figure, use a small brush to paint eight or ten of the drive gear teeth with a stiff mixture of marking raddle, used sparingly, or engineers' blue may be used if preferred. Move the painted gear teeth in mesh with the pinion until a good impression of the tooth contact is obtained. The resulting impression should be similar to Fig. A in Fig. 39.

The illustrations referred to in this section are those shown in Fig. 39 which indicates the tooth bearing impression as seen on the drive gear.

The HEEL is the large or outer end of the tooth.

The TOE is the small or inner end of the tooth.

The FACE top or addendum is the upper portion of the tooth profile.

The FLANK or dedendum is the lower portion of the tooth profile.

The DRIVE side of the drive gear tooth is CONVEX.

The COAST side of the drive gear tooth is CONCAVE.

(a) Ideal Contact

Fig. A shows the ideal tooth bearing impression on the drive and coast sides of the gear teeth. The area of contact is evenly distributed over the working depth of the tooth profile and is located nearer to the toe (small end) than the heel (large end). This type of contact permits the tooth bearing to spread towards the heel under operating conditions when allowance must be made for deflection.

(b) High Tooth Contact

In Fig. B it will be observed that the tooth contact is heavy on the drive gear face or addendum, that is, high tooth contact. To rectify this condition, move the pinion deeper into mesh, that is, reduce the pinion inner race setting distance, by adding shims between the pinion inner bearing outer race and the housing and adding the same thickness of pre-load shims between the pinion bearing spacer, or the shoulder of the pinion shank and outer bearing inner race. This correction has a tendency to move the tooth bearing towards the toe on drive and heel on coast, and it may therefore be necessary after making this change to adjust the drive gear as described in paragraphs (d) and (e).

(c) Low Tooth Contact

In Fig. C it will be observed that the tooth contact is heavy on the drive gear flank or dedendum, that is, low tooth contact. This is the opposite condition from that shown in (b) and is therefore corrected by moving the pinion out of mesh, that is, increase the pinion inner race setting distance by removing shims from between the pinion inner bearing outer race and housing, and removing the same thickness of pre-load shims from between the pinion bearing spacer or the shoulder on the pinion shank and the outer bearing inner race. The correction has a tendency to move the tooth bearing towards the heel on drive and toe on coast, and it may therefore be necessary after making this change to adjust the drive gear as described in (d) and (e).

(d) Toe Contact

Fig. D shows an example of toe contact which occurs when the bearing is concentrated at the small end of the tooth. To rectify this condition, move the drive gear out of mesh, that is, increase backlash, by transferring shims to the drive gear side of the differential from the opposite end.

(e) Heel Contact

Fig. E shows an example of heel contact which is indicated by the concentration of the bearing at the large end of the tooth. To rectify this condition move the drive gear closer into mesh, that is reduce backlash, by removing shims from the drive gear side of the differential and adding an equal thickness of shims to the opposite side.

Note : It is most important to remember when making this adjustment to correct a heel bearing that sufficient backlash for satisfactory operation must be maintained. If there is insufficient backlash the gears will at least be noisy and have a greatly reduced life, whilst scoring of the tooth profile and breakage may result. Therefore, always maintain a minimum backlash requirement of ·004″ (·10 mm.).

Backlash

When adjusting backlash always move the drive gear as adjustment of this member has more direct influence on backlash, it being necessary to move the pinion considerably to alter the backlash a small amount—·005″ (·13 mm.) movement on pinion will generally alter backlash ·001″ (·025 mm.).

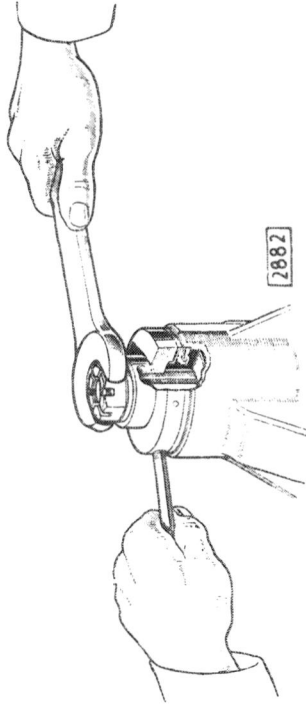

2882

Fig. 38. *Fitting the pinion oil seal using Churchill Tool No. SL.4.*

Drive Gear and Pinion Movement

Moving the drive gear out of the mesh moves the tooth contact towards the heel and raises it slightly towards the top of the tooth.

Moving the pinion out of the mesh raises the tooth contact on the face of the tooth and slightly towards the heel on drive, and towards the toe on coast.

2883	TOOTH CONTACT (DRIVE GEAR)	CONDITION	REMEDY
A	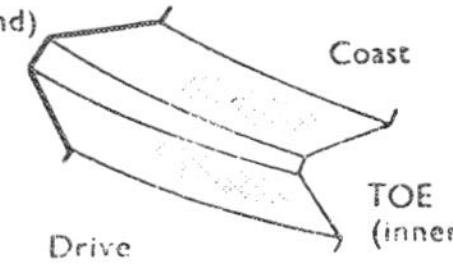	IDEAL TOOTH CONTACT Evenly spread over profile, nearer toe than heel.	o —— o
B	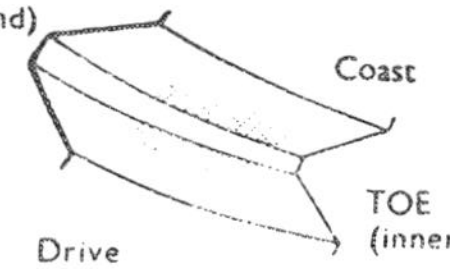	HIGH TOOTH CONTACT Heavy on the top of the drive gear tooth profile.	Move the DRIVE PINION DEEPER INTO MESH. i.e., REDUCE the pinion cone setting.
C	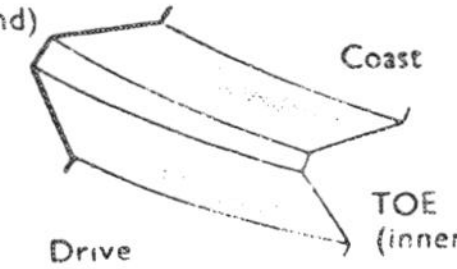	LOW TOOTH CONTACT Heavy in the root of the drive gear tooth profile.	Move the DRIVE PINION OUT OF MESH. i.e., INCREASE the pinion cone setting.
D	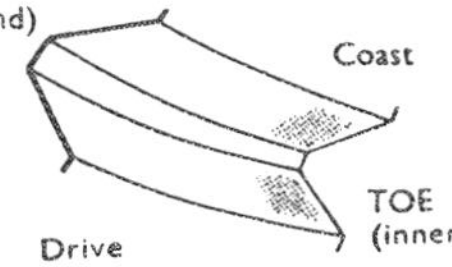	TOE CONTACT Hard on the small end of the drive gear tooth.	Move the DRIVE GEAR OUT OF MESH. i.e., INCREASE backlash.
E	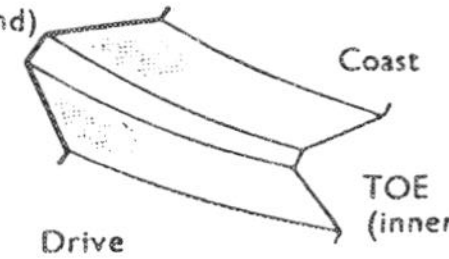	HEEL CONTACT Hard on the large end of the drive gear tooth.	Move the DRIVE GEAR INTO MESH. i.e., DECREASE backlash but maintain minimum backlash as given in "Data".

Fig. 39. Tooth contact indication (Contact markings on the drive gear)

SPECIAL TOOLS

Description		Tool No.
Pinion oil seal replacer	.. ..	SL.4
Pinion bearing cone removal/replacing adaptor	.. .. ..	SL.14–1*
Pinion bearing outer bearing cup replacing adaptor	.. ..	SL.550–4†
Pinion bearing inner bearing cup replacing adaptor	.. ..	SL.550–5†
Differential bearing cone removal adaptor	.. .. ..	SL.14–3*
Differential bearing cone replacing adaptor	.. .. ..	SL.550–1†
Pinion cone setting gauge	.. ..	SL.3
Hub end-float master spacer	.. ..	J.15
Hub end-float dial gauge	.. ..	J.13
Hub extractor (disc wheel hubs)	.. ..	J.1C
Hub extractor (wire wheel hubs)	.. ..	J.7
Hub outer bearing cup removal adaptor	.. .. ..	J.16B*
Hub bearing cup replacing adaptor	.. .. ..	J.20A–1‡

* Use with main tool SL.14—Multi-purpose hand press
† Use with main tool 550—Multi-purpose handle
‡ Use with main tool J.20A—Bearing remover

SUPPLEMENTARY INFORMATION TO

SECTION H "REAR AXLE"

CONVENTIONAL DIFFERENTIAL (FITTED TO THE 3.4 "S" TYPE)

Routine maintenance, removal, service adjustments and refitting procedures of the conventional rear axle are the same as detailed for the axle having the Thornton "Powr-Lok" differential unit. Dismantling and assembling procedures differ and are detailed below.

Dismantling

Bend down the locking tabs of the lock straps and remove the drive gear screws. Withdraw the drive gear from the differential casing by tapping with a rawhide mallet.

Using a small punch, drive out the pinion mate shaft locking pin which is secured by peening over the case, and remove the pinion mate shaft. Fig. 40 indicates the direction in which the pin must be driven out; it is not possible to drift out the pin in the opposite direction.

Rotate the side gears by hand until the pinions are opposite the openings in the differential casing; remove the differential gears taking care not to lose the thrust washers fitted behind them.

Assembling

Assemble the side gears with the thrust washers in position. Insert the differential pinions through the openings in the casing and mesh them with the side gears. Hold the pinion thrust washers on the spherical faces of the pinions whilst rotating the differential gear assembly into its operating position by hand.

Line up the pinions and thrust washers and install the pinion mate shaft in position. Line up the cross hole in the shaft with the hole in the differential case and fit the pinion mate shaft lock pin. Using a punch, peen over some metal of the differential casing to prevent the lock pin working loose and causing extensive damage to the assembly.

Clean the drive gear and differential case contacting surfaces and check for burrs. Align the drive gear attaching screw holes with those in the casing and gently tap the gear home with a hide or lead hammer. Insert the drive gear bolts; fit new locking straps and tighten to a torque of 70—80 lb.ft. (9.7—11.1 kg/m). Lock with the locking straps.

Commencing Chassis No.

	R.H. Drive	L.H. Drive
3.4 "S"		
Standard or Automatic	1B 3576	1B 25511
Overdrive Transmission	1B 3686	1B 25517
3.8 "S"		
Standard or Automatic	1B 53892	1B 77577
Overdrive Transmission	1B 53983	1B 77649

Commencing at the above chassis numbers, "S" Type cars are fitted with a modified final drive unit in which the drive shaft flanges are integral with the drive shafts.

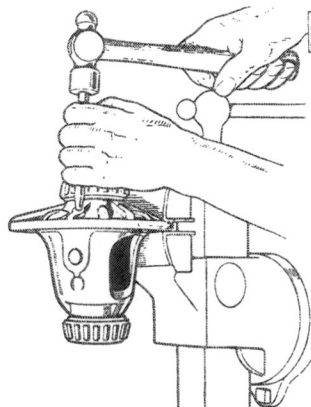

Fig. 40. *Removal of pinion mate shaft locking pin.*

Removal

Unscrew the five bolts securing each drive shaft bearing housing, remove the brake caliper adaptor plates and withdraw the drive shaft assemblies. Note the number of shims fitted between the flange of the bearing housing and the differential case.

Dismantling

Knock back the tab washer and remove the nut from the drive shaft. Press the drive shaft through the bearing housing. Collect the inner bearing, inner race, spacing collar and bearing shims. The outer bearing inner race and oil seal will remain on the drive shaft.

If the outer bearing is to be renewed, the oil seal must also be renewed by withdrawing the bearing from the shaft will damage the seal.

The bearing outer races may be driven from the housing.

Assembling

Press the bearing outer races into the housing ensuring that they are fully home in their recesses. The races must be fitted so that the bearings are opposed.

Fit the inner races in the housing with shims and the spacing collar interposed between them. Fit the drive shaft, tab washer and nut. Tighten the nut securely.

Check the end-float with a dial gauge. This should be .001"—.003" (.025—.076 mm) and is adjusted by adding shims to increase or removing shims to reduce end-float. When correct end-float has been obtained, remove the nut and tab washer; withdraw the drive shaft ensuring that the correct number of shims is retained. Withdraw the outer bearing inner race from the drive shaft and place in position in the housing. Press a new oil seal into the housing; insert the drive shaft. Fit the shims, spacing collar, inner bearing inner race, tab washer and nut. Lock the nut with the tab washer ensuring that the tabs lie as flat as possible against the nut.

Refitting

See "Differential Bearing Preload and Drive Gear Adjustment", page H.25.

SECTION I

STEERING

INDEX

STANDARD STEERING

INDEX (continued)

POWER-ASSISTED STEERING
(FIRST TYPE)

STEERING

A Burman F.3 steering unit is fitted as standard equipment. Burman power-assisted steering is specified as an optional extra and is dealt with at the end of this section.

STANDARD STEERING

DESCRIPTION

The Burman F.3 steering unit is of the high efficiency recirculating ball type in which motion is transmitted from the inner column worm to the rocker shaft by means of a nut on a continuous train of steel balls.

The worm is supported at each end by a loose ball race. Adjustment of the ball races is by means of shims under the end plates at the top and bottom of the steering box.

The rocker shaft is supported in a single bush pressed into the steering box. End float of the rocker shaft is controlled by an adjusting screw and locknut fitted to the top cover plate.

The one piece drop arm is taper splined to the rocker shaft and secured by a large spring washer and nut.

The drop arm and idle lever are connected by an adjustable track rod with a rubber/steel bonded bush at each end. Extensions of the track rod ends are attached to the inner ball joints of steering tie rods. The outer ball joints of the tie-rods are connected to steering arms which are bolted to the stub axle carriers.

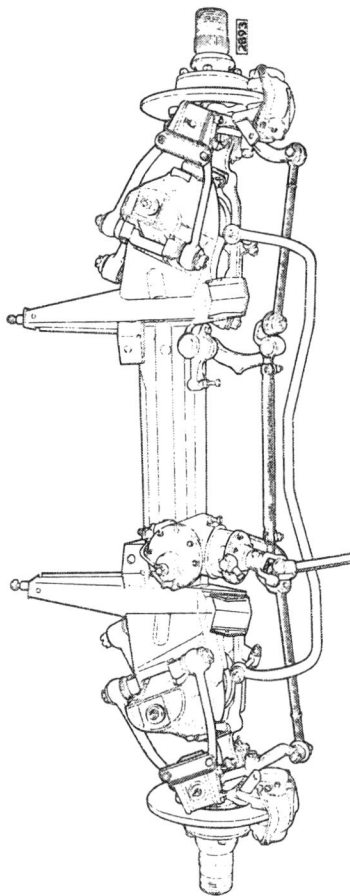

Fig. 1. *Steering layout. (Power steering illustrated.)*

INDEX (continued)

DATA

Type	Recirculating ball
Steering gear ratio at centre of travel ..	20.3 : 1
Number of turns—lock to lock	4.25
Turning circle	33' 6" (10.21 m.)
Diameter of steering wheel	17" (43 cm.)
Front wheel alignment	Parallel to $\frac{1}{8}$" (3.2 mm) total toe-in

ROUTINE MAINTENANCE

Every 6,000 miles (10,000 km.).

Steering Box

(Standard Steering)

The steering box is attached to the front suspension cross member; the filler plug is situated in the top cover and is accessible from the engine compartment on the driver's side of the car. The filler plug has a plain head and should not be confused with the rocker shaft adjustment screw which is threaded externally. Top up the steering box with the recommended grade of lubricant until no more oil will enter.

Steering Idle Lever Housing

The idler housing is pre-packed with grease which only requires replenishing if the idler assembly is dismantled for overhaul.

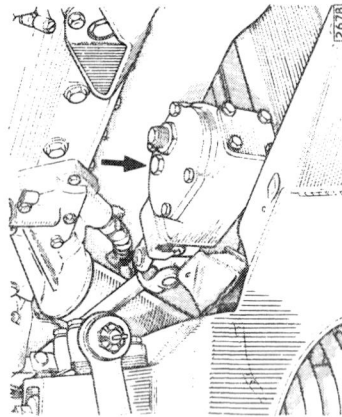

Fig. 2. *Steering box filler plug—standard steering R.H. drive.*

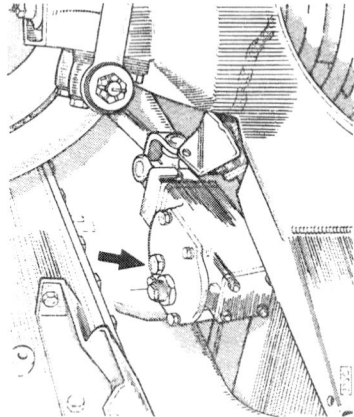

Fig. 3. *Steering box filler plug—standard steering L.H. drive.*

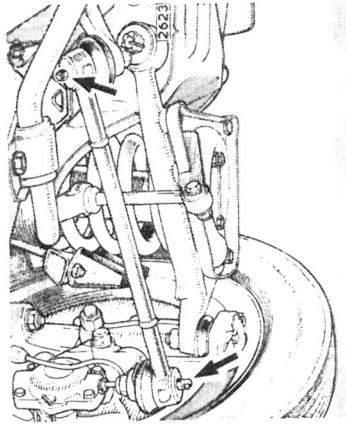

Fig. 4. *Steering tie rod grease nipples.*

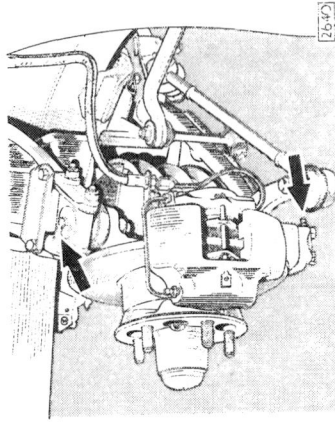

Fig. 5. *Wheel swivel grease nipples.*

This prevents grease from escaping past the seals when too much pressure is applied.

Wheel Swivels

Lubricate the nipples (four per car) fitted to the top and bottom of the wheel swivels.

The nipples are accessible from underneath the front of the car.

Front Wheel Alignment

Check that the car is full of petrol, oil and water. If not, additional weight must be added to compensate for, say, a low level of petrol (the weight of 10 gallons of petrol is approximately 80 lb.—36.0 kg.).

Ensure that the tyre pressures are correct and that the car is standing on a level surface.

With the wheels in the straight ahead position check the alignment of the front wheels with an approved track setting gauge.

The front wheel alignment should be :— Parallel to $\frac{1}{8}$" (3.2 mm.) total "toe-in".

(measured at the wheel rim).

Recheck the alignment after pushing the car forward until the wheels have turned half a revolution (180).

If adjustment is required, slacken the clamp bolt at each end of the central track rod and rotate the rod in the required direction until the alignment of the front wheels is correct. Tighten the clamp bolts and re-check the alignment.

Steering Tie-Rods

Lubricate the ball joints at the ends of the two steering tie-rods with the recommended lubricant. The tie-rods are situated at the rear of the front suspension cross-member. When carrying out this operation examine the rubber seals at the ends of the ball housings to see if they have become displaced or split. In this event they should be repositioned or replaced as any dirt or water that enters the ball joint will cause premature wear.

A bleed hole covered by a circular nylon washer which lifts under pressure indicates when sufficient lubricant has been applied.

Recommended Lubricants

	Mobil	Castrol	Shell	Esso Esso	B.P.	Duckham	Regent Caltex/Texaco
Steering box	Mobilube GX 140	Castrol D	Shell 140 EP	Gear Oil GP 90/140	Gear Oil SAE 140 EP	Nol EP 140	Multigear EP 140
Steering idler housing Steering tie rods	Mobilgrease MP	Castrolease LM	Retinax A	Multi purpose Grease H	Energrease L.2	LB 10	Marfak All purpose

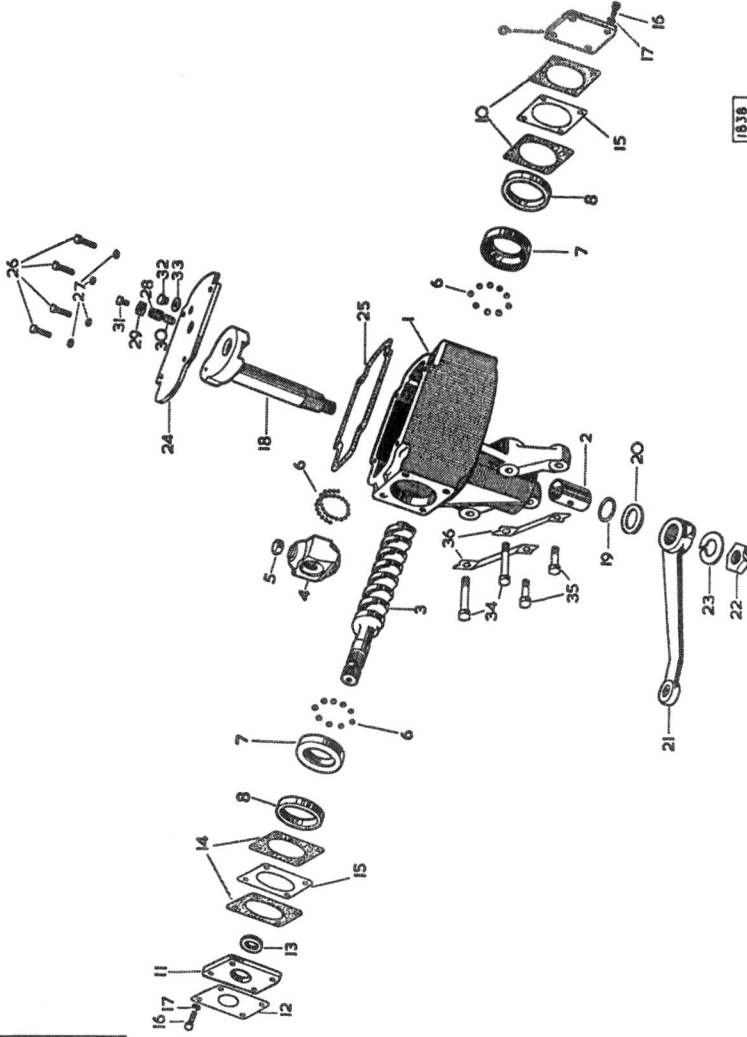

Fig. 6. *Exploded view of the steering unit.*

1.	Steering box	13.	Oil seal	25. Gasket
2.	Trunnion bush	14.	Gaskets	26. Setscrews
3.	Inner column worm	15.	Shims	27. Spring washer
4.	Main nut	16.	Setscrew	28. Rocker shaft adjustment screw
5.	Roller	17.	Spring washer	29. Locking nut
6.	Steel balls	18.	Rocker shaft	30. Spring
7.	Ball race	19.	'O' ring	31. Spring tension bolt
8.	Distance piece	20.	Washer	32. Oil filler plug
9.	End plate (bottom)	21.	Drop arm	33. Washer
10.	Gasket	22.	Nut	34. Bolt (long)
11.	End plate (top)	23.	Spring washer	35. Bolt (short)
12.	Oil seal retainer plate	24.	Cover plate	36. Tab washer

STEERING UNIT

Removal

Turn the road wheels to the straight ahead position and note the position of the slot in the lower column top joint.

Remove the spring clip and retaining ring situated at the lower end of the outer tube inside the car and withdraw the plastic thrust bearing.

Remove the pinch bolt securing the lower column top universal joint to the upper steering column.

Withdraw the upper column by pulling on the steering wheel until the splines clear the socket.

Remove the pinch bolt and detach the lower steering column from the steering unit.

Remove the self-locking nut and washer which secures the track rod end to the drop arm. Drift out the track rod end from the drop arm in which it is a taper fit.

Tap back the tab washers and remove the four bolts attaching the steering unit to the front suspension cross member when the unit can be removed.

Dismantling (Fig. 6)

Remove the four set screws and spring washers securing the rocker shaft cover plate (24). to the steering box. Remove the cover plate and gasket taking care not to dislodge the rocker shaft adjustment screw (30) from the rocker shaft adjustment screw (28). Drain the oil into a suitable receptacle. Remove the roller (5) from the top of the main nut (4).

Remove the nut (22) securing the drop arm (21) to the rocker shaft (18). Observe the line scribed on the drop arm and rocker shaft to ensure correct assembly. Using a suitable extractor, draw the drop arm off the spline on the rocker shaft. (Under no circumstances must the drop arm be hammered off otherwise

indentation will be caused to the ball tracks). Withdraw the rocker shaft. Remove the 'O' ring (19) from the bottom of the box.

Remove the four set screws and spring washers securing the upper end plate and stone guard to the steering box. Remove the retainer plate (12), end plate (11), gasket (14), shims (15), the other gasket and the distance piece (8).

Push the worm shaft upwards and withdraw the outer race of the upper bearing. Collect the ten balls. Unscrew the worm through the worm nut and withdraw from box.

Remove the four set screws and washers attaching the end plate to the bottom of the steering box. Remove the gaskets, shims and distance piece. Withdraw the outer race of the lower bearing and collect the ten balls.

Remove the two set screws and tab washers retaining the transfer tube to the main nut and remove the clip, tube and thirty-one balls (20 balls, high ratio box).

Assembling

Note: When assembling the steering unit carry out adjustment of the worm shaft and rocker shaft end float as described in this section.

Fit the transfer tube and clip to the worm nut and secure with the two setscrews.

Fit the recirculating balls into the unit; use grease to retain the balls in position. Fit the ten ball bearings to the bottom race with grease and assemble to the bottom of the steering box together with the distance piece. Fit the gaskets, shims, distance piece and end plate to the bottom of the steering box and secure with four set screws. Screw the worm shaft into the worm nut until the nut is half way along the worm. Feed the worm shaft carefully into the steering box through the top cover aperture, making sure that the balls in the bottom ball race are all securely in position when the wormshaft makes contact with them.

Fit the ten balls to the top race with grease and assemble to the top of the steering box together with the distance piece.

Fit the shims with a gasket at each side to the top of the steering box. Cover the serrations at the top of the wormshaft with a piece of brown paper or adhesive tape to protect the oil seal when sliding the end plate over the worm shaft.

Assemble the end plate, oil seal and oil seal retainer plate and carefully slide over the wormshaft.

Remove all traces of adhesive tape from the splines. Refit the stone guard and secure the end plate with four setscrews.

Adjust wormshaft end-float as follows :—

The wormshaft bearings should be adjusted to a pre-load of .002" to .003" (.05 — .08 mm) by means of the shims and gaskets at each end of the steering box. The shims are .005" (.13 mm) thick; the gaskets .003" (.08 mm) thick.

Eliminate, or reduce to a minimum, the end-float of the wormshaft by removing shims as necessary. Check that the wormshaft turns freely.

Remove a shim or gasket to obtain the required pre-load. Always maintain a minimum, of two gaskets at each end of the steering box, one at each side of the shim pack.

Adjust rocker shaft end-float as follows :—

When adjusting the end-float, the rocker shaft must be at the centre of its travel.

Unscrew the bolt (31) and extract the spring (30). Slacken the lock nut (29) securing the adjuster screw (28) in the cover plate. Screw down the adjuster screw by hand until it contacts the rocker shaft so that all end-float is eliminated.

Enter the rocker shaft into its bore in the steering box and engage the slotted extension with the top portion of the main nut. Fit the gasket and secure the cover plate with the four setscrews.

Hold the adjuster screw firmly and tighten the locknut. Test the freedom of movement of the rocker shaft throughout its travel; if tightness exists in the centre, it will be necessary to re-adjust the end-float.

Refit the spring and retaining bolt.

This operation may also be carried out with the unit on the car and the front wheels in the straight ahead position.

Refitting

Ensure that the road wheels are still in the straight ahead position.

Fit the lower steering column to the steering unit and fit this assembly to the front suspension cross member.

Attach the drop arm to the rocker shaft ensuring that the scribed line on the rocker shaft matches the appropriate line (right hand drive or left hand drive) on the drop arm. Attach the track rod to the drop arm.

Set the steering wheel so that the spokes are in the horizontal position and the trafficator switch cancels evenly on each side.

Connect the lower steering column top joint to the upper steering column (having noted the position of the slot on removal).

Depress the pot joint fully and then raise ¼" (6.3 mm). Secure with the pinch bolt.

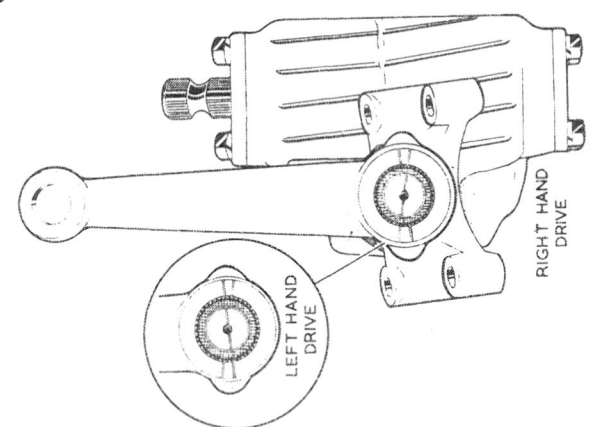

Fig. 7. *Showing the alignment of the drop arm/rocker shaft marks both right-hand and left-hand steering*

LEFT HAND DRIVE — RIGHT HAND DRIVE

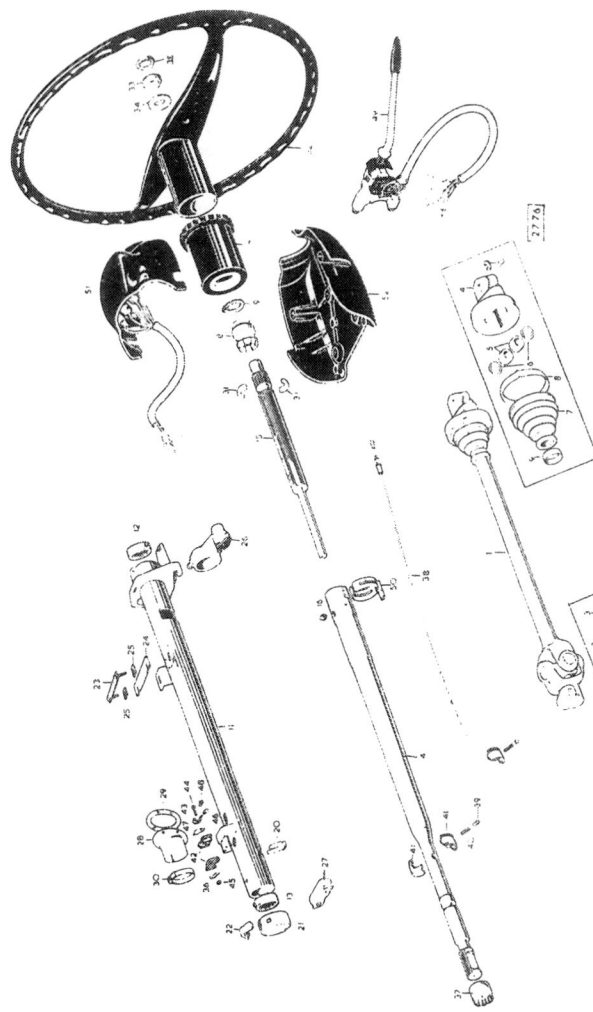

Fig. 8. *Exploded view of steering column.*

1. Lower steering column
2. End yoke
3. Journal assembly
4. Socket
5. Nylon roller
6. Circlip
7. Gaiter
8. Clip
9. Clip
10. Distance washer
11. Outer tube
12. Plastic bearing (top)
13. Plastic bearing (bottom)
14. Inner column
15. Shaft, inner shaft
16. Stop button
17. Locknut
18. Split collet
19. Circlip
20. Thrust bearing
21. Retaining ring
22. Spring clip
23. Screw plate
24. Spacer
25. Shim
26. Lock and ignition switch
27. Stone guard
28. Bracket
29. Gasket
30. Clip
31. Split cone
32. Steering wheel (17" dia.)
33. Nut
34. Special washer
35. Locking nut
36. Earth contact
37. Slip ring
38. Horn wire
39. Contact
40. Spring
41. Rotor
42. Contact holder
43. Contact
44. Bolt
45. Nut
46. Rubber sleeve
47. Insulating strip (fibre)
48. Dowlet eyelet
49. Direction indicator/headlamp flasher switch
50. Striker
51. Upper switch cover
52. Lower switch cover

STEERING WHEEL
Removal (Fig. 8).

Withdraw four cheese-headed screws from the underside of the steering wheel centre and detach the horn switch cover. Undo the four screws securing the horn ring to the steering wheel and remove the horn ring. Unscrew the locknut (35) and the nut (33) securing the steering wheel to the inner column.

Withdraw the steering wheel and collect the two halves of the split cone (31).

Refitting

Secure the split cone (31) in position in the inner column shaft grooves, ensuring that the narrowest portion of the cone is towards the top of the column.

Slide the steering wheel onto the column shaft splines so that the two spokes are horizontal when the road wheels are pointing straight ahead.

Push the steering wheel fully home onto the split cone. Fit the plain washer, the securing nut and the locknut. Refit the horn ring and cover.

STEERING COLUMN
Removal

Disconnect the battery.

Turn the road wheels to the straight ahead position. If retained in this position this will facilitate refitting.

Disconnect the cables from the flashing indicator/headlamp flasher switch at the snap connectors.

If overdrive is fitted, disconnect the cables from the overdrive switch at the snap connectors. Remove the two screws and washers securing the switch upper cover to the lower cover below the steering wheel.

Lift off the upper cover and remove the three bulb holders from the back. Note the location of the bulbs for reference when refitting.

Disconnect the four cables from the inhibitor switch located on a bracket attached to the steering column. Note the location of the cables for reference when refitting. (Automatic transmission cars only).

On right hand drive cars, unscrew the ratchet adjustment on the gear selector adjustment rod and lift out the ball joint and the crank lever.

Unscrew the jubilee clip securing the bottom of the steering column to the mounting bracket inside the car.

On left hand drive cars, it will be necessary to remove the four bolts attaching the bracket to the bulkhead as the control rod which operates the bowden cable passes through the bracket. (Automatic transmission cars only).

Remove the pinch bolt securing the inner column to the top universal joint of the lower column. Mark the location of the inner column splines in relation to the joint for reference when refitting.

Remove the two bolts, nuts and washers securing the steering column to the upper mounting bracket. Collect any shim plates located between the column and body brackets.

Withdraw the steering column as an assembly.

Dismantling

Withdraw four cheese headed screws from the underside of the steering wheel centre and detach the horn ring cover.

Remove the three screws and spring washers securing the horn ring to the steering wheel.

Remove the locknut (35), nut (33) and plain washer (34) securing the steering wheel (32) to the shaft (15) inside the inner column (14). Withdraw the steering wheel and extract the split cone (31).

Remove the steering wheel locking nut (17). Remove the two screws and plain washers securing the lower switch cover (52) to the outer tube (11).

Remove the spring clip (22) and retaining ring (21) situated at the lower end of the outer tube (11). Withdraw the plastic bearing (13). Withdraw the inner column (14).

Remove the two screws, serrated and plain washers securing the flashing indicators striker plate (50) in position.

Remove the circlip from the end of the horn wire contact nipple and remove the washer, spring and rubber.

Unscrew the stop button (16), retaining the inner shaft (15) in position.

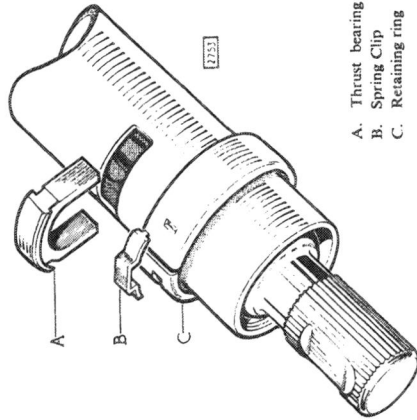

Withdraw the inner shaft.

Slide off the horn pick-up ring by lifting the serrations slightly and remove the bottom half of the rubber rotor (41). Withdraw the horn wire (38) and remove the top half of the rubber rotor (41).

Remove the two screws and serrated washers securing the flashing indicator switch (49) to the outer tube (11).

If the car is fitted with automatic transmission unclip the circlip and slide it together with the washer along the automatic transmission control rod.

If the car is fitted with an overdrive remove the two nuts, bolts and serrated washers securing the overdrive switch to the outer tube (11).

Slacken the screw securing the upper control rod to the lower control rod.

Remove the two bolts, nuts, washers and spacers securing the selector quadrant to the outer column bracket.

Remove the bolt, plain and serrated washers securing the quadrant selector pointer to the outer column bracket.

On early cars compress one of the ends of the starter/reverse inhibitor switch control rod and withdraw the washer together with the control rod.

Remove the bolt securing the starter/reverse inhibitor switch.

On later cars, remove the bolt securing the starter/reverse inhibitor switch and withdraw the switch from the control rod.

Withdraw the upper and lower automatic transmission control rods from their respective bushes.

On left-hand drive cars it will be necessary to slide the lower cranked control rod off the outer tube complete with the column support bracket after slackening the jubilee clip before withdrawing the lower control rod from the support bracket.

Remove the nut and bolt securing the earth contact (36).

Remove the nut (45) and bolt (44) securing the two rubber contact holders (42), fibre insulating strip (47) and contact (43).

Remove the plastic bearings (12 and 13), by depressing the retaining lugs from the inner tube.

Fig. 9. Showing the components of the upper inner steering column retainer.

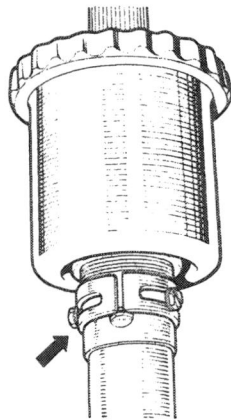

A. Thrust bearing
B. Spring Clip
C. Retaining ring

Fig. 10. Centralising the flashing indicator switch striker peg.

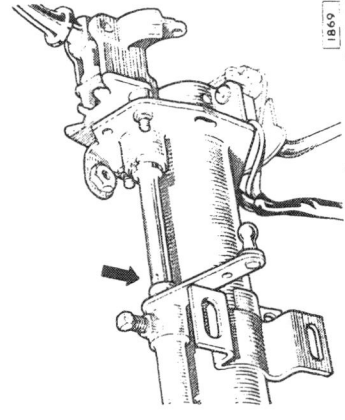

Fig. 11. Align the flat on the end of the upper control rod with the securing screw in the lower control rod.

Reassembling

Replace the two plastic bearings (12 and 13).

Replace the two rubber contact holders (42) in the bracket at the lower end of the outer tube (11).

The contact (39) which touches the horn pick-up ring should face towards the top of the outer tube.

Secure the contact (43) and fibre insulating strip (47) together with bolt (44) and nut (45).

Replace the earth contact (36).

If the car is fitted with automatic transmission, slide the lower control rod through the lower bush.

On left-hand drive cars thread the lower control rod through the hole in the steering column bottom support bracket.

Refit the support bracket, secure with jubilee clip and pass the lower control rod through the supporting bush on the side of the outer column.

Refit the selector quadrant securing it to the outer tube bracket by two nuts, bolts, spacers and washers. The spacers are to fit between the selector quadrant and the bracket.

Pass the upper control rod through the top bush on the outer column.

Slide the plain washer and circlip over the upper control rod.

Align the flat on the end of the upper control rod with the hole for the securing screw in the end of the lower control rod (Fig. 11).

Refit the reverse starter inhibitor switch to the bracket on the outer tube and secure finger tight.

Join the upper and lower control rods together, fit the plain washer and secure the upper control rod with a circlip.

Secure the upper and lower control rods with a screw.

Refit the quadrant selector pointer to the bracket on the outer tube and secure finger tight.

Refit the reverse inhibitor switch control rod through the grommet in the crank, fit the washer and open the ends of the control rod.

Refit the two screws and serrated washers securing the flashing indicator switch to the outer tube.

If the car is fitted with an overdrive secure the overdrive switch to the outer tube bracket with two bolts, serrated washers and nuts.

The top half of the rubber rotor (41), fitted to the inner column has the horn wire passing through it.

Thread the horn wire (38) through the inner column (14) and fit the top half of the rubber rotor (41) with the groove to the bottom of the column.

Fit the bottom half of the rubber rotor (41) and slide the horn slip ring (37) over both halves of the rotor with the serrations in the pick-up ring towards the bottom of the column.

Gently knock the serrations into the groove in the rubber rotors until the horn pick-up ring is secure.

Slide the inner shaft (15) over the horn wire into the inner column (14) so that the slot in the shaft serrations aligns with the stop button hole in the inner column.

Screw the stop button (16) into position until the inner shaft binds on the button.

Slacken the stop button until the inner shaft moves freely.

The striker plate (51) should be fitted with the striker peg towards the bottom of the column and on the opposite side to the stop button (16).

Turn the inner column until the striker retaining bolts are in the vertical position.

Set the striker peg so that it is just below the horizontal axis position.

On early cars refit the two screws, serrated and plain washers securing the flashing indicator striker plate in position.

On later cars, pass the two fixing screws through the switch clamp; attach the spring washer and locknut to the lower screw and the distance piece and washer to the top screw. Feed the two screws through the column bracket and secure to the indicator switch. Tighten the top screw fully.

Attach a spring balance to the steering wheel; tighten the bottom screw until the steering wheel will just turn with a pull of 5 ozs. (141.7 grammes) registered on the balance. Turn the locknut towards the switch carrier bracket and lock the screw. Two thicknesses of distance pieces are available to compensate for any variation in the bore of the outer tube.

Grade A 0.188" (4.7 mm)
Grade B 0.166" (4.2 mm)

Slide the split collet (18) onto the inner shaft (15) with the serrations towards the bottom of the column.

Slide the inner column (14) into the outer tube (11) and ensure that the earth contact (36) is not damaged.

Mount the lower switch cover (52) onto the outer tube and secure.

Refit the upper switch cover and secure.

If the car is fitted with automatic transmission move the hand control to the 'D' position in the selector quadrant and move the pointer along the elongated slot in the outer tube bracket until the pointer coincides with the 'D' on the switch cover.

Tighten the nut and bolt securing the pointer.

Slide the rubber grommet, with the flange towards the top of the column, over the horn wire together with the spring, larger diameter foremost and the washer. Retain this assembly on the horn wire by means of a circlip.

Refitting

Ensure that the front road wheels are in the straight ahead position.

Lower assembly into position ensuring that the striker plate (50) is between the two cancelling arms on the flashing indicator switch.

Engage the splines of the upper steering column with the internal splines of the lower steering column socket (4).

On left-hand drive cars fitted with automatic transmission secure the bottom support bracket on the column to the body with four bolts.

Refit the two nuts and washers securing the upper steering column top mounting bracket to the body.

Secure the splined end of the upper steering column (14) in the lower steering column yoke (4) by means of the pinch bolt.

Depress the pot joint fully and then raise ¼" (6.35 mm). Tighten the pinch bolt.

Tighten the jubilee clip (30) securing the upper steering column to the lower mounting bracket.

On right-hand drive automatic transmission cars, refit the ratchet adjustment on to the ball joint on the crank lever.

To ensure correct operation of the reverse light and starter cut-out inhibitor switch it is important that the following instructions be adhered to :-

(a) On all automatic transmission cars slacken the starter/reverse light inhibitor switch securing bolt.

(b) Move the gear selector lever until the gear indicator is in the 'D' position.

(c) Move the starter/reverse light inhibitor switch until the hole in the lever is in line with the hole in the switch base plate.

(d) Place a piece of wire through the two holes and tighten the nut securing the switch to the upper steering column (see fig. 12).

(e) Remove the wire.

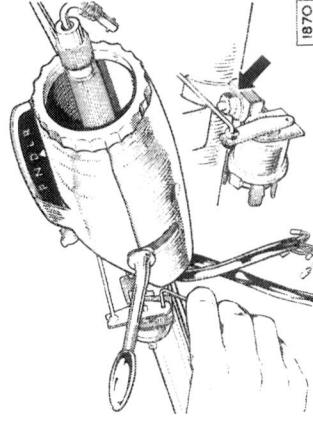

Fig. 12 Setting the starter/reverse inhibitor switch.

Fit three bulbs in the upper switch cover (51).

Fit the upper switch cover (51) and secure to the lower switch cover (52) with two screws.

If an overdrive switch is fitted connect the wires from the overdrive switch to the snap connectors.

Refit the wires from the flashing indicator/headlamp flashing switch to the snap connector.

Screw the steering wheel locking nut (17) into position on the inner column (15).

Place the split cone (31) into the groove on the inner shaft (15) with tapered end towards the top of the shaft.

Fit the steering wheel (32) onto the inner shaft with the two holes on the steering wheel centre boss towards the top.

Refit the plain washer (34) nut (33) and locknut (35).

Secure the horn ring to the steering wheel centre boss with three screws and spring washers.

Ensure that the horn wire nipple makes good contact with the horn ring pick-up.

Clip the horn ring contact cover into position on the horn ring.

LOWER STEERING COLUMN (Fig. 8).

Removal

Turn the road wheels to the straight ahead position and note the position of the slot in the lower column top joint.

Remove the pinch bolt securing the lower column top universal (pot) joint to the upper steering column.

Collect the spacing washer fitted in the slot of the universal joint.

Remove the spring clip (B Fig. 9).

Rotate and withdraw the retaining ring (C Fig. 9).

Withdraw the thrust bearing (A Fig. 9).

Withdraw the inner column (15) by means of the steering wheel, until the splines clear the universal joint.

Detach the lower universal joint from the universal joint.

Dismantling and Re-assembly

Remove the two clips (9 and 8) retaining the rubber gaiter and part the socket (4) from the lower column (1).

Examine and replace if necessary the two nylon rollers (5) retained by circlips (6).

Repack the socket with a suitable grease.

Refit the socket to the lower column and retain the gaiter (7) with two new clips.

Refitting

Refitting is the reverse of the removal procedure but the following points should be noted.

Set the road wheels in the straight ahead position.

Set the steering wheel so that the spokes are horizontal and the motif is upright.

Refit the lower column with the slot in the top universal joint in the same position as noted on removal.

Fit a new spring clip and tighten securely by pinching the raised portion with a pair of pincers.

Before tightening the top universal joint pinch bolt depress the upper half of the joint fully and raise it ⅛" (6 mm).

STEERING IDLER ASSEMBLY

Removal

Remove the self-locking nut and washer securing the track rod end to the idler lever. Drift out the track rod end from the idler lever in which it is a taper fit.

Remove the four bolts and spring washers attaching the steering idler bracket to the front suspension cross member, when the steering idler assembly can be detached.

Dismantling

Prise out the end cap from the top of the idler bracket.

Tap back the tab washer and remove the nut at the top of the idler shaft. The idler lever and the idler spindle can now be withdrawn.

Remove the washers, felt seal, abutment ring and abutment washer.

Remove the inner races of the taper roller bearing.

Assembling

Thoroughly clean out the inner races of the taper roller bearings and the idler housing. Re-pack the housing and bearings with one of the recommended greases.

Fit the lower bearing, abutment ring, seal retainer and a new felt seal.

Place the large washer over the idler spindle and pass the spindle upwards into the mounting bracket. Fit the upper bearing, 'D' washer, tab washer and nut. Tighten the nut to a torque of 5 lb. ft. (.69 kg.m.).

If a torque wrench is not available, tighten the nut until the rotation of the idler shaft by the idler arm displays a tendency to bind and then slacken back the nut one flat; lock the nut by means of the tab washer and refit the end cap.

Refitting

Ensure that the idler lever is in the straight ahead position (see Fig. 1) when refitting the track rod end to the lever.

STEERING ARM

Removal

Raise the car by placing a jack under the front suspension cross member and remove the road wheel.

Remove the self-locking nut and plain washer securing the tie rod to the steering arm. Drift out the tie rod ball pin from the steering arm in which it is a taper fit.

Unscrew the centre self-locking nut securing the stub axle shafts and steering arm to the carrier and remove the wired bolt attaching the end of the steering arm to the carrier. The steering arm can now be removed.

Refitting

Refitting is the reverse of the removal procedure. Use new locking wire to secure the steering arm attachment bolt.

TIE ROD

The tie rod ball joints cannot be dismantled and if worn a complete tie rod assembly must be fitted.

Removal

Remove the self-locking nuts and plain washers securing the tie rod to the steering arm and track rod end.

Tap the tie rod ball pins out of the steering arm and track rod end in which they are a taper fit.

Refitting

Refitting is the reverse of the removal procedure.

TRACK ROD

The track rod ends incorporate rubber/steel bonded bushes. If the bushes show signs of deterioration they should be replaced.

Removal

Remove the self-locking nuts and washers from the inner ball joint of each tie rod. Tap the ball pin out of each track rod in which they are a taper fit.

Remove the self-locking nuts and washers securing the track rod ends to the drop arm and idler lever.

Tap the track rod ends out of the drop arm and idler lever in which they are a taper fit.

Dismantling

To remove the track rod ends, slacken the clamp at each end of the centre tube; unscrew each end from the tube noting that one end has a left-hand thread and the other a right-hand thread.

Assembling

When refitting the track rod ends to the centre tube, screw in each end **an equal number of turns**. The final setting of the track rod length must be carried out after the track rod has been refitted, as described under the heading "Front Wheel Alignment".

FRONT WHEEL ALIGNMENT

Check that the car is full of petrol, oil and water. If not, additional weight must be added to compensate for, say, a low level of petrol (the weight of 10 gallons of petrol is approximately 80 lbs.—36.0 kg.).

Ensure that the tyre pressures are correct and that the car is standing on a level surface.

With the wheels in the straight ahead position check the alignment of the front wheels with an approved track setting gauge.

The front wheel alignment should be: .0" to ⅛" (3.18 mm.) total "toe-in" (measured at the wheel rim).

Re-check the alignment after pushing the car forward until the wheels have turned half a revolution (180°).

If adjustment is required slacken the clamp bolt at each end of the central track rod and rotate the rod in the required direction until the alignment of the front wheels is correct. Tighten the clamp bolts and re-check the alignment.

LOCK STOP ADJUSTMENT

The lock stop bolts are screwed into the idler bracket and are retained in position by locknuts. The stops are set at the factory to allow 38° travel of the drop arm and idler lever each side of the central (straight ahead) position.

Normally, the lock stop bolts should not require adjustment but if attention is found to be necessary the adjustment should be carried in the following manner.

Slacken the locknuts and screw in the lock stop bolts as far as possible. Turn the steering until the steering unit is at the end of its travel on that lock. Screw out the lock stop bolt until the head contacts the abutment on the idler lever. Screw out the stop bolt a further two turns and tighten the locknut. Repeat for the other lock.

ACCIDENTAL DAMAGE

The following dimensional drawings are provided to assist in assessing accidental damage. A component suspected of being damaged should be removed from the car, cleaned off, and the dimensions checked and compared with those given in the appropriate illustration.

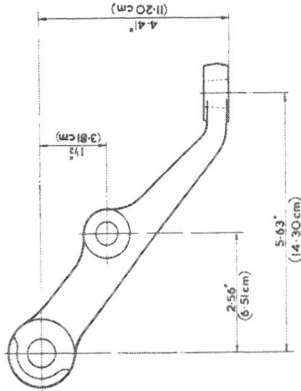

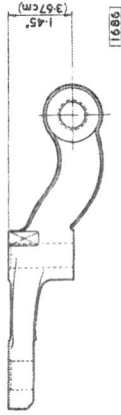

Fig. 13. *Steering arm.*

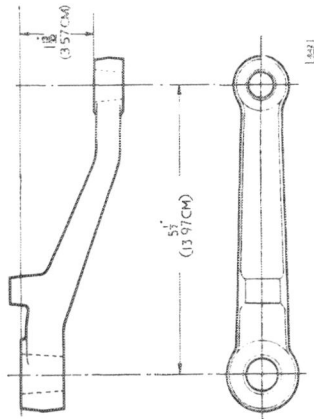

Fig. 14. *Steering idler lever.*

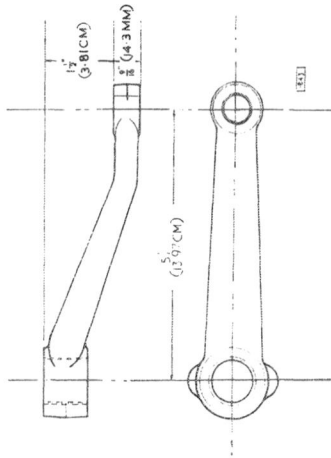

Fig. 15. *Steering drop arm.*

POWER-ASSISTED STEERING

(Fitted as an optional extra)

(FIRST TYPE)

DESCRIPTION

The power-assisted steering system consists of a Hobourn-Eaton roller type pump driven off the rear of the dynamo shaft, an oil reservoir with a replaceable filter element and an hydraulically assisted worm and re-circulating ball type steering box.

These parts are connected by flexible hoses as follows:—

Reservoir to inlet side of pump.

Outlet side of pump to inlet pipe connection attached to the steering box.

Outlet at top of steering box to reservoir.

The pump supplies a continuous flow of oil through the system while the engine is running and the steering is in the straight ahead position. Pressure is only created in the system when the steering column is rotated and is proportional to the effort applied to the steering wheel.

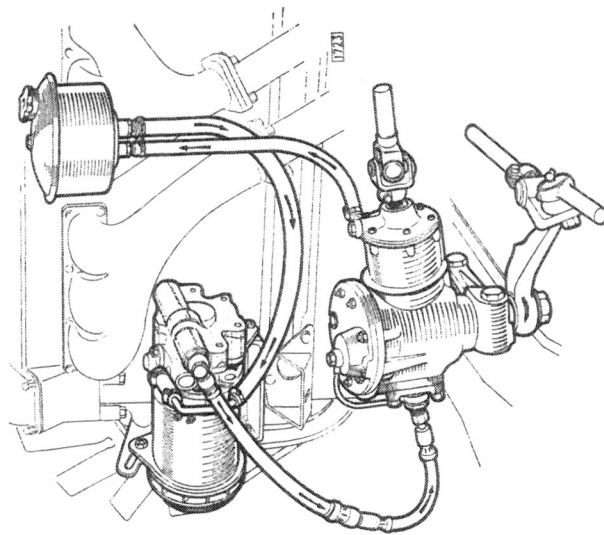

Fig. 16. *Layout of the Power-Assisted Steering (First Type).*

The Steering Unit (Fig. 17).

The steering box is of the re-circulating ball worm and nut type in which hydraulic assistance is applied to a piston forming part of the nut (G).

The piston works within a cast iron cylinder pressed into the aluminium steering box casing, hydraulic pressure being admitted to one side or other of the piston, depending on which steering lock is applied.

Admission of oil to the appropriate pressure chamber is controlled by a selector valve (M) co-axially mounted within the hollow rear end of the wormshaft (P). The valve extends rearwards through the steering box top cover, and forms the input shaft to which the lower end of the steering column is directly connected.

Rotary movement of the valve relative to the wormshaft opens and closes ports in the wormshaft and thus directs oil to the side of the piston in operation for the steering lock required.

When steering wheel effort is at a minimum centralisation of the valve within the wormshaft is effected by the action of an interlock ball (K) which is loaded by a coil spring (Q) located at the bottom of the valve. The interlock ball operates in specially shaped mating holes in the valve and wormshaft.

To obtain a high mechanical efficiency the internal sealing in the box is obtained by the use of sealing sleeves instead of rubber 'O' rings.

DATA

Steering Unit (First Type):

Make	..	..	..	..	..	..	..	Burman
Type	..	..	..	..	..	..	..	Hydraulically assisted worm and recirculating ball

Turning circle	..	..	..	..	..	..	..	33' 6" (10.21 m.)

Oil Pump:

Make	..	..	..	..	..	..	..	Hobourn-Eaton
Type	..	..	..	..	..	..	..	Eccentric rotor
Location	..	..	..	..	..	..	..	Rear of generator

Operating pressure	..	..	..	..	..	..	800—850 lb. per sq. in. (5.24—59.76 kg./cm².)

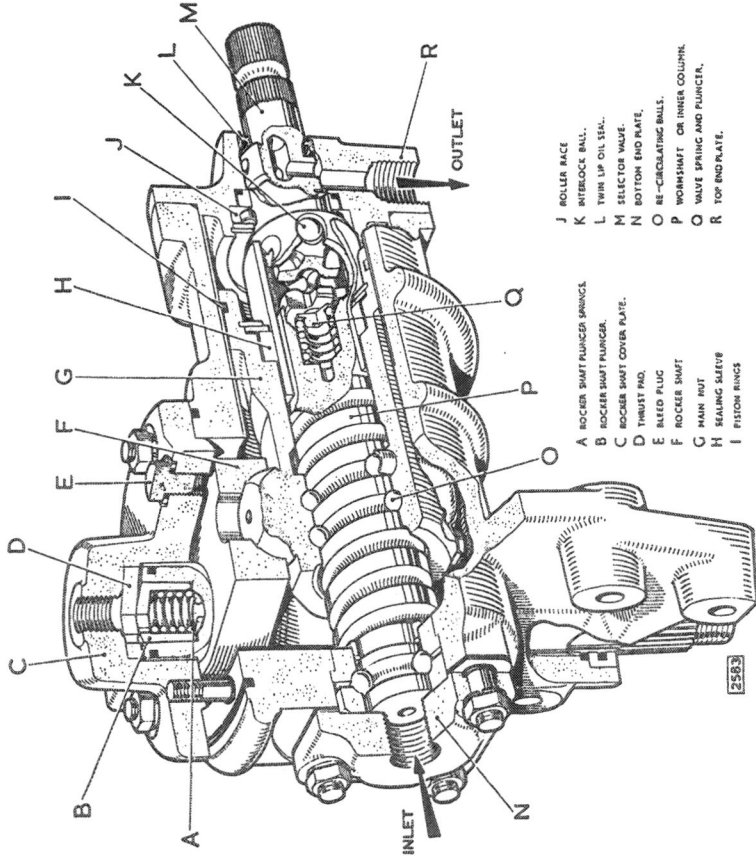

Fig. 18. *Showing the oil feed groove and port at each side of the valve.*

The Valve (Fig. 18).

The valve is of cylindrical form and has a central longitudinal passage which is closed at each end. An interrupted flange formed on the outside of the valve working between stops on the wormshaft, limits the rotary movement of the valve within the wormshaft. This prevents overloading of the valve and permits normal steering in the event of the hydraulic assistance not being available. At each side of the valve symmetrical oil feed grooves and ports are machined, the port drillings communicating with the central passage.

Note: A limited amount of axial movement of the valve (input shaft) may be noticed when turning the steering but this movement is quite normal.

The Rocker Shaft

The rocker shaft (F, Fig. 17) is of the normal spring-loaded type but when the steering is on lock the spring loading is augmented by the hydraulic pressure existing in the system.

Note: It may be noticed when turning the road wheels on lock that there is a certain amount of end movement of the rocker shaft. This is quite normal and is inherent in the design of the steering unit.

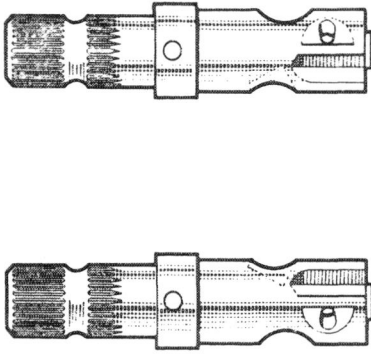

Fig. 17. *Sectional view of the power-assisted steering unit (First Type).*

A ROCKER SHAFT PLUNGER SPRINGS.
B ROCKER SHAFT PLUNGER.
C ROCKER SHAFT COVER PLATE.
D THRUST PAD.
E BLEED PLUG.
F ROCKER SHAFT.
G MAIN NUT.
H SEALING SLEEVE.
I PISTON RINGS.
J ROLLER RACE.
K INTERLOCK BALL.
L TRIM LIP OIL SEAL.
M SELECTOR VALVE.
N BOTTOM END PLATE.
O RE-CIRCULATING BALLS.
P WORMSHAFT OR INNER COLUMN.
Q VALVE SPRING AND PLUNGER.
R TOP END PLATE.

INLET

OUTLET

ROUTINE MAINTENANCE

Every 3,000 miles (5,000 km.)

Checking the Steering Reservoir Oil Level

The oil reservoir is attached to the left-hand wing valance. It is important that absolute cleanliness is observed when replenishing with oil as any foreign matter that enters may affect the hydraulic system.

Clean the area around the filler cap and then remove the cap by turning anti-clockwise.

Check the level of oil and top up if necessary with the recommended grade. The level of oil must be just above the filter element located in the reservoir.
Important. If the oil level is allowed to fall appreciably, the power assistance to the steering will be affected.

Every 6,000 miles (10,000 km.)

Steering Idle Lever Housing

The idler housing is pre-packed with grease which only requires replenishing if the idler assembly is dismantled for overhaul.

Steering Tie Rods

Lubricate the ball joints at the ends of the two steering tie-rods with the recommended lubricant. The tie-rods are situated at the rear of the front suspension cross-member. When carrying out this operation examine the rubber seals at the ends of the ball housings to see if they have become displaced or split. In this event they should be repositioned or replaced as any dirt or water that enters the ball joint will cause premature wear.

A bleed hole covered by a circular nylon washer, which lifts under pressure, indicates when sufficient lubricant has been applied. This prevents grease from escaping past the seals when too much pressure is applied.

Every 12,000 miles (20,000 km.)

Oil Reservoir Filter

Every 12,000 miles (20,000 km.) renew the paper filter element in the oil reservoir.
Unscrew the bolt securing the top cover.
Lift off the cover and collect the spring and retainer plate. Lift out the filter element.

When fitting the new element, ensure that it is located in the support plate at the bottom of the reservoir. Refit the retainer plate, spring and top cover. Tighten the central bolt.

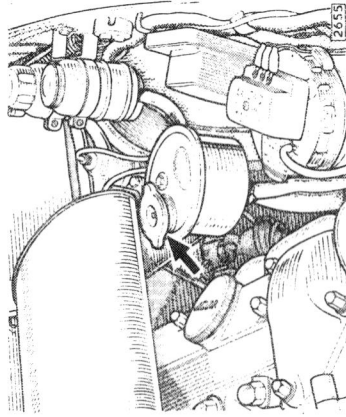

Fig. 19. *Power steering oil reservoir.*

Recommended Lubricants

	Mobil	Castrol	Shell	Esso	B.P.	Dackham	Regent Caltex/Texaco
Steering reservoir	ATF 210	Castrol T.Q.F.	Shell Donax T7	Esso Glide	Autran B	Q-matic	Texamatic Type F
Steering idler housing Wheel swivels Steering tie rods	Mobilgrease MP	Castrolease LM	Retinax A	Esso Multi-purpose Grease H	Energrease L.2	LB 10	Marfak All-purpose

OPERATION

(a) Steering in straight ahead position

Oil direct from the pump enters the steering box via the bottom end cover and passes up through a longitudinal hole in the wormshaft the valve being in its central position permits free flow of oil through the steering box to the outlet in the top cover. In this condition the oil is at low pressure and no thrust is applied to the piston.

(b) Steering on lock

Rotation of the steering wheel causes the valve to move relative to the wormshaft by an amount proportional to the effort applied to the steering wheel; the amount is determined by the resistance of the road wheels to turning. On either side of the worm a milled slot communicates with the oil feed grooves in the valves; the larger milled slot controls the flow to the chamber above the piston, and the smaller slot, together with a longitudinal groove controls the flow to the chamber below the piston. A sleeve pressed on to the outside of the worm and valve assembly effects a seal between the upper and lower chambers and also acts as a retainer for the interlock ball.

The relative movement of the valve to wormshaft restricts or completely closes the return port in the valve, which causes pressure to build up in the chamber on the side of the piston on which it is required to exert hydraulic pressure. Immediately steering wheel movement ceases and the car is held on a constant lock, the valve tends to centralise in the wormshaft by the combined action of the valve spring and interlock ball and the reduction in resistance to turning of the wormshaft due to the hydraulic assistance being applied to the main nut piston. A state of balance then exists between the effort at the steering wheel and the normal self-centering action of the front road wheels. On returning to the straight ahead position the hydraulic condition described in paragraph (a) is restored.

THE STEERING UNIT

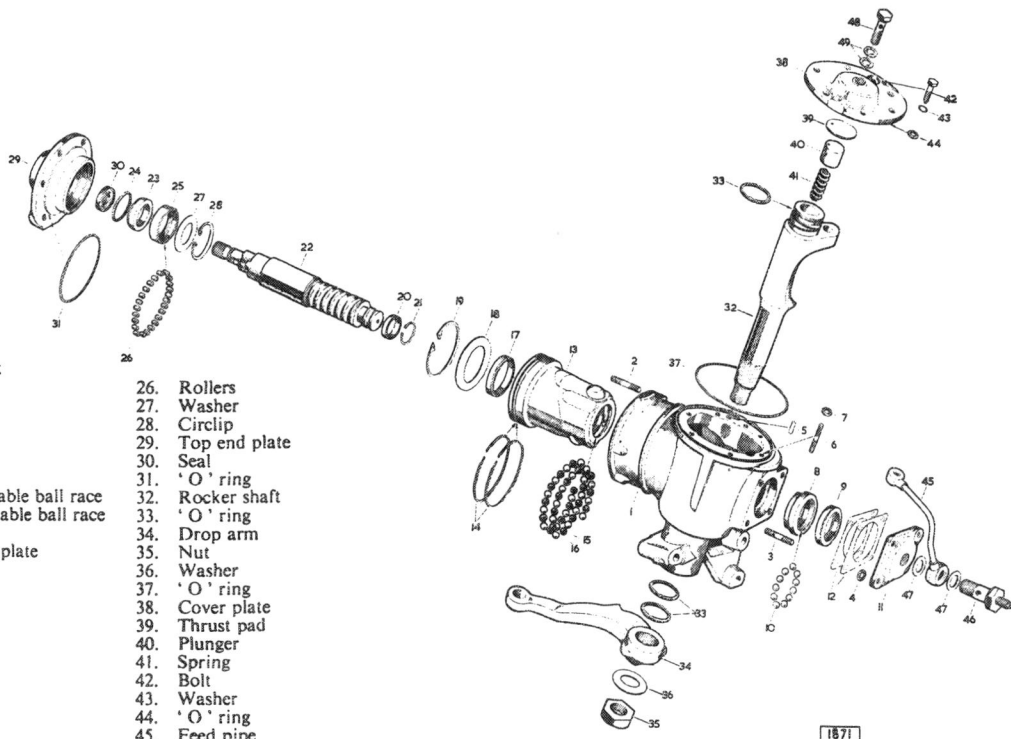

1. Steering box
2. Stud.
3. Stud
4. 'O' ring
5. Dowel
6. Stud
7. 'O' ring
8. Inner adjustable ball race
9. Outer adjustable ball race
10. Balls
11. Bottom end plate
12. Shim
13. Main nut
14. Piston ring
15. Balls (small)
16. Balls (large)
17. Sleeve
18. Washer
19. Circlip
20. Sleeve
21. Sleeve
22. Inner column assembly
23. Packing piece
24. 'O' ring
25. Roller race
26. Rollers
27. Washer
28. Circlip
29. Top end plate
30. Seal
31. 'O' ring
32. Rocker shaft
33. 'O' ring
34. Drop arm
35. Nut
36. Washer
37. 'O' ring
38. Cover plate
39. Thrust pad
40. Plunger
41. Spring
42. Bolt
43. Washer
44. 'O' ring
45. Feed pipe
46. Connection for feed pipe
47. Copper washer
48. Banjo bolt
49. Copper washer

Fig. 20. *Exploded view of the power-assisted steering unit*

REMOVAL

Remove the bolt securing the reservoir return hose banjo to the top end cover of the steering unit and drain the oil into a clean container. Undo the union securing the hose from the pump to the feed pipe adaptor on the lower end of the steering box.

Note the position of the slot in the lower steering column top joint.

Remove the spring clip and retaining ring situated at the lower end of the outer tube inside the car and withdraw the plastic thrust bearing.

Remove the pinch bolt securing the lower column top universal joint to the upper steering column.

Refitting will be simplified if, before removal of the lower column, the road wheels and steering wheels are set in the straight ahead position. If these are then left undisturbed during the dismantling operations refitting can be carried out without any further adjustment of steering wheel or indicator cancelling device.

DISMANTLING (Fig. 20).

Tap back the tab washer and unscrew the nut (35) and with a suitable extractor withdraw the drop arm (34) from the rocker shaft. With the steering box held in a vice by the mounting boss and a suitable tray placed underneath it to catch the oil, remove the feed pipe assembly from the lower end cover and rocker shaft cover. Remove the six nuts and the setscrew securing an 'O' ring. By pushing up on the rocker shaft the setscrew will be held in position on the cover by an 'O' ring. By pushing up the rocker shaft the cover may now be removed. Extract the domed plunger (40) and the coil spring. Collect the thrust pad (39) from the counter bore in the cover. Empty out the oil from the box into the tray.

Remove the four nuts securing the bottom end plate (11) and withdraw the cover and outer ball race. The eleven loose balls may be released by pushing on the inner column.

Remove the six nuts securing the top cover and withdraw the end cover complete with the worm and nut assembly. (The rocker shaft will have to be lifted to disengage the ball on the nut assembly from the socket in the rocker arm before withdrawing the worm and nut assembly.) The rocker shaft may now be withdrawn.

Remove the end cover and roller race assembly and collect the twenty-four loose rollers. Unwind the main nut from the worm-shaft and collect the forty-six re-circulating balls.

Note: **Twenty-two of these balls are exactly $\frac{7}{32}''$ (5.55 mm.) diameter whilst the other twenty-two are undersize by 0.0007" (.017 mm.). The smaller balls are black in colour on assembly whilst the remainder are bright, and it is MOST IMPORTANT that they should be reassembled alternately with the larger balls.**

Remove the spring circlip and sealing sleeve from the lower end of the worm.

From the upper end cover and roller race assembly, remove the circlip and washer. The roller race can then be tapped out of position against a piece of wood. If this does not dislodge the roller race, immerse the end cover in boiling water and repeat the treatment. The sealing sleeve and 'O' ring will come out with the roller race and the oil seal may now be pushed out.

From the main nut assembly remove the piston rings; do not stretch too far or breakage will occur. Remove the internal circlip, washer and sealing sleeve.

ASSEMBLING

When assembling the steering unit it is advisable, owing to the high pressure (800–850 p.s.i.) existing in the system to renew all the rubber 'O' ring seals and the lip seal.

(1) (a) Press the lower inner ball race into the bottom end of the box.

   (b) Fit lower sealing sleeve inside the inner ball race and check for freedom of rotation.

   (c) Assemble the lower sealing sleeve and spring circlip on to the lower end of valve assembly.

(d) Over the tube pressed on to the worm and assembly, fit the sealing sleeve and check for freedom of rotation.

Fit the sealing sleeve in top end cover on to worm and check for freedom of rotation.

Assemble the sealing sleeve, retaining washer and circlip into the main nut.

Assemble the piston rings into groove on main nut.

Assemble the forty-four balls into the main nut making sure that they are placed alternatively, large and small, around the grooves. (See note on page 124 under "Dismantling").

Screw the worm and valve assembly into the nut, taking care not to dislodge any balls from the nut.

Screw the worm and valve assembly into the nut, taking care not to dislodge any balls from the nut.

Fit the washer and circlip.

The End Cover

Press oil seal into recess in end cover with the spring showing towards the inside of the box.

Fit the 'O' ring and sealing sleeve to top end cover and push in the roller race. Insert the twenty-four rollers and pack with petroleum jelly.

Fit the washer and circlip.

Position the piston rings in the groove on the main nut so that the slots are opposed at 180°. By means of a special ring clamp, (a steering box liner is ideal) which will fit into the recess in the end of the box, insert the rocker shaft and worm nut assembly into the box. It will be necessary to turn the rocker shaft on to the far lock and lift the rocker shaft on to the socket in the rocker arm assembly will fit into the socket in the rocker arm. Fit the eleven steel balls into the lower ball-race (with petroleum jelly). Fit the end plate, outer ball race and shim so that the worm assembly rotates freely without end float. Then remove .0025" (0.63 mm.) thick shim to obtain the correct axial pre-load. Always tighten down the end cover evenly.

Wrap tape around serrations on the valve. Fit 'O' ring to spigot on upper end cover and fit end cover over the studs with the oil outlet facing upwards (for L.H.D. cars) or downwards (for R.H.D. cars).

Remove tape when cover is right home and tighten down evenly.

Rocker Shaft Adjustment

The rocker shaft itself rises and falls slightly in travelling to either lock and it is necessary that the adjustment for the end float should be carried out at the highest point of the rocker arm travel.

Assemble the original thrust washer and plunger (omitting the spring) into the top cover and fit the top cover but leave off the 'O' rings at this stage. Bolt down the top cover evenly. Rotate the input shaft by hand to check the assembly for tightness as the rocker passes over the highest point on each lock. If no tightness is felt, remove the top cover and replace the thrust washer by one 0.005" thicker. Carry on in this manner until tightness is felt, then remove the washer and fit one 0.005" thinner.

Rocker Shaft Assembly

(a) Remove top cover complete with distance washer.

(b) Remove and dismantle plunger.

(c) Fit 'O' ring to top cover studs.

(d) Fit 'O' rings to top cover face.

(e) Assemble 'O' ring to top of rocker shaft and check fit within top cover bush.

(f) Insert spring (2) and plunger in rocker shaft.

(g) Refit top cover complete with distance washer.

Note: Care must be taken to tighten top cover evenly or the top bush may be damaged.

To Complete Assembly

(a) Fit drop arm, ensuring line on rocker shaft and drop arm are in alignment.

(b) Fit nut and secure with new lock washer.

(c) Fit the pipe between the bottom cover and the rocker shaft cover with a new copper washer on each side of the banjo connections.

Refitting

Refitting is the reverse of the removal procedure but special attention must be given to the following points.

Whilst reconnecting the upper steering column to the universal joint on the lower column care must be taken to ensure that the horn contact (43 Fig. 8) does not catch the horn slip ring (37 Fig. 8) on the inner column. The contact may be prised up with a screwdriver whilst the column is brought into position.

Should the road wheels or steering wheel have become disturbed during the removal operation, then these must be set in the straight ahead position before connecting the lower column jaw to the input shaft. Check that the indicators cancel evenly on each side of the straight ahead position. An adjustment can be made by removing the switch cover, turning the steering wheel until the adjustment screws on the trigger are visible, slackening the screws and moving the trigger in the required direction. Should the adjustment screws be already hard up against the ends of the slots then the adjustment must be made by re-positioning the lower column jaw in relation to the input shaft of the steering box.

Bleed the system as described on page I.28.

Adjust the lock stops as described on page I.43.

REPLACEMENT OF TOP END PLATE OIL SEAL

Remove the steering box as described on page I.8.

Remove the six nuts securing the upper end cover and carefully withdraw the end cover and roller race assembly. Remove the circlip and washer and collect the twenty-four loose rollers. The roller race can then be tapped out of position against a piece of wood. If this does not dislodge the roller race immerse the end cover in boiling water and repeat the treatment. The sealing sleeve and 'O' ring will now come out with the roller race and the oil seal may be pushed out.

Press the oil seal into the recess in the end cover with the spring showing towards the inside of the box.

Fit the new 'O' ring and sealing sleeve to the top end cover and push in the roller race (if necessary, immerse in boiling water).

Fit the twenty-four loose rollers with petroleum jelly. Fit the washer and circlip.

Fit the 'O' ring to the spigot on the top cover and fit the 'O' ring to the spigot on the top cover using Tool No. 8243 to protect the oil seal as the top cover is pushed over the serrations on the valve (Fig. 28). Fit the top cover with the oil outlet facing downwards (for R.H. Drive cars) or upwards (for L.H. Drive cars). If a service tool is not available, wrap the valve serrations in cellulose tape.

Remove the tool when the cover is right home and tighten down evenly.

Bleed the system as described on page I.28.

THE RESERVOIR

Dismantling, Inspection and Re-assembling

Remove the pump inlet hose from beneath the reservoir and allow the oil to drain into a clean container.

Thoroughly clean the exterior of the reservoir assembly.

Remove the screw, cover, spring and spring seat. Then remove the filter element. (Loosen the screw and remove the reservoir cover and filter).

Examine the filter and renew if damaged.
Place the filter on the reservoir stud.
Install the reservoir cover and a new gasket and secure it with the screw.

Caution: Check to ensure that the cover is installed flush with the reservoir body.

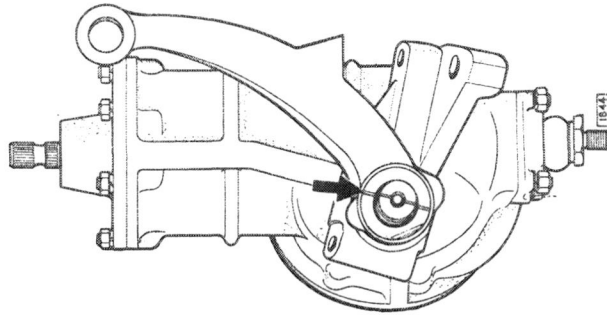

Fig. 21. *Showing the alignment of the drop arm/rocker shaft marks on the steering unit.*

BLEEDING THE SYSTEM

The system requires bleeding only when any part of the steering system has been disconnected. The procedure is as follows :—

(1) Fill the reservoir to the top of the filter element with the recommended grade of oil.

(2) Start the engine and allow to idle. Whilst the engine is idling pour more oil into the reservoir until the level reaches to the top of the filter element. Check the hose connection for leaks.

(3) Increase the engine speed to 1,000 r.p.m. and turn the wheels in each direction five or six times.

(4) Re-check for leaks and oil level in reservoir after road test.

STEERING IDLER ASSEMBLY

Removal

Remove the self-locking nut and washer securing the track rod end to the idler lever. Drift out the track rod end from the idler lever in which it is a taper fit.

Remove the two bolts and steering lock stop bolt attaching the steering idler bracket to the front suspension cross member when the steering idler assembly can be detached.

Dismantling

Prise out the dust cap from the top of the idler bracket.

Tap back the tab washer and unscrew the nut at the top of the idler shaft.

The idler arm and shaft can now be withdrawn and the washers, felt seal, etc., removed. Remove the inner races of the taper roller bearings.

Assembling

Thoroughly clean out the inner races of the taper roller races and the idler housing. Repack the housing and bearings with one of the recommended greases.

Fit the lower bearing, distance piece, seal retainer and a new felt seal.

Place the large washer over the idler shaft and pass the shaft upwards into the housing.

Fit the upper bearing. "D" washer, tab washer and nut. Tighten the nut to a torque of 5lb./ft. If a torque wrench is not available tighten the nut until rotation of the idler shaft by the idler arm feels "sticky" and then slacken back the nut one flat ; lock the nut by means of the tab washer and refit the dust cap.

Refitting

Refitting is the reverse of the removal procedure but it is important to ensure that the idler lever is in the straight ahead position as illustrated in Fig. 1 before fitting the track rod end to the lever.

Reset the lock stop as described on page 1.43.

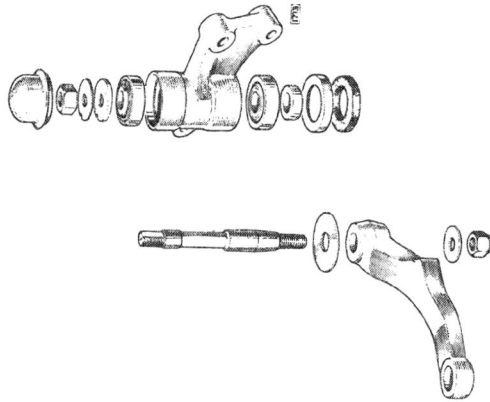

Fig. 22. Exploded view of the power-assisted steering idler lever assembly.

FAULT FINDING

1. High Steering Effort

(a) Insert the pressure gauge in the union at the base of the steering box.

Pressure should be 800-850 lb. per sq. in. (56.24-59.76 Kg/cm²) with the engine running at 1,000 r.p.m. and with stops against the wheels. If the pump is not delivering oil under pressure, it may be due to one of the following causes :—

(i) Low oil level. Add oil as necessary.

(ii) Stuck valves, dirt wedged in valves. Remove valves from pump and check for free operation.

(iii) Valving surfaces scored by abrasive matter. Replace all scored or worn parts. Clearance of relief valve should not exceed .0015" (.038 mm).

(iv) Worn pumping elements. End clearance should not exceed .0025" (.063 mm.).

2. Noise (the installation is moisy at engine idling speed).

(a) Look for hoses rubbing against the chassis or body metal ; isolate hoses.

(b) This may be caused by entrapped air in the system.

(i) Check reservoir to pump inlet connections for leaks.

(ii) Bleed the system.

(iii) Refit reservoir to correct level.

(c) Inspect oil seal ; lips may have been damaged due to faulty installation; replace oil seal.

(d) Flow control valve stuck closed, excessive internal pressure build up causing noise.

Free up valve, if necessary use crocus cloth to remove burrs in bore and on valve.

(e) Some noise may be expected when wheels are against stops. This is caused by the relief valve. It is undesirable to hold wheels in this position.

(f) When the oil level is low on a cold morning start, some noise may be caused by funnelling of the oil, allowing air to enter the system. This will stop when the oil heats.

3. Oil Leaks

(a) Shaft seal leakage. Replace oil seal.

(b) Reservoir gasket. If leakage is indicated by excessive oil around reservoir, replace gasket.

(c) Oil flowing out of air vent usually indicates a clogged filter.

Replace the element.

POWER-ASSISTED STEERING
(SECOND TYPE)

DESCRIPTION

The power-assisted steering system consists of an oil reservoir, roller type pump driven off the rear of the dynamo shaft, and an hydraulically assisted worm and re-circulating ball type steering box. These parts are connected by flexible hoses as follows :—

Reservoir to inlet side of pump.

Outlet side of pump to inlet pipe connection attached to the steering box.

Outlet at top of steering box to reservoir.

The pump supplies a continuous flow of oil through the system while the engine is running and the steering is in the straight ahead position. Pressure is only created in the system when the steering column is rotated and is proportional to the effort applied to the steering wheel.

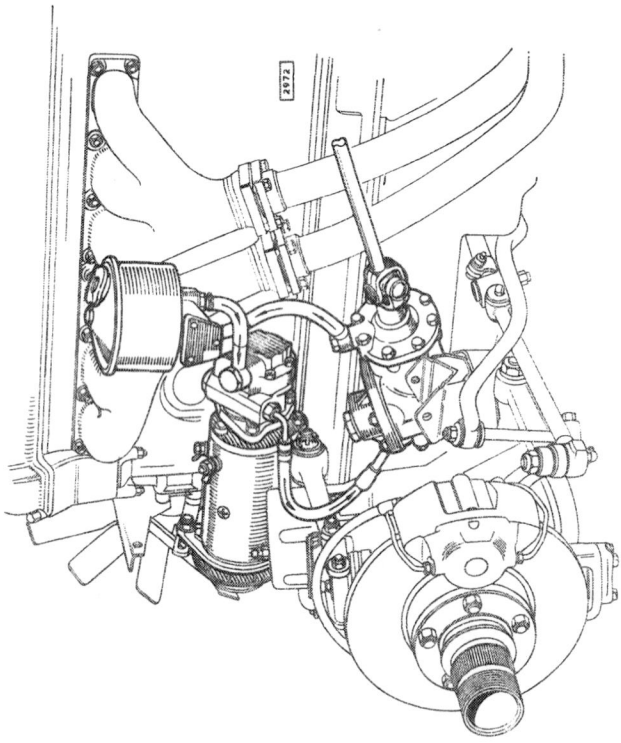

Fig. 23. *Power-Assisted Steering (Second Type).*

1. Rocker shaft
2. Dowel
3. Top cover
4. Adjusting screw
5. Dowel
6. Air bleed plug
7. Sealing sleeve
8. Selector valve
9. Operating ring
10. Torsion bar
11. Seal
12. Top end plate
13. Roller race
14. Main nut
15. Piston rings
16. Sealing ring
17. Inner column or wormshaft
18. Re-circulating balls
A. Inlet
B. Outlet

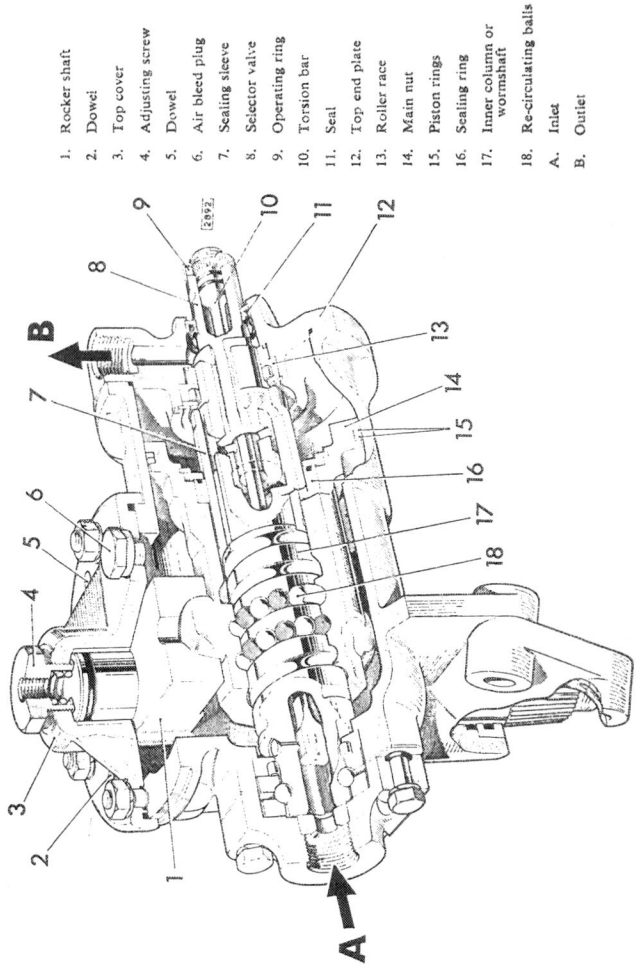

Fig. 24. *Sectional view of the power assisted steering unit (Second Type).*

The Steering Unit (Fig. 24).

The steering box is of the re-circulating ball worm and nut type in which hydraulic assistance is applied to a piston forming part of the nut (14). The piston works within a cast iron cylinder pressed into the aluminium steering box casing, hydraulic pressure being admitted to one side or the other of the piston, dependent upon which steering lock is applied.

Admission of oil to the appropriate pressure chamber is controlled by a selector valve (8) co-axially mounted within the hollow rear end of the wormshaft (17). The valve extends rearwards through the steering box top cover and forms the input shaft to which the lower end of the steering column is directly connected.

Rotary movement of the valve relative to the wormshaft opens and closes ports in the wormshaft directing oil to the side of the piston in operation for the steering lock required.

When steering wheel effort is at a minimum, the centralisation of the valve within the wormshaft is effected by the action of a torsion bar keyed to the valve by means of a serrated operating ring and locked in the wormshaft at the other end by a retaining pin.

To obtain a high mechanical efficiency the internal sealing in the box is obtained by the use of sealing sleeves instead of rubber 'O' rings.

DATA

Steering Unit:

Make	..	Burman
Type	..	Hydraulically assisted worm and recirculating ball
Steering gear ratio at centre of travel	..	16.7 : 1
Number of turns—lock to lock	..	3.0
Turning Circle	..	33' 6" (10.21 m.)

Oil Pump:

Make	..	Hobourn-Eaton
Type	..	Roller
Location	..	Rear of generator
Operating pressure	..	1000 lb. per sq. in (70.31 kg./cm²).

The Valve (Fig. 25).

The valve is cylindrical, one end of which is serrated while the other end is relieved to form valve ports. An interrupted flange, formed around the body of the valve, limits the rotary movement of the valve within the wormshaft. This prevents the overloading of the valve and permits normal steering in the event of hydraulic assistance not being available.

The torsion bar passes through the centre of the valve and it is locked in position by means of a serrated operating ring which is aligned with the serrations on the valve body. To prevent fluid loss from the valve there are two 'O' rings fitted to the serrated end of the torsion bar.

The valve ports consist of four faces, two of which are drilled to provide port drillings communicating with the central passage while the other two faces provide pressure relief.

The Rocker Shaft

The rocker shaft is of the normal pre-loaded type with an adjustable spherical ended bolt running in contact with the rocker shaft. A groove is machined in the rocker shaft which allows the rocker shaft to rise when on either lock.

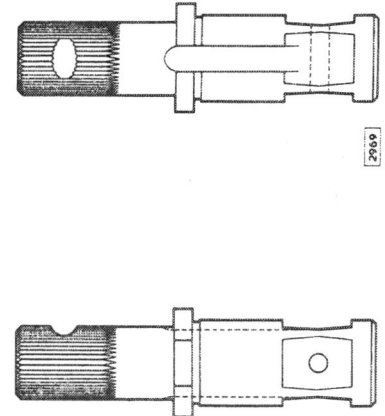

Fig. 25. The valve.

Clean the area around the filler cap and then remove the cap by turning anti-clockwise.

Check the level of oil and top up if necessary with the recommended grade. The level of oil must be just above the filter element located in the reservoir.

Important: If the oil level is allowed to fall appreciably, the power assistance to the steering will be affected.

Steering Idle Lever Housing

The idler housing is pre-packed with grease which only requires replenishing if the idler assembly is dismantled for overhaul.

Every 6,000 miles (10,000 km.)

Steering Tie-Rods

Lubricate the ball joints at the ends of the two steering tie-rods with the recommended lubricant. The tie-rods are situated at the rear of the front suspension cross member. When carrying out this operation examine the rubber seals at the ends of the ball housings to see if they have become displaced or split. In this event they should be repositioned or replaced as any dirt or water that enters the ball joint will cause premature wear.

Every 12,000 miles (20,000 km.)

Oil Reservoir Filter

Every 12.000 miles (20.000 km.) renew the paper element in the oil reservoir filter.

Unscrew the bolt securing the top cover. Lift off the cover and collect the spring and retainer plate. Lift out the filter element.

When fitting a new element, ensure that it is located in the support plate at the bottom of the reservoir. Refit the retainer plate, spring and top cover. Tighten the central bolt.

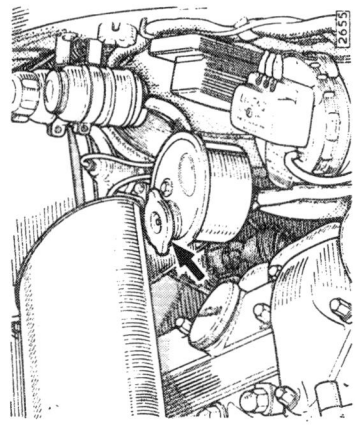

Fig. 26. Power steering oil reservoir.

ROUTINE MAINTENANCE

Every 3,000 miles (5,000 km.)

Checking the Steering Reservoir Oil Level

The oil reservoir is attached to the left-hand wing valance. It is important that absolute cleanliness is observed when replenishing with oil as any foreign matter that enters may affect the hydraulic system.

THE STEERING UNIT

(Second Type)

REMOVAL (Fig. 27)

Remove the bolt (50) securing the reservoir return hose banjo to the rocker shaft cover of the steering unit and drain the oil into a clean container. Undo the union securing the hose from the pump to the feed pipe adaptor on the lower end of the steering box.

Note the position of the slot in the top joint of the lower steering column.

Remove the spring clip and retaining ring situated at the lower end of the outer tube inside the car and withdraw the plastic thrust bearing.

Remove the pinch bolt securing the lower column top universal joint to the upper steering column.

Refitting will be simplified if, before removing the lower column, the road wheels and the steering wheel are set in the straight ahead position. If this state is preserved during dismantling, refitting can be carried out without any further adjustment of either the steering wheel or indicator cancelling device.

Remove the self-locking nut securing the track rod end to the drop arm. Drift out the track rod end from the drop arm in which it is a taper fit. Remove the three setscrews and one long bolt securing the steering unit to the front suspension cross member and remove the unit.

DISMANTLING (Fig. 27)

Tap back the tab washer (38) and unscrew the nut (37).

Withdraw the drop arm (36) from the rocker arm using a suitable extractor.

Secure the steering box in a vice and, with a suitable tray placed underneath in which to collect the oil, remove the feed pipe assembly (47) from the bottom end cover (18) and rocker shaft cover (41). Remove the six nuts and the setscrew securing the rocker shaft cover to the steering box (the setscrew is held in position on the cover by an 'O' ring).

Press the rocker shaft upwards to remove the rocker shaft cover.

Remove the four nuts securing the bottom cover (18) and withdraw the cover together with the outer ball race (16 and 17).

The eleven balls may be released by pushing on the inner column.

Remove the six nuts securing the top cover (31) and withdraw the cover complete with the worm and nut assembly (lift rocker shaft to disengage the ball on the nut assembly from the socket in the rocker arm to withdraw).

Detach the cover and roller race assembly consisting of twenty-four loose rollers. Unwind the main nut from the wormshaft and collect the thirty-seven recirculating balls.

Note: Nineteen of these balls are exactly $\frac{9}{32}''$ (7.14 mm.) diameter whilst the remaining eighteen balls are smaller by .0007" (.017 mm.). As a means of identification, the larger balls are bright whilst the smaller balls are dull. It is MOST IMPORTANT that the smaller balls are assembled alternately with the larger balls.

Remove the circlip (30) and washer (29) from the top cover roller race assembly. The roller race can then be tapped out of position against a piece of wood. If this does not dislodge the roller race, immerse the end cover in boiling water and repeat the treatment. The sealing sleeve and 'O' ring will come out with roller race thus permitting the removal of the oil seal.

Remove the piston rings (22) from the main nut assembly.

Avoid stretching too far or breakage will occur.

Recommended Lubricants

	Mobil	Castrol	Shell	Esso	B.P.	Duckham	Regent Caltex/Texaco
Steering reservoir	Mobil Fluid 200	Castrol T.Q.	Shell Donax T6	Esso Automatic Transmission Fluid	Energol Automatic Transmission Fluid Type 'A'	Nolmatic	Texamatic Fluid
Steering idler housing Wheel swivels Steering tie rods	Mobilgrease MP	Castrolease LM	Retinax A	Esso Multi-purpose Grease H	Energrease L.2	LB 10	Marfak All-purpose

OPERATION

(a) **Steering in straight ahead position**

Oil direct from the pump enters the steering box via the bottom end cover (Fig. 24) and passes up through a longitudinal hole in the wormshaft; the valve being in its central position permits free flow of oil through the steering box to the outlet in the top cover. In this condition the oil is at low pressure and no thrust is applied to the piston.

(b) **Steering on lock**

Rotation of the steering wheel causes the valve to move relative to the wormshaft by an amount proportional to the effort applied to the steering wheel; the amount is determined by the resistance of the road wheels to turning. On either side of the worm a milled slot communicates with the oil feed grooves in the valves: the larger milled slot controls the flow to the chamber above the piston, and the smaller slot, together with a longitudinal groove controls the flow to the chamber below the piston. A sleeve pressed on to the outside of the worm and valve assembly effects a seal between the upper and lower chambers and also acts as a retainer for the interlock ball.

The relative movement of the valve to wormshaft restricts or completely closes the return port in the valve, which causes pressure to build up in the chamber on the side of the piston on which it is required to exert hydraulic pressure. Immediately steering wheel movement ceases and the car is held on a constant lock, the valve tends to centralise in the wormshaft by the combined action of the torsion shaft and the reduction in resistance to turning of the wormshaft due to the hydraulic assistance being applied to the main nut piston. A state of balance then exists between the effort at the steering wheel and the normal self-centring action of the front road wheels. On returning to the straight ahead position the hydraulic condition described in paragraph (a) is restored.

Key to Fig. 27

1. Steering box
2. Bolt
3. Bolt
4. Spring washer
5. Stud
6. 'O' ring
7. Nut
8. Spring washer
9. Dowel
10. Dowel
11. 'O' ring
12. Stud
13. 'O' ring
14. Nut
15. Washer
16. Roller race
17. Rollers
18. Adjustable ball race
19. Ball
20. Bottom end plate
21. Sealing sleeve
22. Shims
23. Inner column and main nut
24. Piston ring
25. Balls (Black finish)
26. Balls (Bright finish)
27. Packing piece
28. 'O' ring

29. Roller race
30. Rollers
31. Washer
32. Circlip
33. Top end plate
34. Seal
35. 'O' ring
36. Rockershaft
37. 'O' ring
38. Drop arm
39. Nut
40. Washer
41. 'O' ring
42. 'O' ring
43. Rocker shaft cover
44. Adjusting screw
45. Spring
46. Shim (copper)
47. Shim (steel)
48. Air bleed plug
49. Washer
50. Feed pipe assembly
51. Connector
52. Copper washer
53. Banjo bolt
54. Copper washer
55. Dust seal

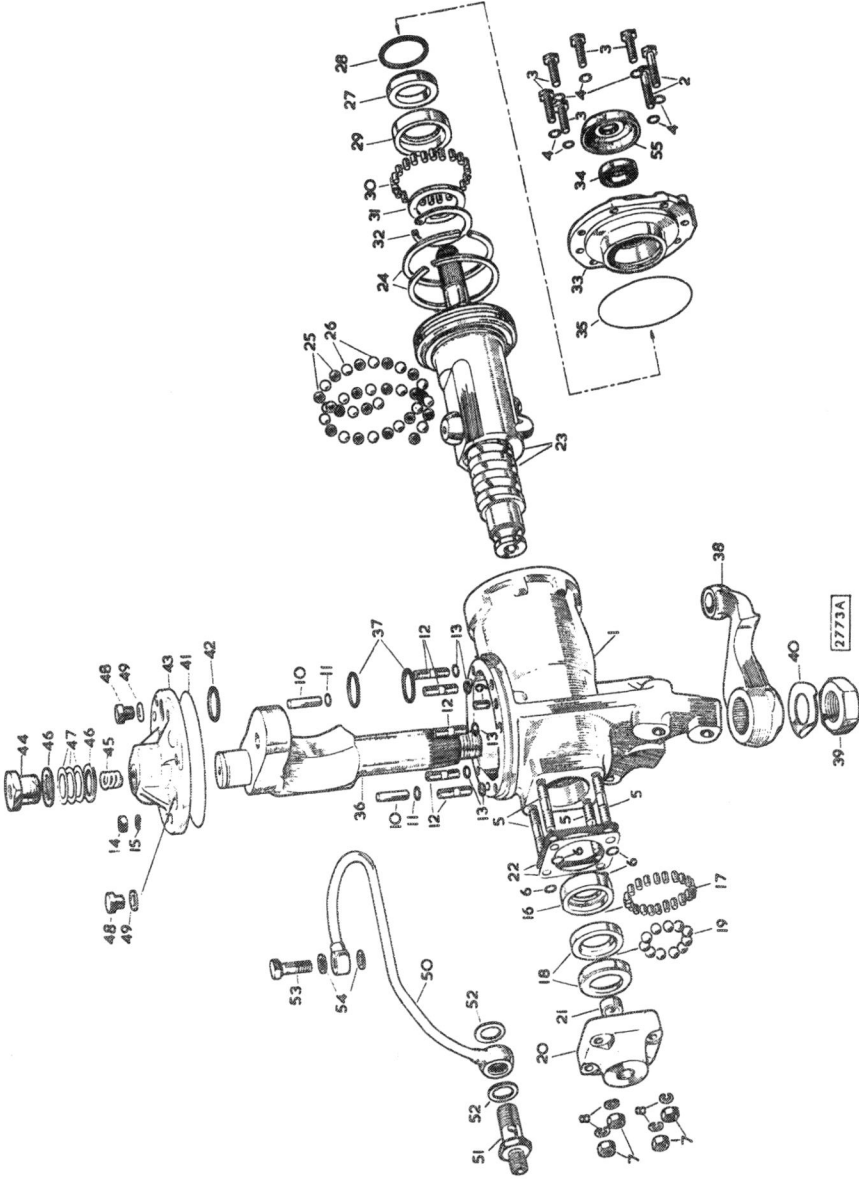

Fig. 27. The Steering Unit—Second Type (Power-Assisted)

2773A

ASSEMBLING (Fig. 27)

When assembling the steering unit, it is advisable, owing to the high pressure. (1,000 lb. per sq. in. or 70.31 kg.cm²) existing in the system, to renew all the rubber 'O' ring seals and the lip seal.

Assemble the piston rings (22) into the groove on the main nut, using a piston spring expander.

Refit the thirty-six balls into the nut and ensure that they are placed alternately, large and small, around the grooves.

Screw the worm assembly into the nut taking care not to dislodge any balls from the nut.

Position the piston rings in the groove on the main nut so that the ring gaps are opposed at 180°. Insert the rocker shaft and worm nut assembly into the box using a piston ring compressor.

Remove the piston ring compressor sleeve and fit the roller race complete with twenty-one rollers and the inner ball race to the lower end of the worm shaft. Tap the inner ball race and roller race into the bottom of the box and fit the eleven steel balls round the wormshaft (with petroleum jelly) and press the shaft up into the box until the balls locate on the inner ball race.

Fit the outer ball race (16), end plate (18) and shims (20) so that the worm assembly rotates freely without end float.

Remove one shim .0025" (0.06 mm) thick to obtain the correct axial preload.

Always tighten the cover down evenly.

The Top Cover

Press the oil seal (32) into the recess in the top cover (31) with the spring showing towards the inside of the box.

Fit the 'O' ring and sealing sleeve to the top cover, push in the roller race and insert the twenty-four rollers (packed with petroleum jelly).

Fit the washer (29) and circlip (30).

Fit the 'O' ring to the spigot on the top cover and fit the cover using Tool No. 8243 to protect the oil seal as the top cover is pushed over the serrations on the valve (Fig. 28).

Fit the top cover with the oil outlet facing downwards. If a service tool is not available, wrap the valve serrations in tape.

Remove the tool when the cover is right home and tighten down evenly.

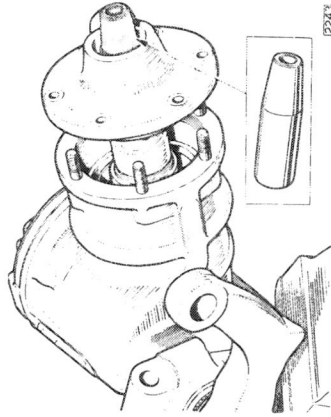

Fig. 28. *Fitting the top cover with special sleeve (Tool No. 8243)*

The Rocker Shaft Cover

Assemble the rocker shaft cover (41) complete with the rocker shaft adjuster bolt (42) and locknut (43). Fit the cover and ensure that the 'O' rings are in place on the steering box studs. Locate the cover on the dowel (9) and tap securely home before bolting the cover down evenly.

Rocker Shaft Adjustment

In travelling to either lock the rocker shaft rises and falls axially to a small degree and it is therefore necessary to check the end-float at the highest points as indicated in Fig. 29.

Adjustment of the rocker shaft end-float is by means of shims (47 Fig. 27) located under the head of the adjusting screw. The shims are available in .005", .010" and .015" (.13, .25 and .38 mm.) sizes. Note that a copper sealing washer (46) must be fitted to each side of the shim pack.

(a) Place the rocker shaft in position in the steering box.

(b) Fit the rocker shaft cover complete with existing shims, copper washer and adjusting screw (45) but omit the spring at this stage. Tighten the cover down fully.

(c) Rotate the input shaft to check the assembly for tightness as the rocker arm passes over the highest point on each lock.

(d) If no tightness is felt reduce the thickness of the shim pack by removing a shim at a time.

Repeat this procedure until tightness is felt as the rocker arm passes over the "high" points and then insert a .005" (.13 mm.) shim to give an end-float of 0 to 005" (0 to .13 mm.). Re-check the assembly for tightness.

Remove the adjusting screw and refit the spring.

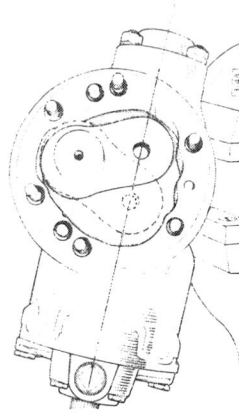

Fig. 29. *Showing the two highest positions of the rocker shaft.*

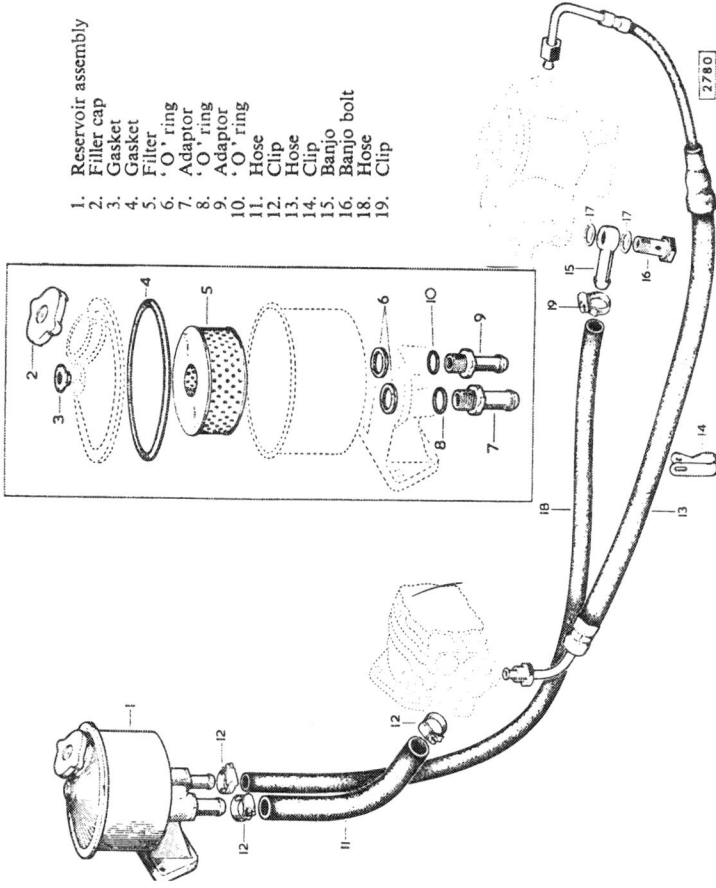

1. Reservoir assembly
2. Filler cap
3. Gasket
4. Gasket
5. Filter
6. 'O' ring
7. Adaptor
8. 'O' ring
9. Adaptor
10. 'O' ring
11. Hose
12. Hose
13. Clip
14. Hose
15. Banjo
16. Banjo bolt
17. Hose
18. Clip

Fig. 30. *Exploded view of the reservoir assembly.*

STEERING (Power-Assisted)

To Complete Assembly

Fit the drop arm, ensuring that the lines on the rocker shaft and drop arm are in alignment (Fig. 21).

Fit the nut and secure with a new lock washer.

Fit the pipe between the bottom cover and the rocker shaft cover with a new copper washer on each side of the banjo connections.

REFITTING

Refitting is the reverse of the removal procedure but special attention must be given to the following points.

Replace the spacer between the steering unit and cross beam. Refit the steering column plastic thrust bearing and clamp ring to retain the inner column in position.

When reconnecting the upper steering column to the universal joint on the lower column ensure that the horn contact does not catch the horn slip ring on the inner column.

The contact may be prised up with a screw-driver whilst the column is being brought in to position.

If the road wheels or the steering wheel have been disturbed during the removal of the steering unit, they must be set in the straight ahead position before connecting the lower column jaw to the input shaft.

Check that the indicators cancel evenly on each side of the straight ahead position. An adjustment can be made by removing the switch cover, turning the steering wheel until the adjustment screws on the trigger are visible, slackening the screws and moving the trigger in the required direction. If the adjustment screws are already hard up against the ends of the slots adjustment must be made by repositioning the lower column jaw in relation to the input shaft of the steering box.

Bleed the system as described on page 1.42.

Adjust the lock stops as described on page 1.45.

REPLACEMENT OF TOP COVER OIL SEAL

Remove the steering unit as described on page 1.35.

Remove the seven bolts (3) securing the top cover (31) and carefully withdraw the cover and the roller race assembly. Remove the circlip (30) and washer (29). Collect the twenty-four loose rollers (28). The roller race can then be tapped out of position against a piece of wood. If this does not dislodge the roller race, immerse the top cover in boiling water and tap out. The

sealing sleeve and 'O' ring will now come out with the roller race and the oil seal may be pushed out.

Press the oil seal into the recess in the top cover with the spring showing towards the inside of the box.

Fit a new 'O' ring and sealing sleeve to the top cover and push in the roller race (if necessary, immerse the cover in boiling water).

Fit the twenty-four loose rollers and pack with petroleum jelly.

Fit the washer and circlip.

Fit the 'O' ring to the spigot on the top cover and fit cover using Tool No. 8243 to protect the oil seal when sliding the top cover over the serrations on the valve input shaft (Fig. 28). Fit the cover with the oil outlet facing downwards. (If a service tool is not available, wrap the valve serrations in tape.)

Remove the tool when the cover is right home and tighten down evenly.

Bleed the system as described below.

THE RESERVOIR (Fig. 30)
Dismantling, Inspection and Re-assembling

Remove the pump inlet hose (11) from beneath the reservoir and allow the oil to drain into a clean container.

Thoroughly clean the exterior of the reservoir assembly.

Remove the screw, cover, spring and spring seat.

Remove the filter element (5).

Examine the filter and renew if damaged.

Place the filter on the reservoir stud.

Install the reservoir cover together with new gaskets and secure with the screw.

Caution: Check to ensure that the cover is installed flush with the reservoir body.

BLEEDING THE SYSTEM

The system requires bleeding only when any part of the steering system has been disconnected. The procedure is as follows :—

(a) Fill the reservoir to the top of the filter element with the recommended grade of oil.

(b) Start the engine and allow to idle. While the engine is idling pour more oil into the reservoir until the level reaches to the top of the filter element.

Check the hose connections for leaks.

(c) Increase the engine speed to 1,000 r.p.m. and turn the wheels in each direction five or six times.

(d) Re-check for leaks and oil level in reservoir after road test.

STEERING IDLER ASSEMBLY (Fig. 31)

Removal

Remove the self-locking nut (13) and washer (12) securing the track rod end to the idler lever (10).

Drift out the track rod end from the idler lever in which it is a taper fit.

Remove the two setscrews and one long bolt attaching the steering idler bracket (7) to the front suspension cross beam when the steering idler assembly can be detached.

Dismantling

Prise out the end cap (1) from the top of the idler bracket (7).

Tap back the tab washer (3) and remove the nut (2) at the top of the idler shaft.

The idler lever (10) and the idler spindle (5) can now be withdrawn.

Remove the washers, felt seal, abutment ring and abutment washer.

Remove the inner races of the taper roller bearing.

Assembling

Thoroughly clean out the inner races of the taper roller races and the idler housing. Repack the housing and bearings with one of the recommended greases.

Fit the lower bearing (9), abutment ring (14), seal retainer (11) and a new felt seal (15).

Place the large washer (8) over the idler spindle (5) and pass the spindle upwards into the mounting bracket (7).

Fit the upper bearing (6), 'D' washer (4), tab washer (3) and nut (2).

Tighten the nut to a torque of 5 lb. ft. (.69 kg.m).

If a torque wrench is not available tighten the nut until the rotation of the idler shaft by the idler arm displays a tendency to bind and then slacken back the nut one flat; lock the nut by means of the tab washer and refit the end cap.

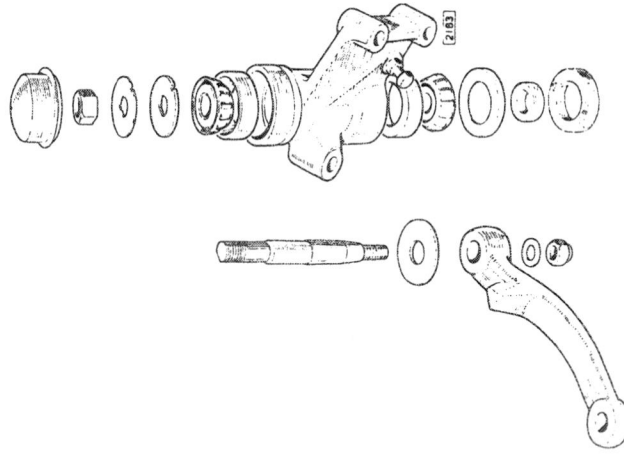

Fig. 31. *Exploded view of the idler assembly.*

1. End cap
2. Nut
3. Tab washer
4. 'D' washer
5. Idler spindle
6. Bearing
7. Mounting bracket
8. Abutment washer
9. Bearing
10. Idler lever
11. Seal retainer
12. Plain washer
13. Self locking nut
14. Abutment ring
15. Felt seal
16. Set screw
17. Nut

LOCK STOP ADJUSTMENT (Fig. 32)

There are two lock stop bolts, one located in the idler bracket and one in the steering box bracket as shown.

The bolts are retained in position by lock-nuts and they are set to permit 35° travel of the drop arm and idler lever either side of the central (straight ahead) position.

Normally, the lock stop bolts should not require adjustment but, if attention is found to be necessary, the adjustment should be carried out in the following manner :—

Slacken the locknuts and screw in the lock stop bolts as far as possible. Turn the steering until the steering unit is at the end of its travel on the lock.

Screw out the lock stop bolt two turns and tighten the locknut.

Repeat for the opposite lock.

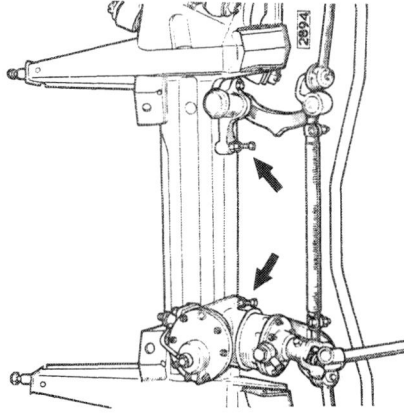

Fig. 32. *Location of the lock stops.*

FAULT FINDING

1. High Steering Effort

(a) Insert the pressure gauge in the union at the base of the steering box (see Fig. 33). Pressure should be 1,000 lb. per sq. in. (70.3 kg. per sq. cm.) with engine at 1,000 r.p.m. and wheels against stops. If the pump is not delivering oil under pressure, it may be due to one of the following causes :—

(i) Low oil level. Add oil as necessary.

(ii) Stuck valves, dirt wedges in valves. Remove valves from pump and check for free operation.

(iii) Valving surfaces scored by abrasive matter. Replace all scored or worn parts. Clearance of relief valve should not exceed .0015″ (.038 mm.).

(iv) Worn pumping elements. End clearance should not exceed .0025″ (.063 mm.).

2. Noise (the installation is noisy at engine idling speed).

(a) Look for hoses rubbing against the chassis or body metal ; isolate hoses.

(b) This may be caused by entrapped air in the system.

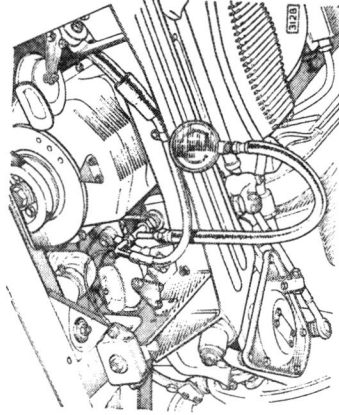

Fig. 33. *Insert the pressure gauge in the union at the base of the steering box*

(i) Check reservoir to pump inlet connections for leaks.

(ii) Bleed the system.

(iii) Refill reservoir to correct level.

(c) Inspect oil seal ; lips may have been damaged due to faulty installation : replace oil seal.

(d) Flow control valve stuck closed, excessive internal pressure build up causing noise. Free up valve. If necessary use crocus cloth to remove burrs in bore and on valve.

(e) Some noise may be expected when wheels are against stops. This is caused by the relief valve. It is undesirable to hold wheels in this position.

(f) When the oil level is low on a cold morning start, some noise may be caused by funneling of the oil, allowing air to enter the system. This will stop when the oil heats.

3. Oil Leaks.

(a) Shaft seal leakage. Replace oil seal.

(b) Reservoir gasket. If leakage is indicated by excessive oil around reservoir, replace gasket.

(c) Oil flowing out of air vent may indicate a clogged filter.

Replace element.

ACCIDENTAL DAMAGE

Dimensioned drawings are provided below to assist in assessing accidental damage. A component suspected of being damaged should be removed from the car, cleaned off and the dimensions checked and compared with those given in the appropriate illustration.

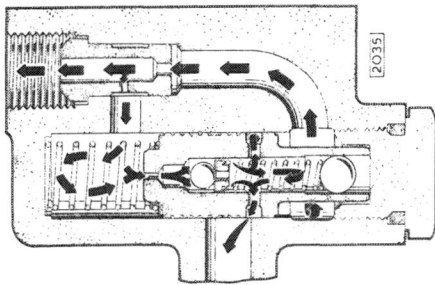

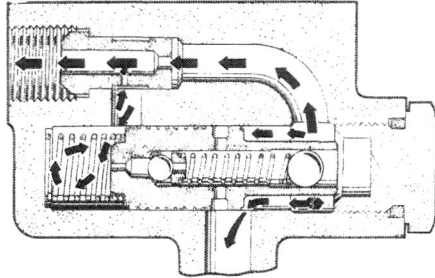

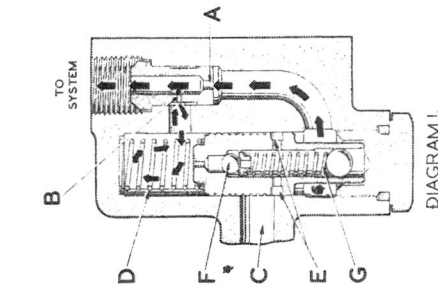

5½″
(13·97 cm)

1½″
(3·81 cm)
9/16″
(14·3 mm)

1685

Fig. 34. *Power-assisted steering drop arm.*

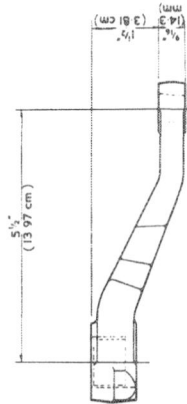

5½″
(13·97 cm)

1½″
(3·81 cm)

1687

Fig. 35. *Power-assisted steering idler lever.*

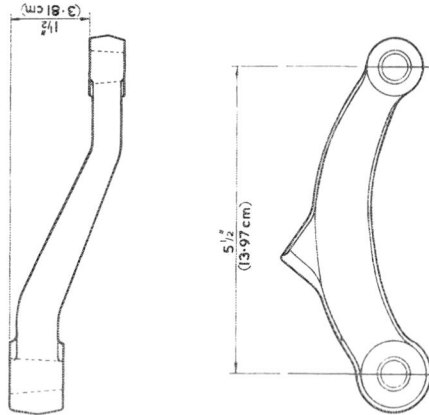

TO
SYSTEM

B

A

D
F
C
E
G

DIAGRAM I.

DIAGRAM 2.

DIAGRAM 3.

2035

Fig. 36. *Diagram showing the operation of the combined flow and relief valve.*

THE PRESSURE PUMP

The pressure pump which provides hydraulic pressure in the system is a Hobourn-Eaton unit of the roller type and incorporates a combined flow and relief valve. The pump is attached to the rear of the dynamo and is driven from the dynamo shaft by means of a rubber coupling.

OPERATION (Fig. 36)

Pressure from the pump is caused by six rollers in slots in a circular carrier, keyed to the pump drive shaft. These rollers circulate inside an eccentrically mounted cam ring. Owing to the eccentricity, the gap between each pair of rollers widens and narrows during the cycle, drawing oil from the inlet side of the pump and forcing it to the flow control valve.

When the pump comes into operation, and oil flow commences, a drop in pressure caused by the primary orifice (A, Dia. 1.) occurs. Oil at this lower pressure passes through the secondary orifice (B) and enters the chamber containing spring (D) (Condition in Diagram 1).

This pressure difference increases with the oil flow causing the control valve to move against the spring (D) and when a pre-determined flow has been reached, the valve uncovers the by-pass hole (C) leading to the intake side of the system.

Any further increase in flow causes the by-pass hole to be uncovered further and thus a constant flow is maintained (Condition in Diagram 2).

Should the line pressure become excessive, the ball (F) in the valve moves against the spring (G) and oil flow from the chamber containing spring (D) by-passes through the annular holes (E) (Condition in Diagram 3).

When this occurs, a further pressure drop caused by the secondary orifice (B) causes the valve to move up to its normal by-pass position irrespective of the oil flow conditions. As soon as the line pressure drops the ball valve closes and the pressure difference is restored bringing the constant oil flow back to normal.

Inspect the rollers paying particular attention to the finish on the ends. Replace if scored, damaged or out of round, refit the rollers.

Check the end float of the carrier and rollers in the pump body using feeler gauges and a straight edge across the cam surface. If the end float exceeds .002" (.051 mm.) the rollers and carrier should be replaced.

Refit the flow control spring in the valve bore. The spring should be replaced if the tension is not between 8–9 lb. (3.63–4.08 kg.) at 0.82" (20.8 mm.).

Refit the valve in the pump body with the exposed ball bearing end entering last. Ensure that the valve is not sticking. Refit the cap sealing ring and tighten the cap to 30–35 lb. ft. (4.15–4.84 kg. m.).

Fit new sealing rings to the pump body joint face. Fit the thrust washer to the cover and refit the pump cover to the body, secure with the six cover screws and tighten evenly to a torque of 18 lb. ft. (2.49 kg. m.).

Important: Check the drive shaft rotation for freedom after tightening the cover screws. There must be no binding.

Fit a new rubber gasket at the adaptor on the cover. Refit the adaptor, adaptor bolt and fibre washer. The plug in the top of the cover housing is used only for sealing purposes and if removed, care should be taken to ensure that an airtight seal is obtained when the plug is replaced.

REFITTING

To install, place the flexible coupling assembly in the slot on the dynamo shaft. Align the slot in the pump driving shaft with the driving tongue on the flexible coupling and push the pump home on to the mounting studs. Secure with nuts and spring washers. Connect the pressure and intake hoses to the pump.

Bleed the system as described on page 1.28.

Fig. 38. *The roller carrier and rollers in position.*

Remove the snap ring (16) from the drive shaft (15) and withdraw the six rollers (14) and roller carrier (13).

Remove the drive pin (15a) and withdraw the drive shaft from the pump body.

Remove the cam (11) from the cam locking peg (10).

Drift the oil seal (2) from the body if worn or damaged ensuring that the drive shaft bushing is not damaged.

Remove the valve cap (8), valve seal (7) and flow control valve (6) and flow control spring (5).

Caution: Place the parts where they will not be damaged.

Wash all parts in a suitable solvent and dry with a lint-free cloth or compressed air.

Check the pump body and cover for wear and replace either part if the faces or bushes are scored or worn.

Grease the lip of a new oil seal and assemble the seal with the lip towards the roller assembly. An arbor press is generally employed with a 1 3/32" (30.95 mm.) diameter piece of steel bar used as a piloting tool. Press the seal in until it is fully home but ensure that the seal is not squashed.

Refit the cam locking peg. Inspect the cam for wear and replace if worn or damaged. Refit the cam with the slot over the locking peg. Ensure that the cam is seated correctly.

Insert the drive shaft from the seal side of the body, ensuring that there are no sharp edges on the shaft to damage the oil seal lip.

Refit the drive pin to the shaft and having inspected the roller carrier, fit it in to position as shown in Fig. 38. Ensure that the correct face of the carrier slots are driving the rollers. Refit the snap ring.

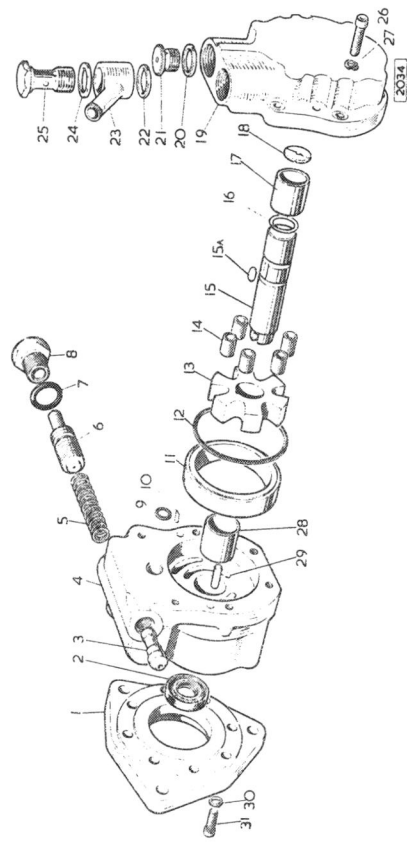

Fig. 37. *Exploded view of the oil pump.*

REMOVAL

Disconnect the hoses at the pump unions and place the hose ends in a raised position to prevent oil drainage. Alternatively, allow the oil to drain into a clean container.

Remove the nuts and lock washers securing the pump to the dynamo and remove the pump. If the flexible coupling comes away with the pump, withdraw the coupling from the slot in the pump shaft.

DISMANTLING, INSPECTION AND ASSEMBLY (Fig. 37)

Note: Thoroughly clean the exterior of the pump ensuring that dirt does not enter the inlet and outlet holes.

Hold the pump in a vice, using soft jaws.

Remove the adaptor screw (25), fibre washer (24), adaptor (23) and gasket (22).

Remove the six screws securing the cover (19) to the pump body (4).

Remove the pump from the vice, remove the cover from the pump body vertically to prevent the loss of parts.

Remove the sealing rings (9 and 12) from the grooves in the pump body. Remove the thrust washer (18) from the bearing hole in the cover.

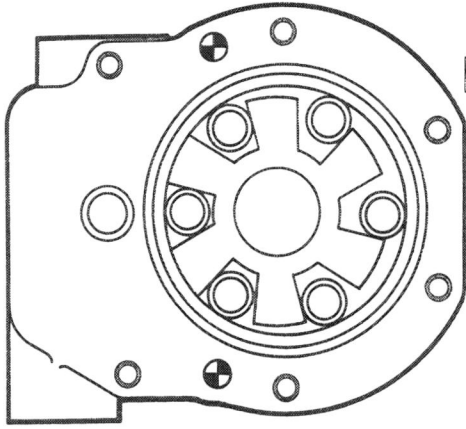

FRONT SUSPENSION

INDEX

INDEX (continued)

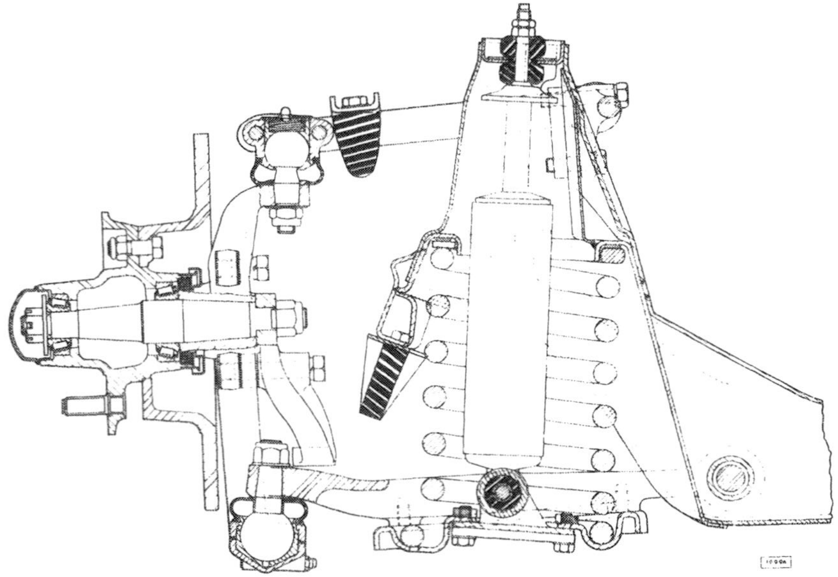

FRONT SUSPENSION

Fig. 1. *Sectional view of the front suspension assembly.*

DESCRIPTION

The assembly comprises a fabricated pressed steel cross-member to which are attached the wishbones, stub axle carriers, coil springs and hydraulic dampers. The steering unit and idler assembly, together with the track rod and tie rods, are also attached to this cross-member.

The coil springs are housed in "turrets" at each end of the suspension cross-member and are retained at the lower ends by seat pans bolted to the lower wishbone.

Each coil spring is controlled by a telescopic direct acting hydraulic damper which is mounted in the centre of the spring. The top of the damper is attached directly to the cross-member turret; the bottom of the damper is bolted to a mounting bracket which in turn is attached to the coil spring seat pan.

The upper wishbone levers are steel forged and are mounted at the fulcrum shaft end on rubber/steel bonded bushes. The outer ends of the wishbone levers are bolted to the upper wishbone ball joint which in turn is attached to the stub axle carrier.

The lower wishbone is a one-piece forging, the inner ends of which are mounted on rubber/steel bonded bushes. The outer end of the lower wishbone is bolted to the lower ball joint which in turn is attached to the stub axle carrier.

The wheel hub is supported on two tapered roller bearings the inner races of which fit on a shaft located in a tapered hole bored in the stub axle carrier.

An anti-roll bar fitted between the two lower wishbones, is attached to the chassis side members by rubber insulated brackets.

The front suspension assembly is attached to the body underframe at four points. The two longitudinal members are attached to brackets at the front end of the chassis side members via flat rubber/steel bonded mountings. The transverse member is attached to the chassis side members via two "V" shaped rubber/steel bonded mountings.

DATA

Type	Independent—Coil spring
Dampers	Telescopic hydraulic
Castor angle	$0° \pm \frac{1}{2}°$
Camber angle	$\frac{1}{4}° \pm \frac{1}{4}°$ positive
Swivel inclination	$3\frac{1}{2}°$
Coil springs :	
No. of coils (approximately)	$6\frac{1}{4}$
Diameter of Bar	.61" (15.478 mm.)

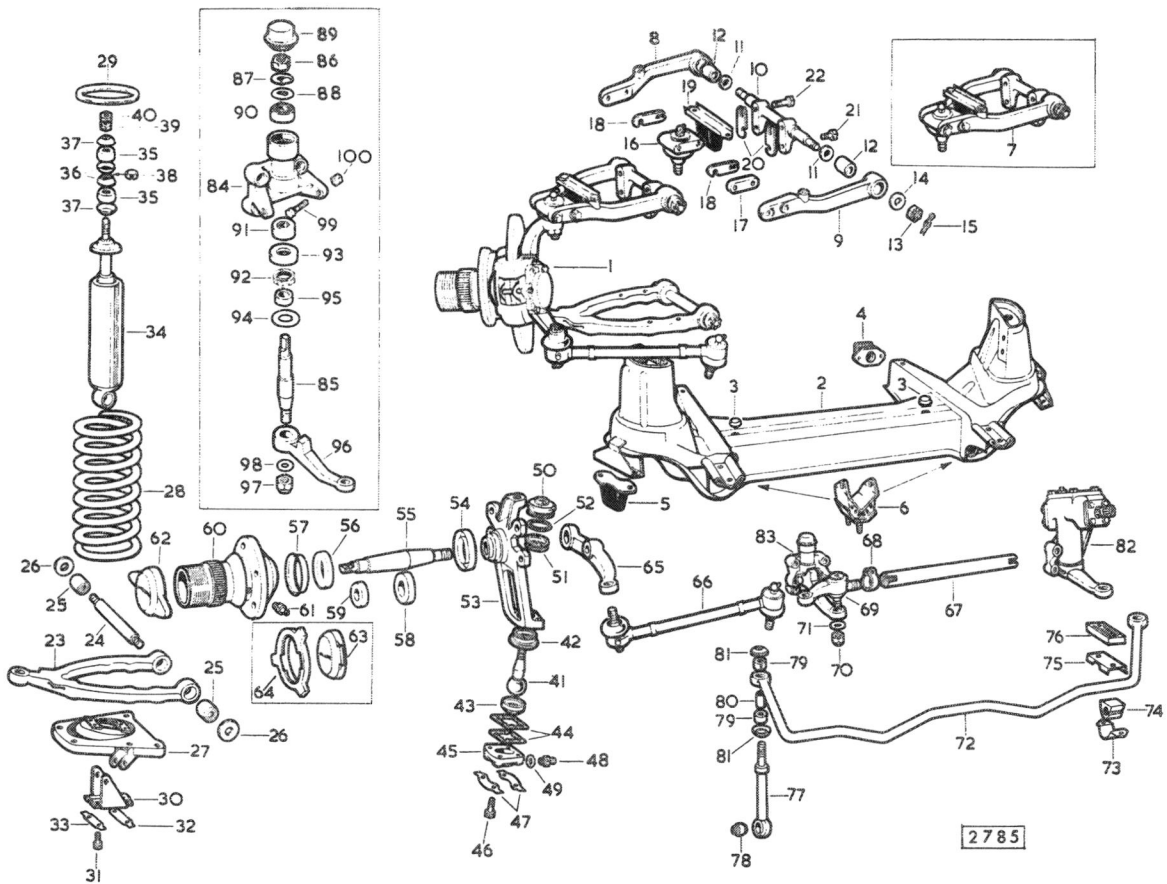

Fig. 3. *Front suspension assembly.*

1. L.H. front suspension assembly	34. Front shock absorber
2. Front suspension cross member	35. Rubber buffer
3. Rubber plug	36. Inner washer
4. Rubber mounting	37. Outer washer
5. Bump stop	38. Spacing collar
6. Rubber mounting	39. Nut
7. Upper wishbone	40. Locknut
8. Upper wishbone lever	41. Ball pin
9. Upper wishbone lever	42. Spigot
10. Fulcrum shaft	43. Railko socket
11. Distance washer	44. Shim
12. Rubber bush	45. Cap
13. Slotted nut	46. Bolt
14. Special washer	47. Tab washer
15. Split pin	48. Grease nipple
16. Ball joint	49. Washer
17. Distance piece	50. Rubber gaiter
18. Shim	51. Plastic insert
19. Rebound stop	52. Ring
20. Shim (camber angle)	53. Stub axle carrier
21. Bolt (short)	54. Water deflector
22. Bolt (long)	55. Stub axle shaft
23. Lower wishbone lever	56. Oil seal
24. Fulcrum shaft	57. Water deflector
25. Bush	58. Inner bearing
26. Special washer	59. Outer bearing
27. L.H. seat assembly	60. L.H. front hub
28. Front suspension coil	61. Grease nipple
29. Packing ring	62. L.H. hub cap
30. Bracket	63. L.H. hub cap
31. Set screw	64. Tool for removing/fitting hub caps
32. Tab washer	65. L.H. tie rod lever
33. Tab washer	66. L.H. outer tie rod

67. Tie rod tube
68. Clamp
69. End assembly
70. Nut
71. Special washer
72. Anti-roll bar (heavy duty)
73. Bracket
74. Rubber bush
75. Keeper plate
76. Packing block
77. Link
78. Rubber bush
79. Rubber pad
80. Distance tube
81. Retaining washer
82. Steering box
83. Steering idler
84. Bracket
85. Idler spindle
86. Nut
87. Tab washer
88. "D" washer
89. End cap
90. Bearing
91. Bearing
92. Felt seal
93. Retainer
94. Abutment washer
95. Abutment ring
96. Idler lever
97. Nut
98. Washer
99. Set screw
100. Nut

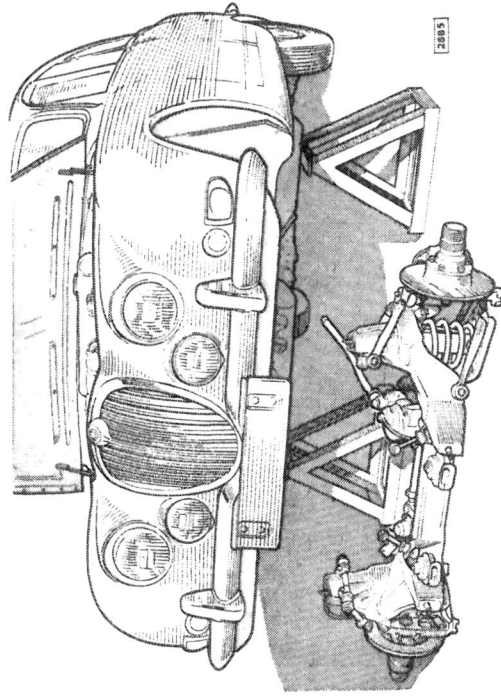

Fig. 4. *Removal of the front suspension assembly.*

FRONT SUSPENSION ASSEMBLY

There are two methods of removing the front suspension assembly. One method entails supporting the body on stands and drawing out the assembly, less the road wheels, on a jack. This method is suitable when this operation is to be carried out on the floor.

The other method which can be carried out on a lift or on the floor entails removing the radiator and raising the body by means of lifting tackle attached to a cross-bar placed under the chassis side members; the assembly is then rolled out from underneath the body on its road wheels.

Removal with car on floor

Jack up under the front suspension cross-member until the road wheels are clear of the ground. Remove the road wheels.

Support the weight of the car under the front jacking sockets by means of blocks not less than 16″ (40 cm.) in height, leaving the jack in position under the front cross-member.

Remove the two bolts securing the front suspension rear mountings to the chassis side members.

Disconnect the two engine stabilisers at the upper attachment points.

Remove the four nuts and bolts securing the front mountings to the brackets at the front ends of the chassis side members.

Disconnect the two anti-roll bar mountings from the body underframe members.

Disconnect the flexible brake hoses at the brackets on the body.

Remove the clamping bolt securing the steering column universal joint to the steering box shaft.

Lower the front suspension cross-member assembly on the jack until the front suspension assembly can be drawn forward.

Removal with car on lift

Mark the bonnet with the positions of the hinges to facilitate reassembly. Remove the four set bolts and washers attaching the bonnet to the hinges. Remove the bonnet.

Drain the radiator.

Remove the coil by unscrewing the two nuts and removing the shakeproof washers. Hang the coil on the side of the cylinder head.

Remove the four nuts and washers attaching the fan cowl to the radiator. The drain tap

---

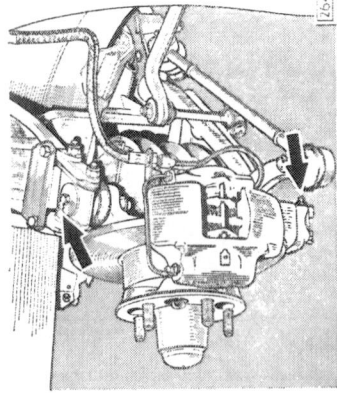

Fig. 2. *Wheel swivel grease nipples.*

ROUTINE MAINTENANCE

Front Suspension

The front suspension wishbone levers and anti-roll bar are supported in rubber bushes which do not require any attention.

Front Shock Absorbers

The front shock absorbers are of the telescopic type and no replenishment with fluid is necessary or provided for.

Every 6,000 miles (10,000 km.)

Wheel Swivels

Lubricate the nipples (four per car) fitted to the top and bottom of the wheel swivels.

The nipples are accessible from underneath the front of the car.

A bleed hole is incorporated in the ball joint. The bleed hole is covered by a circular nylon washer which lifts, under pressure allowing grease to escape and indicating when sufficient lubricant has been applied.

Front Wheel Alignment

Check that the car is full of petrol, oil and water. If not, additional weight must be added to compensate for, say, a low level of petrol (the weight of 10 gallons of petrol is approximately 80 lb.—36.0 kg.).

Ensure that the tyre pressures are correct and that the car is standing on a level surface. With the wheels in the straight ahead position check the alignment of the front wheels with an approved track setting gauge.

The front wheel alignment should be:
Parallel to $\frac{1}{8}$″ (3.2 mm.) total "toe-in" (measured at the wheel rim).

Recheck the alignment after pushing the car forward until the wheels have turned half a revolution (180°).

If adjustment is required slacken the clamp bolt at each end of the central track rod and rotate the rod in the required direction until the alignment of the front wheels is correct. Tighten the clamp bolts and recheck the alignment.

Every 12,000 miles (20,000 km.)

Front Wheel Bearings

Removal of the wheel nave plates will expose a grease nipple in the wheel bearing hubs (on cars fitted with wire wheels it will be necessary to remove the roadwheel). Lubricate sparingly with the recommended grade of lubricant. Always thoroughly clean the grease nipple before applying the grease gun.

On disc wheel hubs a bleed hole is provided in the hub cap to indicate when sufficient lubricant has been applied.

On wire wheel hubs an indication that sufficient lubricant has been applied will be the escape of grease past the nut in the bore of the splined hub.

Wheel Bearing Adjustment

Every 12,000 miles (20,000 km.) check the end-float of the front wheel bearings and adjust if necessary as detailed on page J.12.

remote control rod bracket at the top of the radiator should be refitted. Hang the fan cowl on the fan.

Slacken off the clips and remove the top and bottom hoses from the radiator.

Withdraw the set bolts and remove the shakeproof and plain washers, spacers and mounting rubbers from the top radiator mountings.

From underneath the car, unscrew the self-locking nut and withdraw the plain washer, spacer and mounting rubber from each of the lower radiator mountings. Collect the other two mounting rubbers when the radiator has been removed.

Tilt the top of the radiator towards the cylinder head and withdraw the radiator from its mountings, taking care not to foul the fan.

Place a bar under the chassis side members and attach the lifting tackle to it.

Remove the two bolts securing the front suspension rear mountings to the chassis side members.

Remove the four nuts and bolts securing the front mountings to the brackets at the front ends of the chassis side members.

Disconnect the two anti-roll bar mountings from the body underframe members.

Disconnect the flexible brake hoses at the brackets on the body.

Remove the clamping bolt securing the steering column universal joint to the steering box shaft.

Raise the front of the car by means of the lifting tackle and roll out the front suspension assembly from underneath the body.

Refitting

When refitting the front suspension cross-member assembly ensure that the brake discs are in the straight ahead position and that the steering wheel spokes are in the three and nine o'clock positions with the horn ring at the bottom.

After the front suspension assembly has been completely refitted it will be necessary to "bleed" the brake hydraulic system as described in Section L—"Brakes".

SUSPENSION BUMP STOP

A progressive type of bump stop, which consists of a tapered rubber block, is attached to a welded bracket beneath each suspension cross-member turret.

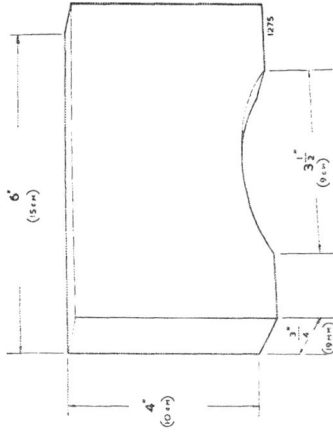

Fig. 5. Showing a fibre block which can be made up and used to support the upper wishbone levers when carrying out certain operations on the front suspension assembly.

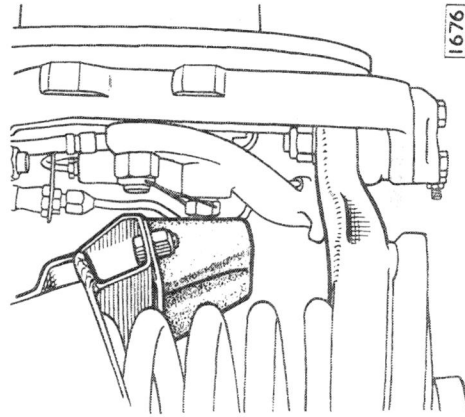

Fig. 6. The bump stop.

HYDRAULIC DAMPERS

The telescopic hydraulic dampers are of the sealed type with no provision for adjustment or "topping up" with fluid. Therefore, in the event of a damper being unserviceable a replacement must be fitted.

Before fitting a damper to a car it is advisable to carry out the following procedure to "bleed" any air from the pressure chamber that may have accumulated due to the damper having been stored in a horizontal position. Hold the damper in its normal vertical position with the shroud uppermost and make several short strokes (not extending more than halfway) until there is no lost motion and finish by extending the damper to its full length once or twice. Do not extend the damper fully until several short strokes have been made first. After the operation of "bleeding" the hydraulic dampers should be kept in their normal upright position until they are fitted to the car.

Removal

Removal of the hydraulic dampers will be facilitated if the wishbone levers are kept approximately horizontal, by either interposing a fibre packing piece between the upper wishbone levers and the cross-member turret (as illustrated in Fig. 9) or by placing a support under the brake drum and partly lowering the jack to compress the spring.

Jack up the car under the front suspension cross-member until the wheels are clear of the ground. Remove the road wheel.

Remove the locknut and nut from the top mounting of the damper and withdraw the outer washer, rubber buffer and inner washer; note the difference between the inner and outer washer.

Note : A distance piece is fitted to the damper top mounting hole which may become displaced during the removal of the damper.

Bend back the tab washers on the four setbolts attaching the hydraulic damper mounting bracket to the coil spring seat. Remove the bolts when the damper can be withdrawn.

Refitting

Refitting is the reverse of the removal procedure. Ensure that the distance piece is in position in the top mounting hole in the cross-member turret.

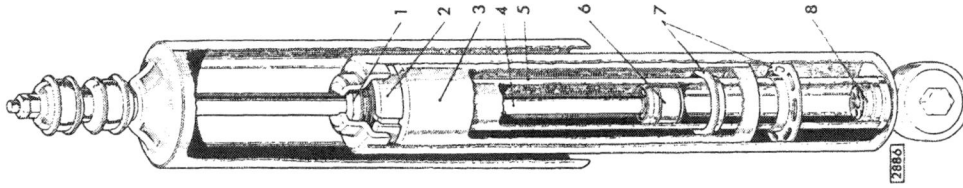

1. Cone seal
2. Piston rod guide
3. Drain sleeve
4. Piston rod
5. Pressure tube
6. Piston assembly
7. Fluid baffles
8. Base valve

Fig. 7. Sectioned view of the hydraulic damper.

COIL SPRINGS

The coil springs are marked with coloured paint on the middle coil (which may be covered by tape) to denote springs of the same static load. It is, therefore, important that the two front springs fitted to a car are of the same colour code. (see page J.11).

Removal

Remove the hydraulic damper as described on page J.9.

Insert a suitable coil spring compressor (Churchill Tool No. JD6A) through the centre of the spring and compress the spring sufficiently to relieve the load on the spring seat pan screws.

Detach the anti-roll bar link arm from the bracket welded to the rear edge of the spring pan by withdrawing the nut and bolt.

Remove the six setscrews and spring washers which secure the seat pan to the lower wishbone.

Release the coil spring compressor until the load of the spring is completely relieved. Completely unscrew the compressor when the coil spring and seat pan can be removed.

Note : On some cars a packing piece may be fitted at the top of the spring, see "Coil Spring Packing Pieces" overleaf.

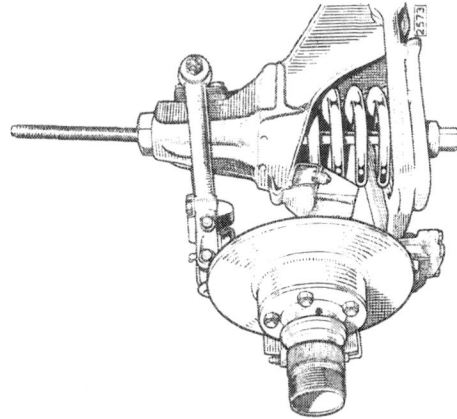

Fig. 8. *Showing the compressor (Churchill Tool No. JD64) in position prior to the removal of the coil spring.*

Fig. 9. *Removal of the hydraulic damper.*

Dismantling

Extract the grease seal. Withdraw the inner races of the taper roller bearings. If new bearings are to be fitted the outer races can be drifted out, grooves being provided in the abutment shoulders in the hub.

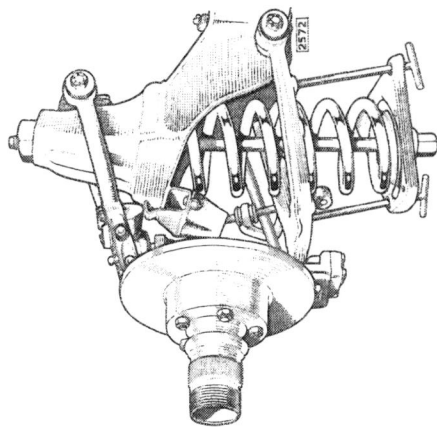

Refitting

Refitting is the reverse of the removal procedure. Alignment of the seat pan holes with the tapped holes in the lower wishbone will be facilitated if 8" (20 cm.) long pilot studs (threaded 3/8" U.N.F.) are fitted as illustrated in Fig. 11.

Coil Spring Packing Pieces

Packing pieces may be fitted above the coil springs of some cars, their purpose being to accommodate the manufacturing variations in the length of the springs which are graded into three groups and identified by a colour patch on its middle coil as previously mentioned. Disregard the stripe of paint along the full length of the spring.

Colour Code of Spring		Thickness of Packing
Red	.. .. ..	1/4" (6.4 mm.)
Yellow	.. .. ..	1/8" (3.2 mm.)
Purple	.. .. ..	None fitted

Commencing chassis numbers 1B 4000 (3.4 "S") and 1B 54357 (3.8 "S"), the right hand spring on **right hand drive cars only** has 1/8" (3.2 mm.) more packing than the left hand spring to equalise the standing height of the car side for side.

Packing rings are fitted in accordance with the following details:—

Colour		Packing Required R.H. Spring	L.H. Spring
Red	.. ..	3/8" (9.5 mm.)	1/4" (6.4 mm.)
Yellow	.. ..	1/4" (6.4 mm.)	1/8" (3.2 mm.)
Purple	.. ..	1/8" (3.2 mm.)	None

WHEEL HUBS

Removal

Firmly apply the handbrake, jack up the front of the car and remove the roadwheel.

Remove the caliper from the front stub axle carrier, noting the shims fitted at the mounting points for centralisation of the disc within the caliper.

Remove the split pin retaining the hub nut. On cars with disc wheels the split pin is accessible after prising off the end cap; cars with wire spoke wheels are provided with holes in the side of the hub through which the split pin can be withdrawn.

Remove the slotted nut and plain washer from the end of the stub axle shaft. The hub can now be withdrawn by hand.

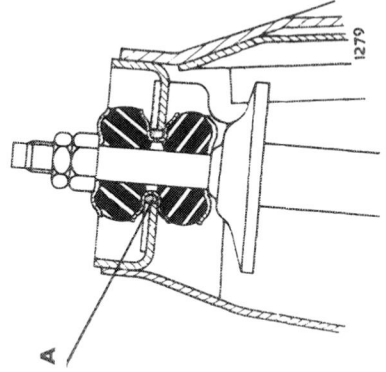

Fig. 10. *When refitting an hydraulic damper ensure that the distance piece "A" is in position.*

Fig. 11. *Refitting the coil spring with a compressor; the pilot studs facilitate alignment of the seat pan securing bolt holes.*

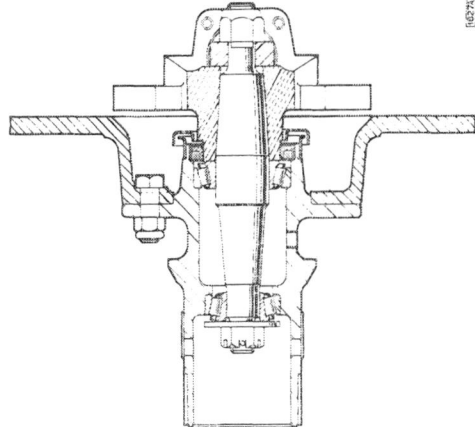

Fig. 12. *Sectional plan view of the disc brake hub arrangement.*

Refitting

Refitting is the reverse of the removal procedure but it will be necessary to re-lubricate the bearings as detailed in "Routine Maintenance" at the beginning of this section and adjust the end-float of the hub bearings as described under the next heading.

Refit the brake caliper, ensuring it is correctly aligned and bleed the brake hydraulic system as detailed in Section L "Braking System".

Bearing End-float Adjustment

The correct end-float of the wheel bearings is ·003" to ·005" (·07 mm. to ·13 mm.). On cars with disc brakes it is particularly important that the end-float does not exceed ·005" (·13 mm.) otherwise the brakes may tend to drag and not function correctly.

The wheel bearing end-float can be measured with a dial indicator gauge, mounted with the plunger against the end of the hub. If a gauge is not available proceed as follows :—

Tighten the hub nut until there is no end-float, that is, when the rotation of the hub feels slightly restricted.

Slacken back the hub nut between one and two flats dependent upon the position of the split pin hole relative to the slots in the nut.

Temporarily attach the road wheel and check that the wheel spins freely.

If satisfactory, fit a new split pin and turn over the ends.

Fig. 13. Exploded view of the wheel hub.

STUB AXLE CARRIERS

Removal

Jack up under the lower wishbone lever and remove the road wheel.

Remove the caliper from the stub axle carrier, noting the shims fitted at the mounting points, and remove the front wheel hub complete with disc brake as described on page J.11.

Remove the self-locking nut and plain washer securing the upper ball joint to the stub axle carrier. Drift out the ball pin from the stub axle carrier in which it is a taper fit.

Remove the split pin, nut and plain washer which secure the ball joint to the lower wishbone.

Drift out the ball pin from the lower wishbone, in which it is a taper fit, when the stub axle carrier can be removed.

Refitting

Refitting is the reverse of the removal procedure.

LOWER WISHBONE

Removal

Remove the coil spring as described on page J.10.

Remove the stub axle carrier as described above.

Withdraw the split pin, slotted nut and washer from one end of the lower wishbone fulcrum shaft. The shaft can now be drifted out.

Fitting the Rubber/Steel Bushes

Drift out or press out the bush from the wishbone eye. Press the new bush into the eye, ensuring that the bush projects from each side by an equal amount. Fitting of the bush will be facilitated if a lubricant, made up of twelve parts of water to one part of liquid soap, is used.

Refitting

Refitting is the reverse of the removal procedure. When refitting the fulcrum shaft the car should be in the normal riding position before the units at each end of the shaft are fully tightened. Omitting to carry out this procedure will result in undue torsional loading of the rubber bushes with possible premature failure.

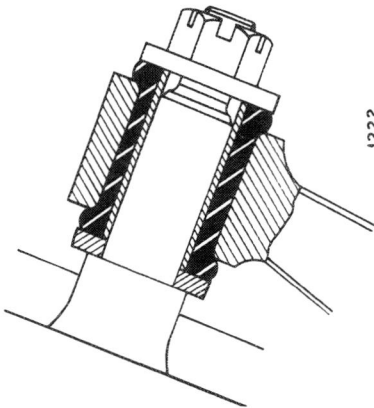

Fig. 14. Section through one of the lower wishbone rubber/steel mounting bushes.

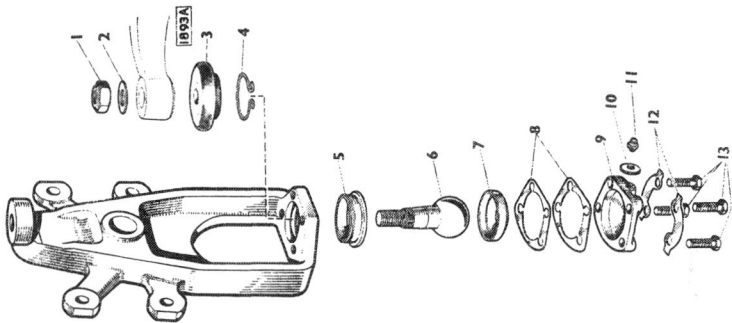

Fig. 15. Exploded view of the lower wishbone ball joint.

LOWER WISHBONE BALL JOINT

Removal

Remove the stub axle carrier complete with the lower wishbone ball joint as described on page J.13.

Dismantling

Release the wire clip and remove the rubber gaiter. Withdraw the retainer from the top of the ball pin.

Later cars have a plastic ring securing the rubber gaiter to the ball pin spigot with a further insert between the ball pin spigot and gaiter.

Tap back the tab washers and unscrew the four setscrews securing the ball pin cap to the stub axle carrier.

Remove the cap, shims, ball pin socket, spigot and ball pin.

Reassembling

Reassembling is the reverse of the dismantling procedure but, if necessary, re-shim the ball joint to obtain the correct clearance of ·004"—·006" (·10 mm.—·15 mm.).

Note: Shims should not be removed to take up excessive wear in the ball pin and socket; if these parts are badly worn, replacements should be fitted.

Adjustment of the Ball Joint

The correct clearance of the ball pin in its socket is ·004"—·006" (·10 mm.—·15 mm.). Shims for adjustment of the ball joint are available in ·002" (·05 mm.) and ·004" (·10 mm.) thicknesses.

To adjust the ball pin clearance to the correct figure, remove shims one by one until, with ball cap fully tightened, the ball is tight in its socket. Fit shims to the value of ·004"—·006" (·10 mm.—·15 mm.) which should enable the shank of the ball pin to be moved by hand.

Refitting

Refit the stub axle carrier complete with the lower wishbone ball joint as described on page J.13.

UPPER WISHBONE

Removal

Jack up under the lower wishbone and remove the road wheel.

Remove the two bolts, nuts and plain washers securing the ball joint to the upper wishbone levers. Note the relative positions of the packing piece and shims as these control the castor angle. Alternatively, remove the self-locking nut and drift out the ball joint from the stub axle carrier. Tie-up the stub axle carrier to the suspension cross-member so that the flexible brake hose does not become extended.

Remove the four set bolts which secure the upper wishbone fulcrum shaft to the suspension cross-member turret. Note the relative positions of the shims as these control the camber angle. The upper wishbone assembly can now be removed.

Dismantling

Remove the nuts, bolts and distance pieces securing the rebound stop bracket to the upper wishbone levers.

Extract the split pin and remove the slotted nuts and plain washers which secure the wishbone levers to the fulcrum shaft. The wishbone levers can now be removed from the fulcrum shaft.

Fitting the Rubber/Steel Bushes

Drift out or press out the bush from the wishbone eye. Press the new bush into the eye, ensuring that the bush projects from each side by an equal amount. Fitting of the bush will be facilitated if a lubricant, made up of twelve parts of water to one part of liquid soap, is used.

Reassembling

The reassembly of the upper wishbone assembly is the reverse of the dismantling procedure but the slotted nuts securing the wishbone levers to the fulcrum shaft must not be tightened until the upper wishbone assembly has been refitted and the full weight of the car is on the suspension. Omitting to carry out the procedure will result in undue torsional loading of the rubber bushes with possible premature failure.

Refitting

Refitting is the reverse of the removal procedure.

Fig. 16. Section through one of the upper wishbone rubber/steel mounting bushes.

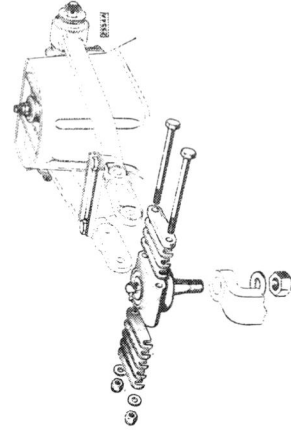

UPPER WISHBONE BALL JOINT

The upper wishbone ball joint cannot be dismantled and, if worn, the complete assembly must be replaced.

Removal

Jack up the car under the lower wishbone and remove the road wheel.

Remove the two bolts, nuts and plain washers securing the ball joint to the upper wishbone levers. Note the relative positions of the packing piece and shims as these control the castor angle.

Remove the self-locking nut and plain washer which secure the ball joint to the stub axle carrier.

The ball joint can now be drifted out of the stub axle carrier in which it is a taper fit.

Note: When carrying out the above operation do not allow the flexible brake hose to become extended; tie up the stub axle carrier to the cross-member turret.

Refitting

Refitting is the reverse of the removal procedure. Ensure that the packing piece and shims are refitted in their original positions otherwise the castor angle will be upset.

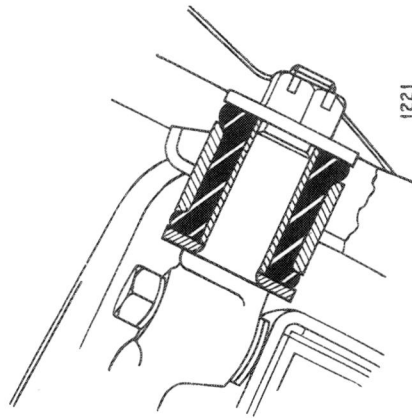

Fig. 17. Removal of the upper wishbone ball joint.

CASTOR ANGLE ADJUSTMENT

Check that the car is full of petrol, oil and water. If not, additional weight must be added to compensate for, say, a low level of petrol (the weight of 10 gallons of petrol is approximately 80 lb.—36·0 kg.).

Ensure that the tyre pressures are correct and that the car is standing on a level surface.

Using an approved gauge, check the castor angle.

Castor angle .. .. $0 \pm \frac{1}{2}°$

Note : The two front wheels must be within a $\frac{1}{2}°$ of each other.

Adjustment is effected by either transposing the shims from the rear of the upper wishbone ball joint to the front, or transposing the packing piece and shim(s).

To decrease negative castor or increase positive castor transpose shims from the rear to the front; the holes in the shims are slotted and therefore it will only be necessary to slacken the two bolts securing the upper wishbone members to enable the shims to be removed.

To increase negative castor or decrease positive castor, transpose the packing piece and shims as necessary. As the holes in the packing piece are not slotted it will be necessary to remove the two bolts after first having placed a support under the brake disc or lower wishbone.

The shims are $\frac{1}{16}$″ (1·6 mm.) thick and it should be noted that $\frac{1}{16}$″ (1·6 mm.) of shimming will alter the castor angle by approximately $\frac{1}{4}°$.

The front of the car should be jacked up when turning the wheels from lock to lock during checking.

If any adjustment is made to the castor angle, the front wheel alignment should be checked and, if necessary, re-set.

Note : A packing piece and 8 shims must be always fitted between the wishbone levers and the upper ball joint; their relative positions may, of course, not always be the same.

CAMBER ANGLE ADJUSTMENT

Check that the car is full of petrol, oil and water. If not, additional weight must be added to compensate for, say, a low level of petrol (the weight of 10 gallons of petrol is approximately 80 lb.—36·0 kg.).

Ensure that the tyre pressures are correct and that the car is standing on a level surface.

Line up the front wheel being checked parallel to the centre line of the car. Using an approved gauge, check the camber angle. Rotate the wheel being checked through 180° and re-check.

Camber angle .. $\frac{1}{2} \pm \frac{1}{2}°$ positive.

Note : The two front wheels must be within a $\frac{1}{2}°$ of each other.

Adjustment is effected by removing or adding shims at the front suspension top wishbone bracket; the holes in the shims are slotted and it is therefore only necessary to slacken the setscrews securing the bracket to enable the shims to be removed. Inserting shims decreases positive camber; removing shims decreases negative camber or increases positive camber. Remove or add an equal thickness of shims from each position, otherwise the castor angle will be affected. Shims for the adjustment of camber are available in $\frac{1}{32}$″ (·8 mm.) $\frac{3}{64}$″ (1·2 mm.) and $\frac{1}{16}$″ (1·6 mm.) thicknesses and it should be noted that $\frac{1}{16}$″ (1·6 mm.) of shimming will alter the camber angle by approximately $\frac{1}{4}°$.

Check the other front wheel in a similar manner. If any adjustment is made to the camber angle, the front wheel alignment should be checked and, if necessary, reset.

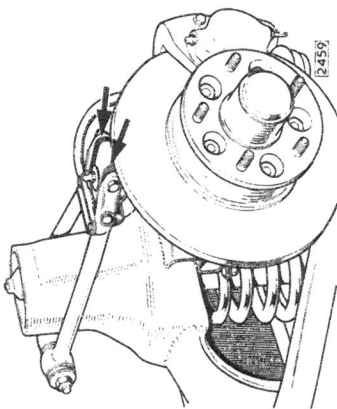

Fig. 18. *The castor angle is adjusted by means of the shims and packing piece indicated by the arrows.*

ANTI-ROLL BAR (Fig. 3)

Removal

Raise the car on a lift to enable work to be carried out underneath. Remove the four bolts from the anti-roll bar support brackets (73) on the chassis side members.

Remove the self-locking nut and remove the bolt attaching the link arm (77) to the coil spring seat. Repeat for the other side.

To separate the anti-roll bar (72) from the link arms (77) remove the self-locking nuts, upper cup washers and rubbers. Care should be taken to replace the distance tube (80) when refitting.

The anti-roll bar bracket rubbers (74) are split to enable them to be removed.

Fitting the Link Arm Bush

Drift out or press out the bush from the link arm eye.

Press the new bush into the eye, ensuring that the bush projects from each side by an equal amount. Fitting of the bush will be facilitated if a lubricant, made up of twelve parts of water to one part of liquid soap, is used.

Refitting

Refitting is the reverse of the removal procedure. It is important when attaching the support brackets to the chassis side members, to have the full weight of the car on the road wheels.

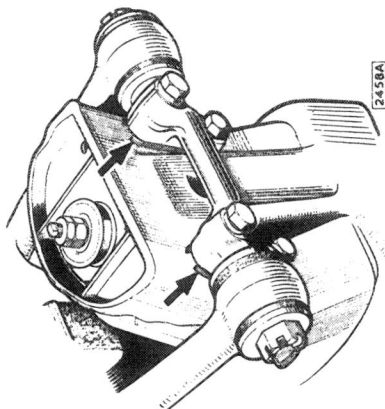

Fig. 19. *The front wheel camber is adjusted by means of the shims indicated by the arrows. Remove or add an equal thickness of shims from each position.*

ACCIDENTAL DAMAGE

The dimensioned drawings are provided to assist in assessing accidental damage. A component suspected of being damaged should be removed from the car, cleaned off and the dimensions checked and compared with those given in the appropriate illustration.

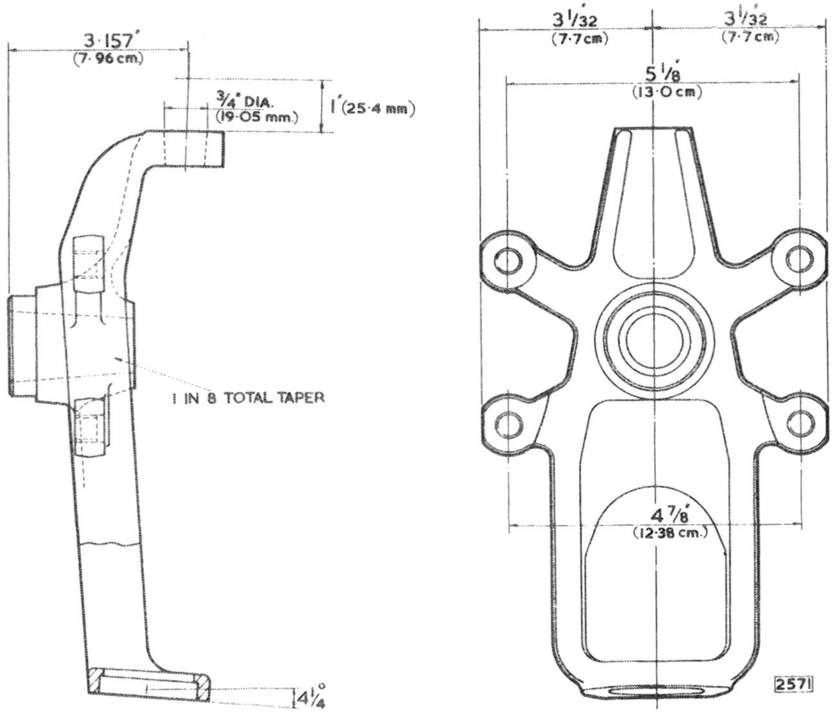

Fig. 20. *Stub axle carrier.*

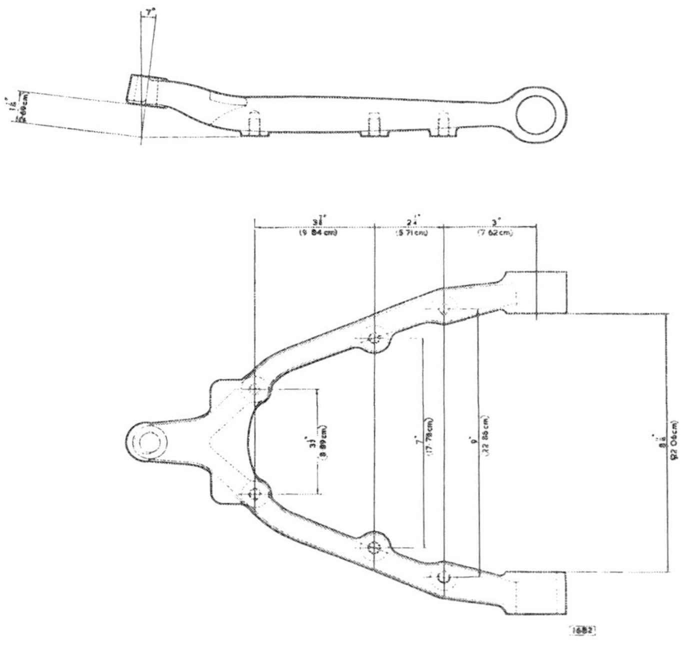

Fig. 21. *Lower wishbone.*

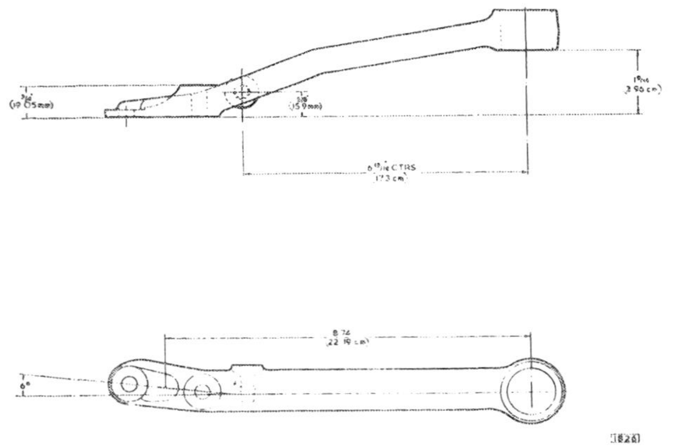

Fig. 22. *Upper wishbone.*

SECTION K

REAR SUSPENSION

REAR SUSPENSION

INDEX

REAR SUSPENSION

DESCRIPTION

The rear wheels are located in a transverse plane by two tubular links of which the top link is the half shafts universally jointed at each end. The lower link is pivoted at the wheel carrier and at the crossbeam adjacent to the differential casing. To provide maximum rigidity in a longitudinal plane the pivot bearings at both ends of the lower link are widely spaced. The suspension medium is provided by four coil springs enclosing telescopic hydraulic dampers, two being mounted on either side of the differential casing. The complete assembly is carried in a fabricated steel crossbeam. The crossbeam is attached to the body by four "Vee" rubber blocks and is located by radius arms. The radius arm pivots are rubber bushes mounted on each side of the car between the lower link and a mounting point on the body structure.

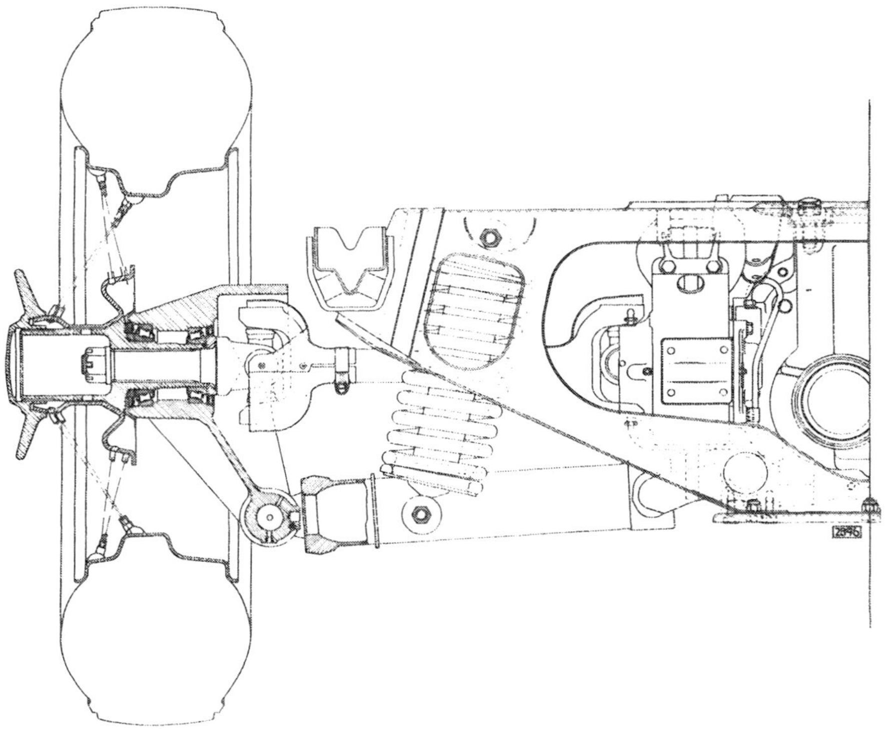

Fig. 1. *Sectioned view of rear suspension.*

DATA

Rear Road Spring:

Free length	11·395" (28·9 cm.)
Identification colour	Red/Yellow
No. of coils (approx.)	8¼
Wire diameter	·475" (12·06 mm.)
Dampers	Telescopic

Road Wheel Movement from mid-laden position:

Full Bump	3¾" (9·52 cm.)
Full Rebound	3¾" (9·52 cm.)
Track	58" (147·32 cm.)
Rear Wheel Camber	see page K.14

ROUTINE MAINTENANCE

Every 6,000 miles (10,000 km.)

Outer Pivot Bearings

A grease nipple is located in the centre of the rear wishbone outer pivot. Lubricate sparingly with the recommended grade of lubricant. A bleed hole is provided, opposite the grease nipple, to indicate when an excess of lubricant has been applied. Always ascertain that the bleed hole is clear before carrying out this operation.

Inner Pivot Bearing

Two grease nipples are provided, one at each end of the wishbone fork. Lubricate sparingly with the recommended grade of lubricant.

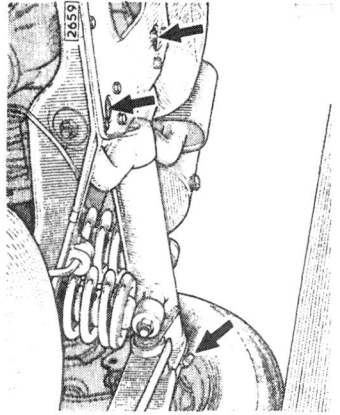

Fig. 2. Inner and outer pivot bearing grease nipples.

Recommended Lubricants

	Mobil	Castrol	Shell	Esso	B.P.	Duckham	Regent Caltex/Texaco
Wishbone Pivots	Mobilgrease MP	Castrolease Medium or LM	Retinax A	Esso Multi-purpose Grease H	Energrease L2	LB10	Marfak Multi-purpose 2

Removal

Slacken the two clamp bolts which secure the muffler boxes to the rear silencers.

Remove the four nuts and washers retaining the muffler mounting rubbers to the underside of the car.

Withdraw the mufflers.

Remove the locking wire from the radius arm safety strap and securing bolt.

Unscrew the two bolts securing the safety strap to the body floor.

Remove the radius arm securing bolt and spring washer and remove the safety strap.

Withdraw the radius arm from the mounting post on the body.

Place a stout piece of wood approximately 9" × 9" × 1" (22·8 cm. × 22·8 cm. × 25·4 mm.) between the rear suspension tie plate and the jack.

Jack up the rear of the car and place two chassis stands of equal height under the body forward of the radius arm mounting posts. Place blocks of wood between the chassis stands and the body to avoid damage.

Remove the rear road wheels.

Disconnect the flexible brake pipe at the connection on the body.

Remove the split pin, washer and clevis pin securing the handbrake cable to the handbrake caliper actuating levers mounted on the suspension cross beam.

Slacken the locknut and screw the outer handbrake cable screw out of the adjuster block.

Remove the four bolts and self-locking nuts securing the mounting rubbers at the front of the cross beam to the body frame. Remove the six self-locking nuts and four bolts securing the rear mounting rubbers to the cross beam.

Remove the four self-locking nuts and bolts securing the propeller shaft to the differential pinion flange.

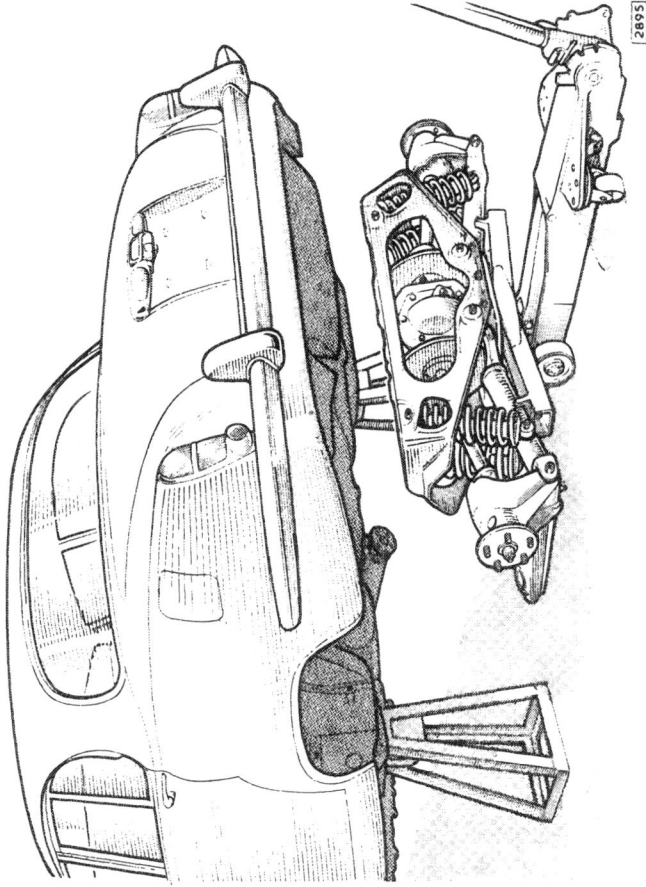

Fig. 3. Removal of the rear suspension assembly from the car.

Lower the rear suspension unit on the jack and withdraw the unit from under the car as shown in Fig. 3.

Refitting

Refitting is the reverse of the removal procedure.

Check all mounting rubbers for deterioration.

Bleed the braking system as described in Section L. "Brakes."

If the radius arms have been removed the rear suspension should be at the normal riding height before tightening the radius arm securing nuts on the rear suspension wishbone. Refit the radius arms as described on page K.7.

IMPORTANT

The following removal and refitting operations are described assuming the rear suspension is removed from the car. If it is possible for the operations to be carried out with the rear suspension in position on the car the fact will be noted in the text.

ROAD SPRING AND HYDRAULIC DAMPER ASSEMBLY

Removal

The road spring and hydraulic damper assembly may be removed from the car with the rear suspension assembly in position.

Remove the two self-locking nuts and washers securing the two hydraulic dampers to the wishbone.

Support the appropriate wishbone and drift out the hydraulic damper mounting pin, Fig. 6.

Remove the self-locking nut and bolt securing each hydraulic damper to the cross beam.

Withdraw the hydraulic damper and road spring assembly.

Refitting

Refitting is the reverse of the removal procedure.

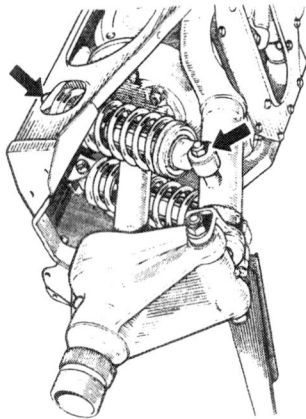

Fig. 4. Hydraulic damper mounting points (Wire Wheels).

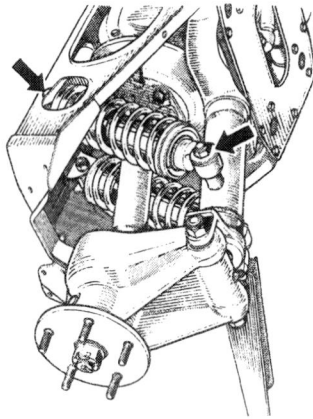

Fig. 5. Hydraulic damper mounting points (Disc Wheels).

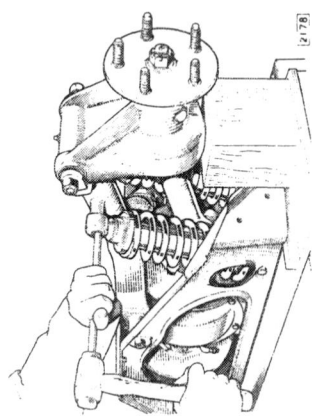

Fig. 6. Drifting out the hydraulic damper mounting pin.

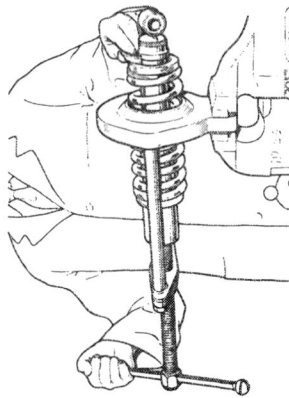

Fig. 7. Removing the rear road spring from the hydraulic damper with Churchill Tool J.11A in conjunction with SL.14.

HYDRAULIC DAMPERS

The telescopic hydraulic dampers are of the sealed type with no provision for adjustment or "topping-up" with fluid. Therefore, in the event of a damper becoming unserviceable a replacement must be fitted.

Before fitting a damper to a car it is advisable to carry out the following procedure to "bleed" any air from the pressure chamber that may have accumulated due to the damper having been stored in the horizontal position. Hold the damper in its normal vertical position with the shroud uppermost and make several short strokes (not extending more than half way) until there is no lost motion. Finish by extending the damper to its full length once or twice. Do not extend the damper fully until several short strokes have been made first. After the operation of "bleeding," the hydraulic dampers should be kept in their normal upright position until they are fitted to the car.

Removal

Remove the road spring and hydraulic damper as described on page K.6.

Utilizing a suitable press, Fig. 7, compress the road spring until the split collet can be removed from under the road spring retainer.

Carefully release the pressure on the road spring and withdraw the hydraulic damper.

Refitting

Compress the road spring, utilizing Churchill Tool No. J.11A and SL.14, sufficiently to allow the hydraulic damper to be passed through the road spring. Fit the packing ring, spring and split collet. Ensure that the split collet and spring are seating correctly. Release the pressure on the road spring.

Refit the road spring and hydraulic damper assembly as described on page K.6.

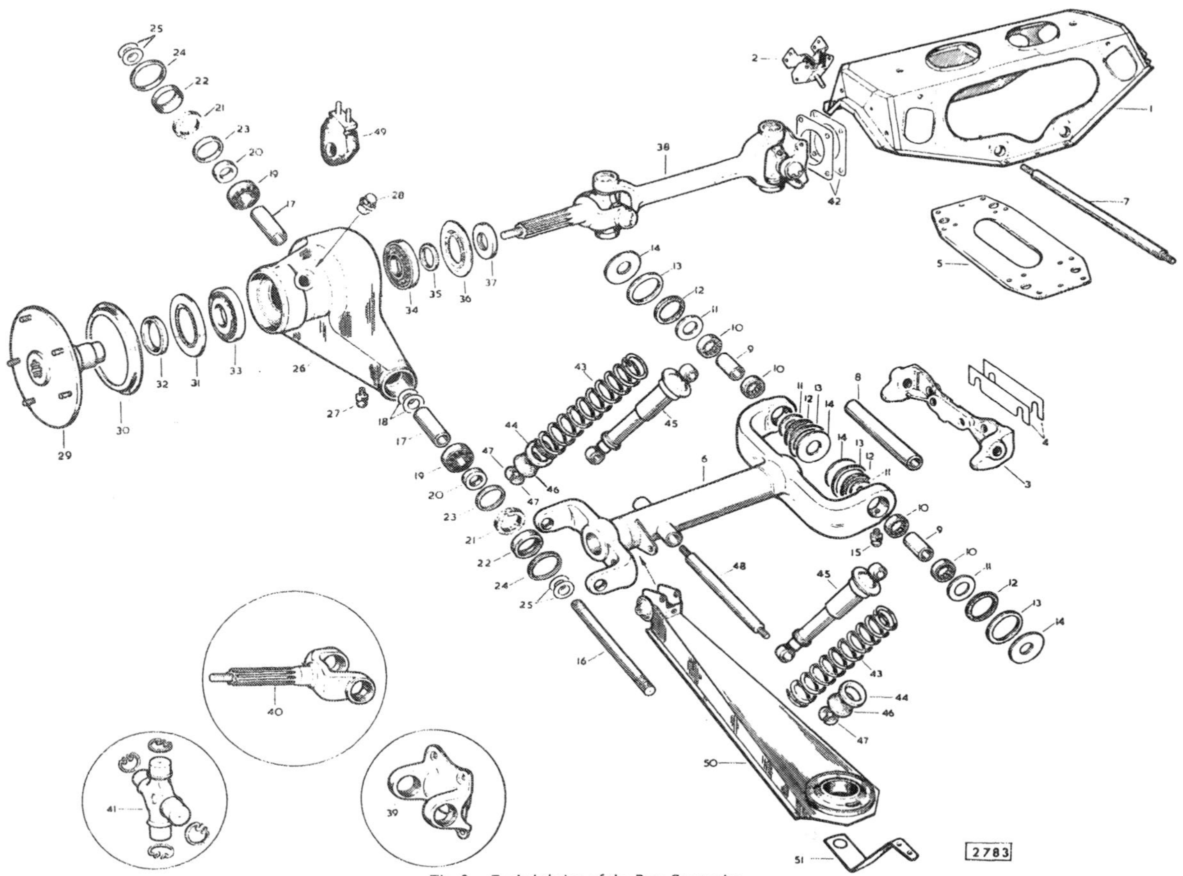

1. Rear suspension cross member
2. Rubber mounting
3. Inner fulcrum mounting
4. Shim
5. Bracing plate
6. R.H. wishbone
7. Fulcrum shaft
8. Distance tube
9. Bearing tube
10. Needle bearing
11. Thrust washer
12. Sealing ring
13. Retainer
14. Thrust washer
15. Grease nipple
16. Fulcrum shaft
17. Sleeve
18. Shim
19. Bearing
20. Seating ring
21. Oil seal
22. Container
23. Spacer
24. Retaining washer
25. Shim
26. Hub carrier

27. Grease nipple
28. Grease retaining cap
29. Rear hub
30. Water thrower
31. Oil seal (outer)
32. Seating ring
33. Outer bearing (taper roller)
34. Inner bearing (taper roller)
35. Spacer
36. Oil sealer (inner)
37. Seating ring
38. Half shaft
39. Flange yoke
40. Splined yoke
41. Journal
42. Shim
43. Rear suspension coil spring
44. Packing ring
45. Rear suspension damper
46. Seat
47. Retainer
48. Mounting shaft
49. Bump stop
50. Radius arm
51. Safety strap

2783

Fig. 9. *Exploded view of the Rear Suspension.*

RADIUS ARM

Removal

Remove the locking wire from the radius arm safety strap and securing bolt.

Unscrew the two bolts securing the safety strap to the body floor.

Remove the radius arm securing bolt and spring washer and remove the safety strap.

Withdraw the radius arm from the mounting post on the body.

Remove one of the self-locking nuts securing the hub bearing assembly fulcrum shaft to the wishbone.

Drift out the fulcrum shaft from the wishbone and hub assembly as described on page K.10.

Remove the self-locking nut and bolt securing the radius arm to the wishbone and remove the radius arm.

Examine the radius arm mounting rubbers for deterioration.

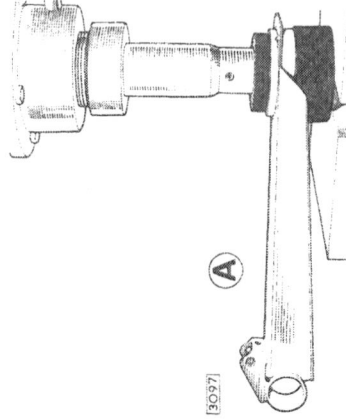

Refitting

Refitting is the reverse of the removal procedure.

If the rubber bushes need replacing they can be removed and refitted by means of Churchill Tool Number J.21.

When replacing the large rubber bush the two holes should be in the longitudinal position in the radius arm.

When fitting the smaller bush ensure that an equal amount of the steel centre sleeve protrudes from each side of the radius arm.

When refitting the hub bearing assembly shaft refer to page K.12.

Refit the safety strap into position, refit the spring washer and radius arm securing bolt.

Refit the two bolts and nuts securing the safety strap to the body.

Tighten the radius arm securing bolt to 46 lb. ft. (6.36 kg.m.) and pass the locking wire through the hole in the head of the bolt and secure round the safety strap.

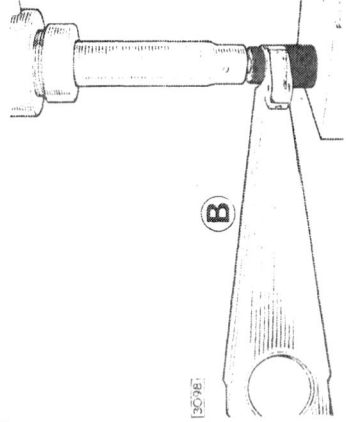

A. Removing large rubber bush

B. Removing small rubber bush

C. Refitting large rubber bush

Fig. 8. *Showing removal of radius arm bushes (Churchill Tool No. J.21).*

WISHBONE

Removal

Remove the hydraulic dampers as described on page K.6.

Remove the six self-locking nuts and bolts securing the tie plate to the cross beam.

Remove the eight self-locking nuts and bolts securing the tie plate to the inner fulcrum wishbone mounting brackets and remove the tie plate, see Fig. 11.

Remove one of the self-locking nuts securing the hub bearing assembly fulcrum shaft to the wishbone and drift out the fulcrum shaft, see Fig. 6.

Separate the hub carrier from the wishbone. If any shims are fitted between the wishbone and hub assembly note the amount and position of the shims as it is essential to replace the exact amount in the correct position. To facilitate refitting, slide a dummy fulcrum shaft Churchill Tool No. J.14 through the hub carrier.

Place a piece of sticky tape over each of the hub carrier assembly oil seal tracks to prevent them becoming displaced.

Remove the self-locking nut securing the radius arm to the wishbone. Withdraw the special thin headed bolt and remove the radius arm from the wishbone.

Remove the self-locking nut securing the wishbone fulcrum shaft to the cross beam.

Drift the inner fulcrum shaft out of the wishbone and inner fulcrum mounting bracket.

Withdraw the wishbone assembly and collect the four outer thrust washers, inner thrust washers, oil seals and oil retainers.

Examine the oil seals for deterioration.

Remove the two bearing tubes.

There is no need to remove the spacer fitted between the inner fulcrum mounting bracket unless the mounting bracket is to be replaced. To remove the spacer, tap out of position. To remove the needle rollers gently tap the needle cages out of the wishbone using a suitable drift. Remove the needle roller spacer.

Refitting

If the needle rollers have been removed from the larger fork of the wishbone lever press one roller cage into position, with the engraving on the roller cage facing outwards.

Insert the roller spacing tube and press in the other roller cage.

Repeat for the other side.

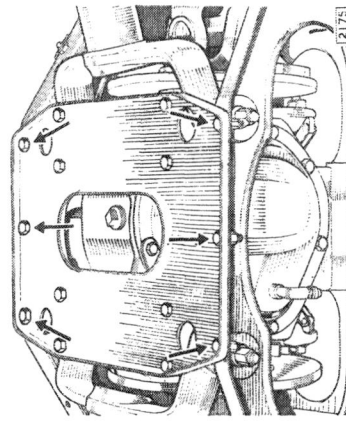

Fig. 10. *Showing the six bolts which secure the tie plate to the cross beam.*

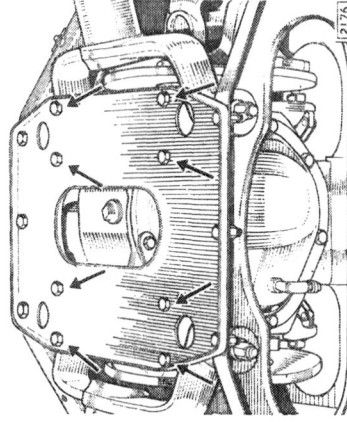

Fig. 11. *Showing the eight bolts which secure the tie plate to the inner fulcrum mounting bracket.*

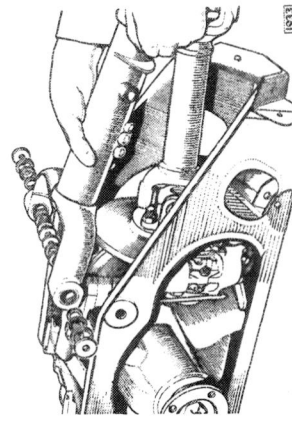

Fig. 12. *Showing the wishbone inner fork and components.*

Insert the bearing tubes. Smear the four outer thrust washers, inner thrust washers, oil seals and oil seal retainers with grease and place into position on the wishbone, see Fig. 12.

Offer up the wishbone to the inner fulcrum mounting bracket with the radius arm mounting bracket towards the front of the car. Align the holes and spacers. Press a dummy shaft Churchill Tool No. J.14 through each side of the cross beam and wishbone.

The dummy shafts locate the wishbone, thrust washers, cross beam and inner fulcrum mounting bracket and facilitate refitting of the fulcrum shaft.

Smear the fulcrum shaft with grease and gently tap the shaft through the cross beam, wishbone and inner fulcrum mounting bracket. As the fulcrum shaft is tapped into position the short dummy shafts will be displaced from the opposite side. It will be found advantageous to keep a slight amount of pressure exerted on the dummy shafts as they emerge from the cross beam. This will reduce the tendency for the dummy shafts to be knocked out of position and allow a spacer or thrust washer to be displaced. If a washer or spacer becomes displaced it will be necessary to remove the fulcrum shaft, dummy shafts and wishbone and then repeat the operation.

When the fulcrum shaft is in position tighten the two self-locking nuts to 55 lb. ft. (7·60 kg.m.) with a torque wrench.

Refit the eight bolts and self-locking nuts securing the tie plate to the inner fulcrum wishbone mounting bracket, see Fig. 11.

Refit the six bolts and self-locking nuts securing the tie plate to the cross beam, see Fig. 10.

Refit the radius arm to the wishbone as described on page K.7.

Remove the two pieces of sticky tape holding the oil seal tracks in position.

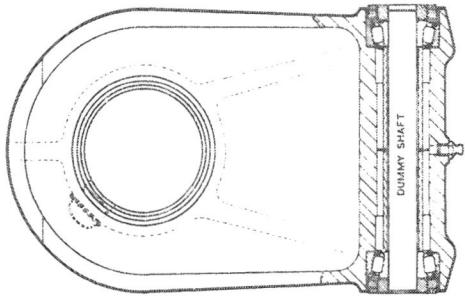

Fig. 13. *Showing the dummy shaft in position in the hub carrier.* [2217]

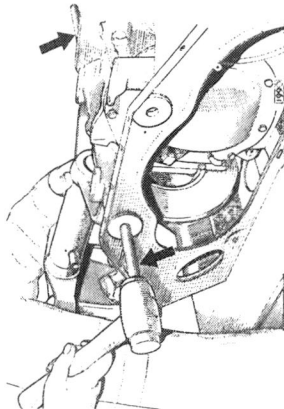

Fig. 14. *Tapping the dummy shafts into position at the wishbone inner fulcrum.*

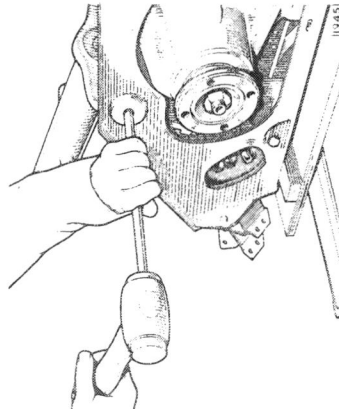

Fig. 15. *Drifting the inner fulcrum shaft into position and displacing the dummy shafts.* [1945]

---

seal tracks and spacers. Smear the fulcrum shaft with grease and gently tap the fulcrum shaft into position and displace the dummy shaft.

It will be found advantageous to apply a small amount of pressure on the dummy bar against the fulcrum shaft to prevent the bar being knocked out of position and allowing a spacer to be displaced. If a spacer is displaced it may be necessary to repeat the operation.

Slide the fulcrum shaft through the wishbone and hub carrier. Using feeler gauges check the amount of clearance between the hub carrier and the wishbone lever, see Fig. 19. If necessary fit sufficient shims between the hub carrier and the wishbone to centralize the hub carrier and the wishbone to centralize the hub carrier. Tighten the nuts on the fulcrum shaft to 55 lb. ft. (7·60 kg.m.).

Check the rear suspension camber angle as described on page K.14.

Refit the hydraulic dampers as described on page K.7.

Refit the rear suspension as described on page K.6.

Re-lubricate the wishbone fulcrum shafts as described in "Routine Maintenance" at the beginning of this section.

WISHBONE OUTER PIVOT

Removal

Support the hub carrier and wishbone.

Remove one of the self-locking nuts securing the outer fulcrum shaft.

Drift out the fulcrum shaft, Fig. 16, and collect the shims, if any, between the hub carrier and the wishbone.

Separate the hub carrier and wishbone.

Dismantling

Remove the oil seal track and prise out the oil seals.

Remove the inner races of the tapered roller bearings by tapping out with the aid of a drift in the grooves provided.

Remove the spacers and shims.

Reassembly

Refit the inner races for the tapered roller bearings.

Fit the spacers and a known quantity of shims, this is necessary to obtain the correct bearing adjustment as described in the following paragraphs.

Fit the tapered roller bearings and oil seal tracks.

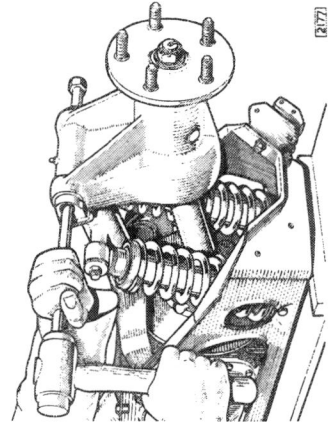

Fig. 16. *Drifting out the wishbone outer fulcrum shaft.* [2177]

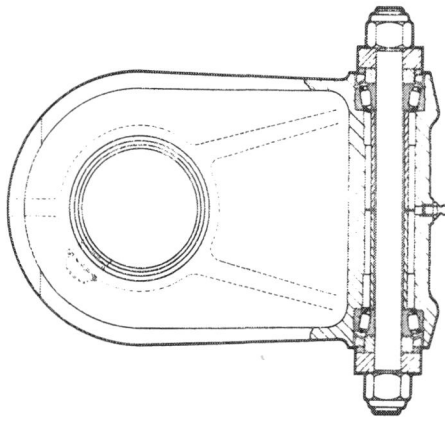

Fig. 17. *Section through hub carrier and wishbone showing outer fulcrum shaft in position.* [2185]

Offer up the wishbone to the hub assembly. Using a dummy shaft, Churchill tool No. J.14, line up the wishbone hub assembly oil

Bearing Adjustment

If it is necessary to adjust the tapered roller bearings it will be necessary to extract the hub from the rear axle half shaft as described in Section H "Rear Axle."

Bearing adjustment is effected by shims fitted between the two fulcrum shaft spacer tubes. The correct bearing adjustment is ·000"—·002" (·00 mm.—·05 mm.) pre-load.

Shims are available in sizes of ·004" (·101 mm.) and ·007" (·17 mm.) thick and 1⅛" (28·67 mm.) diameter.

A simple jig should be made consisting of a piece of plate steel approximately 7" × 4" × ⅜" (17·7 cm. × 10·1 cm. × 9·5 mm.). Drill and tap a hole suitable to receive the outer fulcrum shaft. Place the steel plate in a vice and screw the fulcrum shaft into the plate and slide an oil seal track onto the shaft. Place the assembly into position on the fulcrum shaft minus the oil seals and with an excess of shims between the spacers. Place an inner wishbone fork outer thrust washer onto the fulcrum shaft so that it abuts the oil seal track. Fill the remaining space on the shaft with washers and secure with a nut. Tighten the nut to 55 lb. ft. (7·60 kg.m.).

Press the hub carrier assembly towards the steel plate using a slight twisting motion to settle the rollers onto the bearing surface. Maintain a steady pressure against the hub carrier and using a feeler gauge measure the amount of clearance between the large diameter washer and the machined face of the hub carrier.

Pull the hub carrier assembly towards the large diameter washer slightly rotating the carrier to settle the rollers onto the bearing surface. Maintain a steady pressure against the hub carrier and using feeler gauges measure the amount of clearance between the large diameter washer and the machined face of the hub carrier.

Subtract the one measurement from the other which gives the amount of end float present in the bearings.

Remove sufficient shims to obtain a reading of ·000"—·002" (·00 mm.—·05 mm.) preload.
Example:—

Correct preload ·000"—·002" (·00 mm.—·05 mm.).

Mean ·001" (·02 mm.).

Assume the bearing end-float to be ·010" (·35 mm.).

Therefore ·010" + ·001" = ·011" (·25 mm.

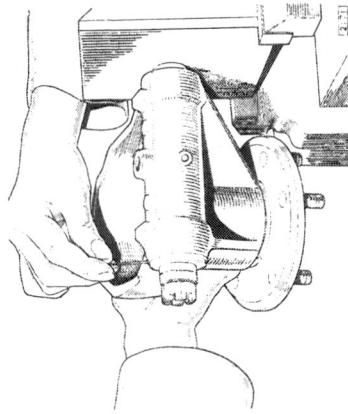

Fig. 18. *Measuring the amount of clearance between the hub carrier and large diameter washer to determine the end float in the bearings.*

+ ·02 mm. = ·27 mm.) to be removed to give correct preload.

Refit the hub carrier to the half shaft as described in Section H "Rear Axle."

Fit new oil seals with the lips inwards and place the fulcrum shaft into position in the hub carrier.

Offer up the hub carrier to the wishbone. Chase the dummy shaft through the wishbone with the fulcrum shaft.

Using feeler gauges measure the gap between the oil seal track and the wishbone. Shims of ·004" (·101 mm.) thickness by 1⅛" (22·2 mm.) diameter should be used.

Repeat for the other end and shim as necessary to centralize the hub carrier in the wishbone fork. The above procedure is to prevent the wishbone fork ends from closing inwards. Tighten the nuts on the fulcrum shaft to 55 lb. ft. (7·60 kg.m.).

Refitting

To facilitate refitting, slide a dummy shaft Churchill tool No. J.14 through the hub carrier before offering up the wishbone to the hub carrier.

Refitting is the reverse of the removal procedure.

Re-lubricate the bearings as described in "Routine Maintenance" at the beginning of the section.

INNER FULCRUM WISHBONE MOUNTING BRACKET

Removal

Remove the eight bolts and self-locking nuts securing the tie plate to the inner fulcrum wishbone mounting bracket.

Remove the six bolts and self-locking nuts securing the tie plate to the cross beam.

Remove one self-locking nut and drift out the inner fulcrum shaft.

Withdraw the forks of the wishbone from between the cross beam and inner fulcrum wishbone mounting bracket.

Collect the oil seal retainers, oil seals, inner and outer thrust washers and bearing tubes.

Remove the lock wire from the two setscrews which secure the inner fulcrum wishbone mounting bracket to the differential unit.

Remove the spacer between the inner fulcrum mounting bracket.

Remove the two setscrews and note the

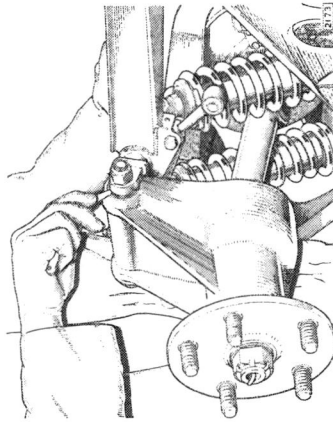

Fig. 19. *Using feeler gauges to measure the clearance between the hub carrier oil seal tracks and wishbone fork.*

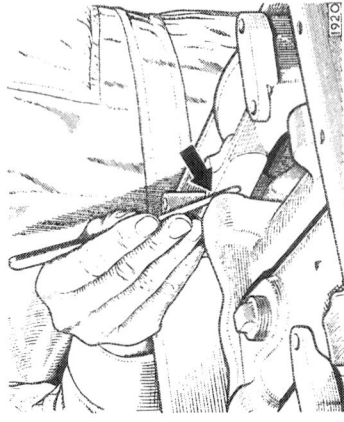

Fig. 20. *Measuring the clearance between the inner fulcrum mounting bracket and the differential casing.*

amount of shims between the bracket and the differential.

Remove the inner fulcrum wishbone mounting bracket.

Refitting

When refitting the inner fulcrum wishbone mounting bracket, replace the same amount of shims between the differential casing and the bracket.

Shims are available in sizes of ·005" (·13 mm.) and ·007" (·18 mm.) thickness.

Hold the inner fulcrum wishbone mounting bracket in position between the cross beam.

Insert the fulcrum shaft through the cross beam and bracket. Screw the inner fulcrum bracket securing setscrews in two or three threads, enough to locate the bracket.

Insert the required amount of shims and tighten the two setscrews securing the inner fulcrum wishbone mounting bracket to the differential casing. Secure the two setscrews with locking wire.

Withdraw the inner fulcrum shaft from the cross beam and fulcrum bracket.

Offer up the wishbone to the inner fulcrum mounting bracket complete with bearing tubes, needle roller bearing and spacers, inner and outer thrust washers, oil seals and oil seal retainers. Ensure that the radius arm mounting bracket is towards the front of the car.

Align the holes and spacers. Press a dummy shaft through each side of the cross beam and wishbone.

The dummy shafts locate the wishbone, spacers, cross beam and inner fulcrum mounting bracket and facilitate refitting of the fulcrum shaft.

Smear the fulcrum shaft with grease and gently tap the shaft through the cross beam, wishbone and inner fulcrum mounting bracket. As the fulcrum is tapped into position the short dummy shafts will be displaced from the opposite side. It will be found advantageous to keep a slight amount of pressure exerted on the dummy shafts as they emerge from the cross beam. This will reduce the tendency for the dummy shafts to be knocked out of position and allow a spacer or thrust washer to be displaced. If a washer or

spacer becomes displaced it will be necessary to remove the fulcrum shaft, dummy shafts and wishbone and then repeat the operation.

When the fulcrum shaft is in position tighten the two self locking nuts to 55 lb. ft. (7·60 kg.m.) with a torque wrench.

Refit the eight bolts and self-locking nuts securing the tie plate to the inner fulcrum wishbone mounting bracket.

Refit the six bolts and self-locking nuts securing the tie plate to the cross beam.

Refit the rear suspension unit as described on page K.6.

REAR WHEEL CAMBER ANGLE—ADJUSTMENT

To check the camber of the rear wheels the car must be standing on a level surface with the tyre pressures set correctly. Owing to the variations in the camber angle with different suspension heights it is necessary to lock the rear suspension in the mid-laden position by means of two setting links (Churchill tool No. J.25) as shown in Fig. 23. To fit the setting links hook one end in the lower hole of the rear mounting and depress the body until the other end can be slid over the hub carrier fulcrum nut. Repeat for the other side.

With the car in this condition the camber angle should be: −¼° negative ±¼°.

Note: The two rear wheels must be within a ¼° of each other.

If the reading is incorrect it will be necessary to add or subtract shims between the half shaft and the brake disc. One shim ·020" (·05 mm.) will alter the rear camber angle by approximately ¼°.

Jack up the car on the appropriate side and remove the rear road wheel.

Remove the self-locking nut and washer securing the forward road spring and hydraulic damper assembly to the wishbone mounting pin. Drift the mounting pin through the wishbone until the assembly is free from the pin.

Remove the self-locking nut and bolt securing the top of the road spring and hydraulic damper assembly to the cross beam and remove the assembly.

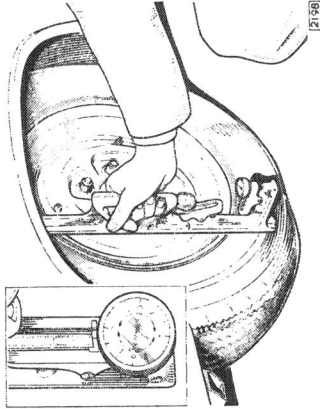

Fig. 21. *Checking the rear wheel camber angle.*

Fig. 22. *The rear wheel camber angle is adjusted by means of shims indicated by the arrow.*

Unscrew the four self-locking nuts securing the half shaft and the camber shims to the brake disc. Pull the hub and half shaft away from the shims sufficiently to clear the disc mounting studs. Remove or add shims as necessary.

Offer up the half shaft to the four disc mounting studs and secure with four self-locking nuts. Offer up the forward road spring and hydraulic damper assembly to the cross beam and secure with a bolt and self-locking nut.

Align the hydraulic damper and road spring assembly bottom mounting with the mounting pin in the wishbone and drift the pin through the assembly. Replace the plain washer and secure with a self-locking nut.

Replace the road wheel(s) and recheck the camber angle.

Warning: After completing the adjustment do not omit to remove the setting links from the suspension.

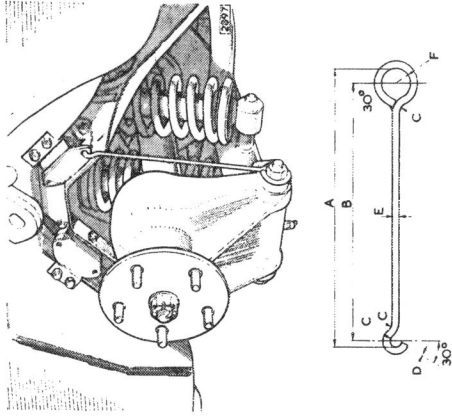

	INCHES	METRIC
A	9 1/32	22·9 cm
B	8 3/16	20·79 cm
C	1/4 RAD	6·3mm
D	1/16	1·5mm
E	9/32	7·1mm
F	19/32 RAD	15·0 mm

Fig. 23. *When checking the rear camber angle the rear suspension must be retained in the mid-laden position by means of the setting links (Churchill Tool No. J.25).*

REAR SUSPENSION

ACCIDENTAL DAMAGE

The dimensional drawings below are provided to assist in assessing accidental damage. A component suspected of being damaged should be removed from the car and cleaned off, the dimensions then should be checked and compared with those given in the appropriate illustration.

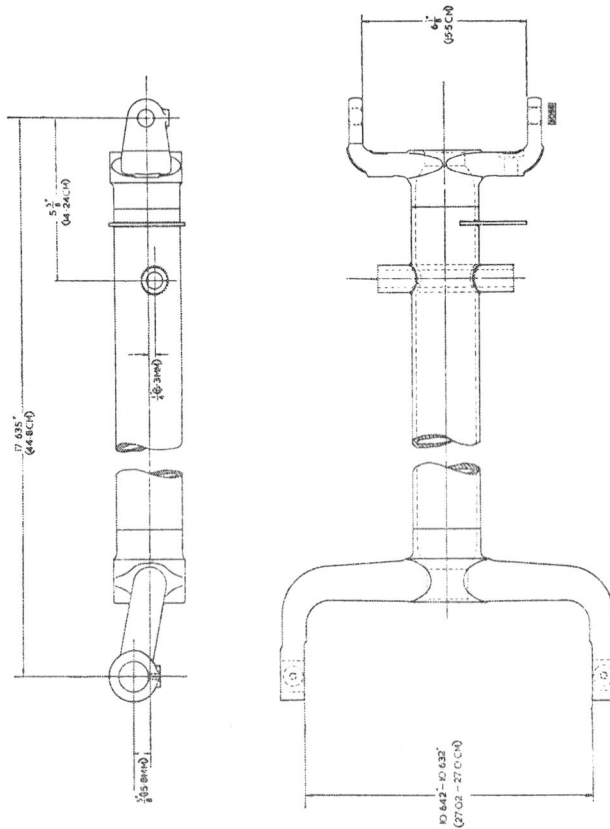

Fig. 24. *The wishbone.*

SPECIAL TOOLS

DESCRIPTION	TOOL NO.
Shock Absorber/Spring Unit Dismantling Tool	J.11 A (Use with S.L.14)
Rear Wishbone Pivot Dummy Shaft	J.14 (2 off per set)
Radius Arm Bush Remover/Replacer	J.21
Rear Suspension Setting Link (for camber checking)	J.25 (2 off per set)

Page K.16

SECTION L

BRAKES

ADDENDA

The following corrections should be noted when using SECTION L.

Page	*Correction*		
L.5	Servo unit type (early cars)	.. ..	Lockheed 6⅞"
	Servo unit type (later cars)	.. ..	Lockheed 8"
	Main friction pad material	..	"M.33" should read "M.59"
L.10, L.13	"Wakefield Rubber Grease H95/59" should read "Girling Red Rubber Grease".		

INDEX

INDEX (continued)

THE BRAKING SYSTEM

DESCRIPTION

Each wheel brake assembly consists of a hub mounted disc, rotating with the wheel and a braking unit rigidly attached to the suspension member.

The braking unit consists of a caliper, straddling the disc, which houses a pair of piston-operated friction pads. The outer piston assembly is housed in a cylinder integral with the caliper while the inner cylinder is bolted to the inner face of the caliper and can be detached. A securing pin, retained by a spring clip, passes through the body of the caliper and the inner cylinder block securing the friction pads and a stop plate. The stop plate serves as a stop for the friction pads when they have reached the end of their life and also excludes dirt from the caliper. A dust seal spigots on the outer face of each piston and a rubber seal is fitted between the piston and the cylinder. A counterbore in the piston accommodates a retractor bush which tightly grips the stem of a retractor pin. This pin forms part of an assembly which is peened into the base of the cylinder bore. The assembly consists of a retractor stop bush, two spring

washers, a dished cap and the retractor pin; it functions as a return spring and maintains a "brake-off" working clearance of approximately ·008" — ·010" (·20 — ·25 mm.) between the pads and the disc throughout the life of the pads.

Handbrake

The mechanical handbrake units are secured by pivot bolts to the top of the rear caliper bodies.

Each handbrake unit consists of two carriers, one on each side of the brake disc, on which friction pads are mounted. The free end of the inner pad carrier is equipped with a pivot seat to which the forked end of the operating lever is attached. A trunnion is mounted within the forked end of the operating lever which carries the end of the adjuster bolt. Located on the shank of the adjuster bolt and in a counterbore in the inside face of the inner pad carrier is the operating lever return spring. This spring is held under load by a nut retained by a spring plate which is riveted to the inside face of the inner carrier. The adjuster bolt passes through the outer pad carrier and seats in the outer carrier.

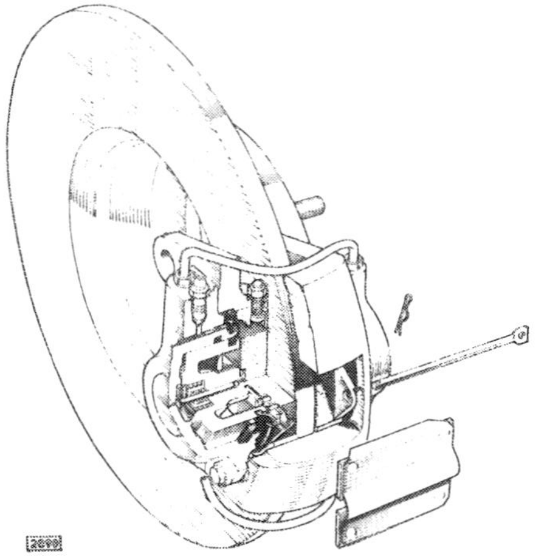

Fig. 1. Sectional view of front disc brake.

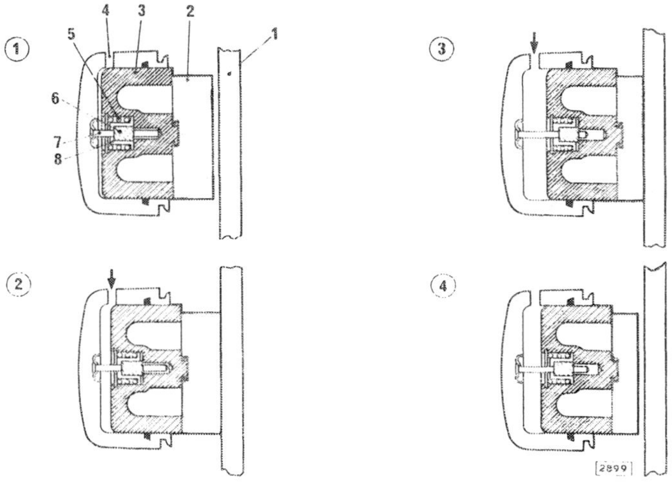

1. Brake OFF
2. Brake ON
   Retractor spring under compression.
3. Brake ON
   Retractor bush drawn along pin by piston as pad wears.
4. Brake OFF
   Bush retains its new position on pin and spring returns to normal position thus restoring correct clearance between pad and disc.

1. Disc
2. Friction pad
3. Piston
4. Fluid connection
5. Retractor spring
6. Retractor bush
7. Retractor pin
8. Spring retainer

Fig. 2. Operation of self-adjusting mechanism.

DATA

		Front	Rear
Make	Dunlop		
Type	Bridge type calipers with quick change pads		
Brake Friction Pads:			
Face dimensions		$2\frac{1}{32}'' \times 2''$	$2\frac{1}{32}'' \times 2''$
		(51·59 × 50·8 mm.)	(51·59 × 50·8 mm.)
Thickness		·438" (11·1 mm.)	·438" (11·1 mm.)
Brake Disc:			
Diameter		11" (279·4 mm.)	$10\frac{3}{8}''$ (263·5 mm.)
Thickness		$\frac{3}{8}''$ (9·5 mm.)	$\frac{1}{2}''$ (12·7 mm.)
Master cylinder bore diameter		$\frac{7}{8}''$ (22·22 mm.)	
Master cylinder stroke		$1\frac{3}{8}''$ (35 mm.)	
Brake cylinder bore diameter — Front		$2\frac{1}{8}''$ (53·97 mm.)	
— Rear		$1\frac{1}{2}''$ (38·1 mm.)	
Servo unit type		Lockheed $6\frac{7}{8}''$	
Main friction pad —material		Mintex M.33	
Handbrake friction pad —material		Mintex M.34	

ROUTINE MAINTENANCE

Every 3,000 miles (5,000 km.)

Brake Fluid Level

The fluid reservoir for the hydraulic brakes is attached to the wing valance on the driver's side of the car.

At the recommended intervals check the level of the fluid in the reservoir and top up if necessary to the level mark above the fixing strap marked "Fluid Level" using only the correct specification of brake fluid.

Do NOT overfill.

The level can be plainly seen through the plastic reservoir.

First, disconnect the two electrical cables from the "snap-on" terminals. Unscrew the filler cap and "top up" if necessary to the recommended level. Insert the combined filler cap and float slowly into the reservoir to allow for displacement of fluid and screw down the cap. Wipe off any fluid from the top of the cap and connect the cables to either of the two terminals.

Note: An indication that the fluid level is becoming low is provided by an indicator pin situated between the two terminals.
First, press down the pin and allow it to return to its normal position; if the pin can then be lifted with the thumb and forefinger the reservoir requires topping up immediately.

Brake Fluid Level Warning Light

A warning light (marked "Brake Fluid—Handbrake") situated on the facia behind the steering wheel, serves to indicate if the level in the brake fluid reservoir has become low, provided the ignition is "on". As the warning light is also illuminated when the handbrake is applied, the handbrake must be fully released before it is assumed that the fluid level is low. If with the ignition "on" and the handbrake fully released the warning light is illuminated the brake fluid must be "topped up" immediately.

As the warning light is illuminated when the handbrake is applied and the ignition is "on" a two-fold purpose is served. Firstly, to avoid the possibility of driving away with the handbrake applied. Secondly, as a check that the warning light-bulb has not "blown"; if on first starting up the car with the handbrake fully applied, the warning light does not become illuminated the bulb should be changed immediately.

Note: If it is found that the fluid level falls rapidly indicating a leak from the system, the car should be taken immediately to the nearest Jaguar Dealer for examination.

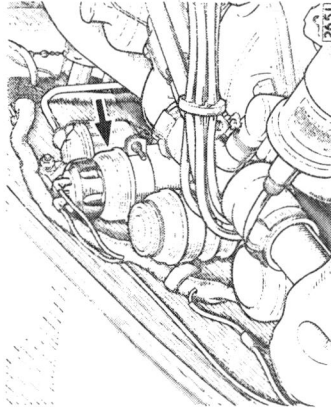

Fig. 3. Brake fluid reservoir—right hand drive.

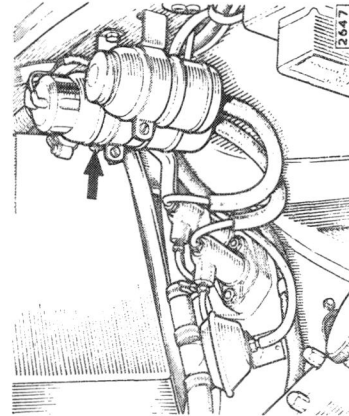

Fig. 4. Brake fluid reservoir—left hand drive.

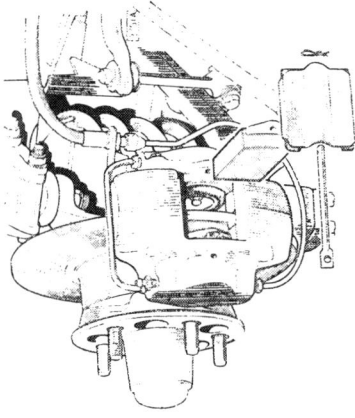

Fig. 5. Friction pad removal.

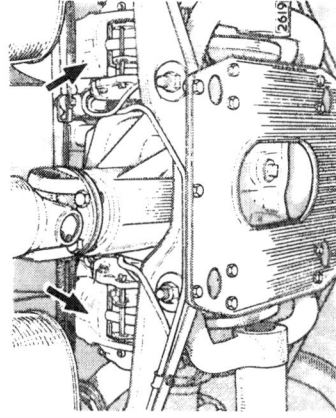

Fig. 6. Location of the rear brake calipers.

Every 6,000 miles (10,000 km.)

Friction Pads—Examination for Wear

At the recommended intervals, or if a loss of braking efficiency is noticed, the brake friction pads (2 per brake) should be examined for wear; the ends of the pads can be easily observed through the apertures in the brake caliper. When the friction pads have worn down to a thickness of approximately ¼" (7 mm.) they need renewing.

Friction Pads—Renewal

To remove the friction pads withdraw the spring clip and extract the pad retaining pin and stop plate.

Insert a hooked implement through the hole in the metal tag attached to the friction pad and withdraw the pad by pulling on the tag.

To enable the new friction pads to be fitted it will be necessary to force the pistons back into the cylinder blocks by means of special tool number 10416.

Before doing this, it is advisable to half empty the brake supply tank, otherwise forcing back the friction pads will eject fluid from the tank with possible damage to the paintwork. When all the new friction pads have been fitted, top up the supply tank to the recommended level.

Insert the new friction pads into the caliper ensuring that the slot in the metal plate attached to each pad engages with the button in the centre of the piston.

Refit the retaining pin through the caliper and brake pads and attach the spring clip. Apply the footbrake a few times to operate the self-adjusting mechanism, so that the normal travel of the pedal is obtained.

Brake Servo Air Cleaner

At the recommended intervals the brake servo air cleaner, which is attached to the right-hand wing valance, should be removed and washed in **methylated spirits.** After drying out re-lubricate the wire mesh with **brake fluid.**

On later cars fitted with the Type 8 brake servo the filter is attached to the unit and should be removed and any light dust blown out with an air line. This filter should never be cleaned in a cleansing fluid.

BLEEDING THE BRAKE HYDRAULIC SYSTEM

"Bleeding" the brake hydraulic system (expelling the air) is not a routine maintenance operation and should only be necessary when a portion of the hydraulic system has been disconnected or if the level of the brake fluid has been allowed to fall. The presence of air in the hydraulic system will cause the brakes to feel "spongy".

During the bleeding operation it is important that the level in the reservoir is kept topped up to avoid drawing air into the system.

Check that all connections are tightened and all bleed screws closed.

Fill the reservoir with brake fluid of the correct specification.

Attach the bleeder tube to the bleed screw on the near side rear brake and immerse the open end of the tube in a small quantity of brake fluid contained in a clean glass jar. Slacken the bleed screw and operate the brake pedal slowly backwards and forwards through its full stroke until fluid into the jar is reasonably free from air bubbles. Keep the pedal depressed and close the bleed screw. Release the pedal.

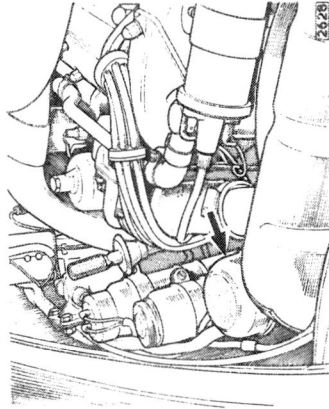

Fig. 7. *The brake servo air cleaner.*

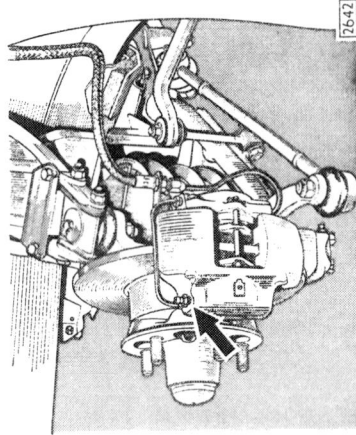

Fig. 8. *Brake bleed nipple.*

Repeat the complete bleeding sequence until the brake fluid pumped into the jar is completely free from air bubbles.

Lock all bleed screws and finally regulate the fluid level in the reservoir.

Apply normal working load on the brake pedal for a period of two or three minutes and examine the entire system for leaks.

Do not use the fluid which has been bled through the system to replenish the reservoir as it will have become aerated. Always use fresh fluid straight from the tin.

BRAKE OVERHAUL—PRECAUTIONS

The brake system is designed to require the minimum of attention and, providing the hydraulic fluid in the reservoir is not allowed to fall below the recommended level, no defects should normally occur. Fluid loss must be supplemented by periodically topping up the reservoir with fluid to the same specification as that in the system. If the recommended brand of brake fluid is not available and it is intended that an alternative approved brand be used the complete system must be drained before the substitution of fluid for another.

The inclusion of air in a system of this type will be indicated by sluggish response of the brakes and spongy action of the brake pedal. This condition may be due to air induction at a loose joint or at a reservoir in which the fluid has been allowed to fall to a very low level. These defects must be immediately remedied and the complete system bled. Similarly, bleeding the system is equally essential following any servicing operation involving the disconnecting of part or whole of the hydraulic system.

The following instructions detail the procedure for renewal of component parts and for complete overhaul of the disc brakes, handbrakes and master cylinder. The units should be thoroughly cleaned externally before dismantling. Brake system fluid should be used for cleaning internal components and, except where otherwise stated in these notes, the use of petrol, paraffin or chemical grease solvents should be avoided as they may be detrimental to the rubber components. Throughout the dismantling and assembling operations it is essential that the work bench be maintained in a clean condition and that the components are not handled with dirty or greasy hands. The precision parts should be handled with extreme care and should be carefully placed away from tools or other equipment likely to cause damage. After cleaning, all components should be dried with clean lint-free rag.

When it is not the intention to renew the rubber components, they must be carefully examined for serviceability. There must be no evidence of defects such as perishing, excessive swelling, cutting or twisting, and, where doubt exists, comparison with new parts may prove to be of some assistance in making an assessment of their condition. The flexible pipes must show no signs of deterioration or damage and the bores should be cleaned with a jet of compressed air. No attempt should be made to clear blockages by probing as this may result in damage to the lining and serious restriction to fluid flow. Partially or totally blocked flexible pipes should always be renewed. When removing or refitting a flexible pipe, the end sleeve hexagon (A, Fig. 9) should be held with the appropriate spanner to prevent the pipe from twisting. A twisted pipe will prove detrimental to efficient brake operation.

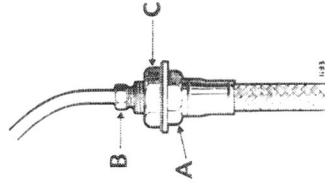

Fig. 9. *Flexible hose connection. Hold hexagon 'A' with spanner when removing or refitting locknut 'C'.*

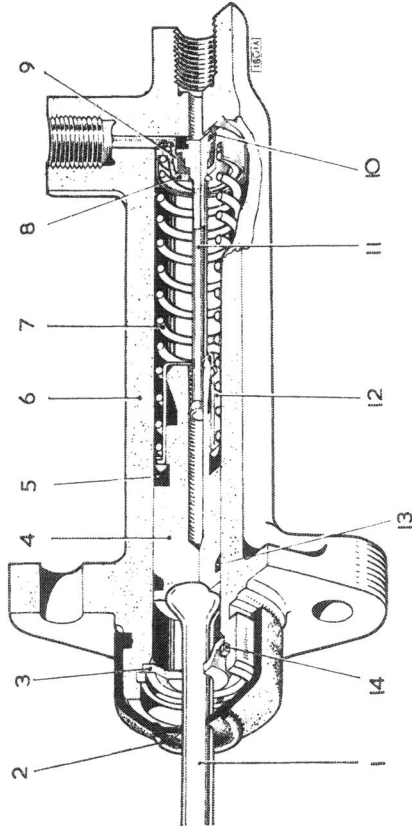

Fig. 10. *Sectioned view of master cylinder.*

1. Push rod
2. Dust excluder
3. Dished washer
4. Piston
5. Cup seal
6. Body
7. Return spring
8. Spring support
9. Valve spring
10. Seal
11. Valve
12. Spring support
13. Sealing ring
14. Circlip

THE MASTER CYLINDER (Fig. 10)

Description

The master cylinder is mechanically linked to the footbrake pedal and, in proportion to the load applied, provides the hydraulic pressure necessary to operate the brakes.

The components of the master cylinder are contained within the bore of the body (6) which has, at its closed end, two 90 opposed pipe connection bosses and, at the other end, a flange provided with two holes for the master cylinder attachment bolts.

A piston (4), sealed at the front and rear, actuated by a ball-ended push rod (1) against a return spring (7) is housed in the central bore of the master cylinder. Located in a blind drilling in the piston is the stem of a valve (11), the head of which is housed in the spring support (8) and acts against the valve spring (9). The forward movement of the push rod and piston is controlled by a dished washer (3) which is retained in the body by a circlip (14).

Operation

On releasing the foot brake, the pedal return spring re-asserts itself withdrawing the push rod (1). This, in turn, permits the piston (4) to follow the push rod under the action of the return spring (7). Before the piston reaches the end of its travel, the shoulder on the end of the valve (11) engages against the spring support (12). As the piston and valve continue to move to the rear together, the head of the valve compresses the valve spring (9) thus permitting hydraulic fluid to flow either to or from the reservoir.

When the brakes are applied, the push rod forces the piston down the bore against the compression of the return spring, freeing the valve. The valve spring re-asserts itself isolating the reservoir.

Further loading of the pedal results in the discharge of fluid under pressure from the outlet connection and by way of the pipe lines to the brake system.

Removal

Drain the brake fluid reservoir and detach the inlet and outlet pipes from the brake master cylinder, by withdrawing the two union nuts

(see Fig. 4). Detach the master cylinder push-rod from the brake pedal by discarding the split pin and withdrawing the clevis pin. Remove the brake master cylinder from the housing situated inside the engine compartment by removing two nuts.

Renewing the Master Cylinder Seals

Ease the dust excluder clear of the head of the master cylinder.

With suitable pliers remove the circlip; this will release the push rod complete with dished washer.

Withdraw the piston and remove both seals.

Withdraw the valve assembly complete with springs and supports. Remove the seal from the end of the valve.

Lubricate the new seals and the bore of the cylinder with brake fluid, fit the seal to the end of the valve ensuring that the lip registers in the groove. Fit the seals in their grooves around the piston.

Insert the piston into the spring support, ensuring that the head of the valve engages the piston bore.

Lubricate the piston with Wakefield Rubber Grease H. 95/59 and slide the complete assembly into the cylinder body taking particular care not to damage or twist the seals. The use of a fitting sleeve is advised.

Position the push-rod and depress the piston sufficiently to allow the dished washer to seat on the shoulder at the head of the cylinder. Fit the circlip and check that it fully engages the groove.

Fill the dust excluder with clean Wakefield H. 95/59 Rubber Grease.

Reseat the dust excluder around the head of the master cylinder.

Master Cylinder Push-rod—Free Travel

When the brake pedal is in the "off" position it is necessary that the master cylinder piston is allowed to return to the fully extended position, otherwise pressure may build up in the system causing the brakes to drag or remain on.

To ensure that this piston returns to the fully extended position clearance is provided between the enlarged head of the push-rod, the piston and dished washer. As this washer also forms the return stop for the brake pedal, no means of adjustment is necessary.

The push-rod clearance will give approximately $\frac{1}{4}''$ (7 mm.) free movement at the brake pedal pad and can be felt if the pedal is depressed gently by hand.

THE BRAKE ASSEMBLY

Assembling

It is MOST IMPORTANT that the end-float of the front wheel bearings and the differential output shafts do not exceed the limits laid down in Sections H and J otherwise the brakes may tend to drag and not function correctly.

Check the disc for true rotation by mounting a dial test indicator so that the needle pad bears on the face of the disc. "Run-out" should not exceed 0·006" (·15 mm.) gauge reading. Manufacturing tolerances on the disc and hub should maintain this truth and in the event of the "run-out" exceeding this value the components should be examined for damage.

Locate the caliper body (complete with cylinder assemblies) in position and secure with two bolts and locking wire.

Check the gap between each side of the caliper and the disc. The difference should not exceed 0·010" (·25 mm.) and shims may be fitted to centralise the caliper.

If not already fitted, fit the bridge pipe connecting the two cylinder assemblies. Connect the supply pipe to the cylinder block and ensure that it is properly secured.

Renewing the Friction Pads (Fig. 14)

Brake adjustment is automatic during the wearing life of the pads. The pads should be checked for wear every 6,000 miles (10,000 km) by visual inspection and measurement.

When wear has reduced the pads to the minimum permissible thickness of ¼" (7 mm.), the friction pad assemblies must be renewed.

It is essential that only one brake unit is dealt with at one time as fluid pressure would otherwise prevent reassembly.

Withdraw the pad retaining pin (14) and stop plate (13).

Lever out the friction pads with a suitable tool.

Return the piston assemblies to the base of the cylinder bores using a piston re-setting tool (Fig. 13).

Insert the new friction pad assemblies in the unit, ensuring that the clip on each pad backing plate engages with the button in the centre of the piston.

Refit the pad retaining pin in order to retain the friction pads.

It is not necessary to bleed the system after changing the pads but it is important that fluid is withdrawn from the reservoir otherwise the displaced fluid from the cylinders will cause an overflow from the reservoir.

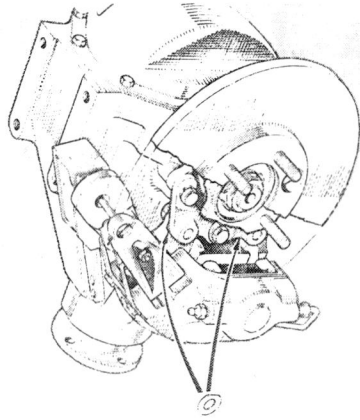

Fig. 12. *Friction pad removal.*

Fig. 11. *Disc brake caliper adjustment shims.*

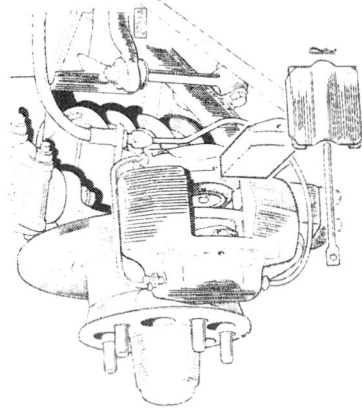

Renewing Cylinder Seals

Withdraw the friction pad assembly as previously described.

Disconnect and blank off the fluid supply pipe.

Remove the caliper.

If it is the inboard piston assembly which is defective it will be necessary to remove the outboard piston and cylinder assembly in order to gain access.

Clean the assemblies thoroughly before proceeding with further dismantling.

Remove the dust cover from the groove around the cylinder block face.

Connect the cylinder block to a source of fluid supply and apply a steady pressure to eject the piston assembly.

Extract the defective seal from the retaining groove in the cylinder bore.

When extracting a seal ensure that the cylinder bore is neither scratched nor scored.

Clean the piston and cylinder thoroughly with trichloroethylene and inspect for damage.

Deep scoring resulting in fluid loss will render the assembly unserviceable. Examine the seal "lead in" chamfer on the piston and remove any burr or sharp corner which may be present.

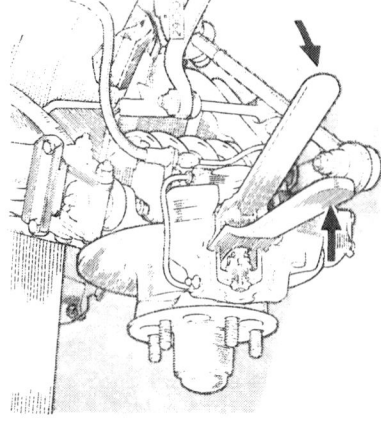

The new seal should be soaked in the correct brake fluid for several minutes before inserting in the wet condition into the retaining groove in the cylinder block.

Smear the cylinder bore and piston barrel with Dunlop Preservative Fluid.

Locate the piston assembly on the end of the retractor pin.

Press the piston assembly into the cylinder bore using a hand press. During this operation ensure that the piston assembly is aligned correctly with the cylinder bore and that pressure is applied slowly and evenly.

Before fitting a new dust cover, inject a small quantity of Dunlop Preservative Fluid behind the piston to prevent corrosion.

On fitting the new dust cover ensure that it is correctly located in the piston barrel groove and on the cylinder face spigot.

Fit the serviced assembly to the caliper and refit the unit to the vehicle.

Reconnect the fluid supply pipeline.

Refit the friction pad assemblies and retain with the retaining pin.

Bleed the brake unit free of air.

Test under maximum pressure and check for fluid leakage at all connections.

Fig. 13. *Resetting the pistons with special tool (Tool No. 10416).*

ing the adjuster bolt (5) inwards and bringing the friction pads closer to the brake disc until the normal running clearance is restored.

FRICTION PAD CARRIERS

Removal

With the car on a ramp, disconnect the handbrake cable from the operating levers on the handbrake mechanisms as follows.

Remove the split pin, withdraw the clevis pin and disconnect the fork end on one lever and withdraw the outer cable from the trunnion on the other lever.

Lift the locking tabs and remove the pivot bolts and retraction plate. Remove the friction pad carriers by moving them rearwards around the disc and withdrawing from the rear of the rear suspension assembly. Repeat for the second handbrake.

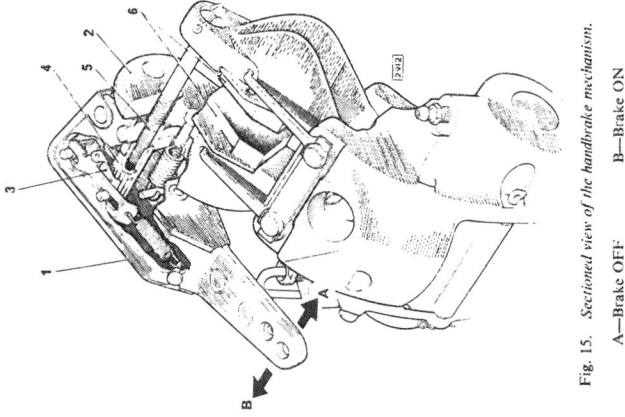

Fig. 15. *Sectioned view of the handbrake mechanism.*

A—Brake OFF   B—Brake ON

THE HANDBRAKE

Operation (Fig. 15)

The self-adjusting handbrakes are attached to the rear brake caliper bodies but form an independent mechanically actuated system carrying its own friction pads. These handbrakes are self-adjusting to compensate for friction pad wear and automatically provide the required clearance between the brake discs and the friction pads.

When the handbrake lever is operated, the operating lever (1) is moved away from the friction pad carrier (2) and draws the friction pads (6) together. Under normal conditions, when the lever is released the pawl (3) in the adjusting mechanism returns to its normal position so that the normal running clearance between the brake disc and the friction pads is maintained.

If the clearance is excessive, the pawl will turn the ratchet nut (4) on the bolt thread draw-

Dismantling

Remove the cover securing bolt, discard the split pin and withdraw the pivot clevis pin. Remove the dust cover and remove the split pin from the screwdriver slot in the adjusting bolt. Unscrew the adjusting bolt from the ratchet nut and withdraw the nut and bolt. Detach the pawl return spring and withdraw the pawl over the locating dowel. Detach the operating lever return spring and remove the operating lever and lower cover plate.

Assembling

Assembly is the reverse of the dismantling procedure.

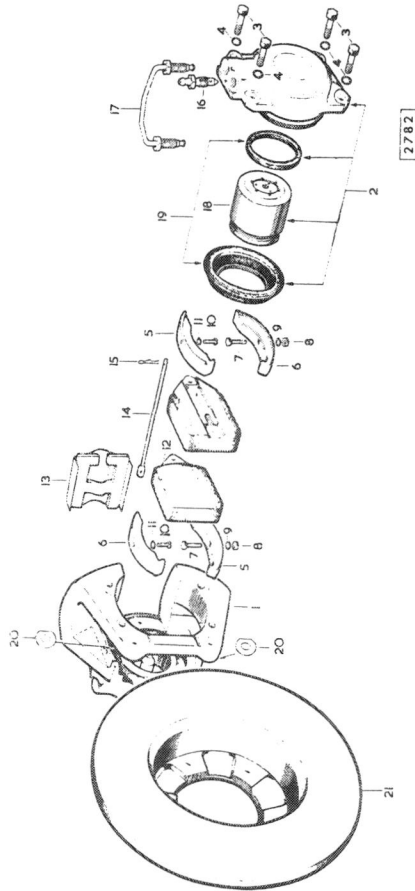

Fig. 14. *Exploded view of front brake caliper.*

1. Caliper and piston assembly
2. Piston and cylinder assembly
3. Bolt
4. Shakeproof washer
5. Pad support R.H.
6. Pad support L.H.
7. Bolt
8. Nut
9. Shakeproof washer
10. Screw
11. Shakeproof washer
12. Friction pad kit
13. Stop plate
14. Retaining pin
15. Clip
16. Bleed screw
17. Bridge pipe
18. Piston
19. Seal kit
20. Shim
21. Disc

FRICTION PAD CARRIERS

Refitting

Refitting is the reverse of the removal procedure but the handbrake should be set as follows:—

Ensure that the handbrake pivot bolts are slack.

Remove the split pin from the head of the adjuster bolt and slacken the bolt until there is approximately ¼" (6·35 mm.) free movement between the head and outer pad carrier.

Pull the inner and outer pad carriers away from the disc bending the brass retraction fingers until there is 1/16" (1·6 mm.) clearance between each pad and the disc.

Take up the free movement at the adjuster bolt by tightening until the bolthead is in light contact with the outer pad carrier seating.

Fit a new split pin to lock the adjuster bolt.

Pull and release the handbrake lever repeatedly until the ratchet ceases to operate, which will indicate that the correct adjustment has been obtained.

With the handbrake applied reasonably hard, tighten the pivot bolts and secure with the tab washer.

Note: It is ESSENTIAL that the brass retraction fingers are in good condition, i.e. not badly distorted. The ends which fit into the pad carriers must be inserted fully into the holes to avoid the possibility of twisting the fingers.

Reconnect the handbrake compensator linkage to the operating levers and check the cable adjustment as follows:—

HANDBRAKE CABLE ADJUSTMENT

The handbrake cable adjustment linkage (Fig. 17) is situated to the rear of the front suspension assembly on the driver's side of the car.

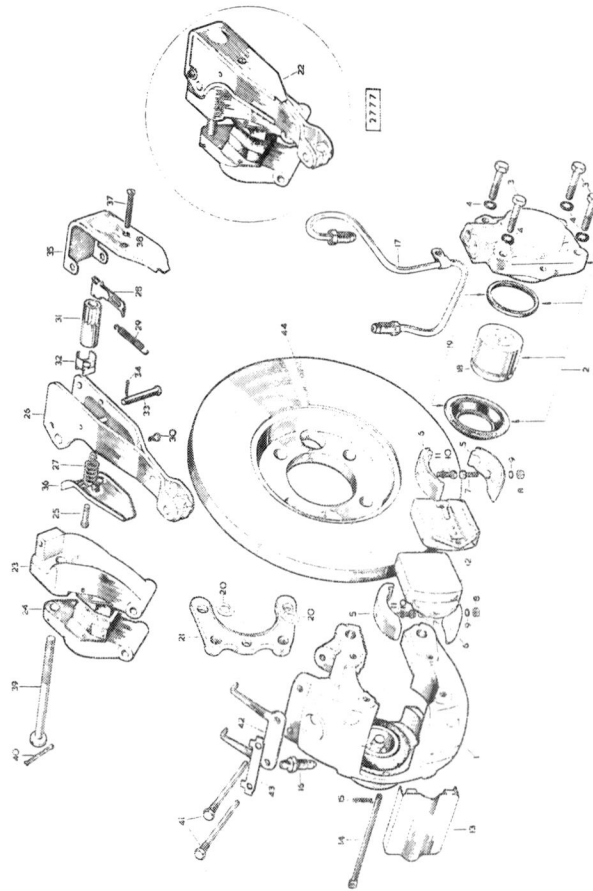

Fig. 16. Exploded view of the rear brake caliper.

1. Caliper and piston assembly	23. R.H. Inner pad carrier
2. Piston and cylinder assembly	24. R.H. Outer pad carrier
3. Bolt	25. Anchor pin
4. Shakeproof washer	26. Operating lever
5. Pad support R.H.	27. Return spring
6. Pad support L.H.	28. Pawl
7. Bolt	29. Tension spring
8. Nut	30. Anchor pin
9. Shakeproof washer	31. Adjusting nut
10. Screw	32. Friction spring
11. Shakeproof washer	33. Hinge pin
12. Friction pad kit	34. Split pin
13. Stop plate assembly	35. Protection cover
14. Pin	36. Protection cover
15. Clip	37. Bolt
16. Bleed screw	38. Washer
17. Bridge pipe assembly	39. Bolt
18. Piston	40. Split pin
19. Seal kit	41. Bolt
20. Shim	42. Retraction plate
21. Adaptor plate	43. Tab washer
22. R.H. Handbrake mechanism assembly	44. Disc

Fully release the handbrake control in the car and slacken the locknut at the rear of the adjustment linkage. Ensure that levers at the calipers are in the "fully off" position by pressing toward the caliper and adjust the length of the cable to a point just short of where the caliper levers start to move; no attempt should be made to place the cable under tension otherwise the handbrake may bind.

HANDBRAKE FRICTION PADS—RENEWING

With the friction pad carriers removed, withdraw the old pads by slackening the nuts in the outer face of each carrier and utilizing a hooked tool in the hole of each pad securing plate. Fit new pads, short face upwards, ensuring that each pad locates the head of the retaining bolt. Fit new retraction fingers and assemble the carrier to the main calipers, leaving the pivot bolts slack.

Pull and release the handbrake lever repeatedly until the ratchet ceases to operate, which will indicate that the correct adjustment has been obtained.

With the handbrake applied reasonably hard, tighten the pivot bolts and secure the tab washer.

Note: It is recommended that new retraction fingers are fitted when replacing the handbrake pads.

Reconnect the handbrake compensator linkage to the operating levers and check the handbrake cable adjustment.

Fig. 17. *Handbrake components.*

1. Handbrake assembly
2. Cover plate
3. Shaft assembly
4. Shaft housing
5. Rubber seal
6. Lever
7. Cable abutment bracket
8. Cable
9. Trunnion
10. Adjusting nut
11. Spring
12. Clip
13. Clevis pin
14. Return spring
15. Handbrake warning light switch
16. Bracket
17. Bracket
18. Blanking plate
19. Gasket
20. Rubber plug

HANDBRAKE CABLE

Removal

From under the car, remove the rear end of the handbrake cable from the operating levers on the handbrake mechanism as follows. Remove the split pin, withdraw the clevis pin and disconnect the fork end at one operating lever and withdraw the outer cable end from the trunnion on the other lever.

Remove the pinch bolt and withdraw the forward end of the outer cable from the trunnion adjacent to the rear suspension radius arm mounting post. Remove the two small bolts and self-locking nuts securing the top half of the nylon intermediate cable support. Disconnect the front end of the cable from the cross shaft linkage on the chassis side member, adjacent to the rear of the front road wheel by removing the front brass nut.

Withdraw the main handbrake cable.

Refitting

Refitting is the reverse of the removal procedure and the cable should be adjusted as described previously.

Fig. 18. *Brake fluid level and handbrake warning light.*

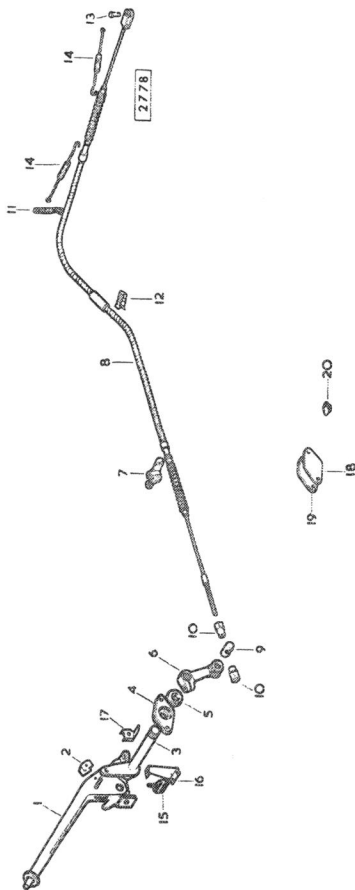

THE BRAKE FLUID LEVEL AND HANDBRAKE WARNING LIGHT

Description (Fig. 18)

The brake fluid level and handbrake warning light, situated in the side facia panel, will indicate after the ignition has been switched on whether the brake fluid in the reservoir is at a low level or the handbrake has not reached the fully off position. This is effected by two switches, one in the top of the fluid reservoir and a second on the handbrake lever, being in circuit with a single warning lamp which is included in the ignition circuit.

When the ignition is switched on and while the handbrake remains applied, the warning light will glow but will become extinguished when the handbrake is fully released with the brake fluid in the reservoir at a high level.

Should the warning light continue to glow after the handbrake has been fully released, it indicates that the brake fluid in the reservoir is at a very low level and the cause must be immediately determined and eliminated. Should the brake fluid be at a high level, the cause of the handbrake remaining on must be investigated.

Handbrake Warning Light Switch—Setting

A bracket mounted interrupter switch is attached to the floor assembly adjacent to the handbrake lever and a small right angle bracket on the latter contacts and depresses the plunger of the interrupter switch when the handbrake is placed in the "off" position.

Should the warning light fail to extinguish when the handbrake is in the "off" position, ensure the handbrake is moving the full length of its travel or the switch bracket has not become inadvertently off set before examining the leads for short circuiting or resetting the interrupter switch.

Examine the handbrake for full travel and the interrupter switch brackets for misplacement. Apply the handbrake and switch on the ignition when the warning light should glow, depress the plunger in the centre of the interrupter switch while observing the warning light; should the light still continue to glow, check the brake fluid in the reservoir which may be at a low level and top up to the correct level. When the warning light ceases to glow re-position the switch in the mounting bracket by slackening and tightening the two nuts on the threaded shank of the switch.

THE VACUUM RESERVOIR AND CHECK VALVE

Description (Fig. 19)

The vacuum reservoir is incorporated in the vacuum line between the inlet manifold and the vacuum servo unit and it is located together with a stone guard in the front section of the right-hand front roadwheel arch. Its purpose is to provide a reserve of vacuum in the event of braking being required after the engine has stalled.

A vacuum check valve is fitted in the bottom end of the front face of the vacuum reservoir with the bottom-most connection communicating with the inlet manifold while the second connection communicates directly with the vacuum port of the vacuum servo unit thus any reduction of pressure inside the reservoir is conveyed to the vacuum servo unit.

Included in the inlet port of the check valve is a flat rubber spring-loaded valve and when there is a depression in the inlet manifold the valve is drawn away from its seat against its spring loading thus the interior of the reservoir becomes exhausted. When the depression in the reservoir becomes equal to that of the inlet manifold the valve spring will return the valve to its seat thus maintaining the highest possible degree of vacuum in the reservoir.

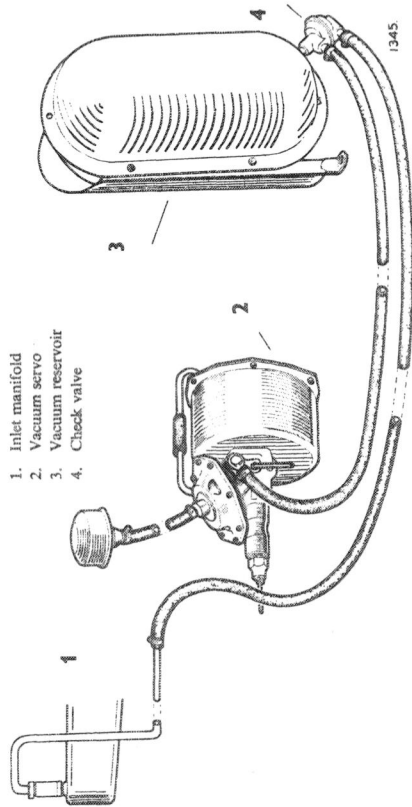

1. Inlet manifold
2. Vacuum servo
3. Vacuum reservoir
4. Check valve

Fig. 19. *Layout of vacuum servo system with reservoir.*

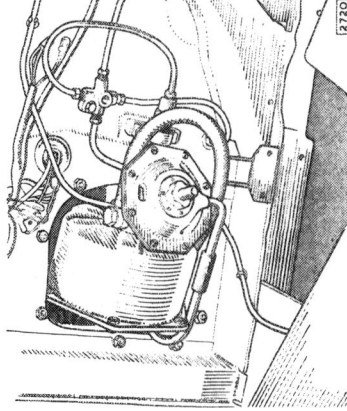

Fig. 20. *Location of the servo unit.*

THE LOCKHEED $6\frac{7}{8}''$
VACUUM SERVO UNIT

A vacuum servo is incorporated in the braking system, and provides the driver with a degree of assistance when applying the brakes, so reducing the effort required on the brake pedal; it is installed in the line between the master cylinder and the four brake calipers.

The unit (refer to Fig. 21) consists mainly of a control-valve, a slave cylinder and booster piston, and power for its operation is provided by atmospheric pressure and by vacuum from the inlet manifold.

When the brake pedal is depressed, fluid pressure created by the master cylinder causes the control-valve to create a pressure-differential across the booster-piston, and this boosts the pressure within the slave cylinder and the brake line, so assisting the driver's effort.

In order to smooth out fluctuations of manifold depression, and to provide a safety factor in the event of braking being required when the engine is stopped, a vacuum reservoir is interposed between the manifold and the servo.

A safety factor, in the event of vacuum failure, is provided by the design of the servo being such as to permit direct communication between the master cylinder and the wheel cylinders, thus allowing brake applications by unassisted foot pressure on the brake pedal.

When the brakes are "off", vacuum is present throughout the servo.

Removal

Firmly apply the handbrake, jack up the front of the car and remove the right-hand front roadwheel. Detach the vacuum reservoir and stone guard from the front of the roadwheel arch by removing three nuts and bolts. Identify and remove the two pipes from the check valve by slackening one hose clip each. Remove the stone guard from the vacuum reservoir by withdrawing four nuts and bolts. Unscrew the check valve from the bottom of the vacuum reservoir when necessary.

Refitting

The refitting is the reverse of the removal procedure but particular attention must be given to the following points:—

(i) That the rubber hose from the vacuum servo unit is attached to the pipe of the check valve having the two grooves in its body; it is also the pipe nearest the screwed connection.

(ii) That the rubber hose from the inlet manifold is attached to the pipe of the check valve having two annular ribs in its body; this pipe is moulded into the centre of the check valve cap.

Slave-Cylinder

The slave-cylinder body carries the control-valve and is bored to accommodate a spring-loaded piston which, when the servo is in the "released" position, is held against a distance-piece. The piston is sealed by a rubber cup and, when the servo is operating is displaced by the push-rod. This latter passes through a spring loaded rubber cup which is backed by a spigot.

Control-Valve

Important: The control-valve assembly comprises a spring-loaded "two-stage" diaphragm which, when the brakes are applied, is deflected by a piston by means of pressure from the master cylinder. In turn the diaphragm operates two valves located within the cover. A short length of tube protruding from the top of the cover provides for connection of an air filter; this cover also incorporates a pipe which is connected to a further pipe in the vacuum shell end cover, by means of a short length of rubber hose.

Principle of Operation

Operation of the servo is portrayed, in purely diagrammatic form, on Fig. 22; these diagrams follow very closely the layout of the actual unit.

When the brakes are "off", the servo is at rest (Fig. 22A), the valve piston and the diaphragm are in the normal position, valve "3" is open and valve "4" is closed. From this it will be seen that whatever degree of vacuum is present within the engine's inlet manifold is also present within chambers "6" and "7" of the servo; additionally the vacuum is transferred to chamber "5", via holes in the centre of the diaphragm assembly and the open valve "3", and to chamber "8", via the connecting pipe. The valves "3" and "4" are maintained in the position illustrated in Fig. 22A due to the pressure of the small conical spring and the pressure differential across valve "4" (there being atmospheric pressure on the spring side).

Upon depressing the brake pedal, initial fluid pressure from the master cylinder passes through connection "Y" and is transmitted to the brakes via the holes in the centre of the slave-cylinder piston and the cup.

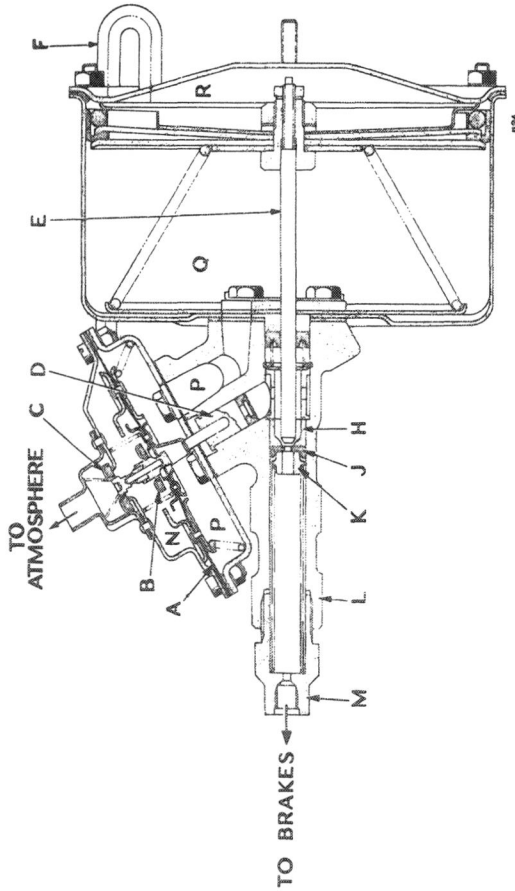

Fig. 21. *Sectional view of 6¾" servo unit.*

A Diaphragm assembly
B Vacuum valve
C Air valve
D Air valve piston
E Servo piston push rod
F Cylinder pipe

H Slave cylinder piston
J Rubber cup
K Spring guide
L Slave cylinder
M Adaptor
N Chamber above diaphragm assembly

P Chamber below diaphragm assembly
Q Chamber inner (vacuum) side of servo piston.
R Chamber outer side of vacuum piston: vacuum when brakes are off, atmospheric pressure when brakes are being applied.

DESCRIPTION (Fig. 21)

Booster-Piston

The booster-piston assembly is contained within the vacuum cylinder shell and consists mainly of a cup held between two plates which are carried on a centre piece; a rubber seal is fitted between the plates to prevent leakage between the chambers on either side of the piston. A felt wick is fitted inside the cup, for lubrication purposes, and is held in place by a barbed retainer and an end stop. The bore of the centre piece is threaded to receive an adjustable push rod which extends into a slave cylinder and is locked by a nut. The whole piston-assembly is under load from a return spring.

Meanwhile, the fluid pressure generated by the master cylinder is also felt upon the rear face of the valve piston which displaces the inner portion of the diaphragm to close valve "3" (Fig. 22B). The initial deflection of the diaphragm is resisted only by a light spring, so ensuring operation of the servo by only a small effort on the brake pedal. The closing of valve "3" isolates the vacuum source from chambers "5" and "8".

Continued effort on the brake pedal causes valve "4" to open, which admits atmospheric pressure to chambers "5" and "8". (Fig. 22C) and partially destroys the vacuum within these chambers.

Note: A progressive entry of atmospheric pressure is ensured by the two stage opening of valve "4". In the first stage, the centre portion only lifts, allowing a restricted passage; in the second stage, the complete valve moves off its seat allowing free flow of the air.

This admission of atmospheric pressure creates a pressure-differential across the faces of the booster piston, which overcomes the spring-loading and displaces the piston, causing the push-rod to move nearer to the cup and to seal off the hole in its centre. Since the push-rod has sealed-off the hole in the cup, there is a locked "line" of fluid to the brakes, and the continued influence of the pressure-differential across the booster piston will boost the pressure within the slave cylinder and the brake line.

The same pressure-differential also exists across the diaphragm and, when the desired degree of braking is reached, the forces due to the master-cylinder pressure acting on the valve piston and the pressure-differential across the diaphragm will balance, and the influence of the small spring which loads the inner portion of the diaphragm will cause the diaphragm to deflect in the direction of the valve piston. Consequently, the valve "4" will close on to its seat and prevent the further entry of air. The control-valve will now be in the condition shown on Fig. 22C, whilst the booster-piston remains in the displaced position.

The spring fitted under the outer portion of the diaphragm ensures that the force required to deflect the diaphragm to this position is substantially equal to that required to displace the booster piston, thus ensuring that the valve "4" does not close until the servo has given maximum performance appropriate to a given effort on the brake pedal.

Greater effort upon the brake pedal increases the thrust on the valve piston, which causes the valve "4" to re-open and admit more air, so increasing the pressure-differential across the booster piston and allowing the piston to perform a greater amount of effort; when the opposing forces on the diaphragm are once more in balance, the valve "4" will again close on to its seat. It will be apparent, therefore, that the diaphragm acts as a "proportioning" device, ensuring that the performance of the servo is progressive.

When the brake pedal is release, pressure is removed from the valve piston, allowing the spring load to push the diaphragm back and thereby cause all four chambers of the servo once more to be communicated with the manifold. The booster-piston and the slave-cylinder piston are now able to return to the "off" position but, since the rearward travel of the slave-cylinder piston is limited by a distance-piece (this part is not shown on the diagrams but may be seen on Fig. 24), a gap exists between this piston and the push-rod when the booster piston has fully returned.

Fluid in the brake line (being under load from the disc brake pistons) is now able to return to the master cylinder via connection "Y".

Removal

Apply the handbrake firmly, open the engine compartment and jack up the front of the car, remove the right-hand roadwheel and drain the hydraulic system at that brake caliper.

Remove the air cleaner hose from the central port in the air valve cover by slackening the hose clip, detach the vacuum reserve tank hose from the large slave cylinder connection by withdrawing a banjo bolt, remove the rigid hydraulic pipe lines from the top and end slave cylinder connections by withdrawing the two union nuts.

Detach the servo unit clamp and support block on the cylindrical body of the slave cylinder from the right-hand wing valance inside the engine compartment by removing two nuts and bolts. Withdraw the vacuum servo unit and supporting cowl from inside the right-hand front roadwheel arch by removing eight setscrews. Detach the vacuum servo unit from the supporting cowl by removing three nuts.

Dismantling

Slide the hose clear of the junction in the connecting pipe. Whilst holding the air-valve cover against the spring load, remove the eight screws and nuts securing it, and take off the cover, the diaphragm, the two springs and the gaskets (Fig. 24 refers).

Remove the four bolts securing the valve-housing to the slave cylinder, lift off the valve-housing and the gasket, and expel the valve piston (this is done by closing the end connection with the thumb and applying a low air pressure at the smaller connection at the side of the cylinder). Ease the rubber cup off the piston.

Scribe a line on the vacuum-cylinder shell to ensure correct alignment of the connecting pipe when reassembling, and mark the positions which the longer bolts securing the end cover occupy. Remove the longer bolts and whilst holding the end cover against the load of the spring, remove the short bolts. Lift off the end cover, withdraw the piston and remove the rubber gasket; disengage the spring from the tabs of the locking plates.

Unscrew the small nut and the large nut (in that order) from the rear end of the booster-piston assembly. Take off the end-stop and disengage the wick-retainer, noting that there are barbs on it which stick into the wick. Take the various parts off the centre-piece and unscrew the push-rod.

W ATMOSPHERIC PRESSURE
X SUCTION FROM ENGINE INLET MANIFOLD
Y CONNECTION FOR LINE FROM MASTER CYLINDER
Z TO BRAKES

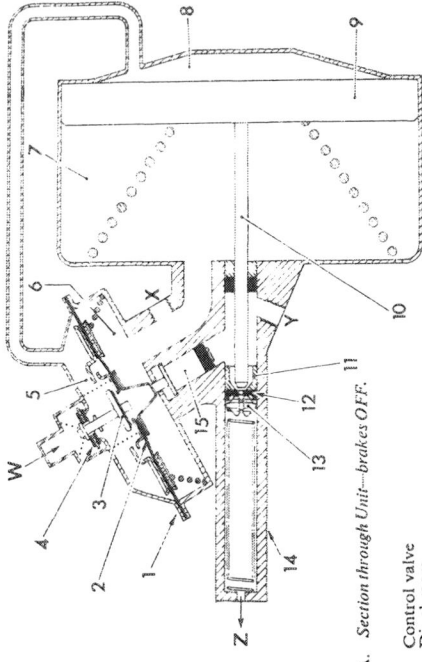

22A. *Section through Unit—brakes OFF.*

1. Control valve
2. Diaphragm
3. Valve (vacuum)
4. Valve (air)
5. Chamber
6. Chamber
7. Chamber
8. Booster piston
9. Slave cylinder piston
10. Push rod
11. Cup
12. Spring guide
13. Slave cylinder
14. Slave cylinder
15. Valve piston

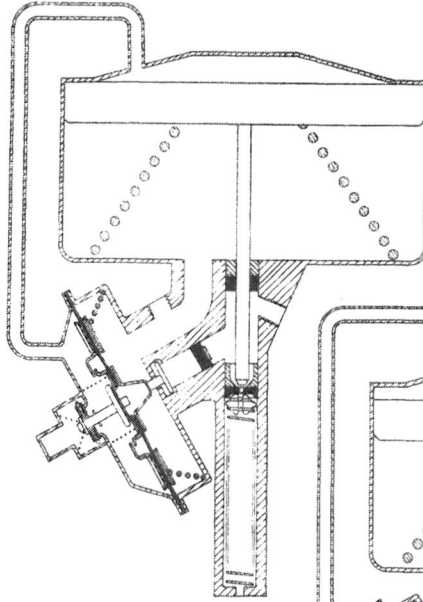

22B. *Initial application—vacuum valve closed.*

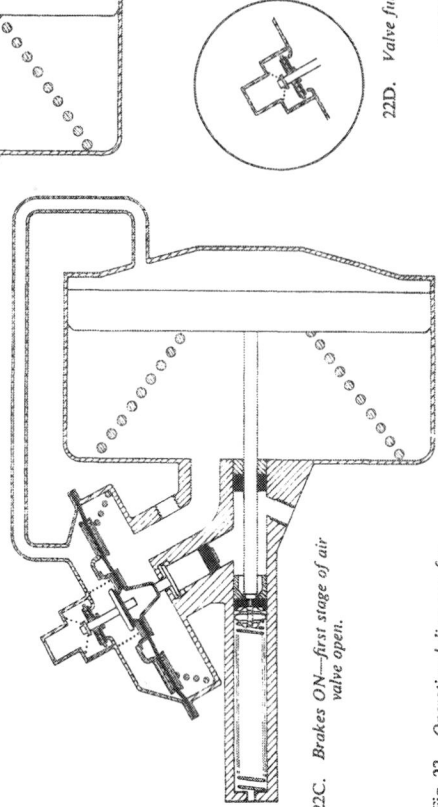

22C. *Brakes ON—first stage of air valve open.*

22D. *Valve fully open.*

2901

Fig. 22. *Operational diagrams for vacuum suspended unit.*

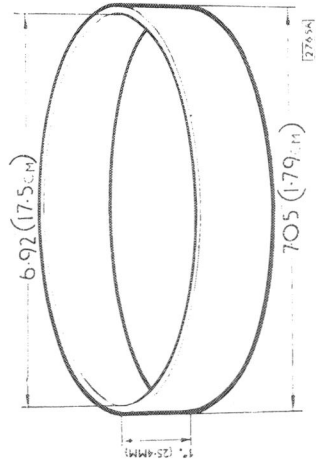

Fig. 24. *Exploded view of the 6¾" servo unit.*

1.	Setscrews	21.	Hose	42.	Abutment plate
2.	Cover and valve assembly	22.	Return spring	43.	Main vacuum cylinder
3.	Return spring	23.	Push rod	44.	Spigot
5.	Diaphragm	24.	Adjuster nut	45.	Spring short
6.	Gasket	25.	Locating washer	46.	Seal
7.	Gasket	26.	Outer piston plate	47.	Spreader
8.	Balance spring	27.	'O' ring	48.	Circlip
8.	Setscrew	28.	Rubber cup	49.	Washer
9.	Lockwasher	29.	Inner piston plate	50.	Distance piece
10.	Valve housing	30.	Piston ring	51.	Piston
11.	Lockwasher	31.	Wick	52.	Seal
12.	Nut	32.	End stop	53.	Spring support
13.	Gasket	33.	Washer (plain)	54.	Main spring
14.	Valve piston	34.	Lockwasher	55.	Spring support
15.	Seal	35.	Nut	56.	Copper washer
16.	Gasket	36.	Copper gasket	57.	Adaptor
17.	Bolt (short)	37.	Nut	58.	Slave cylinder
18.	Bolt (long)	38.	Cover and pipe		
18A.	Tab washer	39.	Gasket		
19.	Nut	40.	Setscrew		
20.	Lockwasher	41.	Lockplate		

Bend back the tabs of the two locking plates inside the vacuum-cylinder shell and unscrew the four bolts which secure the abutment plate and the slave cylinder (Fig. 24). Separate the shell from the cylinder, collect the gasket and withdraw the spigot, cup, cup spreader and spring from the cylinder.

Unscrew the adaptor from the slave cylinder and withdraw the spring retainer, the spring, the spring guide and the copper gasket. Insert a rod into the opposite end of the body and push out the cup and the piston. It is not recommended that the remaining parts be removed from the cylinder, since the bore would possibly be damaged when removing the circlip.

Assembling

The following procedure is recommended for assembling the servo, using new parts as required and new rubber parts (the cup used on the booster piston must be pliable, otherwise it too should be renewed). If the bore of the vacuum-cylinder shell is lightly corroded it may be polished with fine emery cloth or steel wool, but if pitting or scoring is present a new shell should be used. During assembly, Shell "Tellus" 33 lubricant (or equivalent) will be required to lubricate the wick and the booster-piston cup, and also the bore of the vacuum-cylinder shell; the lubricant must not be allowed in contact with any of the rubber parts. Also required, for assembling the booster-piston, will be an assembly ring of the type illustrated in Fig. 23.

Hold the slave-cylinder body upright, with the jointing face at the top, and insert the spring into the bore. Locate the concave side of the cup spreader on the spring, and the grooved side of the cup on the spreader, and push the parts into the bore, followed by the spigot. Locate the gasket on the jointing face and, whilst keeping the spigot depressed, bring the vacuum-cylinder shell into contact with the slave cylinder; fit the abutment plate, two new locking plates and the four bolts to secure the shell to the cylinder. Tighten the four bolts to a torque reading of 150 lb.-in. and bend the ears of the locking plates up to the heads of the bolts.

Place the assembly ring flat on the bench, and position the larger piston plate inside it, with the lip facing downwards. Dip the cup in the specified lubricant, allow the excess to drain off, and place the cup on the plate, with the lip of the cup facing upwards. Place the rubber seal in the chamfer at the centre of the plate, and locate the other piston plate on top of it, with the lip facing downwards.

Lubricate the wick and fit inside the cup, bend the wick retainer into a circular form (with the barbs outermost), locate the retainer inside the wick and push the bent end of the retainer inwards and backwards to engage the fork with the tongue. Place the end stop on the assembled parts, with the raised side downwards and with the cut-away over the bent end of the retainer. Leave the assembly within the assembly ring.

Screw the centre-piece fully on to the push-rod, with the hexagon leading.

Whilst taking care to avoid damage to the high surface finish of the push-rod, hold the centre-piece by the hexagon in the vice, with the threads uppermost, and place the locating washer on it.

Take the piston assembly, still within the assembly ring, and locate it on the centre-piece, with the endstop facing upwards; secure these parts by fitting the backing washer, the shake-proof washer and the large nut. Pass the copper gasket over the push-rod and fit the small nut; adjustment will be effected at a later stage. Remove the assembly from the vice, but do not remove the assembly ring yet.

Lightly smear the bore of the vacuum-cylinder shell with the specified lubricant.

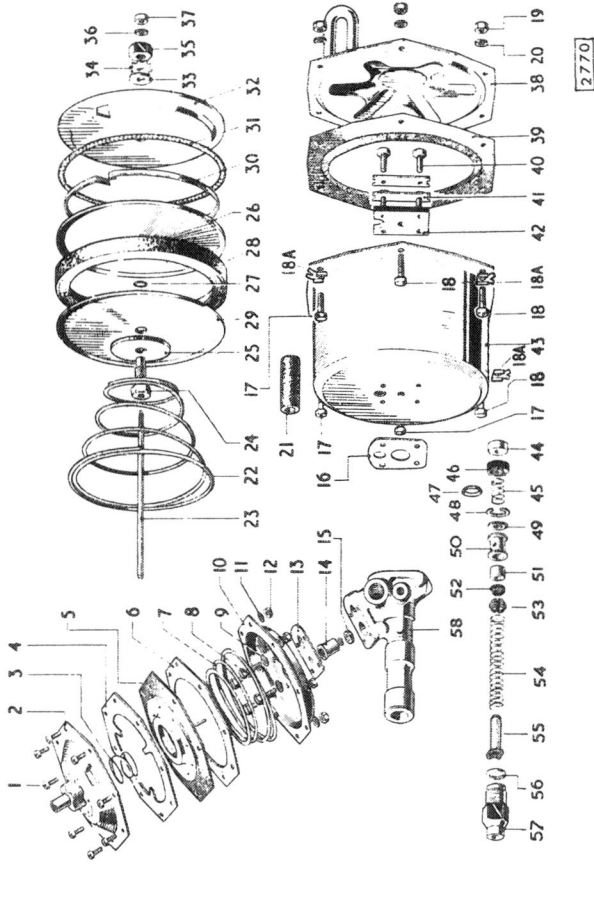

6.92 (17.5cm)

7.05 (17.9cm)

1" (25.4mm)

Fig. 23. *Assembly ring for booster piston. This may be made from a section of scrap vacuum cylinder shell.*

SUPPLEMENTARY INFORMATION
TO
SECTION L "BRAKES"

Hold the assembly with the slave cylinder pointing downwards and engage the small end of the spring behind the tabs of the locking plates inside the shell: place the piston on the large end of the spring and push inwards until the piston enters the shell (removing the assembly ring as it enters). Depress the piston several times and allow to return unassisted to check that there is not excessive friction between the cup and the shell. Pour approximately 20 c.c. of the specified lubricant on to the piston.

Whilst holding the piston within the shell, position the rubber gasket upon the end flange, and place the end cover on the gasket, with the connecting pipe aligned with the mark made prior to dismantling. Hold the end cover down against the spring load and fit the short bolts, nuts and spring washers to secure it; fit the long bolts in the positions marked when dismantling.

Insert the piston into the slave-cylinder, with the recessed side innermost, and push the piston fully home. Use a depth gauge to ascertain dimensions "A" and "B" (see Fig. 25); the difference between these dimensions should be between 0·060" and 0·070" (1·5—1·8 mm.). If the difference is outside these figures, make a note of what it is, then remove the vacuum-shell end cover (ensuring that the spring does not push the booster-piston out of the shell whilst doing so) slacken the nut at the end of the push-rod and adjust the rod "C" to obtain the correct figures.

Note: One turn of the rod produces a difference of 0·035" (0·89 mm.). Hold the push-rod against rotation whilst tightening the locknut, to avoid disturbing the adjustment. Refit the end cover as previously detailed, and re-check the dimensions "A" and "B".

Insert the cup, flat face leading, into the slave cylinder bore, and push it down until it meets the piston. Locate the spring guide in one end of the spring, insert the spring into the bore, with the guide innermost, and locate the spring retainer within the outer end of the spring. Pass the copper gasket over the spring so that it drops squarely into the recess at the end of the bore. Offer up the adapter to the slave cylinder body, and tighten to a torque reading of 37·5—42 lb.ft. (52—58 mkg.).

Ease the cup into the groove on the valve piston, and insert the piston into the recessed bore in the inclined face of the slave cylinder. Place the gasket on the face, followed by the

valve housing, and secure the housing with the four bolts and shakeproof washers. Tighten the bolts to a torque reading of 8·3—11 lb.ft. (1·2—1·5 mkg.).

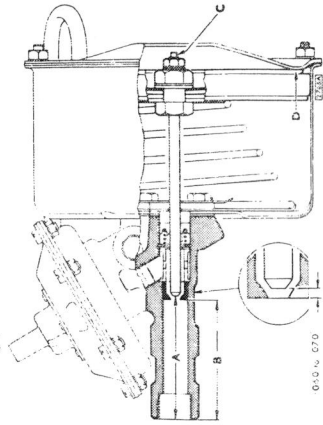

0·060—0·070
(1·5 to 1·8 mm)

Fig. 25. When assembled there must be ·060" to ·070" (1·5—1·8 mm.) clearance between the end of the push rod and the piston. The clearance is adjusted by means of a screwdriver at the point indicated by the arrow.

Place the larger of the two springs within the valve housing so that it is resting on its larger diameter, and position the plain gasket on the flange of the housing. Place the diaphragm on the gasket, with the stem downwards. Position the small spring in the centre of the diaphragm, and the other gasket on the outer edge of this latter part. On top of this place the end cover, with the connecting pipe correctly aligned. Pass two screws through diametrically opposite holes in the cover, the gaskets and the diaphragm, and compress the complete assembly, passing the screws through the holes in the valve housing; fit the shakeproof washers and the nuts. Continue holding the cover against the spring load whilst fitting the remaining six screws, washers and nuts. Slide the hose over the junction in the pipes.

Refitting

Refitting is the reverse of the removal procedure.

Check that the five rubber support grommets are in position in the valance before fitting the unit.

Bleed the system as detailed under "Bleeding the Brake System", page L.8.

VACUUM SERVO UNITS MUST BE STORED IN THE UPRIGHT POSITION, i.e., RESTING ON THE END COVER.

SERVICE TOOLS

Piston re-setting tool .. Part Number 10416

TYPE 8 VACUUM SERVO UNIT

	Commencing Chassis No.	
	R.H. Drive	L.H. Drive
3.4 "S"	1B 2102	1B 25286
3.8 "S"	1B 52037	1B 76292

Commencing at the above chassis numbers the Type "8" eight inch vacuum servo unit is fitted.

Description

Vacuum Chamber and Diaphragm Assembly

The rolling type diaphragm and diaphragm support assembly is housed within the vacuum shell and end cover. The outside edge of the diaphragm is pinched between the vacuum shell and the end cover, thus providing an air-tight joint.

The inside diameter of the diaphragm fits tightly into a groove in the diaphragm support. A push rod, centrally positioned in the diaphragm support, is retained by means of a key which locates in a slot in the base of the diaphragm groove in the diaphragm support; the key is, therefore, retained in position by the tightly fitting diaphragm.

The vacuum shell is attached to a cast iron slave cylinder by means of four setscrews, locked in position by tab washers. The tab washers are formed with lugs which provide location for the diaphragm support return spring interposed between the vacuum shell and the diaphragm support.

Slave Cylinder

The slave cylinder carries the control valve and is bored to accommodate a spring loaded piston which, when the servo is in the "released" position, is held against a distance piece. The piston is sealed by a rubber cup and, when the servo is operated, is displaced by the push rod. The push rod passes through a spring-loaded rubber cup which is backed by a spigot.

Control Valve

The control valve assembly comprises a spring-loaded "two-stage" diaphragm which, when the brakes are applied, is deflected by a piston by means of fluid pressure from the master cylinder. In turn, the diaphragm operates two valves located within the valve cover. To prevent foreign matter entering the unit when the valves are open, an air filter is located within the filter housing.

Principle of Operation

When the brakes are "off", the servo is at rest (Fig. 26), the valve piston and diaphragm are in the normal position; valve "A" is open while valves "B" and "BI" are closed. From this it will be seen that whatever degree of vacuum is present within the engine induction manifold is also present within the vacuum chambers "D" and "E"; additionally, the vacuum is transferred to chamber "C" via holes in the centre of the diaphragm and the valve "A" and to the chamber "F" via the connecting pipe.

Upon depressing the brake pedal, initial fluid pressure from the master cylinder passes through the connection "Y" and is transmitted to the brakes via the hole in the centre of the slave cylinder piston and rubber cup.

Meanwhile, fluid pressure from the master cylinder acts upon the valve piston which displaces the inner portion of the diaphragm to close valve "A" (Fig. 27). The internal deflection of the diaphragm is restricted only by a light spring, thus ensuring operation of the servo by only a small effort on the brake pedal. The closing of valve "A" isolates the vacuum source from chambers "C" and "F".

Continued effort on the brake pedal causes valve "B" to open which admits atmospheric pressure via a restricted passage to chambers "C" and "F" partially destroying the vacuum in these chambers (Fig. 28).

Further effort on the brake pedal causes valve "BI" to open allowing an unrestricted passage of air to chambers "C" and "F" (Fig. 29).

This admission of atmospheric pressure operates a pressure differential across the large booster diaphragm which overcomes the spring loading and displaces the diaphragm support causing the push rod to move forwards towards, the cup to seal off the hole in its centre. Since there is a "locked line" of fluid to the brakes and the continued influence of the pressure differential across the diaphragm will boost the pressure within the slave cylinder and the brake line.

The same pressure differential also exists across the control valve diaphragm and, when the desired degree of braking is obtained, the forces due to fluid pressure from the master cylinder acting on the valve piston and the pressure differential across the control valve diaphragm will balance and the influence of the small spring which loads the centre of the diaphragm will cause the centre of the diaphragm to deflect in the direction of the piston. Consequently, the valves "B" and "BI" will close on their seats thus preventing further entry of air to chamber "F" (Fig. 30).

Greater effort on the brake pedal increases the thrust on the valve piston which causes the valves "B" and "BI" to re-open (Fig. 29) and admit more air so increasing the pressure differential across the booster diaphragm and allow the diaphragm support to perform a greater amount of effort; when the opposing forces on the control valve diaphragm are once more in balance, the valves "B" and "BI" will again close. It will, therefore, be apparent that the control valve diaphragm acts as a proportioning device ensuring that the performance of the servo is progressive.

When the brake pedal is released, pressure is removed from the valve piston allowing the spring load to push the diaphragm back to the position shown in Fig. 26, and thereby cause all four chambers of the servo once again to be in communication with the engine manifold. The diaphragm support and slave cylinder piston are now able to return to the "off" position as shown in Fig. 26 where a gap exists between the push rod end and the slave cylinder piston allowing excess fluid returning from the brake line to return to the master cylinder supply tank.

Removal

The removal sequence is identical to that described on page L.24 for the 6¾″ servo unit.

Dismantling

Mark the position of the valve cover (7) in relation to the housing (15) to assist in their correct replacement, otherwise the air pipe will not connect with the pipe from the vacuum shell.

Disengage the rubber connecting pipe (21) from the valve cover (7), unscrew the eight fixing screws (6). Lift away the cover (7) and remove the two gaskets (8 and 11), two springs (9 and 12) and the diaphragm assembly (10). Removal of the air filter (3) can be effected by bending back the small lugs retaining the air filter cover inwards sufficiently to remove, the air valve cover (1). The small spring (5), sorbo

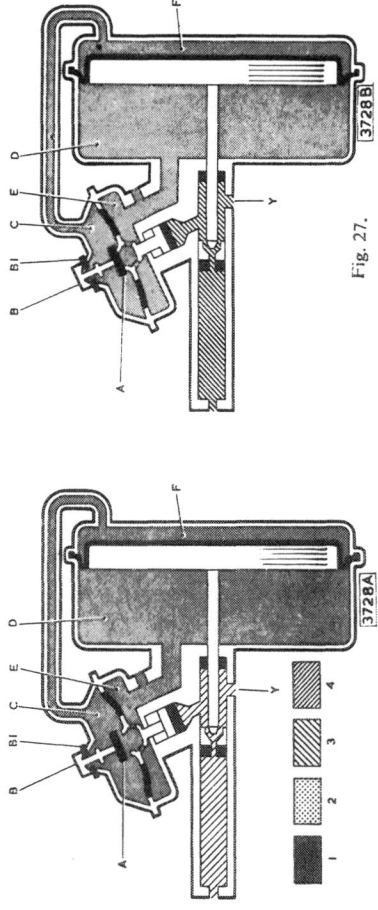

Fig. 26.

1. Vacuum.
2. Atmospheric pressure. 3. Fluid at master cylinder pressure.
4. Fluid at brake line pressure.

Fig. 27.

3728A

3728B

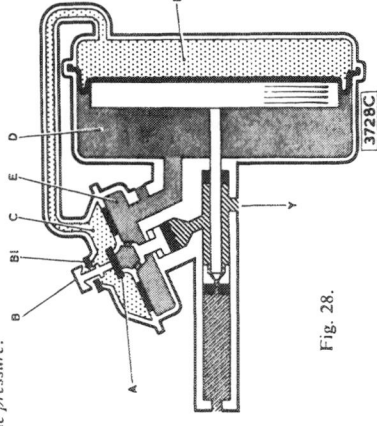

Fig. 28.

3728C

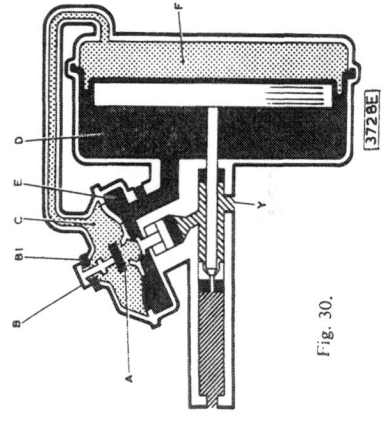

Fig. 29.

3728D

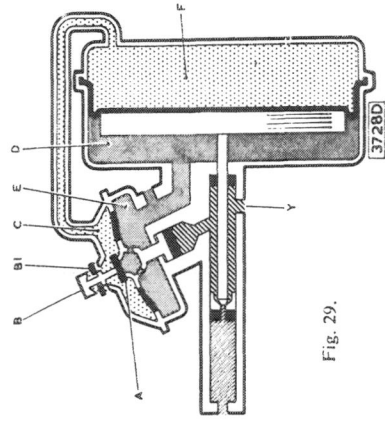

Fig. 30.

3728E

Operation of servo unit.

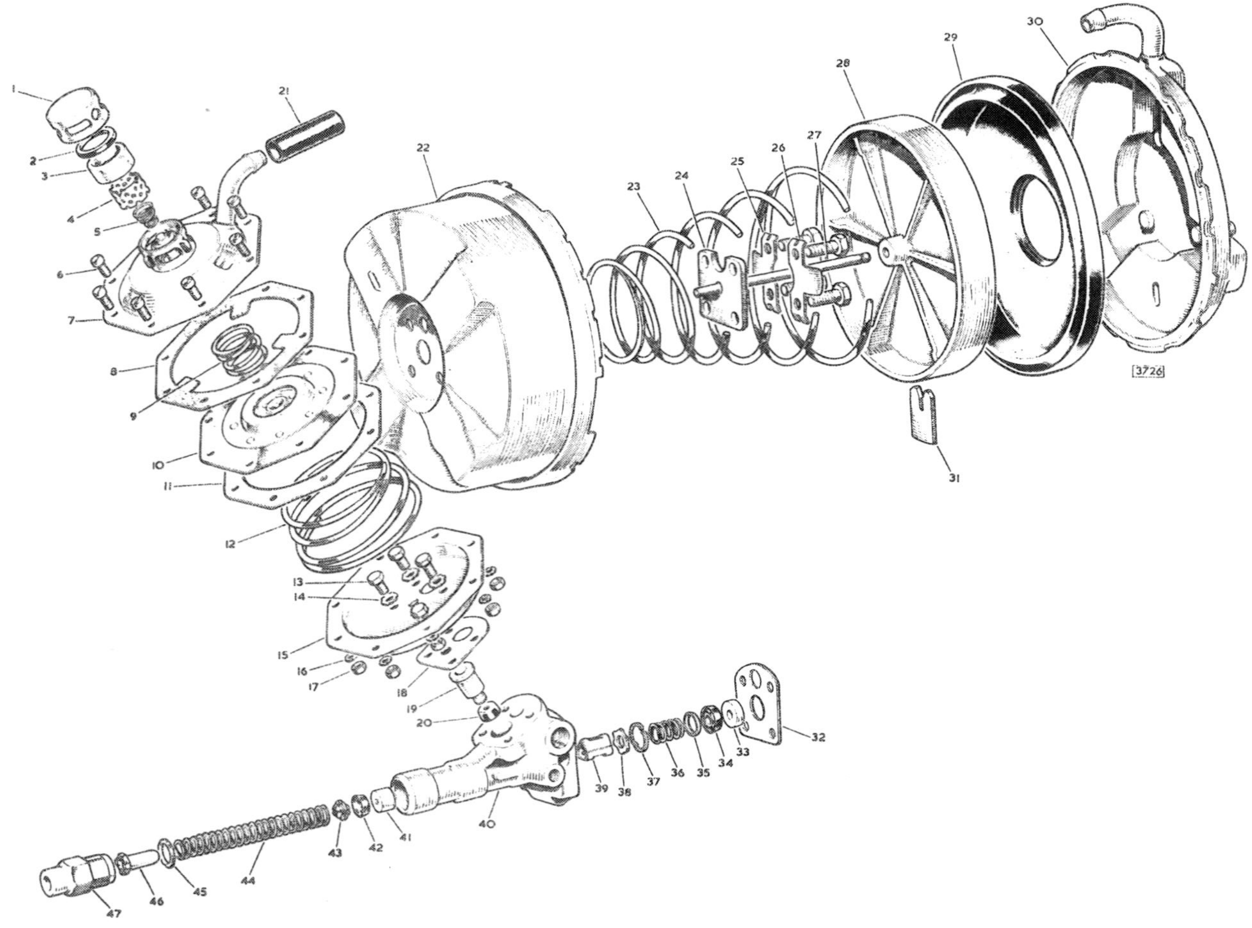

1. Cover
2. Sorbo washer
3. Filter
4. Guide
5. Spring
6. 2 B.A. screw
7. Air valve cover
8. Gasket
9. Spring
10. Diaphragm
11. Gasket
12. Spring
13. Setscrew
14. Shakeproof washer
15. Valve housing
16. Shakeproof washer
17. Nut
18. Gasket
19. Piston
20. Cup
21. Rubber connection
22. Servo shell
23. Spring
24. Abutment plate

25. Locking plate
26. Setscrew
27. Pushrod
28. Diaphragm suppor
29. Diaphragm
30. End cover
31. Key
32. Gasket
33. Push rod guide
34. Gland seal
35. Cup spreader
36. Spring
37. Circlip
38. Washer
39. Distance piece
40. Slave cylinder
41. Piston
42. Cup
43. Spring support
44. Spring
45. Copper washer
46. Spring retainer
47. Adaptor

Fig. 31. *Exploded view of Type 8 8" servo unit.*

washer (2), guide (4) and air filter (3) can now be extracted. Further dismantling of the air control valve cover and diaphragm assembly is not recommended as these are new assemblies reduced to their minimum as spare parts.

Unscrew the four setscrews (13) securing the housing to the slave cylinder body (40) and remove the valve housing and gasket (15).

Apply low air pressure at the fluid inlet connection and eject the valve piston (19) and cup (20). Separate the cup from the piston by easing over the end of the piston.

Holding the slave cylinder (40) in the protected jaws of a vice, fit the special tool (Churchill Tool No. C.2030) to the end cover (30) and secure by fitting three nuts. Remove the end cover by turning anti-clockwise.

Withdraw the diaphragm support (28) complete with the pushrod (27), key (31) and diaphragm (29). Disengage the diaphragm from the diaphragm support, shake the key from the slot in the support and withdraw the pushrod.

Release the return spring (23) from the locking plates (25) inside the vacuum shell: bend back the tabs of the locking plates and remove the four screws (26) securing the abutment plate (24). The vacuum shell (22) and gasket (32) may now be separated from the slave cylinder.

Extract the pushrod guide (33) from the mouth of the slave cylinder bore together with the gland seal (34), cup spreader (35) and spring (36).

Whilst depressing the piston (41) down the bore of the slave cylinder with a length of $\frac{5}{16}''$ (8 mm.) diameter bar, extract the circlip (37). Release the load from the piston and extract the remaining parts from the slave cylinder bore.

Assembling the Servo

Examine all parts before re-assembly and renew any parts showing signs of damage or wear. Inspect the bore of the slave cylinder for signs of rust or scoring: replace if necessary. All rubber cups, seals and diaphragms should be renewed on overhaul.

Clamp the slave cylinder in the protected jaws of a vice. Fit the spring retainer (46) to one end of the spring and insert these two parts (spring retainer leading) into the bore. Follow with the spring support (43), cup (42) (lip leading) and the piston (41). Insert the piston

so that its flat face is in contact with the cup. Push the piston down the bore using a length of $\frac{5}{16}''$ (8 mm.) brass bar: insert the distance piece (39) and washer (38) and secure with the circlip (37). Ensure that the circlip is bedded in the groove provided in the bore.

Remove the brass bar from the bore and insert the spring (36), cup spreader (35), concave face innermost. Fit the gland seal (34) so that its lip is adjacent to the cup spreader (35). Position the push rod guide (33) in the mouth of the bore so that it projects to serve as a spigot when fitting the gasket and vacuum shell.

Locate the gasket (32) over the push rod guide following with the vacuum shell (22) abutment plate (24) and locking plates (25). Line up the holes in the locking plates, abutment plate, vacuum shell and gasket with the tapped hole in the slave cylinder and insert the four setscrews (26). Tighten down the setscrews to a torque of 130 lb in. and bend up the tabs of the locking plates.

Insert the grooved end of the pushrod (27), into the hole in the diaphragm support (28), and secure by fitting the key (31), in the slot in the side of the diaphragm support so that it engages with the groove in the pushrod.

Stretch the smaller diameter of the diaphragm (29), into position on the diaphragm support correctly bedding it into its groove. If the flat face of the diaphragm appears buckled, this indicates that the diaphragm has not been assembled properly and should be checked before proceeding further.

Engage the small end of the return spring (23) with the locking tabs inside the vacuum shell and the larger end against the inside face of the diaphragm support assembly. Lightly smear the outside edge of the diaphragm with brake fluid and, with the special tool (Churchill Tool No. C.2030) attached to the end cover, position the end cover over the diaphragm support assembly.

Press the end cover downwards against the load of the spring so that the pushrod enters the hole in the push rod guide. Continue pressing the end cover down until the lugs on the outside edge of the cover engage the cutaway portions of the servo shell. Lock the end cover in position by turning clockwise to its full extent and remove the special tool.

Ease the cup (20) into the groove in the valve piston (19) so that the lip of the cup faces

away from the shoulder of the piston and insert the piston into the valve bore of the slave cylinder. Fit the gasket (18) and the valve housing (15) and secure with the four setscrews and shake proof washers tightening to a torque of 110 lb in. Insert the spring (12) inside the valve housing with the larger diameter innermost. Place a gasket on the valve housing and fit the diaphragm assembly on the gasket so that the small pushrod engages the depression in the valve piston. Place a further gasket on the diaphragm and locate the spring (9) on the centre of the diaphragm. Fit the air valve cover (7) over the diaphragm; line up the holes and the tube from the air valve cover with the tube from the vacuum shell end cover. Secure the air valve cover with the eight 2 B.A. screws (6), shakeproof washers (16), and nuts (17). Tighten to a torque of 19 lb in.

Examine the air filter for clogging and renew if necessary. Any light dust can be removed with a low pressure air line. Do NOT use cleansing fluid.

Insert the air filter (3), guide (4), sorbo washer (2), and spring (5), with the small end of the spring innermost.

Place the cover (1) over the housing and retain in position by bending the lugs on the housing outwards to engage the holes in the cover.

Push the rubber connection (21) into position to connect the air valve cover with the tube in the vacuum shell end cover.

Refitting

Refitting is the reverse of the removal procedure.

SECTION M

WHEELS AND TYRES

ADDENDA

The following corrections should be noted when using SECTION M.

Page	Correction
M.5	"Construction of the Tyre" should read "Construction of the RS.5 Tyre".

WHEELS AND TYRES

DESCRIPTION

Pressed steel disc wheels are fitted as standard equipment; wire spoke wheels are specified as an optional extra.

Dunlop SP41 tyres (with tubes) are fitted to both types of wheels. (Early cars were fitted with Dunlop RS5 tyres).

DATA

Roadwheel:

Type—standard equipment	Pressed steel disc
—optional equipment	Wire spoke (72 spoke)
Fixing—pressed steel disc	Five studs and nuts
—wire spoke	Centre lock, knock on hub cap
Rim section—disc	5J
—wire spoke	5K
Rim diameter—all types	15″ (381 mm.)

Tyres:

Make	Dunlop
Type (early cars)	R.S.5
Size (early cars)	6·40 × 15
Type (later cars)	SP.41
Size (later cars)	185 × 15 (185 × 380)

IMPORTANT

It is particularly important that tyres of different makes, types, or those having different tread patterns should not be mixed on individual cars as this may adversely affect the handling and steering characteristics.

A car should not, of course, be driven on bald tyres or on tyres which have only part of the tread left showing. Driving with badly worn tyres on wet roads also greatly increases the risk of "aquaplaning", with consequent loss of steering and braking.

The importance of having tyres that are in good condition and of the correct type cannot be over-stressed. The Dunlop RS.5 Road Speed (or Dunlop SP 41) tyres fitted as original equipment are specially produced to suit the performance of the model concerned and a change of make or type of tyre should not be made unless an assurance is given by the tyre manufacturer concerned that the alternative type is suitable for the particular car under maximum performance conditions.

INDEX

INFLATION PRESSURES

PRESSURES SHOULD BE CHECKED WHEN THE TYRES ARE COLD, SUCH AS STANDING OVERNIGHT, AND NOT WHEN THEY HAVE ATTAINED THEIR NORMAL RUNNING TEMPERATURES.

Dunlop RS.5 Tyres

	Front	Rear
Normal driving including use on motorways up to 110 m.p.h. (175 k.p.h.)	28 lbs. per sq. in. (1·97 kg./cm.²)	25 lbs. per sq. in. (1·76 kg./cm.²)
For sustained high speed using the maximum performance of the car	33 lbs. per sq. in. (2·32 kg./cm².)	30 lbs. per sq. in. (2·11 kg./cm².)
Town use or secondary roads with bad surfaces where speeds are limited, minimum pressures are permissible.	25 lbs. per sq. in. (1·76 kg./cm².)	22 lbs. per sq. in. (1·55 kg./cm².)
Fully laden condition	When undertaking a long journey with a full load of passengers and luggage, increase the rear tyre pressures by 4 lbs. per sq. in. (28 kg./cm².).	

Dunlop SP.41 Tyres

	Front	Rear
Normal motoring up to 100 m.p.h. (160 k.p.h.):		
—Not more than 3 persons in car ..	30 lb. per sq. in. (2·1 kg./cm².)	27 lb. per sq. in. (1·9 kg./cm².)
—Car fully laden	30 lb. per sq. in. (2·1 kg./cm².)	30 lb. per sq. in. (2·1 kg./cm².)
Sustained high speeds up to maximum:		
—Not more than 3 persons in car ..	36 lb. per sq. in. (2·5 kg./cm².)	33 lb. per sq. in. (2·3 kg./cm².)
—Car fully laden	36 lb. per sq. in. (2·5 kg./cm².)	36 lb. per sq. in. (2·5 kg./cm².)

TYRES—GENERAL INFORMATION

The Dunlop tyres specified have been specially designed for cars with the high speed range of the Jaguar 3·4 'S' and 3·8 'S' models.

When replacing worn or damaged tyres and tubes it is essential that tyres with exactly the same characteristics are fitted.

Due to the high speed performance capabilities of the Jaguar 3·4 'S' and 3·8 'S' models it is important that repair of damaged or punctured tyres should only be undertaken by a tyre repair specialist.

All tyres which are suspect in any way should be submitted to the tyre manufacturers for their examination and report. The importance of maintaining all tyres in perfect condition cannot be too highly stressed.

CONSTRUCTION OF THE TYRE

One of the principal functions of the tyres fitted to a car is to eliminate high frequency vibrations. They do this by virtue of the fact that the unsprung mass of each tyre—the part of the tyre in contact with the ground—is very small.

Tyres must be flexible and responsive. They must also be strong and tough to contain the air pressure, resist damage, give long mileage, transmit driving and braking forces, and at the same time provide road grip, stability and good steering properties.

Strength and resistance to wear are achieved by building the casing from several plies of cord fabric, secured at the rim position by wire bead cores, and adding a tough rubber tread (Fig. 1).

Part of the work done in deflecting the tyres on a moving car is converted into heat within the tyres. Rubber and fabric are poor conductors and internal heat is not easily dissipated. Excessive temperatures weaken the tyre structure and reduce the resistance of the tread to abrasion by the road surface.

Heat generation, comfort, stability, power consumption rate of tread wear, steering properties and other factors affecting the performance of the tyres and car are associated with the degree of tyre deflection. All tyres are designed to run at predetermined deflections, depending upon their size and purpose.

Load and Pressure Schedules are published by all tyre makers and are based on the correct relationship between tyre deflection, tyre size, load carried and inflation pressure. By following the recommendations the owner will obtain the best results both from the tyres and the car.

Inflating

When inflating the tyre, after refitting to wheel, be sure that the valve core is in the valve and DO NOT EXCEED 40 POUNDS AIR PRESSURE as there is a risk of breaking the bead wires.

If it is found that the bead will not seat properly, deflate, lubricate and centralize tyre before re-inflating. When the tyre bead does not seat properly at the second attempt, the wheel rim circumference is suspect and should be checked with a rim gauge, if available, or replaced with a new wheel.

After the beads have seated properly, reduce pressure to the recommended operating pressure.

Note: Lock the wheel down when using the mounting machine and do not stand over the tyre when inflating it. Check the tyre pressure frequently to be absolutely sure that the pressure never exceeds 40 pounds per sq. in. It is advisable to use an extension pressure gauge with a clip-on chuck and stand well back for maximum safety.

Inflation Pressures

It is important to maintain the tyre pressures at the correct figures, incorrect pressures will affect the steering, riding comfort, and tyre wear.

Effect of Temperature

Air expands with heating and therefore tyre pressures increase as the tyres warm up. Pressures increase more in hot weather than in cold weather and as a result of high speed. These factors are taken into account when designing the tyre and when determining recommended inflation pressures.

Pressures in warm tyres should not be reduced to standard pressures for cold tyres. "Bleeding" the tyres increases their deflections and causes their temperatures to climb still higher. The tyres will be underinflated when they have cooled.

Always ensure that the valve caps are fitted as they prevent the ingress of dirt and form a secondary seal to the valve core.

Nylon Tyres (RS5 type)

Nylon tyres may develop a temporary flat after standing for some time and cooling off, following a long run during which high temperatures have been reached.

These flat spots can be run out quite quickly but it may usually be necessary to approach the speeds and temperatures which have led to the flatting. For example, flats on tyres which have developed after a long fast run will be difficult to remove if the car is then used for local motoring especially if the weather has become colder and wetter.

Before balancing nylon tyres, it is essential to ensure that these flats have been completely run out, otherwise a false balance reading will be obtained.

Tyre Examination

Examine tyres periodically for flints, nails, etc., which may have become embedded in the tread. These should be removed with a blunt screwdriver or a similar instrument.

WHEEL ALIGNMENT AND ITS ASSOCIATION WITH ROAD CAMBER

It is very important that correct wheel alignment should be maintained. Misalignment causes a tyre tread to be scrubbed off laterally because the natural direction of the wheel differs from that of the car.

An upstanding sharp "fin" on the edge of each pattern rib is a sure sign of misalignment and it is possible to determine from the position of the "fins" whether the wheels are toeing in or toeing out.

"Fins" on the inside edges of the pattern ribs—nearest to the car—and particularly on the near side tyre indicate toe in. "Fins" on the outside edges, particularly on the offside tyre, indicate toe out.

With minor misalignment, the evidence is less noticeable and sharp pattern edges may be caused by road camber even when wheel alignment is correct. In such cases it is better to make sure by checking with an alignment gauge.

Road camber affects the direction of the car by imposing a side thrust and if left to follow its natural course the car will drift towards the near side. This is instinctively corrected by steering towards the road centre.

As a result the car runs crab-wise, diagrammatically illustrated in an exaggerated form in Fig. 2. The diagram shows why nearside tyres are very sensitive to too much toe in and offside tyres to toe out. It also shows why sharp "fins" appear on one tyre but not on the other and why the direction of misalignment can be determined by noting the position of the "fins". Severe misalignment produces clear evidence on both tyres.

The front wheels on a moving car should be parallel. Tyre wear can be affected noticeably by quite small variations from this condition. It will be noted from the diagram that even with parallel wheels the car is still out of line with its direction of movement, but there is less tendency for the wear to be concentrated on any one tyre.

The near front tyre sometimes persists in wearing faster and more unevenly than the other tyres even when the mechanical condition of the car and tyre maintenance are satisfactory. The more severe the average road camber the more marked will this tendency be.

Precautions when Measuring Wheel Alignment

1. The car should have come to rest from a forward movement. This ensures as far as possible that the wheels are in their natural running positions.

2. It is preferable for alignment to be checked with the car laden.

3. With conventional base-bar tyre alignment gauges measurements in front of and behind the wheel centres should be taken at the same points on the tyres or rim flanges. This is achieved by marking the tyres where the first reading is taken and moving the car forwards approximately half a road wheel revolution before taking the second reading at the same points. With the Dunlop Optical Gauge two or three readings should be taken with the car moved forwards to different positions—180° road wheel turn for two readings and 120° for three readings. An average figure should then be calculated.

Wheels and tyres vary laterally within their manufacturing tolerances, or as the result of service, and alignment figures obtained without moving the car are unreliable.

TYRE AND WHEEL BALANCE

Static Balance

In the interests of smooth riding, precise steering and the avoidance of high speed "tramp" or "wheel hop" all Dunlop tyres are balance checked to predetermined limits.

To ensure the best degree of tyre balance the covers are marked with white spots on one bead, and these indicate the lightest part of the cover. Tubes are marked on the base with black spots at the heaviest point. By fitting the tyre so that the marks on the cover bead exactly coincide with the marks on the tube, a high degree of tyre balance is achieved (Fig. 3). When using tubes which do not have the coloured spots it is usually advantageous to fit the covers so that the white spots are at the valve position.

Some tyres are slightly outside standard balance limits and are corrected before issue by attaching special patches to the inside of the covers at the crown. These patches contain no fabric, they do not affect the local stiffness of the tyre and should not be mistaken for repair patches. They are embossed "Balance Adjustment Rubber".

The original degree of balance is not necessarily maintained and it may be affected by uneven tread wear, by cover and tube repairs, by tyre removal and refitting or by wheel damage and eccentricity. The car may also become more sensitive to unbalance due to normal wear of moving parts.

If roughness or high speed steering troubles develop, and mechanical investigation fails to disclose a possible cause, wheel and tyre balance should be suspected.

A Tyre Balancing Machine is marketed by the Dunlop Company to enable Service Stations to deal with such cases.

Warning: If balancing equipment is used which dynamically balances the road wheels on the car, the following precaution should be observed.

In the case of the rear wheels always jack **both** wheels off the ground otherwise damage may be caused to the differential.

This is doubly important in the case of cars fitted with a Thornton "Powr-Lok" differential as in addition to possible damage to the differential, the car may drive itself off the jack or stand.

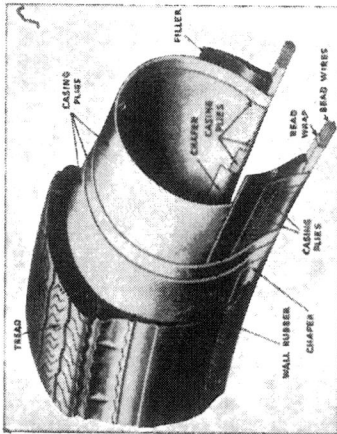

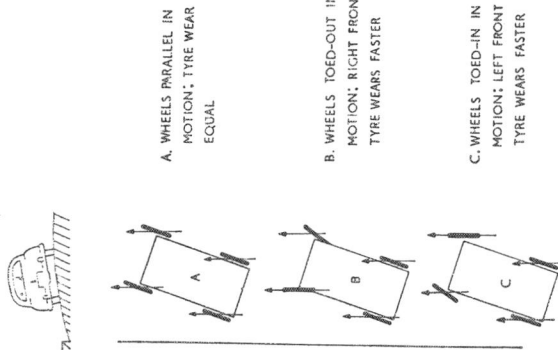

Fig. 1. Tyre construction.

Fig. 2. Exaggerated diagram of the way in which road camber affects a car's progress.

A. WHEELS PARALLEL IN MOTION; TYRE WEAR EQUAL

B. WHEELS TOED-OUT IN MOTION; RIGHT FRONT TYRE WEARS FASTER

C. WHEELS TOED-IN IN MOTION; LEFT FRONT TYRE WEARS FASTER

Dynamic Balance

Static unbalance can be measured when the tyre and wheel assembly is stationary. There is another form known as dynamic unbalance which can be detected only when the assembly is revolving.

There may be no heavy spot, that is, there may be no natural tendency for the assembly to rotate about its centre due to gravity, but the weight may be unevenly distributed each side of the tyre centre line. Laterally eccentric wheels give the same effect. During rotation the off set weight distribution sets up a rotating couple which tends to steer the wheel to right and left alternately.

Dynamic unbalance of tyre and wheel assemblies can be measured on the Dunlop Tyre Balancing Machine and suitable corrections made when cars show sensitivity to this form of unbalance. Where it is clear that a damaged wheel is the primary cause of severe unbalance it is advisable for the wheel to be replaced.

TYRE REPLACEMENT AND WHEEL INTERCHANGING

When replacement of the rear tyres becomes necessary, fit new tyres to the existing rear wheels and, after balancing, fit these wheels to the front wheel positions on the car, fitting the existing front wheel and tyre assemblies (which should have useful tread life left) to the rear wheel positions on the car.

If at the time this operation is carried out the tyre of the spare wheel is in new condition, it can be fitted to one of the front wheel positions in preference to replacing one of the original rear tyres, which have been used on the rear wheel positions, interchanging of part worn tyres from rear to front wheel positions is not recommended the spare.

Note: Due to the change in the steering characteristics which can be introduced by fitting to the front wheel positions, wheels and tyres which have been used on the rear wheel positions, interchanging of part worn tyres from rear to front wheel positions is not recommended.

Fig. 3. *Correct position of the inner tube to outer cover to facilitate wheel balance.*

WIRE SPOKE WHEELS REPAIR AND ADJUSTMENT

DESCRIPTION

Dunlop cross-spoked wire wheels are fitted as optional equipment and the following instructions are issued to assist in the repair and adjustment of the road wheels in the event of damage due to accident or from any other cause.

Cross-spoking refers to the spoke pattern, where the spokes radiate from the well of the wheel rim to the nose or outer edge of the hub shell, and from the tyre seat of the rim to the flanged or inner end of the shell (Fig. 4).

REMOVAL AND DISMANTLING

Detach wheel from car and remove tyre complete from wheel rim.

Remove spoke nipples and detach spokes from rim and centre.

Check wheel rims and centre; renew if damaged beyond normal repair.

Examine spokes and renew as necessary.

REBUILDING

Place the wheel centre and the rim on a flat surface with the valve hole upwards in the 6-o'clock position.

Note: All spoking operations commence in this position, and the valve hole is always the starting point for all rebuilding operations.

With the valve hole in the 6-o'clock position, fit one A, B, C, and D spoke to produce the pattern as shown in Fig. 4.

Having established the correct pattern remove the A and B spokes and proceed as follows:—

(1) Attach the D spoke to the rim, and screw up the nipple finger tight; leave the C spoke loosely fitted without a nipple attached.

(2) Attach all the D spokes with the nipples finger tight.

(3) Insert all the C spokes through the hub shell without nipples.

Page M.8

(4) Attach all the B spokes as paragraph 2 above.

(5) Attach all the A spokes as paragraph 2 above.

(6) Attach the nipples and finger tighten all C spokes.

(7) Tighten the two C spokes and the two D spokes on each side of the valve hole until the ends of the spokes are just below the slot in the nipple heads.

(8) Tighten the four C and D spokes diametrically opposed to the valve hole (12-o'clock position).

(9) Mark around the wheel until all the C and D spokes are similarly tightened.

(10) Follow with all A and B spokes as in paragraphs 7, 8 and 9 above.

(11) Work around the wheel with a spoke spanner and tighten all nipples until some resistance is felt. Diametrically opposed spokes should be tightened in sequence.

The wheel is now ready for trueing and adjustment.

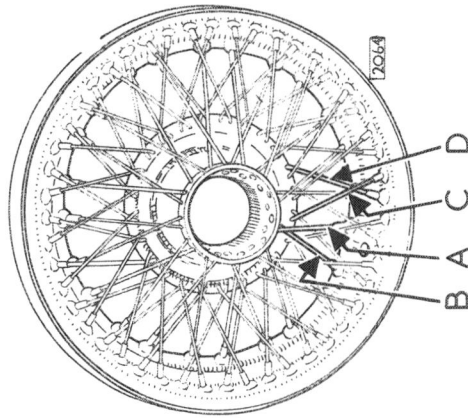

Fig. 4. *Showing the spoke arrangement.*

Page M.9

TRUEING

Wheels can be out of truth in a lateral or radial direction, or in a combination of both.

As a general rule, lateral out of truth should be corrected first.

The wheel to be trued must be mounted on a free-running trueing stand before any adjustment can be carried out.

Lateral Correction

Mount the wheel on the trueing stand. Spin the wheel, and holding a piece of chalk near the wall of the rim flange, mark any high spots. Tighten the A and B spokes in the region of the chalk marks and slacken the C and D spokes in the area.

Note: Throughout the trueing operations, no spoke should be tightened to such an extent that it is impossible to tighten it further without risk of damage. If any spoke is as tight as it will go, all the other spokes should be slackened.

Radial Correction

When lateral out of truth has been corrected, spin the wheel on the trueing stand, and, with the chalk, mark the high spots on the horizontal tyre seat. Tighten **all** spokes in the region of the chalk marks, or if the spokes are on the limit of tightness, slacken all the remaining spokes.

CHECKING FOR DISH

The term "dish" defines the lateral dimension from the inner face of the flanges of the wheel centre to the inner edge of the wheel rim. To check "dish" place straight edge across the inner edge of the wheel rim and measure the distance to the inner face of the wheel centre flange (Fig. 5). This dimension should be $3\frac{7}{16}'' \pm \frac{1}{16}''$ (87.3 mm. $\pm$ 1.58 mm.).

Adjustment for "Dish"

If the "dish" is in excess of the correct dimension $3\frac{7}{16}'' \pm \frac{1}{16}''$ (87.3 mm. $\pm$ 1.58 mm.) tighten all A and B spokes, and slacken all C and D spokes by a similar amount.

When the "dish" dimension is less than the given tolerance slacken all A and B spokes and tighten all C and D spokes by a similar amount.

It will be necessary after completing the "dish" adjustments to repeat the lateral and radial trueing procedure until the wheel is not more than ·060″ (1·5 mm.) out of truth in either direction.

It is important that after the wheel trueing operation is completed that all spokes should be tensioned uniformly, and to a reasonably high degree.

Correct tension can be closely estimated from the high pitched note emitted when the spokes are lightly tapped with a small hammer.

If a spoke nipple spanner of the torque recording type is used, a normal torque figure should be in the order of 60 lb.ins. (0·7 kg./m.).

RADIAL PLY (SP.41) TYRES—SERVICE PROCEDURE

It is not always realised that fitting radial ply tyres can introduce a different "feel" to driving the car. Therefore, it is essential to establish whether a condition should be assessed as abnormal or as something which must be accepted as inherent in radial ply tyre operation. For instance, while SP tyres offer considerable advantage in road roar suppression and ride comfort at open road speeds, some increase in harshness at town speeds and an increase in parking effort is inevitable.

Lurch

This is a rolling kind of progress where the front or back or both ends of the car appear to lurch from side to side, sometimes accompanied by a slight up and down movement. These low frequency lateral and vertical oscillations may occur at speeds between 10 and 30 m.p.h. (16 and 48 k.p.h.).

To some extent, it is inherent in the very flexible casing construction of radial ply tyres. More severe cases may be associated with poor suspension damping possibly due to weakened shock absorbers or with severe lateral or radial run-out of the tyre/wheel assemblies. The assemblies should be checked for truth starting with the end of the car which appears most affected.

Heavy Steering at Low Speeds

This condition, particularly at parking speeds, must be accepted as normal. If it appears to be abnormal, points to check are tyre pressures, castor angles and excessive friction in wheel swivels, ball joints and steering box.

Surge

This is a fore and aft hesitancy causing uneven progress. It is related to engine speed and, if present, will occur at town speeds mainly in the intermediate gears although it may be felt in top gear at times.

This again is partly inherent in radial ply construction because of the low torsional rigidity of the casing permits some degree of wind-up between the tread and the rim. If the condition is not aggravated by other factors, it can in some circumstances be an advantage in providing a more gradual getaway on very slippery surfaces.

Surge may be more noticeable as the result of a torque disturbing factor such as misfiring, incorrect carburation, momentary fuel starvation or excessive flexibility in the engine mountings. Similarly, it may be aggravated by excessive run-out of the driving tyre/wheel assemblies but, in this case, other forms of vibration will also be present.

Steering Niggle

This normally refers to a marked torsional oscillation of the steering wheel at speeds of up to 50 m.p.h. (80 k.p.h.) something similar to a condition caused by excessive dynamic un-balance.

Balance the tyre/wheel assembly statically and dynamically and, if the fault is not corrected, proceed as detailed under the heading "High Speed Vibrations".

Pull

This is a greater than usual tendency for the car to steer to one side when the steering wheel is released. A marked steering effort is required to hold the car on a straight course.

Check the camber, castor angles and make sure that they are equal on both sides of the car. Equal camber angles are important and they must not vary by more than $\frac{1}{4}$° on each side.

If steering angles are suspected as being the cause of the trouble, fit two cross ply tyres to the front wheels for testing purposes only. If the pull persists on cross-ply tyres, then the cause is mechanical.

If the steering angles are correct and the excessive pull does not occur with cross-ply tyres, refit each SP tyre in turn to the near side front wheel with a cross ply tyre on the off front

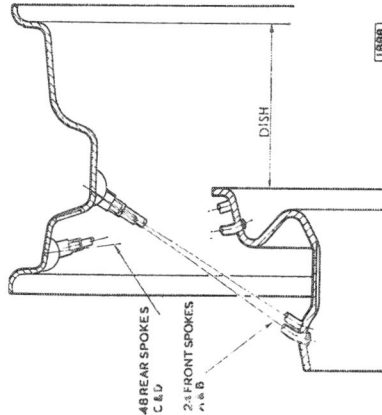

48 REAR SPOKES
C & D

24 FRONT SPOKES
A & B

DISH

1666

Fig. 5. Location for measuring the dish and the 'A', 'B', 'C' and 'D' spokes.

SECTION N

BODY & EXHAUST SYSTEM

wheel and re-test on a flat road. If one SP tyre is found to be the offender it may have to be replaced, but, if both tyres are similarly affected, it should be possible to cancel this by reversing one tyre on its rim.

It will be clear that pull to the near side will be increased by road camber and pull to the off side reduced or cancelled by it. This can often be used to make an unacceptable condition acceptable by reversing or interchanging tyres without the need to replace.

High Speed Vibrations

These are vibrations which occur throughout the car at speeds around 80 m.p.h. (130 k.p.h.). They are similar to vibrations caused by out of balance cross ply/tyre/wheel assemblies.

The vibrations may result from out of balance in the propeller shaft or play in its joints. Out of true brake discs, untrue wheels, fatigued engine mountings and general engine vibration may also be factors other than the tyres themselves.

Check all tyre/wheel assemblies for static and dynamic balance and correct as necessary. Check for lateral and radial run-out and concentricity of the tyre fitting line. An assembly run-out which exceeds ·060″ (1·5 mm.) on the tread may be suspect.

Should the run-out exceed ·060″ (1·5 mm.) strip the tyre from the wheel, check the wheel separately at both bead seats for wobble and lift and refit the tyre to give minimum run-out (by rotating the tyre relative to the wheel in 90° steps). At this stage, road-test the car and, if the vibrations are still unacceptable, check the spare wheel as above and use to replace the worst wheel. If the measurements of the suspect wheel exceed S.M.M. & T. tolerances, the owner may negotiate with his agent on the basis of whether the car is sufficiently new to justify a claim or whether the wheel is out of truth from wear and tear.

If the wheel is well within the tolerance and an assembly run-out of or below ·060″ (1·5 mm.) cannot be achieved, the tyre may be exchanged at an appropriate allowance.

INDEX

BODY

INDEX (continued)

BODY AND EXHAUST SYSTEM

BODY

THE INSTRUMENT PANEL

Opening

Detach the earth lead from the battery.

Withdraw the picnic tray to the full extent and locate the two spring steel retaining clips attached to the tray at the back edge. Press both clips inwards to the centre of the car and complete the withdrawal of the tray.

Remove the ignition key and cigar lighter for safe keeping. Hinge the centre instrument panel downwards on its bottom edge, after withdrawing the thumb screws situated in each top corner.

Removal

The instrument panel can be removed completely by detaching the earth lead from the battery, identifying and removing the leads from the instruments, cigar lighter and switches, removing the electrical harness together with the clips from the instrument panel and withdrawing two bolts from the extended portion of each hinge, accessible through the newspaper tray beneath.

Refitting

Refitting is the reverse of the removal procedure, but particular attention must be given to the following point:—

That the leads are refitted in accordance with their colour coding, utilising the wiring diagram as a reference.

Closing

Closing is the reverse of the opening procedure but particular attention must be given to the following points:—

(i) That the leads are replaced in accordance with their colour coding, utilising the wiring diagram as a reference.

(ii) That the clips securing the main harness to the instrument panel will in no way foul any of the switch or instrument terminals, otherwise a direct short will occur when the battery is connected.

THE SCREEN RAIL (1, Fig. 1)

Removal

Remove the screen rail from the base of the windscreen after detaching four nuts, two adjacent to each end and two central.

The nut situated above the glove-box is accessible through a hole in the glove-box top panel, the centre nuts being exposed when the instrument panel is lowered.

Disconnect the two cables connected to the maplight and withdraw the rail.

Refitting

Refitting is the reverse of the removal procedure.

THE SIDE FACIA PANEL (2, Fig. 1)

Removal

Detach the earth lead from the battery.

Withdraw the picnic tray (3) and lower the instrument panel.

Remove the screen rail.

Remove the upper cover from the steering column nacelle by withdrawing the two sunken screws from beneath.

Identify and withdraw the flasher warning light harness through the loop bracket attached to the hidden face of the instrument panel by disconnecting the snap connectors.

Detach the fabric covered steering column cover, located above the parcel tray, by removing the two drive screws.

Release the two nuts, now exposed, securing the steering column assembly to the body to the full extent of the thread and lower the column to the newspaper tray.

Remove the fabric sheeting, attached with upholstery solution between the demister ducts and the facia panel.

Remove the setscrew securing the facia panel to the bracket attached to the body below the screen pillar.

Withdraw the two facia panel securing screws located in the instrument panel aperture.

Disconnect the speedometer drive cable, warning lights and electrical leads and remove the facia panel.

Refitting

Refitting is the reverse of the removal procedure. Ensure that the flasher warning lights are replaced in their correct holders.

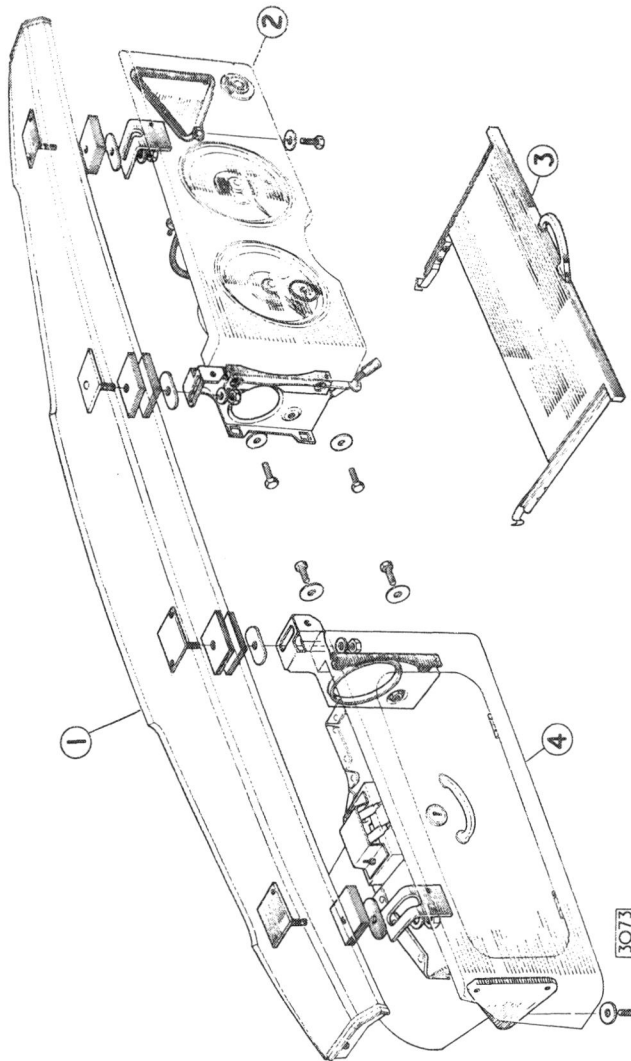

Fig. 1. Showing the attachment points for the side facia panel, glove box and screenrail.

THE GLOVE-BOX (4, Fig. 1)

Removal

Detach the earth lead from the battery.

Withdraw the picnic tray (3) and lower the instrument panel.

Remove the screen rail.

Remove the fabric sheeting, attached with upholstery solution between the demister ducts and the glove-box.

Remove the setscrew securing the glove-box to the bracket attached to the body below the screen pillar.

Remove the two glove-box securing screws located in the instrument panel aperture.

Disconnect the illumination lamp cables and remove the glove-box.

Refitting

Refitting is the reverse of the removal procedure.

THE CONSOLE

Removal

Remove the right-hand and left-hand parcel tray crash rolls after removing the two domed nuts securing each roll to the tray.

Remove the heater control perforated guard cover from the parcel tray after withdrawing four nylon retaining pins.

Pull off the heater control lever knob. Disconnect the control lever at the forward end, unscrew the lever pivot pin, note the washer fitted between the lever and bracket.

Withdraw the lever through the escutcheon plate.

Disconnect the control cables from the right and left-hand front air outlet ducts, located under the parcel tray, by releasing the locking screw securing the inner cable to the air duct operating spindles.

Remove the locknuts securing the outer cables to the air duct brackets. Disconnect the cables and collect the loose adaptors.

Unscrew the outer cables from the centre finisher and withdraw the complete assemblies.

Withdraw the parcel tray centre finisher after removing the domed nuts, plain nuts and large washers from the two outer studs and plain nuts with small washers from the two inner studs.

Remove the heater control button escutcheon, retained by two pegs to the control panel and unscrew the two round-headed screws now exposed.

Lift the parcel tray trimming above the console panel and withdraw the two round-headed thumb screws located in recesses in the tray.

Withdraw the control panel facia and disconnect the three rubber pipes connected to the control button unit. Note the location of the individual pipes for reference when refitting.

Disconnect the rear heater duct pipe from the front junction. Access is gained through the control panel aperture.

Remove the two hexagon headed drive screws securing the rear end of the console to the gearbox tunnel.

Unscrew the gear control knob and tapered nut (not applicable to automatic transmission).

Slide the console rearwards to clear the front clip fixings and lift over the gear control lever to remove.

Refitting

Refit in the reverse order.

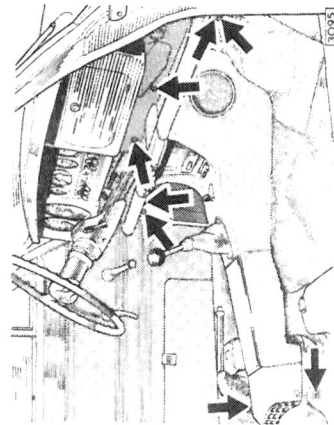

Fig. 2. Showing the attachment points for the console.

Insert the fingers under the nose of the bonnet and lift the safety catch upwards when the bonnet may be raised.

The bonnet is automatically retained in the fully open position by the action of the hinge springs.

Mark the positions of the hinge brackets on the bonnet to facilitate refitting.

Remove the four setscrews and washers securing the bonnet to each hinge and lift off the bonnet.

Refitting

Position hinges on marks made before removal. Refitting is the reverse of the removal procedure.

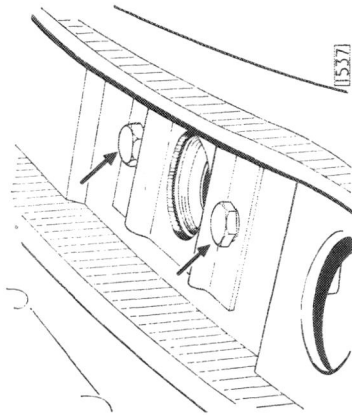

Fig. 3. Showing the two setscrews which secure the bonnet catch striker plate.

BONNET LOCK

Removal

To remove the bonnet catch slacken the locknut at the top of the peg. Insert a screwdriver into the slot in the peg and unscrew the peg complete with locknut, two washers and spring.

Remove the two setscrews securing the closing plate which joins the front wings under the radiator grille opening. Withdraw the plate.

Remove the setscrew, plain and serrated washers securing the top of the radiator grille. Unscrew the five nuts, plain and serrated washers and plates securing the radiator grille to the body. Remove the radiator grille.

Slacken the nut securing the bonnet release cable and withdraw the cable from the release lever.

Remove the two setscrews securing the striker plate, catch plate and base plate to the body.

Remove the striker, catch and base plates, spacers and spring.

Withdraw the release cable from the outer casing if it is to be renewed.

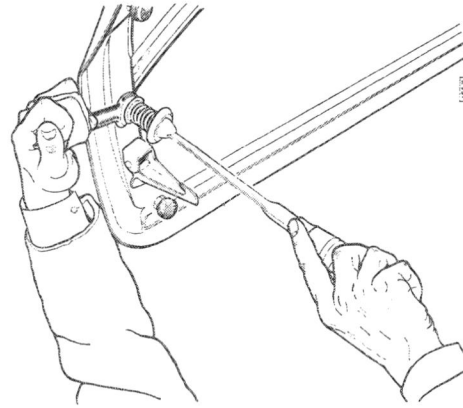

Fig. 4. Adjusting the bonnet lock pegs.

BONNET

Removal

To open the bonnet pull the control knob situated under the facia panel on the right-hand side. This will release the bonnet which will still be retained by the safety catch.

Adjustment of the Bonnet Lock (Fig. 4)

Slacken the locknut on the striker peg and rotate the peg with a screwdriver, until there is approximately $\frac{1}{16}$" (1·5 mm.) movement between the catch plate and the peg. This is to ensure that the catch plate will fully engage with the peg.

Tighten the locknut on the striker peg.

Refitting

Refitting is the reverse of the removal procedure.

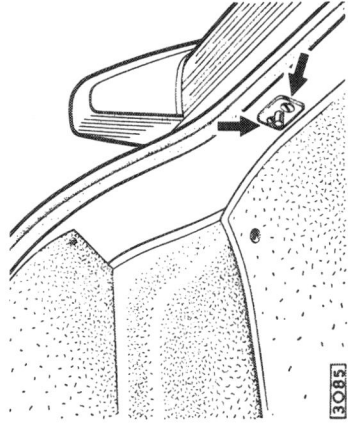

CHROME STRIPS ON BONNET

Removal

The chrome strips along the sides of the bonnet are secured with clips. Straighten the prongs of the clips and withdraw the chrome strip.

To remove the chrome strips in the centre of the bonnet, insert a screwdriver between the strip and the top of the bonnet and prise off the strip taking care not to damage the paintwork.

Refitting

When refitting the chrome strip to the centre of the bonnet renew the clips.

Refitting is the reverse of the removal procedure.

RADIATOR GRILLE (Fig. 5 and 6)

Removal

Remove the two setscrews together with the shakeproof washers securing the closing plate between the front wings under the radiator grille opening and remove the plate.

Remove the setscrew, plain and serrated washers and securing the top of the radiator grille to the body.

Unscrew the five nuts, plain and serrated washers and plates securing the grille to the body.

Dismantling

Remove the two setscrews and washers securing the chrome centre strip to the bottom of the radiator grille.

Remove the setscrew and washer from the top of the grille and detach the centre strip.

Remove four nuts and washers and detach the surround from the grille.

To remove the badge from the centre strip withdraw two drive screws and the fixing strip.

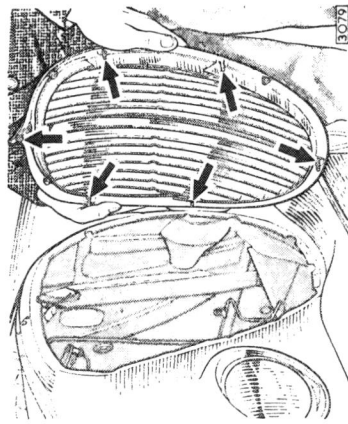

Fig. 5. Showing the mounting points for the radiator grille.

Fig. 6. Removal of the centre chrome strip on the radiator grille.

Reassembling

Reassembling is the reverse of the dismantling procedure.

Refitting

Refitting is the reverse of the removal procedure.

LUGGAGE COMPARTMENT LID AND HINGES

Removal

Open the luggage compartment lid and disconnect the electrical connections in the reverse lamp.

Remove the setscrew securing the earth wire to the luggage compartment lid.

Withdraw the harness from the luggage compartment lid. Remove the two metal straps securing the reverse lamp cable to the left hand hinge. Mark the positions of the hinges on the luggage compartment lid.

Remove the eight setscrews, plain and serrated washers and remove the luggage compartment lid.

Mark the positions of the hinges on the body and remove the eight setscrews, plain and serrated washers securing the hinges to the body.

Remove the luggage compartment lid hinges.

Refitting

Refitting is the reverse of the removal procedure.

Boot Lock Adjustment

Slacken the four setscrews securing the boot lid lock striker to the boot lid. Move the striker in the elongated holes until the lock operates correctly and does not rattle. Tighten the retaining setscrews.

Fig. 7. Showing the screws for adjustment of the boot striker.

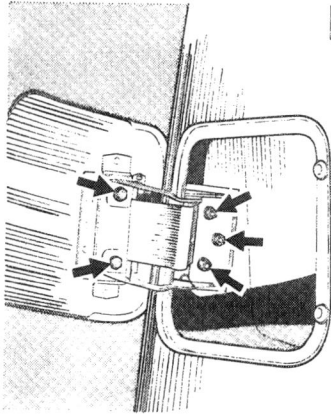

Fig. 8. Removal of fuel filler lid (L.H.).

PETROL FILLER LIDS

Removal

Unscrew the three setscrews and washers securing the hinge to the petrol filler compartment and detach the assembly from the body.

Remove the two setscrews and washers securing the lid to the hinge.

Refitting

Refitting is the reverse of the removal procedure. When refitting the lid retain it by screwing the setscrews finger tight in the elongated holes, then align the lid to fit into the recess of the body panel. Tighten the setscrews securely.

MASCOT

Removal

Raise the bonnet and remove the two setscrews, plain and serrated washers securing the mascot to the bonnet.

Refitting

Refitting is the reverse of the removal procedure.

FRONT BUMPER

Removal

Remove two bolts, nuts and washers from the outer mountings and two setscrews and washers from the inner mountings and detach the bumper from the support brackets.

Refitting

Refitting is the reverse of the removal procedure. Adjust the position of the bumper to the body by utilising the slotted holes in the support brackets before finally tightening the bolts.

FRONT BUMPER OVER-RIDERS

Removal

Remove the two nuts, plain and serrated washers securing the over-riders to the front bumper.

Remove the over-riders and beading.

Refitting

When refitting the over-riders replace the beading between the over-riders and the bumper.

Refitting is the reverse of the removal procedure.

REAR BUMPER

Removal

Remove the eight setscrews, plain and serrated washers from the outer and inner mountings and detach the bumper.

Remove the nuts and washers securing the bumper mounting rubbers to the wings and re-inforcement panels.

Refitting

Refitting is the reverse of the removal procedure.

REAR BUMPER OVER-RIDERS

Removal

Remove the two nuts, plain and serrated washers securing the over-riders to the rear bumper.

Remove the over-riders and beading.

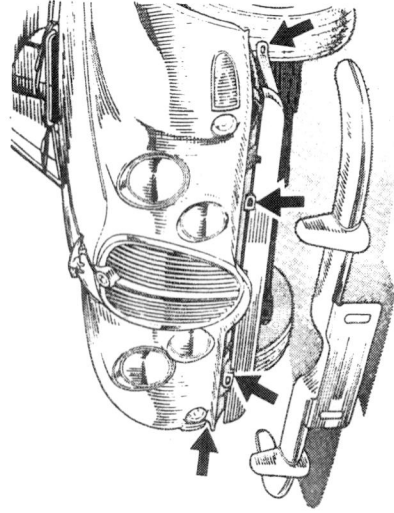

Fig. 9. *Showing the mounting points for the front bumper.*

Refitting

When refitting the over-riders replace the beading between the over-riders and the bumper.

Refitting is the reverse of the removal procedure.

WINDSCREEN

Removal

Prise off the two chrome finisher pieces securing the ends of the chrome finisher which encircles the windscreen.

Prise off the chrome finisher from the windscreen rubber. Extract one end of the rubber insert and withdraw completely.

Run a suitable thin bladed tool around the windscreen to break the seal between the rubber and the windscreen aperture flange.

Strike the glass with the flat of the hand from inside the car, starting in one corner and working towards the bottom.

Repeat this process around the complete windscreen.

Withdraw the windscreen.

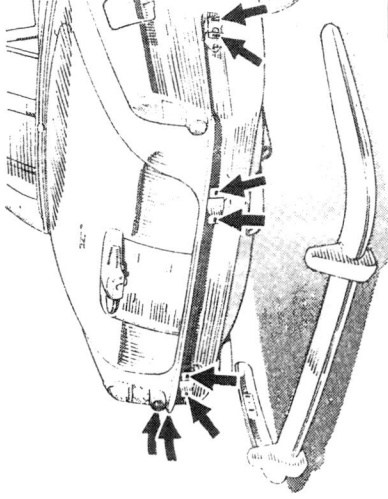

Fig. 10. *Showing the mounting points for the rear bumper.*

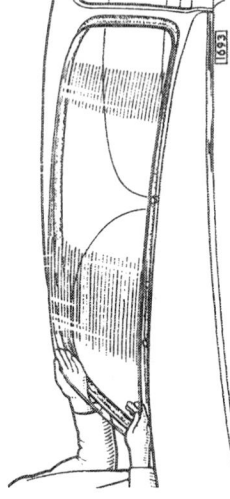

Fig. 11. *Removal of the windscreen.*

Refitting

Remove the old sealer from the windscreen "flange. Examine the windscreen rubber for cuts. If the windscreen was of the toughened glass type it is recommended that the windscreen rubber should be replaced. This is because small particles may have been impregnated in the rubber and could break the screen again. If, however, the windscreen was not broken by a projectile the windscreen aperture flange should be examined for a bump in the metal. If this is found the bump should be filed away otherwise the glass may break again.

The rubber should be attached to the windscreen aperture with the flat side of the rubber towards the rear and the joint in the rubber preferably at the bottom.

Using the special tool (A Fig. 14) insert the screen into the rubber along the bottom edge first (Fig. 13). It is important that the glass should be fitted equally. DO NOT fit one end and then try to fit the other. Using the special tool (B Fig. 14) insert the rubber sealing strip with the rounded wide edge to the outside.

Using a pressure gun filled with a sealing compound, and fitted with a copper nozzle (so that the glass will not be scratched) apply the nozzle of the gun between the metal body flange and the rubber and fill with sealing compound. Repeat the operation between the glass and the rubber. Remove excess sealing compound with a rag soaked in white spirit. DO NOT USE THINNERS as this will damage the paintwork.

Fit the chrome strip on top of the windscreen rubber and bend to suit contour if necessary. Coat the inside of the chrome strip with a layer of Bostik 1251 and allow to become tacky. Place the chrome strip on the rubber over the rubber sealing strip and lip the rubber sealing strip with a hook (A Fig. 14) lip the rubber of the chrome finisher. Fit the two centre chrome clips and lip the rubber over the edges of the clips.

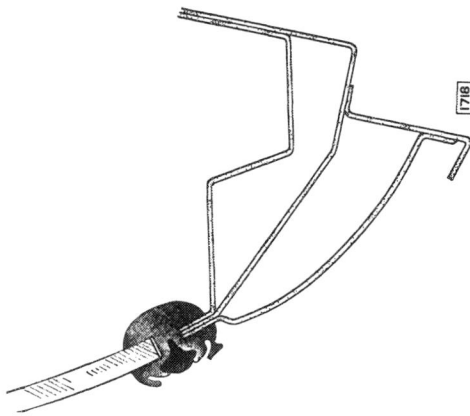

Fig. 12. Section through windscreen glass and sealing rubber.

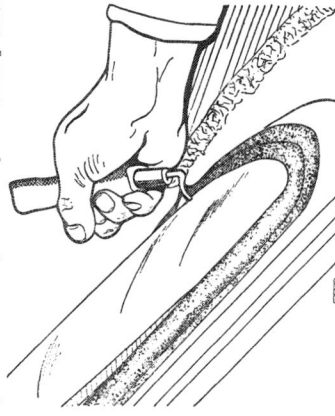

Fig. 13. Using the special tool ('A', Fig. 14) for lifting the windscreen rubber over the glass.

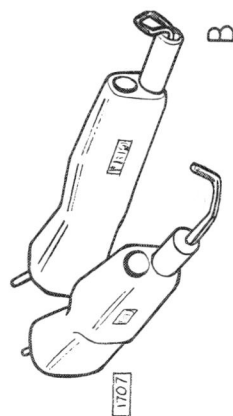

Fig. 14. The two special tools (Churchill Tool No. JD23) used when refitting a windscreen.

Prise off the two chrome finisher pieces securing the ends of the chrome finisher which encircles the windscreen.

REAR GLASS

Removal

Prise off the chrome finisher from the back light rubber. Extract one end of the rubber insert and withdraw completely.

Run a suitable thin bladed tool around the back light to break the seal between the rubber and the back light aperture flange.

Strike the glass with the flat of the hand from inside the car, starting in one corner and working towards the bottom.

Repeat this process around the complete rear light glass.

If the car is fitted with a heated back light it is necessary to disconnect the two electrical connections in the boot (Fig. 86-87, page P.70) and care should be taken when removing the back light not to break the two wires which pass through the holes in the sealing rubber.

Refitting

Remove the old sealer from the back light flange. Examine the sealing rubber for cuts. If the back light is of the toughened glass type (i.e. all cars except those fitted with heated back lights) it is recommended that the back light rubber should be replaced.

This is because small particles of glass may have been impregnated into the rubber and could break the glass again. If, however, the back light was not broken by a projectile the back light aperture flange should be examined for a bump in the metal. If this is found the bump should be filed away otherwise the glass may break again.

The rubber should be attached to the back light aperture with the flat side of the rubber facing the inside of the car.

If the car is fitted with a heated back light, pierce the sealing rubber in the two appropriate positions Fig. 16 to take the wires which lead to the element.

Using the special tool (A. Fig. 14) insert the back light into the rubber along the bottom edge first. On cars fitted with heated back light feed electrical wires through holes in

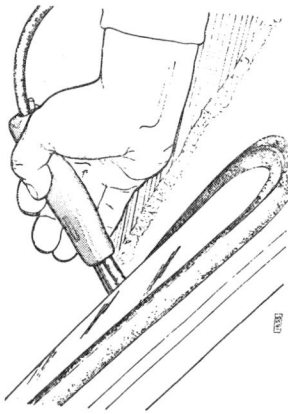

Fig. 15. Using the special tool ('B', Fig. 14) for inserting the rubber sealing strip in the windscreen sealing rubber.

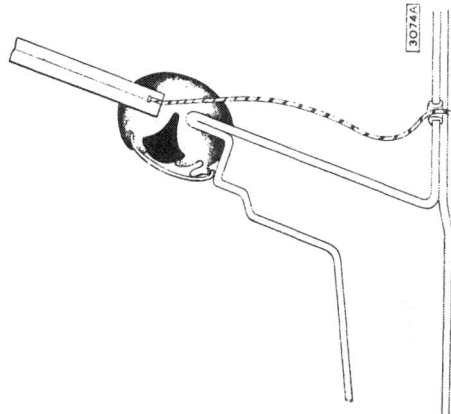

Fig. 16. Section through rear light glass and sealing rubber.

sealing rubber first (Fig. 16). It is important that the glass is fitted equally DO NOT fit one end and then try to fit the other. Using the special tool (B. Fig. 14) insert the rubber sealing strip with the rounded wide edge to the outside.

Using a pressure gun filled with a sealing compound and fitted with a copper nozzle (so that it will not scratch the glass) apply the nozzle of the gun between the metal body flange and the rubber and fill with sealing compound between the rubber and the glass (Fig. 18). Also apply sealing compound between the rubber and the glass (Fig. 17).

Remove excess sealing compound with a rag soaked in white spirit. DO NOT USE THINNERS as this will damage the paintwork.

Check that there is a small gap between the sealing rubber edge and the depression for the rear glass aperture. This is necessary to allow the chrome finishing strip to seat on the sealing rubber. If the rubber bends hard onto the depression at certain points insert a small length of stiff piping cord $\frac{1}{8}$" (3.1 mm.) diameter at the required positions, this will facilitate the fitting of the chrome strip. Fit the chrome on top of sealing rubber and bend to suit contour if necessary. Coat the inside of the chrome strip with a layer of Bostik adhesive and allow to become tacky. Place the chrome strip onto the sealing rubber and with a hook (A, Fig. 14) lift the lip of the rubber over the chrome finisher. Fit the two centre chrome clips and lip the rubber over the edges of the clips.

Reconnect the heated backlight cables if fitted.

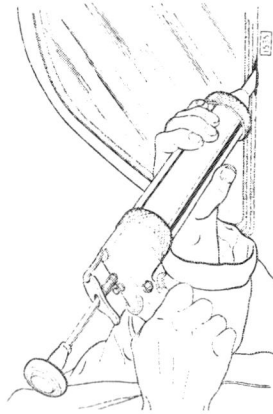

Fig. 17. *Using a gun to inject sealing compound between the rubber and the rear glass aperture.*

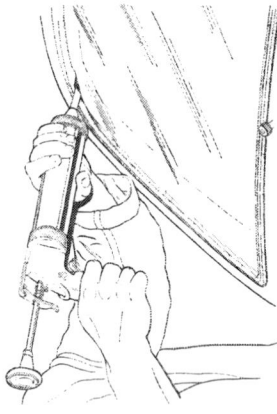

Fig. 18. *Using a gun to inject sealing compound between the surround rubber and the glass.*

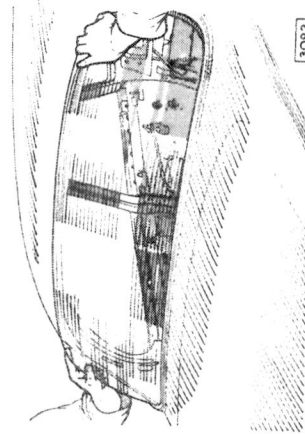

Fig. 19. *Removal of the rear glass.*

FRONT DOORS AND HINGES

Removal

Remove the split pin and clevis pin on the door check strap bracket situated on the door hinge pillar.

Mark the positions of the hinges on the door.

Remove the six bolts securing the hinges to the door side and remove the door.

Remove the scuttle side casing by unscrewing the three drive screws. Remove the two screws securing the aperture cover plate. Unscrew the door courtesy light switch from the bottom hinge recess. Pull out the electrical connection at the rear of the switch.

To remove the hinges unscrew the four cross headed screws and two bolts inside the hinge recess.

Refitting

Refitting is the reverse of the removal procedure.

REAR DOORS AND HINGES

Removal

Remove the split pin and clevis pin on the door check strap bracket on the door hinge pillar. Remove the door trim casing as described on page N.18.

Mark the positions of the hinges on the door.

Remove the three bolts securing the bottom hinge to the rear door and remove the four cross headed screws securing the top hinge to the door.

Remove the rear door.

Remove the four cross headed screws securing the hinges to the rear door side of the centre pillar.

Remove the two cross headed screws from the front door side of the centre pillar.

Withdraw the hinges.

Refitting

Refitting is the reverse of the removal procedure.

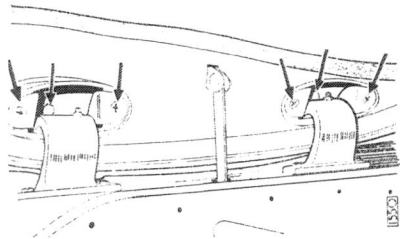

Fig. 20. *Showing the screws which secure the front door hinges.*

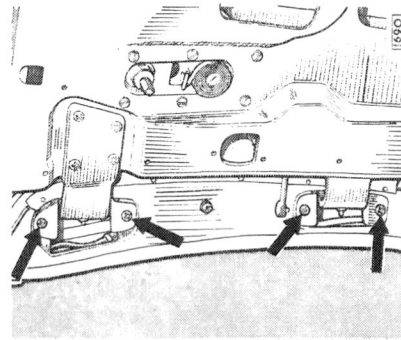

Fig. 21. *Showing the screws which secure the rear door hinges.*

Collect all the packing pieces.

Care should be taken to replace the same number of packing pieces under their respective screws.

Remove the two bolts, serrated and plain washers securing the two legs of the window frame to the door. Collect the wooden packing pieces.

Unclip the weather strip from the door frame, this is secured by four clips. Withdraw the window frame from the door frame and collect the rubber packing piece between the light frame and door frame above the door lock.

Slide the glass out of the retaining channel.

Refitting

Refit the four clips securing the weather strip to the outer inside edge of the door frame.

Clip the weather strip in position.

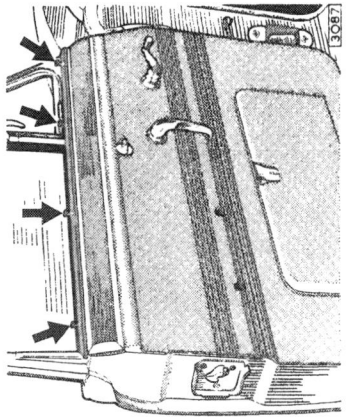

Fig. 28. Showing the window regulator arm and channel.

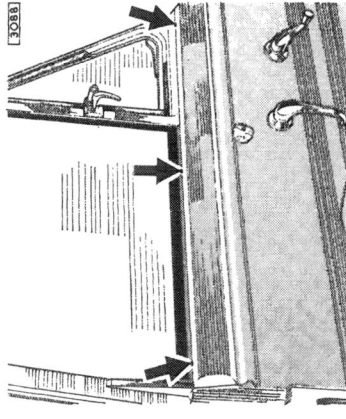

Fig. 25. Showing the location of the interior door handle retaining pin.

Fig. 26. Removal of the door trim casing. Arrow "A" shows the location of the map pocket tensioner anchorage.

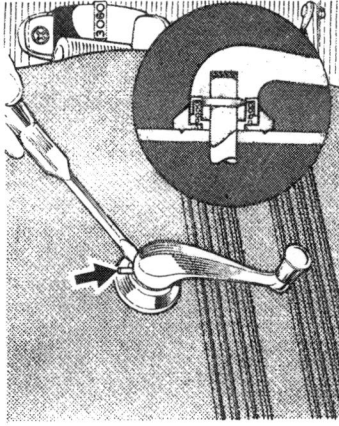

Fig. 27. The location of the window frame securing screws.

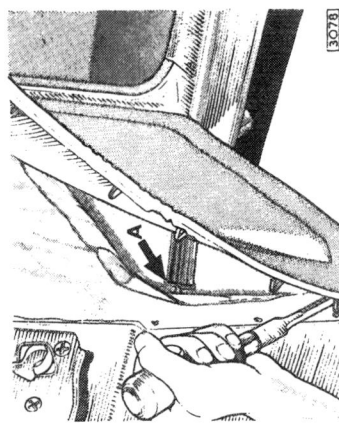

Fig. 22. Showing the four screws which secure the wood capping to the waist rail.

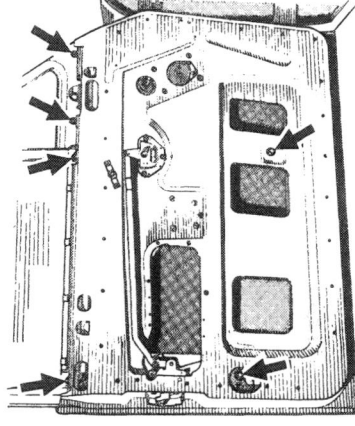

Fig. 23. Showing the three screws which secure the waist rail to the door frame.

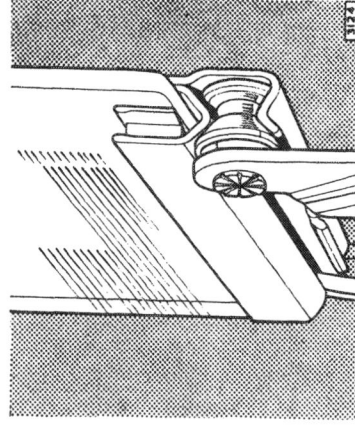

Fig. 24. Showing the screws which secure the top of the door trim casing.

FRONT AND REAR DOOR TRIM CASINGS

Removal

Remove the four chrome screws and washers securing the wood capping to the waist rail.

Ensure that the winding window is completely closed. Insert a screwdriver between the handle and the spring cap and press the cap inwards, into the escutcheon. This will expose the retaining pin which can then be pushed out. Remove the handle, spring cap and escutcheon.

Repeat operation to remove the door handle.

Remove the central retaining screw and detach the locking turn button.

Remove two screws and washers and detach the arm rest (front doors only).

Carefully remove the door casing fabric, where secured by adhesive solution to the door inner panel at the bottom of the window aperture.

Locate the drive screws, five on the rear doors and six on the front doors, securing the top of the casing to the door panel through the moulded rubber strip. Ease away the strip locally and remove the drive screws.

Insert a thin bladed screwdriver between the door casing and the door panel. Prise off the casing which is secured by spring clips to the door.

On the front doors it will be necessary to unhook the two tensioning devices, which retain the map pocket in the closed position, from the door inner panel before the casing can be completely detached.

Refitting

Refitting is the reverse of the removal procedure.

FRONT AND REAR DOOR WINDOW FRAMES AND GLASS

Removal

Remove the door trim casing as described previously.

Pull off the clear plastic sheet which is stuck to the door frame with upholstery solution.

Remove the four round headed screws, serrated and plain washers securing the window frame to the top of the door panel.

FRONT NO DRAUGHT VENTILATOR

Removal

Remove the trim casing from the front door as described on page N.18.

The no draught ventilator adjustment and securing mechanism is visible through a small aperture in the door frame.

Remove the locknut, nut and washer securing the spring against the quadrant on the N.D.V. post.

Remove the pin and segment on the N.D.V. post.

Remove the two screws securing the front N.D.V. hinge to the window frame.

Turn the N.D.V. catch to allow it to open.

Withdraw the N.D.V. from the window frame.

Refitting

Care should be taken not to leave any parts between the inner and outer door frames.

Refitting is the reverse of the removal procedure.

REAR NO DRAUGHT VENTILATOR

Removal

Remove the nut, screw and fibre washer securing the rear N.D.V. bracket to the catch arm which operates the N.D.V.

Open the N.D.V.

Remove the five screws securing the rear N.D.V. light hinge to the window frame.

Refitting

Refitting is the reverse of the removal procedure.

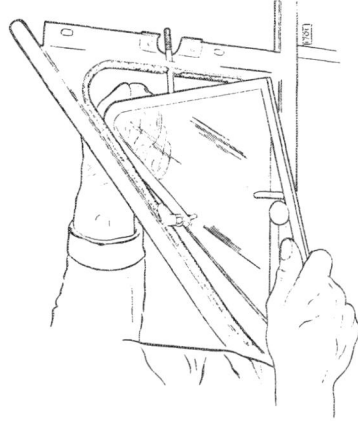
Fig. 31. Removal of front N.D.V. glass from frame.

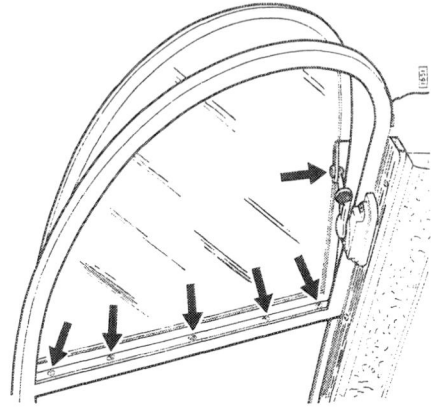
Fig. 32. Removal of rear N.D.V.

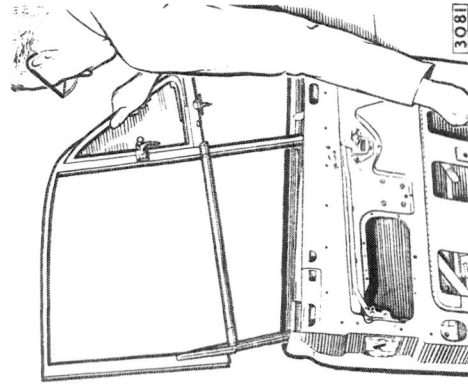
Fig. 29. Removal of the front window frame.

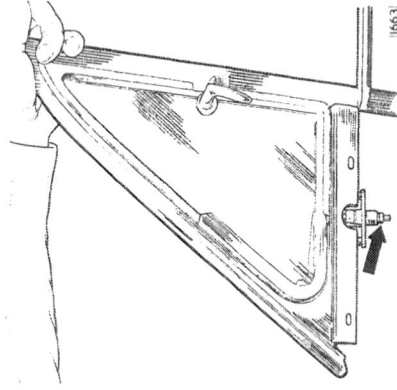
Fig. 30. Showing the front N.D.V. light adjustment nut.

BODY AND EXHAUST SYSTEM

Place a layer of seating compound on the section of the door frame on to which the no draught ventilator window in the window frame seats.

Place the door glass into position on the window winding mechanism slide channel, and slide the glass into position between the door frame.

Insert the window frame into the door frame. On the rear door it is necessary to wind up the window glass approximately one third of its maximum height before inserting the window frame. Insert the rubber packing piece. Refit all screws and bolts finger tight.

Insert the four round headed screws, serrated and plain washers which secure the window frame to the door. Replace the various packing pieces under the round headed screws.

Refit the bolt, serrated and plain washers securing the window frame bracket furthest away from the door hinge and replace the fibre packing pieces.

Refit the bolt, serrated and plain washer securing the window frame bracket nearest the door hinge and replace any wooden packing pieces.

The window frame should clear the front screen pillar by ¹⁄₁₆" (1·5 mm.).

When the correct clearance has been achieved tighten the four round headed screws and two bolts securing the window frame to the door.

Remove any excess sealing compound from the bottom of the no draught ventilator.

Refit the door trim casing and wooden capping straps as described.

FRONT WINDOW REGULATOR

Removal

Remove the door casing and window glass as described on pages N.18 and N.19.

Remove the felt placed over the window regulator spindle.

Remove the four screws and serrated washers securing the window regulator to the door frame.

Remove four screws and serrated washers securing the window regulator spring to the door frame.

Withdraw the window regulator mechanism from the door frame.

Refitting

Refitting is the reverse of the removal procedure.

REAR WINDOW REGULATOR

Removal

Remove the door casing and window as described on pages N.18 and N.19.

Remove the piece of felt placed over the window regulator spindle.

Remove the four screws and serrated washers securing the window regulator mechanism to the door frame.

Withdraw the window regulator mechanism from the door frame.

Refitting

Refitting is the reverse of the removal procedure.

FRONT SEATS AND SEAT RUNNERS

Removal

Remove the cushion from the front seat.

Slide the seat fully rearward.

Remove the two bolts and plain washers securing the front of the seat runners to the body floor.

Slide the seat forwards and remove the two bolts and washers securing the rear of the seat runners to the body floor.

Remove the two bolts and washers securing the seat front support plate to the body floor ; remove the seat.

Refitting

Refitting is the reverse of the removal procedure.

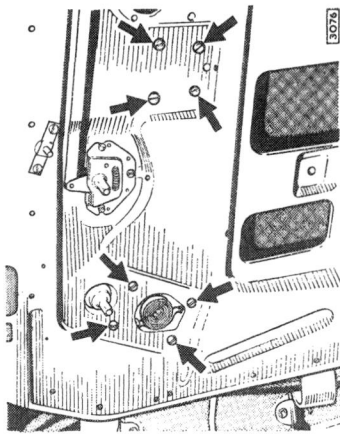

Fig. 33. *Showing the screws securing the window winding mechanism to the door frame.*

REAR SEAT AND SQUAB

Removal

Lift the rear seat cushion upward, off the two locating pins on the rear seat and remove the rear seat.

Remove the two round headed screws, serrated and plain washers securing the bottom of the rear seat squab to the back of the seat pan.

Lift the rear seat squab to disengage the three retaining hooks and withdraw the squab.

Refitting

Refitting is the reverse of the removal procedure.

POLISHED WOOD CAPPINGS

Removal of Upper and Lower Capping on the Door Pillar

Insert a thin bladed screwdriver between the trim casing and the centre door pillar. Prise off the trim casing and pull downwards to release the tongue on the casing from behind the upper capping.

Knock the wooden capping downward with the hand and remove. The wooden capping is secured by two clips.

Note: If seat belts are fitted it will be necessary to detach the anchorage plate by withdrawing the fixing bolt before removing the capping.

Removal of Capping Rail at the Side of the Windscreen

Release the two nuts with serrated and plain washers securing the screen rail to the body.

Remove the screw securing the bottom of the side capping rail to the body.

Pull away the sealing rubber around the top of the front door aperture. This will expose the screw securing the side capping rail.

Withdraw the side capping rail downwards to clear the cant rail joint.

Refitting

Refitting is the reverse of the removal procedure.

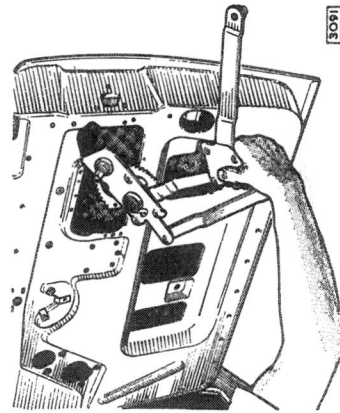

Fig. 34. *Removal of window winding mechanism.*

COURTESY LIGHT AND CAPPING

Removal

Pull off the courtesy light glass from the two plastic prongs securing it to the lamp body.

Remove the two screws securing the wooden base to the body of the car.

Disconnect the positive battery lead.

Disconnect the electrical connections at the rear of the courtesy light.

Refitting

Refitting is the reverse of the removal procedure.

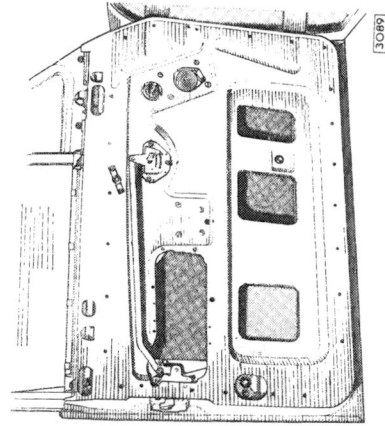

Fig. 35. *Showing the door lock in position.*

3089

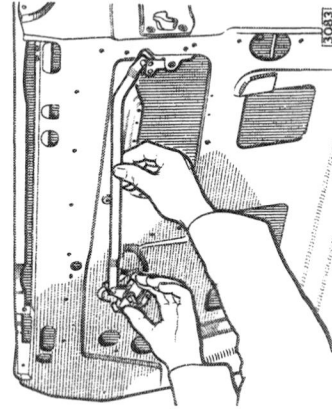

Fig. 36. *Showing the door lock being placed in position.*

3083

REMOVAL OF LOCK MECHANISM

Front Doors (Fig. 37)

Release the spring clip holding the top of the spring loaded handle/lock connecting link (A) to the dowel on the plunger operating lever (B). This is accessible through an aperture in the inner door panel.

Rear Doors

Remove the starlock washer holding the bottom of the handle/lock connecting link (C) to the cross-shaft (D), together with the plain washer behind the link.

Removing Remote Opening Control

Remove the wire clip, plain washer and the waved washer fitted between the connecting link and the latch operating lever (E). Detach the connecting link.

Remove the three screws (F) securing the remote control to the inner door panel.

Removing Turn-Button Locking Control

Detach the control wire by removing the large spring clip which secures it to the locking lever (G).

To facilitate removal, place control in the locked position.

Remove the two screws (H) together with the shakeproof washers to release the turn-button control.

Removing Lock Unit

Remove the three countersunk screws (I) which pass through the dovetail plate (J) and the door into the lock.

On the front doors it is advisable to remove the lower glass run channel bolt with its plain washers, shakeproof washers and packing pieces in order to press the lock inwards so that the projecting latch passes inside the shut face.

Remove the lock through the large aperture in the inner door panel.

Removing Outside Handle Base-Plate

Remove the two bolts (K) together with the shakeproof washers from the inside of the door.

Removing Striker Unit

Do not disturb the three fixing screws (L) unless making an adjustment or fitting a new striker unit.

Removing Outside Handle

When fitting a replacement, remove nut (M) together with the two washers, and remove the handle.

Fit the replacement handle noting that a packing washer (N) is fitted to the front fixing stud.

On the rear doors it is desirable that the lower screw, in the shut face of the door, be removed to facilitate the removal of nut (O) together with the shakeproof and plain washers.

In the case of front doors, it is preferable to remove the window frame complete with glass to facilitate the removal of nut (O) which is in the extreme inside top rear corner of the door unless a suitable spanner is available.

To remove the window frame, lower the glass and release the bolts thus freeing the bottoms of the front and rear glass channels.

Remove the four screws from the window sill and lift the window frame out of the top of the door.

Raise the glass: slide it towards the lock face to release from the window regulator lifting arms and remove the glass.

Important: All window frame mounting points have spring washers, plain washers and packing pieces. Ensure that all these components are replaced in the reverse order.

Refitting

If the outside handle has been removed this should be refitted before installing the lock mechanism. It is, however, advisable to fit the window frame and glass later.

When the outside handle has not been removed, ensure that the winding window is raised to its full extent.

On front doors attach the bottom of the spring-loaded handle lock connecting link (A) to the dowel on the cross-shaft (D). The link is fitted with a plain washer and it is retained with a new starlock washer.

Insert the lock through the upper aperture in the inner door panel so that the latch projects through the aperture in the shut face of the door.

On front doors it is advisable to remove the lower glass run channel bolt together with the plain washers, shakeproof washers and packing pieces in order to facilitate this operation.

The dovetail plate (J) should then be placed in position over the latch bezel and secured by means of the three countersunk screws (I) which pass through the shut face into the lock.

The base-plate assemblies are stamped LH (left hand) and RH (right hand).

On rear doors the top of the handle lock connecting link (C) should be located on the lug on the plunger operating lever (B).

Then the washer (P) should be fitted and retained by the plunger-bolt (Q) and the lock-nut (R).

Each base-plate assembly should be held in position inside the door in order that the clearance between the end of push button plunger bolt (Q) and the lock contactor (S) may be checked through the aperture in the inner door panel.

This clearance should be $\frac{5}{32}$" (·79 mm.).

To adjust, release the locknut (R). screw the plunger bolt (Q) in or out as required and retighten the locknut.

The base-plate assemblies are secured from the inside of the doors by means of two bolts (K). with shakeproof washers, which pass into the back of the outside handle.

When connecting the push-button mechanism to the lock unit, ensure that the cross shaft (D) is moved **downwards** into the locked position.

Important: On the rear doors provision is made for the plunger operating lever (B) to be pegged in the locked position prior to connecting the handle/lock link (C). This is done by inserting a short length of $\frac{1}{8}$" dia. (3·18 mm.) rod through the **rectangular** hole (T) in the base-plate assembly.

To compensate for variations in fitting, the links (A and C) are provided with three holes which are at the top end on the front links and at the bottom end on the rear links.

It will be observed that one of these holes, usually the centre one, can be used to give the correct setting.

The rear link is fitted with a plain washer underneath and it is retained by a starlock washer.

The front link is finally secured by a spring clip.

To check the locking action, remove the rod from the rectangular hole in the rear base-plate assembly, depress the push-button and check that the plunger bolt (Q) clears the lock contactor (S).

Alternatively, raise the cross-shaft (D) into the unlocked position and check that the plunger passes squarely behind the lock contactor to come into contact with it when the push button is operated.

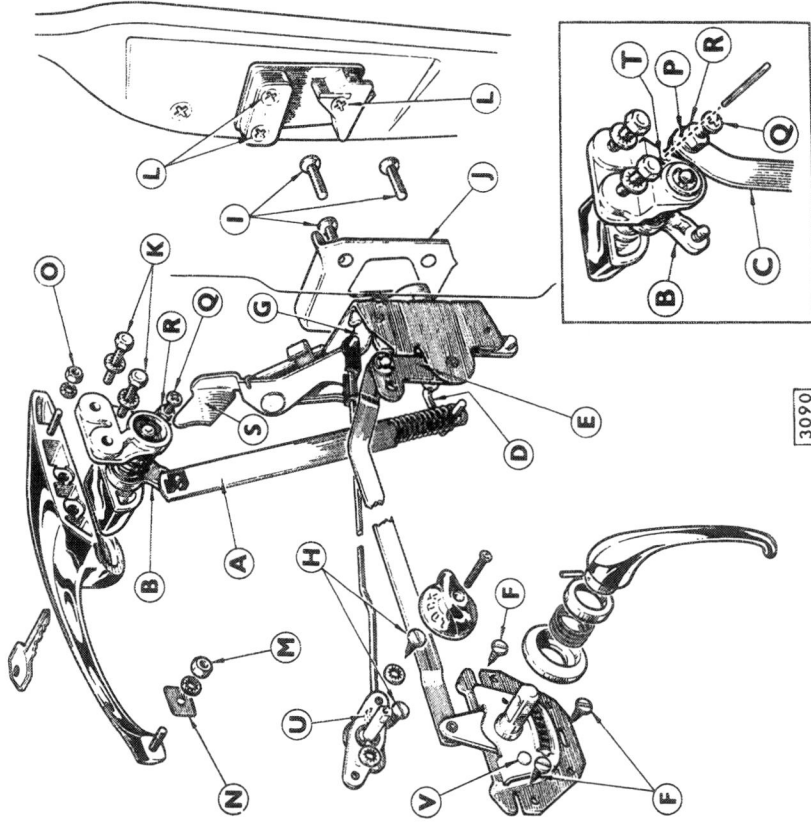

3090

Fig. 37. The door lock mechanism.

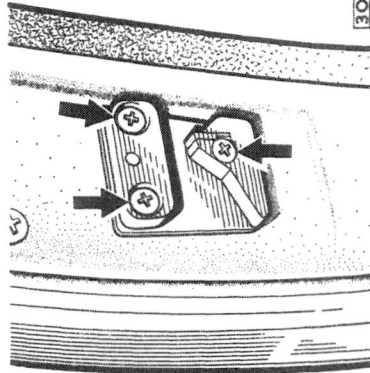

3086

Fig. 38. Location of the door striker plate securing screws

When refitting the turn button locking control it is important to ensure that each turn button is in the unlocked position when inclined vertically and in the locked position when rotated into the horizontal plane.

In the locked position the control wire is moved **away** from the lock passing over the top of the control.

Note: The front controls may be identified by a larger front stop (U).

To refit, the control wire is threaded through the aperture provided for the control in the inner door panel and it is connected to the locking lever (G) with the large spring clip provided.

The control is retained by two screws (H) together with shakeproof washers, which should be adjusted to provide a uniform turn button position before tightening.

When refitting the remote opening controls it must be noted that these controls are handed and a riveted stop (V) ensures that the control can only be operated in one direction, that is away from the lock face.

Each control should be loosely fitted to the inner door panel with its three screws (F). The dowel on the connecting link is then attached via the hole in the latch operating lever (E) nearest to the lock with a waved washer interposed. The assembly is retained by one or two plain washers and a wire clip.

The remote control is aligned by sliding it through its slots towards the lock until the latch operating lever (E) is in contact with the stop in the lock case and then secured by tightening the three screws (F).

To fit and adjust the striker unit, attach the striker loosely by means of the three screws (L) which pass through the door pillar into an adjustable tapping plate.

Positioning is carried out by a process of trial and error until the door can be closed easily without rattling or movement up or down.

Important: The striker must be retained in the horizontal plane relative to the door axis.

MASTER CHECK FOR CORRECT ALIGNMENT

Front Doors

Fit an inside handle **vertically downwards** on the remote control spindle. Turn the handle rearward to open the door. Fit the appropriate locking turn-button vertically, and rotate anti-clockwise on the right side door or clockwise on the left side door into the horizontally locked position. Close the door while holding the push-button in the fully depressed position. The door will remain locked although the push button may be freely depressed.

Insert the key in the push-button slot and turn in the appropriate direction. Push-button control will then be restored and the door can be opened.

After turning, the key will automatically return to the horizontal position when it can be removed.

Important: The key must be removed from the locking device before closing a door in the locked position.

Rear Doors

Fit the inside handle vertically upwards on the remote control spindle. Turn the handle forward to open the door. Fit the appropriate locking turn-button vertically and rotate anti-clockwise on the right side door, clockwise on the left hand side door into the horizontally locked position.

Close the door. It will then be locked although the push-button may be freely depressed.

To unlock, the turn-button is rotated to its original position when push-button control is restored.

Lubrication

Before fitting the door casing ensure that any moving parts are adequately greased.

After assembly introduce a few drops of thin machine oil into the private lock key slots. These must be lubricated once a month.

The private lock cylinders must not under any circumstances be lubricated with grease.

Features of Lock Operation

When the doors are locked the push-buttons may be freely depressed. No amount of wilful pressure on the outside button will force or damage the lock.

Either front door can be locked from inside or outside irrespective of which door was last used as an exit.

This feature is invaluable in cases of traffic congestion and parking.

If either front door is closed after accidentally setting the turn-button in the locked position, locking is automatically cancelled. This action obviates the risk of locking oneself out of the car.

Front doors however, can be locked from the outside without using the key, a great advantage in inclement weather or under heavy traffic conditions where instant locking is desirable. This is achieved by setting the turn-button in the locked position and, while closing the door, deliberately holding the push-button in the fully depressed position. The key will then be used for unlocking either front door in the usual way.

ACCIDENTAL DAMAGE

The repair of integral construction bodies varies in some degree, dependent upon the extent of the damage, to that of separate body and chassis construction.

Superficial damage can be rectified in a similar manner to that employed on "all steel" bodies which is familiar to all body repairers.

Repairs to rectify extensive damage affecting the main members of the underframe must be carried out so that when the repair is completed the main mounting points for the engine, front and rear suspensions, etc., are in correct relation to each other.

When checking for or rectifying distortion in the main underframe members, reference should be made to the diagrams in the section headed "Checking Body Underframe Alignment" which gives the important dimensions to be observed.

Replacement Body Panels

Where the existing panels or members are badly damaged and it is not possible to effect a satisfactory repair in position, the affected panels will have to be cut out and replacement panels welded in their place.

It will frequently be found advantageous to use only a part of a given panel so that the welded joint can be made in a more accessible position. Great care must, of course, be taken when cutting the mating portions of the panel to ensure that perfect matching is obtained.

For example, if damage to a front wing is confined to the forward end a simpler and quicker repair can be effected by cutting the front wing off between the wheel aperture and the wing valance. If the replacement front wing is then cut to match, a simple butt weld can be made and after cleaning down with a sanding disc and filling with plumber's lead the joint should be invisible.

Any unused portions of replacement panels should be retained as it will often be found that they can be used for some future repair job.

Where a replacement panel to be fitted forms part of an aperture such as for a door or the luggage boot lid, an undamaged door or lid should be temporarily hinged in position and used as a template to assist location while the replacement panel is clamped and welded in position.

Similarly, an undamaged radiator grille can be used as a template to accurately form the aperture when fitting a replacement front wing or wings.

Before any dismantling takes place after accidental damage a check of the underframe alignment should be carried out.

CHECKING BODY UNDERFRAME ALIGNMENT

Checking for Distortion in the Horizontal Plane

The plan view of the body on page N.34 provides the important dimensions for checking for distortion in the underframe. These dimensions can be measured actually on the underside of the body or by dropping perpendiculars from the points indicated by means of a plumb-bob on to a clean and level floor. If the latter method is adopted the area directly below each point should be chalked over and the position at which the plumb-bob touches the floor marked with a pencilled cross.

Checking for Distortion in the Vertical Plane

For checking the underframe for distortion in the vertical plane the side elevation gives the details of the important dimensions from a datum line.

If the relative distance between two points above the datum line is required one dimension should be subtracted from the other.

If the relative distance between a point above the datum line and the straight section of the chassis side member is required, add the dimension "T"—3⅛" (9.7 cm.)—to the dimension above the datum line.

If it is required to check the dimensions from ground level raise up the car at the front and rear and insert four blocks or stands of exactly equal height between the ground and the straight section of the chassis side members.

Do not allow the weight of the car to rest on the blocks, use them only as test pieces.

The distance from the ground to any given check point will be: height of blocks + "T" (3⅛"—9.7 cm.) + distance from datum line to check point.

Body Alignment Jig

The use of the Churchill '700' body alignment jig is recommended. This jig is suitable for the 3.4 'S' and 3.8 'S' models and with the additional adaptors supplied also covers many other Jaguar models.

Full details of this equipment can be obtained from the manufacturers:—
Messrs. V. L. Churchill & Co. Limited.
London Road,
Daventry, Northants.

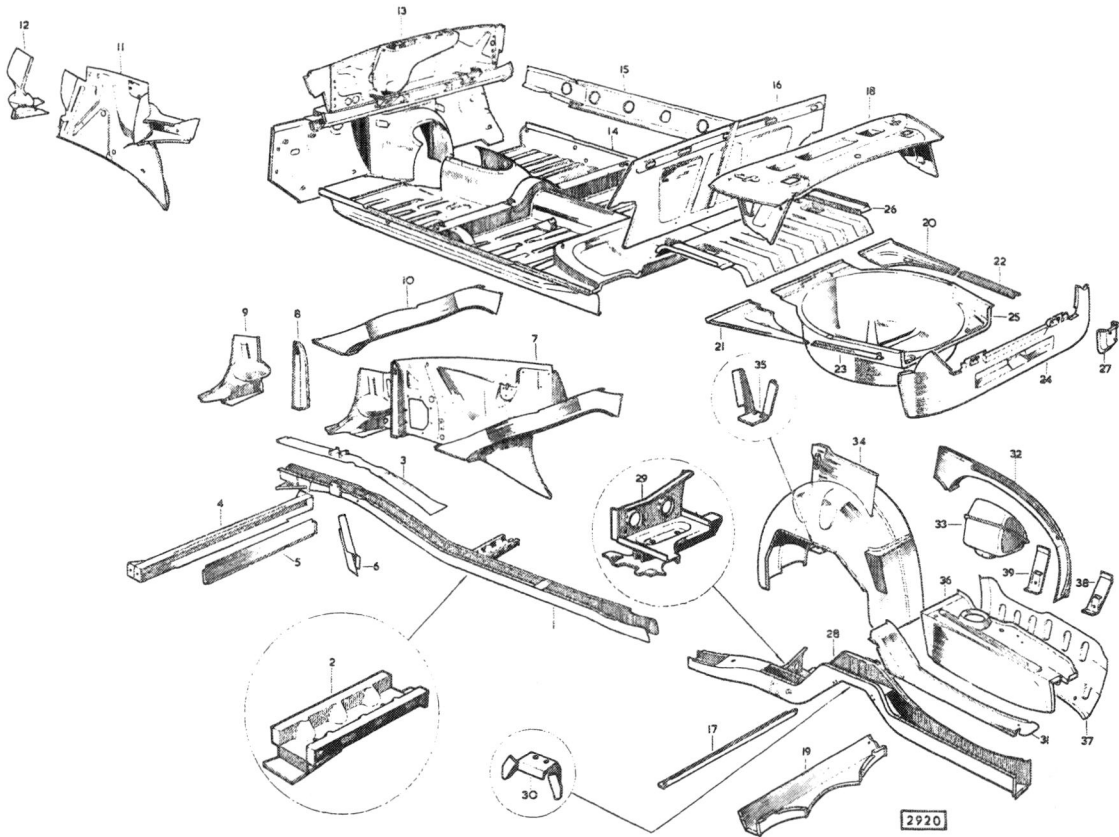

Fig. 39. *Body underframe components.*

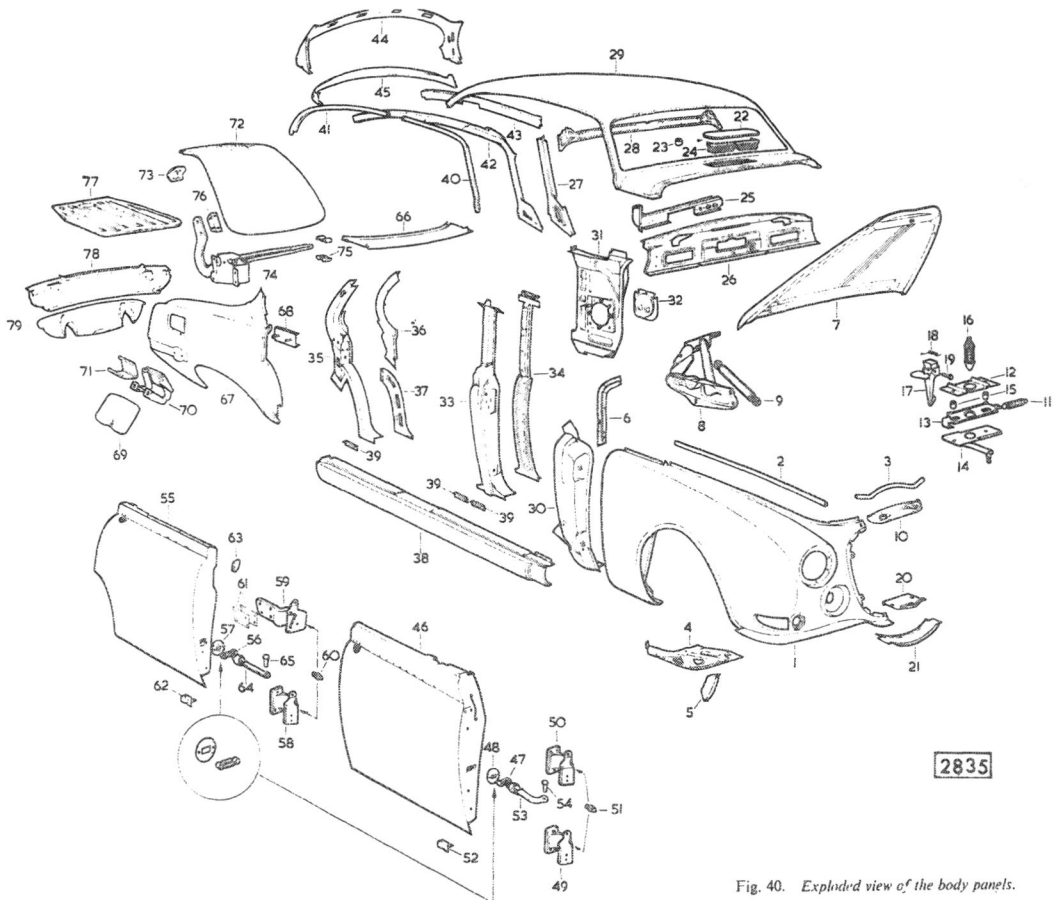

Fig. 40. *Exploded view of the body panels.*

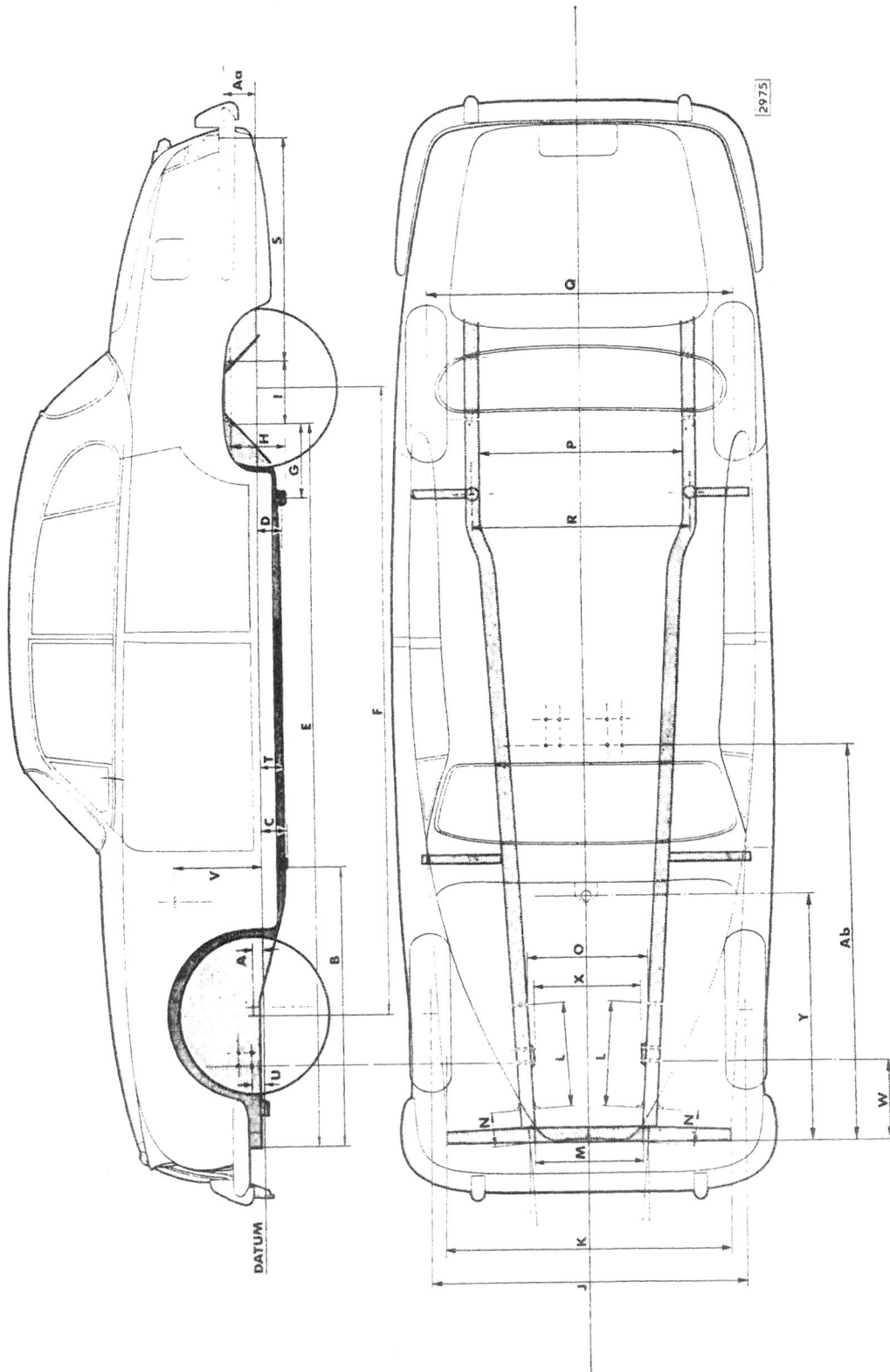

Fig. 41. Underframe alignment diagram.

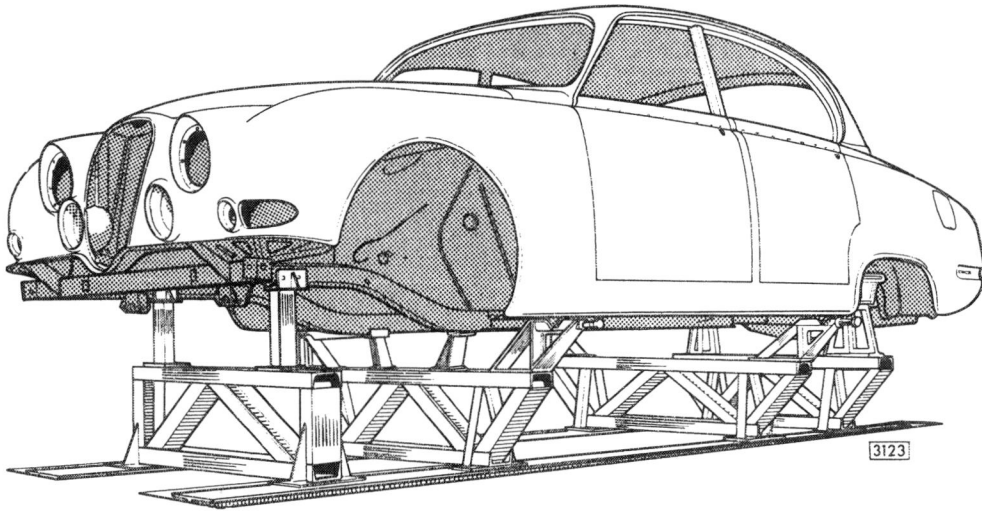

Fig. 42. *The Churchill "700" body alignment jig.*

Symbol	Measurement taken from	Dimension
A	Datum line to centre of tube in chassis side member for front suspension cross-member mounting	2 1/16" (52·4 mm.)
B	Front of jacking tube to front face of subframe cross-member	47 29/32" (121·7 cm.)
C	Datum line to bottom face front jacking tube	4 7/32" (107·16 mm.)
D	Datum line to bottom face rear jacking tube	4 9/64" (105·17 mm.)
E	Forward face of front cross-member to ℄ lower tube of rear suspension frame mounting	123 33/64" (3·14 m.)
F	Centre of front wheel to centre of rear wheel (wheelbase)	107 1/8" (2·72 m.)
G	℄ radius arm body mounting bracket to ℄ lower front tube of rear suspension frame mounting	12 35/64" (31·87 cm.)
H	℄ bottom of radius arm body mounting bracket to ℄ lower front tube of rear suspension frame mounting	9 23/32" (24·69 cm.)
I	℄ of lower front tube to ℄ of lower rear tube rear suspension frame mounting	10 21/32" (27·07 cm.)
J	Front track	55 1/4" (1·403 m.)
K	Outer ends of front cross-member	50" (1·27 m.)
L	Forward face of front suspension mounting bracket to ℄ tube in chassis side for front suspension cross-member mounting	17 27/32" (45·32 cm.)
M	Inner faces of chassis side members at joints with front suspension cross-member mounting brackets	18 23/32" (47·55 cm.)
N	Forward face of front cross-member to forward face of front suspension cross-member mounting bracket (measured along ℄ chassis side member)	5 25/32" (14·68 cm.)
O	Inner faces of chassis side members at ℄ front suspension cross-member mounting tubes	21 1/16" (53·49 cm.)
P	Inner faces of rear chassis side member	35 5/8" (90·49 cm.)
Q	Rear track	54·2" (1·38 m.)
R	Centres of radius arm body mounting	38" (96·52 cm.)
S	℄ lower rear tube of rear suspension frame mounting to rear bumper mounting face	38 41/64" (97·5 cm.)
T	Datum line to underside straight section of chassis side member	3 13/16" (9·7 cm.)
U	Lower forward hole forward engine mounting above datum	2 19/64" (5·83 cm.)
V	Datum line to top face engine mounting bracket (dash centre)	15 57/64" (40·24 cm.)
W	Forward face of front cross-member to ℄ lower front tube forward engine mounting	13 25/32" (35 cm.)
X	Inner faces of support plates engine mounting at ℄ forward lower tube	18 29/64" (46·87 cm.)
Y	Forward face of front cross-member to ℄ of dash centre engine mounting bracket	42 3/16" (1·07 m.)
Z	Forward face of front cross-member to 'O' datum	5 5/8" (14·29 cm.)
Aa	Datum line to ℄ upper bumper mounting bolt hole	5 5/8" (14·29 cm.)
Ab	Forward face of front cross-member to ℄ of forward holes to the rear engine mounting support channel	67 31/32" (1·726 m.)

WELDING METHODS

The following are the principal methods of welding used in the assembly of the body and underframe panels. The instructions given below for breaking the different types of welds should be adhered to when removing a damaged panel as this will facilitate the assembly of the new panel.

Spot Welding

This type of welding is used for the jointing of two or more overlapping panels and consists of passing electric current of high amperage through the panels by means of two copper electrodes.

This results in complete fusion of the metal between the electrodes forming a "spot" weld which is frequently repeated along the length of the panels to be joined. Spot welds can easily be recognised by slight indentation of the metal.

Lap joints on the outer body panels which are spot welded together are usually lead filled and in this case it will be necessary to direct the flame of an oxyacetylene torch on the lead so that the filling can be melted and wiped off by means of a piece of cloth.

Breaking Spot Welds

Spot welds cannot be broken satisfactorily other than by drilling; any attempt to separate the panels by using a chisel will result in the tearing of the metal in the vicinity of the spot welds.

Use a $\frac{3}{16}$" (4·7 cm.) diameter drill and carefully drill out each weld. There is no necessity to drill completely through both panels; if the "spot" is drilled out of one of the panels the weld can be completely broken by inserting a thin sharp chisel between the two panels and tapping lightly with a hammer.

Where possible, drill the spot welds completely out of the panel that is to be left in position on the body. This will allow the new panel to be joined to the mating panel on the body by gas welding through the holes in the overlapping flange. (This does not apply if spot welding equipment is available).

If this is not possible, and the holes have to be drilled out in the damaged panel, new holes can be drilled in the replacement panel and the same type of weld effected.

Gas Welding

This type of welding is carried out by means of oxy-acetylene equipment and is used for the jointing of overlapping panels or the butt-welding of the edges of two panels.

Breaking Gas Welds

Gas welds may be broken either by means of a sharp chisel or by cutting through with a hacksaw; welding can be removed by grinding with a pointed emery wheel.

EXHAUST SYSTEM

EXHAUST SYSTEM

Removal

Remove the two setscrews, together with the spring washers, securing each of the ring brackets (28) to the body.

Slacken the two clips (26) and remove the tail pipe and rear silencer assemblies (18) and (22).

Remove the four setscrews, together with nuts and shakeproof washers, securing the rubber mounting brackets (16 and 17) for the main silencers.

Slacken the two clips (8) and remove the main silencers (14 and 15).

Remove the intermediate exhaust pipes (6 and 7) by slackening clips (8) at the front silencer and removing the securing bolts in the rubber mounting bracket (9).

Slacken the two clips (4) and remove the front silencer (5).

Remove the four nuts, together with the washers, securing each downpipe to the exhaust manifold on the engine and remove the down-pipes (1 and 2) having first separated the down-pipes at the clamping strap adjacent to the flexible pipes (1A and 2A) and having removed the bolt securing the downpipe (2) to the clutch housing.

Collect the sealing rings (3) between the exhaust manifold and the downpipes.

Refitting

Renew the sealing rings (3) when refitting and then proceed in the reverse order.

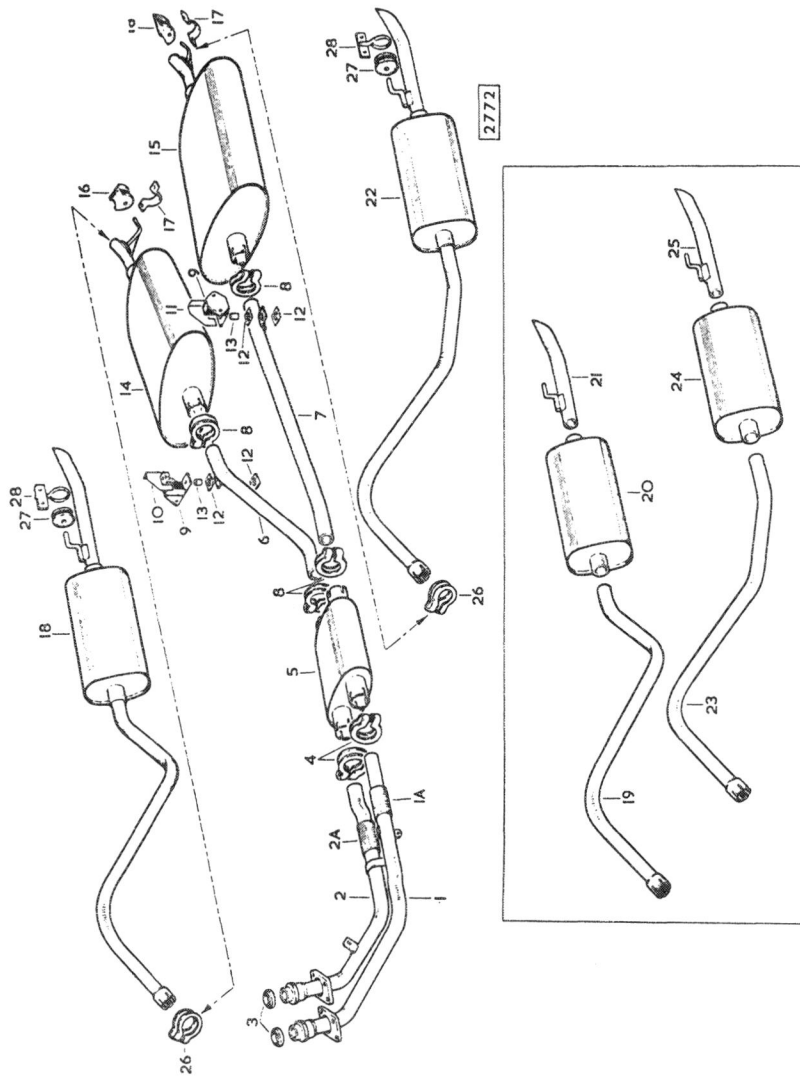

Fig. 43. *Exploded view of exhaust system.*

1. Front downpipe 1A. Flexible pipe
2. Rear downpipe 2A. Flexible pipe
3. Sealing ring
4. Clip
5. Front silencer
6. Intermediate exhaust pipe R.H.
7. Intermediate exhaust pipe L.H.
8. Clip
9. Rubber mounting/intermediate exhaust pipe
10. Heat shield
11. Heat shield
12. Washer (insulating)
13. Distance collar (insulating)
14. Main silencer R.H.
15. Main silencer L.H.
16. Rubber mounting for main silencers
17. Bracket
18. Tail pipe and rear silencer R.H.
19. Rear intermediate pipe R.H.
20. Rear silencer
21. Tail pipe
22. Tail pipe and rear silencer L.H.
23. Rear intermediate pipe L.H.
24. Rear silencer
25. Tail pipe
26. Clip
27. Rubber mounting for tail pipes
28. Ring bracket

Key to Fig. 43. Exhaust system.

SECTION O

HEATING AND WINDSCREEN WASHING EQUIPMENT

INDEX

CAR HEATING AND VENTILATING SYSTEM

INDEX (Continued)

WINDSCREEN WASHING EQUIPMENT

HEATING AND WINDSCREEN WASHING EQUIPMENT

CAR HEATING AND VENTILATION SYSTEM

DESCRIPTION

The car heating and ventilation system consists of a combined heating element and a two-speed electrically driven fan assembly.

Air from the heating and ventilation system is directed:

(a) To the front of the car through outlets (one on the driver's side and one on the passenger's side) below the parcel shelf.

(b) To the rear of the car through an outlet situated on the propeller shaft tunnel cover between the two front seats.

(c) To vents at the base of the windscreen to provide demisting and defrosting.

Fresh air is introduced into the car by pressing the Air button and switching on the fan if required (see also Air Distribution).

Heater Controls

The heater control buttons marked "AIR", "HEAT", "OFF", are situated centrally below the parcel tray (Fig. 1).

These controls operate the air intake vent on the scuttle and the water valve. Operating the "OFF" button automatically cancels the "HEAT" and "AIR" buttons.

The "HEAT" button also cancels the "AIR" button. If it is desired to have the "HEAT" and "AIR" buttons in operation at the same time the "HEAT" button must be pressed first.

The heater control quadrant marked "HOT —COLD" situated centrally in the edge of the parcel shelf regulates the temperature of the air delivered.

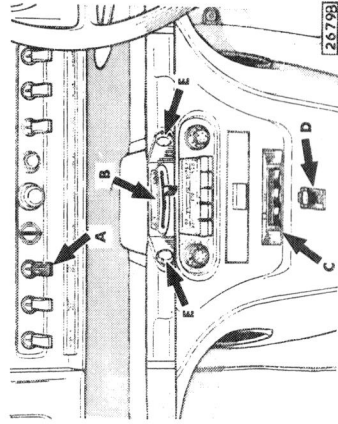

Fig. 1. Heating and ventilating controls. 'A' Fan switch. 'B' Temperature control. 'C' Heater control switches. 'D' Rear outlet control. 'E' Controls for front outlets.

Off

When the "OFF" button is pressed the system is inoperative.

Heat

To obtain hot or warm air in the car, press the "HEAT" button which will open the water valve to supply hot water to the heater element. Before operating the "AIR" button, it is advisable to allow the engine to reach normal operating temperature, particularly in cold weather, to enable hot water to circulate through the heater unit prior to admitting cold air through the scuttle vent.

Adjust the heater control quadrant to give the required temperature.

Air

If cold fresh air is required, press the "AIR" button which will open the scuttle vent and direct the air to the outlets in the car by-passing the heating element. The fan can be switched on if it is desired to increase the circulation. The heater control quadrant should be set at cold.

THE FAN SWITCH

The fan for the heating and ventilating system increases the flow of air through the system and is controlled by a three-position switch (marked "FAN" on the indicator strip) on the instrument panel (Fig. 1).

Lift the switch to the second position for slow speed and to the third position for fast speed, whichever is required.

Operation of the fan is required mainly when the car is stationary or running at a slow speed. At higher speeds it will be found possible to dispense with the fan due to the speed of the car forcing air through the scuttle vent.

AIR DISTRIBUTION

The proportion of air directed to the car or to the windscreen can be controlled by the positions of the front and rear air outlet controls.

The demisting outlets operate whenever the system is working. To obtain the maximum amount of air at the windscreen, both the front and rear outlets should be closed.

The two front outlets are fitted with thumb operated directional controls, one each side of the heater control quadrant. Fully rotating the

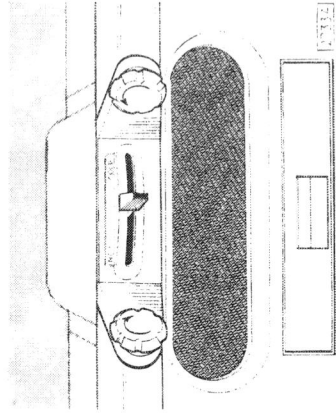

Fig. 2. Rotation of the knobs in the direction indicated by the arrows will progressively re-direct the flow of air out of the front outlets from the feet to the body.

right-hand knob clockwise and the left-hand knob anti-clockwise will cut off the supply of air completely. Reverse rotation of the knobs will progressively re-direct the air flow from the feet to the body.

Operating the lever for the rear outlet (Fig. 1) turns the air supply "ON" or "OFF".

COLD WEATHER

To obtain heating, demising and defrosting:

(a) Depress the button marked "HEAT" and allow a short period to elapse to permit the heater to warm up.

(b) Depress the "AIR" button.

(c) Switch the fan ON at the desired speed.

(d) Open the front and rear outlets as desired.

(e) Adjust the heater control quadrant to give required temperature.

To obtain rapid demisting and defrosting:

(a) Depress the button marked "HEAT" and allow a short period to elapse to permit the heater to warm up.

(b) Depress the "AIR" button.

(c) Switch the fan ON at the desired speed.

(d) Close the front and rear outlets.

(e) Move the heater control quadrant to "HOT".

HOT WEATHER

To obtain ventilation and demisting:

(a) Depress the button marked "AIR".

(b) Switch the fan ON at the desired speed.

(c) Open the front and rear outlets as desired.

(d) Move the heater control quadrant to "COLD".

To obtain rapid demisting:

(a) Depress the button marked "AIR".

(b) Switch the fan ON at the "FAST" position.

(c) Close the front and rear outlets.

(d) Move the heater control quadrant to "COLD".

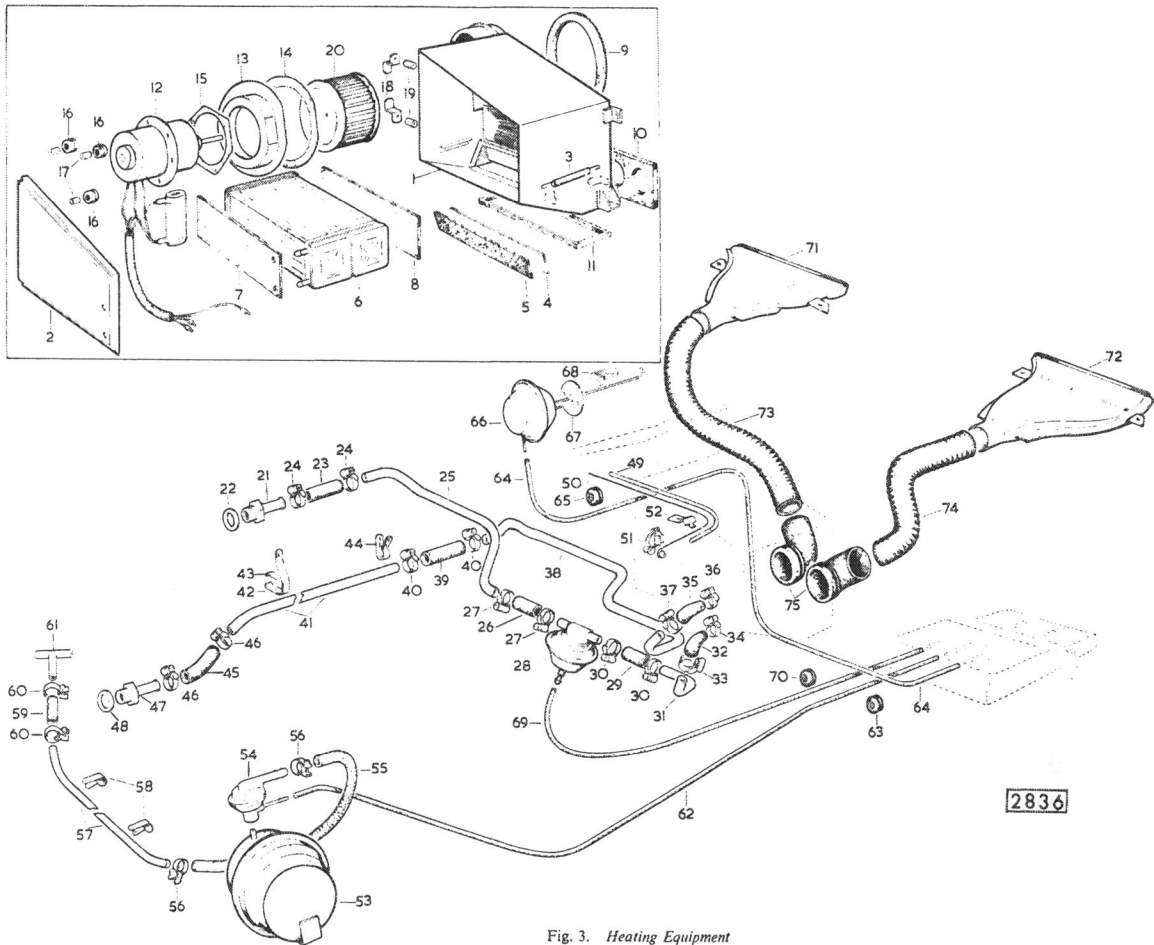

Fig. 3. *Heating Equipment*

2836

1.	Heater case	26.	Hose (feed pipe to vacuum valve)	51.	Clip
2.	Lid	27.	Clip	52.	Clamp
3.	Spring	28.	Valve	53.	Reservac tank
4.	Seal (Polyurethane)	29.	Hose (vacuum valve to feed pipe)	54.	Check valve
5.	Seal (Polyurethane)	30.	Clip	55.	Vacuum hose
6.	Water radiator	31.	Feed pipe	56.	Clip
7.	Seal (Polyurethane)	32.	Elbow hose (feed pipe to radiator)	57.	Front vacuum pipe
8.	Seal (Polyurethane)	33.	Clip	58.	Clip
9.	Seal (Polyurethane)	34.	Clip	59.	Hose (front vacuum pipe to T-piece)
10.	Seal (Polyurethane)	35.	Elbow hose (water radiator to rear return pipe)	60.	Clip
11.	Seal (rubber)			61.	T-piece
12.	Heater motor	36.	Clip	62.	Vacuum hose
13.	Heater motor housing	37.	Clip	63.	Grommet
14.	Seal (Polyurethane)	38.	Rear return pipe	64.	Vacuum hose
15.	Seal (Polyurethane)	39.	Hose	65.	Grommet
16.	Grommet	40.	Clip	66.	Vacuum diaphragm
17.	Distance tube	41.	Front return pipe	67.	Packing
18.	Resistance mounting bracket	42.	Bracket	68.	Clip
19.	Distance tube	43.	Clamp	69.	Vacuum hose
20.	Heater motor fan	44.	Clip	70.	Grommet
21.	Heater feed adaptor	45.	Hose	71.	R.H. demister nozzle
22.	Washer (copper)	46.	Clip	72.	L.H. demister nozzle
23.	Hose (adaptor to feed pipe)	47.	Adaptor	73.	Hose
24.	Clip	48.	Washer (copper)	74.	Hose
25.	Feed pipe	49.	Outer cable	75.	Rubber elbow
		50.	Inner cable		

VACUUM SERVO SYSTEM

Description (Fig. 4)

The vacuum servo system, controlled by the heater buttons (6), includes a vacuum supply tank (1), with non-return valve situated under the left-hand front wing and two servo units one functioning as a heater water tap control (11) and one controlling the opening and closing of the scuttle (fresh air) vent (2).

The vacuum supply tank will provide approximately six complete operations after the ignition is switched off.

Note: In frosty weather it is advisable to close the scuttle vent on leaving the car, to obviate the possibility of the controls freezing up, by pressing the OFF button (6).

The vacuum supply tank is connected by heavy duty hoses and a rigid pipe via a "T"-piece junction to the inlet manifold at the rear of the manifold balance pipe (B).

A small diameter rubber tube from the non-return valve leads directly to the air temperature control panel situated on the console.

Rubber tubing connects the "AIR" control button to the scuttle vent vacuum servo and the "HEAT" control button to the heater water tap vacuum servo.

The servo units are sealed during manufacture and must be replaced if faulty.

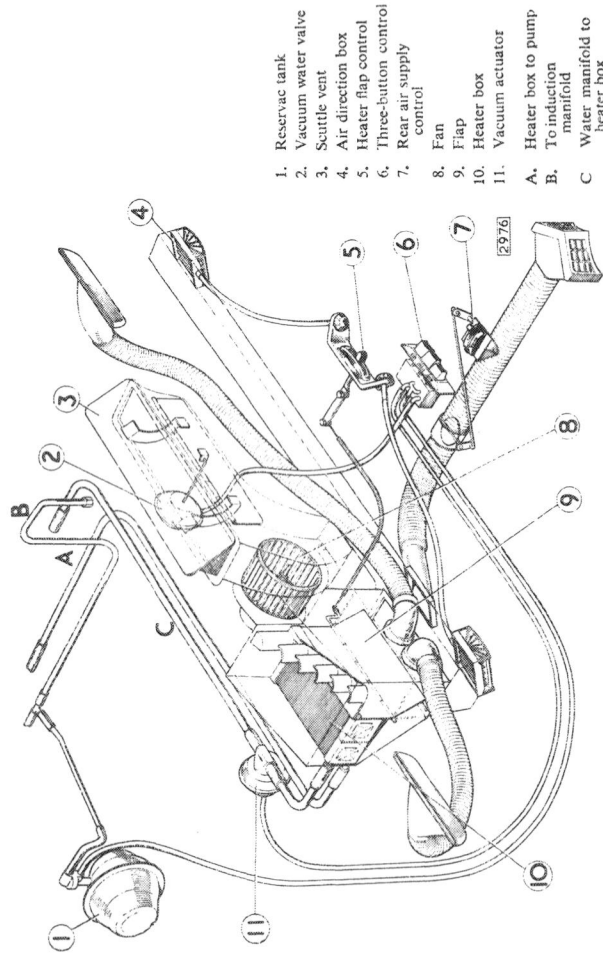

1. Reservac tank
2. Vacuum water valve
3. Scuttle vent
4. Air direction box
5. Heater flap control
6. Three-button control
7. Rear air supply control
8. Fan
9. Flap
10. Heater box
11. Vacuum actuator

A. Heater box to pump
B. To induction manifold
C. Water manifold to heater box

Fig. 4. *Vacuum servo system.*

Refitting

Refit the unit to the bulkhead and attach the operating rod to the inner flap with the spring clip. Secure the unit with the two self-locking nuts and washers. Ensure that the sealing washer is located correctly when refitting.

Locate the shroud panel on the countersunk screws. Seal the top edge of the panel with Bostik 692 sealing compound.

Refit the rubber seal and secure with a good quality adhesive. Renew seal if worn or damaged.

Reconnect the rubber tubing to the vacuum unit, switch on the ignition, start the engine, depress the "AIR" button and raise the vent lid brackets.

Switch the engine OFF.

Refit the vent lid and secure with the two self-locking nuts and washers. Ensure that the two distance pieces are fitted before attaching the lid.

Do not fully tighten the nuts when fitting. Lower the vent by depressing the "OFF" button and position the vent lid correctly in the aperture.

Raise the vent and fully tighten the nuts. Refit the ventilator gauze.

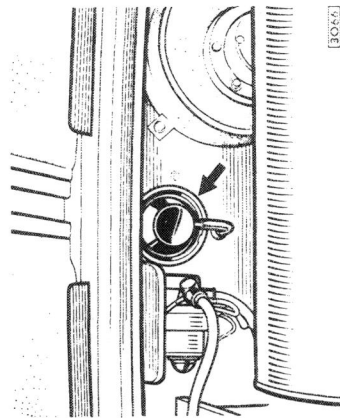

Fig. 5. *Showing location of scuttle vent servo unit*

SCUTTLE VENTILATOR SERVO UNIT

Removal

Start the engine and depress the "AIR" button to open the scuttle vent. Switch off the ignition.

Withdraw the three retaining screws and remove the scuttle ventilator gauze.

Remove the two self-locking nuts and plain washers and lift off the ventilator lid. Retain the two distance pieces attached to the fixing studs.

Remove the rubber seal, secured to the ventilator aperture with an adhesive and release the three countersunk-headed screws now exposed.

Withdraw the shroud panel from the air vent box.

Note: It is not necessary to remove the three countersunk screws completely to enable the shroud to be removed.

Release the spring clip anchoring the operating rod to the inner flap.

Remove the two nuts and washers securing the vacuum unit to the bulkhead and detach the rubber tube.

Withdraw the unit noting the sealing washer located between the unit and the bulkhead.

HEATER TAP SERVO UNIT

Removal

Drain the coolant from the cooling system by opening the radiator tap.

Conserve the coolant if an anti-freeze is in use.

Release the hose clips and withdraw the rubber hoses from the unit. Remove the small rubber tube from the connecting nipple.

Release the nut securing the servo unit to the mounting bracket and withdraw the unit.

Note: The mounting bracket is slotted, it is therefore unnecessary to remove the nut completely unless the unit is to be re-placed.

Refitting

Refitting is the reverse of the removal procedure.

Ensure when fitting that the arrow indicating the direction of the water flow points to the outside of the car.

VACUUM SUPPLY TANK

Removal

Pull off the small rubber tube at the vacuum supply tank non-return valve.

Slacken off the clip and remove the large rubber hose.

Remove the two nuts and washers securing the tank to the cross-member and withdraw the unit.

Remove the non-return valve by turning anti-clockwise.

Note: The non-return valve is a sealed unit and must be replaced if faulty.

Refitting

Refit the non-return valve. The valve is secured to the tank by a tapered thread and must be fully tightened to ensure an air-tight seal.

Refitting the tank is the reverse of the removal procedure.

Fig. 6. Showing location of tap servo unit.

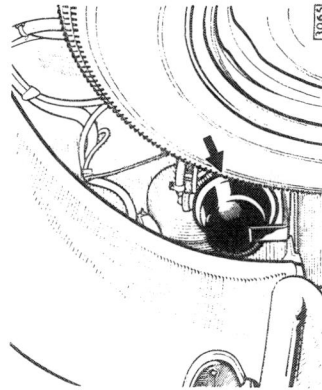

HEATER UNIT

Removal

Drain the radiator and cylinder block. Conserve the coolant if an anti-freeze is in use.

Disconnect the battery earth terminal.

Remove the bonnet as detailed in Section B.

Remove the air cleaner. Roll back the sealing rubber between the carburetter elbow and the air cleaner. Slacken the two wing nuts securing the air cleaner to the brackets on the cylinder head.

Release the air cleaner by pulling it towards the left-hand wing valance.

Slacken off the clips and detach the heater water pipe hoses from the heater unit. Remove the top water pipe from the support clips attached to the heater box.

Remove the electrical cables from the clip on the bulkhead and disconnect at the snap connectors. Disconnect the earth cable by removing the drive screw.

Remove the water valve vacuum unit as detailed previously.

Remove the spring clip securing the control cable to the hot/cold flap control lever and with-draw the cable. Release the bolt and nut securing the outer casing clip.

Remove the five set-screws and washers securing the heater unit to the bulkhead and lift the unit away.

Refitting

Refitting is the reverse of the removal procedure.

Renew the sealing rings if worn or damaged.

Adjust the flap control cable to maintain full movement of the control lever.

HEATER MATRIX

Removal

Drain the radiator and cylinder block. Con-serve the coolant if an anti-freeze is in use.

Slacken off the clips and detach the heater water pipe hoses from the heater unit. Remove the top water pipe from the support attached to the heater box.

Remove the two nuts and washers and with-draw the 2-speed resistance unit from the mounting studs.

Carefully prise away the plastic covered felt covering the front of the unit.

Remove the three hexagon-headed drive screws from the top edge of the front cover and three screws from the bottom edge.

Remove the cover and withdraw the matrix.

Refitting

Refitting is the reverse of the removal procedure.

When refitting the material over the cover plate use a rubber solution.

FAN MOTOR

Removal

Disconnect the battery earth terminal.

Disconnect the wires from the fan motor at the snap-connectors and detach the earth wire from the bulkhead. Remove the resistance unit from the mounting studs.

Carefully prise away the plastic-covered felt on the front of the unit.

Remove the four hexagon-headed set-screws and lockwashers securing the motor mounting bracket to the heater unit.

Withdraw the motor complete with fan, note the plastic foam joint between the mounting bracket and the heater unit.

Remove the fan after slackening off the nut on the spindle.

Note: In order to preserve the balance of the fan assembly care must be taken that the balance pieces are not displaced nor the assembly damaged on removal.

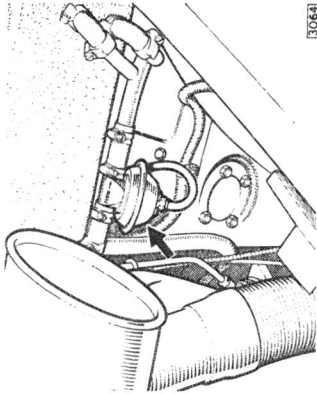

Fig. 7. Showing location of vacuum supply tank.

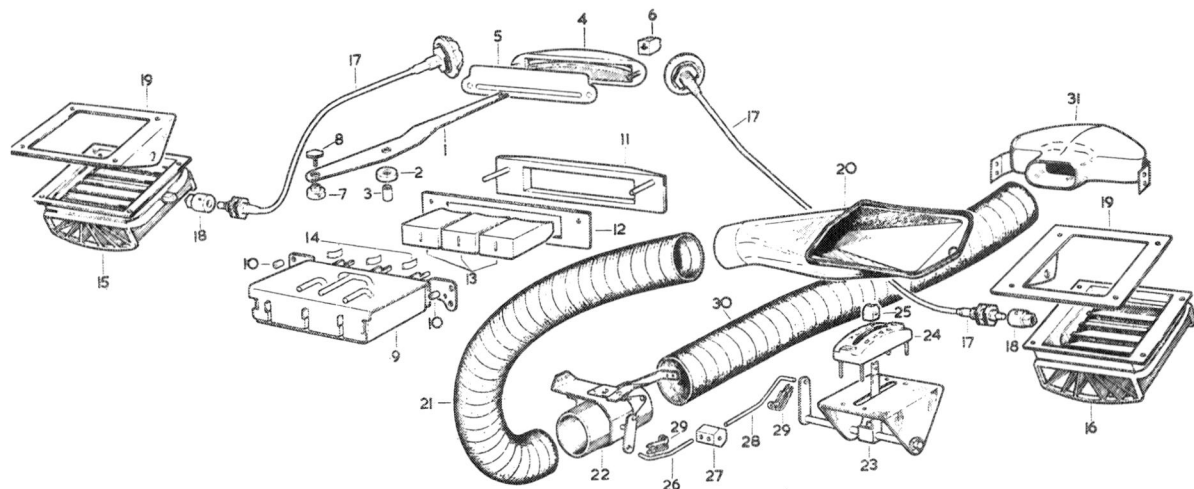

1. Lever (Hot and Cold control)
2. Spacer
3. Distance tube
4. Escutcheon
5. Control plate
6. Control lever knob
7. End fitting
8. Screw
9. Vacuum control box
10. Friction bush
11. Escutcheon
12. Backing plate
13. Vacuum control box knob
14. Clip
15. R.H. air director box
16. L.H. air director box
17. Remote control
18. Adaptor
19. Bracket
20. Collector
21. Hose
22. Butterfly control
23. Lever
24. Escutcheon
25. Control lever knob
26. Front control rod
27. Adjustment slide
28. Rear control rod
29. Clip
30. Hose
31. Air distributor box

Fig. 8.   *Exploded view of heater controls.*

AIR TEMPERATURE CONTROL PANEL

Removal

Disconnect the battery earth lead.

Withdraw the control button escutcheon, retained by two pegs to the control panel and unscrew the two round-headed screws now exposed.

Remove the heater control perforated guard cover from the parcel tray after withdrawing the four nylon retaining pins.

Lift the parcel tray trimming above the control panel and unscrew the two round-headed thumb screws located in recesses in the tray.

Withdraw the control panel facia and disconnect the three rubber pipes connected to the control button unit. Note the location of the individual pipes for reference when refitting.

Note: If a radio is fitted to the car it will be necessary to disconnect the supply feed cable from the fuse holder and the aerial and loud-speaker leads from the control unit before completing the removal of the control panel.

Remove the two wood-screws from the back of the panel and withdraw the control button unit.

Refitting

Refitting is the reverse of the removal procedure.

Ensure that the rubber tubes are connected to the correct junctions as noted on removal.

Remove the three Phillips-headed drive-screws and detach the mounting bracket from the motor unit. Note the plastic foam joint between the mounting bracket and the motor.

Refitting

Refitting is the reverse of the removal procedure.

When reassembling the fan to the motor spindle check that there is at least ⅛" (3·2 mm.) clearance between the fan and motor mounting bracket and that the fan is running true on the spindle.

FAN SWITCH

Removal

Disconnect the battery earth terminal.

Remove the picnic tray. Withdraw the tray to the full extent and locate the two spring steel retaining clips attached to the tray at the back edge. Press both clips inwards to the centre of the car and complete the withdrawal of the tray.

Remove the two thumb-screws securing the instrument panel to the facia.

Remove the three "Lucar" connectors from the fan switch.

Unscrew the chrome bezel securing the switch to the instrument panel and withdraw the switch. Note the location of the wires on the switch before removing.

Refitting

Refitting is the reverse of the removal procedure.

Connect the switch cables to the correct terminals as noted on removal.

HEATER CONTROL LEVER

Removal

Remove the heater control perforated guard cover from the parcel tray after withdrawing four nylon retaining pins.

Pull off the heater control lever knob.

Disconnect the control lever at the forward end, unscrew the lever pivot pin, note the washer fitted between the lever and the bracket.

Withdraw the lever through the escutcheon plate.

Refitting

Refitting is the reverse of the removal procedure.

Ensure that full movement of the lever on the heater unit is maintained when connecting the control cable.

FRONT AIR DIRECTION VENTS

Removal

Remove the four drive-screws securing the vents to the air duct.

Lower the vent and withdraw from the control cable junction.

Note on removal that the assemblies are handed.

Refitting

Refitting is the reverse of the removal procedure.

DIRECTION VENT CONTROL CABLES

Removal

Release the locknuts securing the outer cables to the vent bracket.

Disconnect the cables and collect the loose adaptor.

Unscrew the cable from the centre finisher and withdraw the assemblies.

Note: A thin spanner will be required to remove the outer casing from the finisher.

Refitting

Refitting is the reverse of the removal procedure.

Fig. 9. Cross section of heater unit showing path of hot and cold air.

A. Path of Hot Air
B. Path of Cold Air
C. Cold Position
D. Car Interior
E. To Car Interior
F. Screen Position
G. To Demisters
H. Hot Position

——— A
– – – B

Fig. 10. Windscreen washer water container.

ELECTRICAL CONTACTS
BRASS CONNECTING TUBE
FILLER CAP
PUMP ASSEMBLY
AUXILIARY RESERVOIR
MOTOR UNIT
MOULDED COVER
CONTAINER
PUMP ROTOR SPINDLE
PUMP ROTOR

WINDSCREEN WASHING EQUIPMENT

The windscreen washer is electrically operated and comprises a glass water container mounted in the engine compartment which is connected to jets at the base of the windscreen. Water is delivered to the jets by an electrically driven pump incorporated in the water container.

OPERATION

The windscreen washer should be used in conjunction with the windscreen wipers to remove foreign matter that settles on the windscreen.

Lift the switch lever (marked "Washer") and release immediately when the washer should operate at once and continue to function for approximately seven seconds. Allow a lapse of time before operating the switch for a second time.

Warning

If the washer does not function immediately check that there is water in the container. The motor will be damaged if the switch is held closed for more than one or two seconds if the water in the container is frozen.

The washer should not be used under freezing conditions as the fine jets of water spread over the windscreen by the blades will tend to freeze up.

In the summer the washer should be used freely to remove insects before they dry and harden on the screen.

FILLING-UP

The water should be absolutely CLEAN. If possible use SOFT water for filling the container, but if this is not obtainable and hard water has to be used, frequent operation and occasional attention to the nozzle outlet holes will be amply repaid in preventing the formation of unwelcome deposits.

The correct water level is up to the bottom of the container neck. Do not overfill, or unnecessary splashing may result. Always replace the rubber filler cover correctly after filling, pressing it fully home.

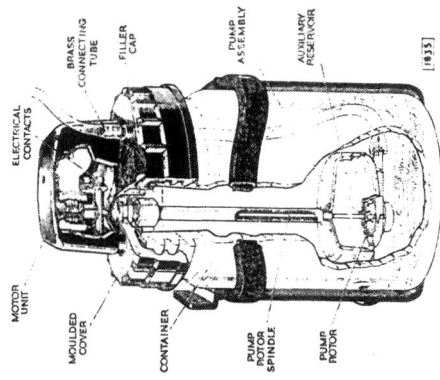

It is not possible to empty the container completely with the pump. **Refilling is necessary when the water level has fallen so that the top of the auxiliary reservoir is uncovered.** About 30 full operations will be obtained from one filling.

When using the washer, an indication of the need to refill the container is given by the behaviour of the unit. The time taken for the auxiliary reservoir to refill increases as the water level in the container falls.

As soon as the water level has fallen to the top of the auxiliary reservoir, the amount of water delivered to the windscreen will decrease with successive operations and the time the unit runs will, in proportion, become less.

If the water level is allowed to fall still further, until it is down to the bottom of the auxiliary reservoir, the automatic action will cease and water will be delivered to the windscreen only as long as the switch is operated. This will continue until the water level has fallen to the inlet orifices, when the pump will be above the water level and no water will be available for delivery to the windscreen.

Do not continue to operate the switch after the available water has been used up, otherwise damage may be caused to the unit.

Refilling the container will restore normal operation of the unit.

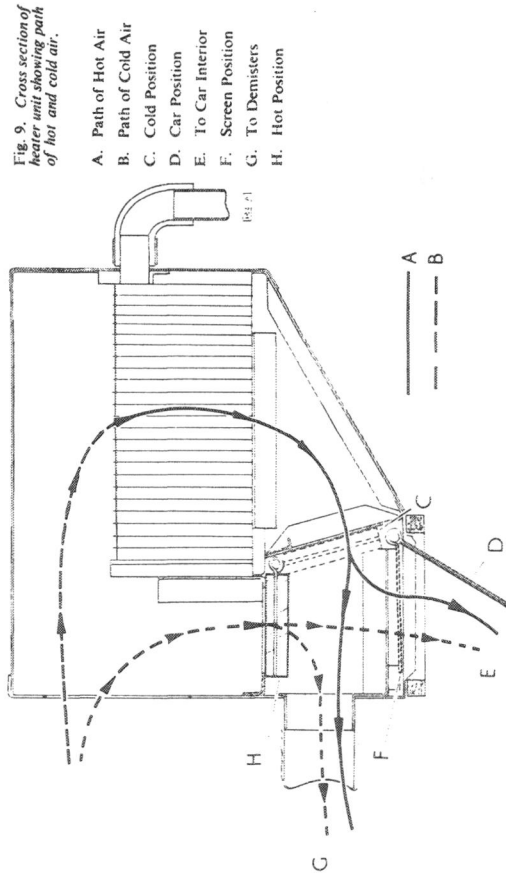

SUPPLEMENTARY INFORMATION
TO SECTION O
" HEATING & WINDSCREEN WASHING EQUIPMENT "

HEATING AND WINDSCREEN WASHING EQUIPMENT

COLD WEATHER

To avoid damage by frost, add denatured alcohol (methylated spirits) as follows:

The underside of the rubber filler cover will be found to form a measure. Two measures of denatured alcohol should be added per container of water. USE NO OTHER ADDITIVES WHATSOEVER.

ADJUSTING THE JETS

With a screwdriver turn each nozzle in the jet holder until the jets of water strike the windscreen in the area swept by the wiper blades. It may be necessary to adjust the nozzles slightly after a trial on the road due to jets of water being deflected by the airstream.

JET NOZZLES

Cleaning

To clear a blocked jet nozzle completely unscrew the nozzle from the jet holder. Clear the small orifices with a thin piece of wire or blow out with compressed air; operate the washer with the nozzle removed. Allow the water to flush through the jet holder and then replace the nozzle.

LUBRICATION

If, after lengthy service, the motor is found to be running slowly, unscrew the moulded cover from the container and apply one or two drops only of thin machine oil to the felt pad situated in the gap between the cover and the motor unit. Do not over-lubricate or excess oil may find its way into the water container when the cover is refitted, with consequent smearing of the windscreen.

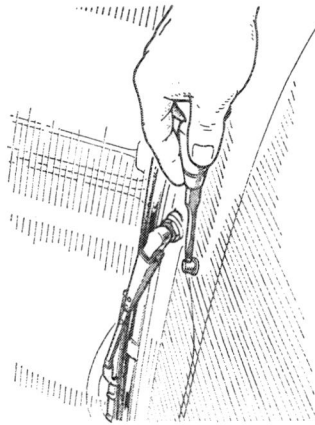

Fig. 11. *Adjusting the windscreen washing jets.*

INTRODUCTION OF LUCAS 5 SJ WINDSCREEN WASHER

	Commencing Chassis No.
	R.H. Drive	L.H. Drive
3.4 "S"	1B 3605	1B 25515
3.8 "S"	1B 53950	1B 77613

Commencing at the above chassis numbers, Lucas 5SJ windscreen washers are fitted to all "S" type cars. The 5SJ unit is electrically operated and comprises a small permanent-magnet motor driving a centrifugal pump through a 3-piece Oldham type coupling, a high density polythene water container mounted inside the engine compartment and connected to two jets at the base of the windscreen.

Operation

The windscreen washer should be used in conjunction with the windscreen wipers to remove foreign matter that settles on the windscreen.

Lift the switch (marked "Washer") when the washer should operate immediately and continue to function until the switch is released.

Warning. If the washer does not function immediately, check that there is water in the container. The motor will be damaged if the switch is operated for more than one or two seconds if the water in the container is frozen. The washer should not be used under freezing conditions as the jets of water spread over the windscreen by the blades will tend to freeze.

Servicing
Testing in Position

(a) Testing with a voltmeter:—

Connect a suitable direct current voltmeter to the motor terminals observing the polarity as indicated on the moulded cover. Operate the switch. If a low or zero reading is given, check the fuse, switch and external connections.

If the voltmeter gives a reverse reading, the connections on the motor must be transposed.

If supply voltage is registered at the terminals but the unit fails to operate, an open circuit winding or faulty brush gear must be suspected. Dismantle the motor as described below.

(b) Checking the external tubes and nozzles:—

If the motor operates, but little or no water is delivered to the screen, the external tubes and nozzles may be blocked.

Remove the external plastic tube from the short connection on the container cover and, after checking that the connector tube is clear, operate the switch. If a jet of water is ejected, check the external tubes and nozzles for blockage. If no water is ejected, proceed as described below.

(c) Testing with an ammeter:—

Connect a suitable direct current ammeter in series with the motor and operate the switch. If the motor does not operate but the current reading exceeds that given in "Data", remove the motor and check that the pump impeller shaft rotates freely.

If the shaft is difficult to turn, the water pump unit must be replaced. If the shaft turns freely, the fault lies in the motor which must be dismantled and its components inspected.

Dismantling

(a) Disconnect the external tube and electrical connections and remove the cover from the container.

(b) Remove the self-tapping screw which retains the motor on the cover and pull away the motor unit. Take care not to lose the intermediate coupling which connects the armature coupling to the pump spindle coupling.

(c) Remove the armature coupling from the armature shaft as follows:—

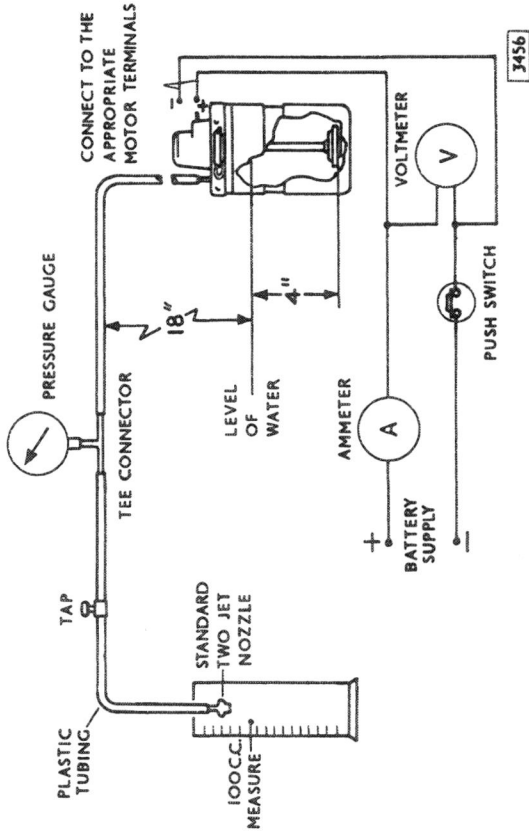

Fig. 12. *Performance testing the windscreen washing equipment.*

Test Data

Nominal voltage of unit	..	..	..	..	..	..	12
Maximum current consumption	..	..	..	..	..	2 amp.
Resistance between commutator segments	..	..	..	2·8—3·1 ohms.
Maximum delivery—water pressure	..	..	..	4·5 p.s.i. (0·32 kg./sq. cm.)
Container capacity	..	..	..	..	..	2¼ pints (1·1 litres)
Usable quantity of water	..	..	..	..	..	2 pints (1 litre)
Diameter of nozzle orifice	..	..	..	..	0·25″—0·28″ (6·3—7 mm.)

Hold the armature shaft firmly with a pair of snipe nosed pliers and, using a second pair of pliers, pull off the coupling.

(d) Remove the two self-tapping screws from the bearing plate. The bearing plate and rubber gasket may now be removed. Remove the two terminal nuts and brushes can now be removed and the armature withdrawn. Take care not to lose the bearing washer which fits loosely on the armature shaft.

(e) The pole assembly should not normally be disturbed. If, however, its removal is necessary, make careful note of its position relative to the motor housing. The narrower pole piece is adjacent to the terminal locations. Also note the position of the pole clamping member. When fitted correctly it locates on both pole pieces but, if fitted incorrectly, pressure is applied to one pole piece only.

If the motor has been overheated or if any part of the motor housing has been damaged, a replacement motor unit must be fitted.

Bench Testing

(a) Armature:—

If the armature is damaged, or if the windings are loose or badly discoloured, a replacement armature must be fitted.

The commutator must be cleaned with a fluffless, petrol-moistened cloth, or, if necessary, by polishing with a strip of very fine glass paper.

The resistance of the armature winding should be checked with an ohm meter. The resistance between commutator segments should be shown under "Data".

If the carbon is less than ⅟₁₆" (1·59 mm.) long, a new brush set must be fitted. Check that the brushes bear firmly against the commutator.

Re-assembling

Reverse the dismantling procedure to re-assemble the unit. The following points should be observed:—

(a) Make sure that the bearing recess in the motor housing is filled with Rocol Molypad molybdenised grease. Remove excess grease from the face of the bearing boss.

(b) Check that the pole piece assembly does not rock and that the pole pieces are firmly located on the circular spigot. Ensure that both the pole piece and the clamping member are the right way round (see paragraph "e". Dismantling).

(c) Before replacing the motor unit on the cover, ensure that the armature coupling is pushed fully home. Also check that the intermediate coupling is in place.

Performance Testing

Equipment required:—

D.C. supply of appropriate voltage.

D.C. Voltmeter, first grade, moving coil.

0—3 amp. D.C. Ammeter.

0—15 lb/sq. in. (0—1 kg./sq. cm.) pressure gauge.

Pushbutton with normally open contacts.

Two-jet nozzle (Lucas No. 295005).

On-Off tap.

100 c.c. capacity measure.

4' 6" (1·27 m.) length of plastic tubing.

(a) Connect up the equipment as shown in Fig. 12. The water level in the container must be 4" (101·6 mm.) above the base of the pump assembly. The pressure gauge and nozzle must be 18" (457·2 mm.) above the water level.

(b) Open the tap.

(c) Lift the switch for approximately 5 seconds and check the voltmeter reading which should be the same as the supply voltage. On releasing the switch, immediately close the tap to ensure that the plastic tubing remains charged with water.

(d) Empty the measuring cylinder.

(e) Open the tap and operate push switch for precisely 10 seconds after which period release the switch and close the tap.

During the 10 second test the current and pressure values should be in accordance with those given in "Data" and at least 35 c.c. of water should have been delivered.

ELECTRICAL
AND
INSTRUMENTS

ADDENDA

The following corrections should be noted when using SECTION P.

Page	Correction
P.42	**Traffic Hazard Warning Light Bulb—Replacement.**
	Remove the chrome bezel and withdraw the bulb from the bulb holder.
	Heated Backlight Warning Light Bulb—Replacement.
	Remove the chrome bezel and withdraw the bulb from the bulb holder.
P.65	Under "AIRIEL MOUNTING" reference to "Fig. 76" should read "Fig. 75" and "Fig. 77" should read "Fig. 76."
P.66	Alter reference "Fig. 78" to read "Fig. 77" and three references to "Fig. 77" to read "Fig. 76."

INDEX

INDEX (continued)

ELECTRICAL AND INSTRUMENTS

BATTERY

The BV.11A battery is of the "clean top" pattern, in which small holes are provided over each intercell connector to enable the prongs of a heavy discharge tester to be inserted for testing purposes.

Battery type	..	..	..	BV.11A
Voltage	..	..	..	12
Number of plates per cell	..	..	11
Capacity at 10 hour rate	..	..	60
Capacity at 20 hour rate	..	..	67

ROUTINE MAINTENANCE

Wipe away any foreign matter or moisture from the top of the battery, and ensure that the connections and the fixings are clean and tight.

About once a month, or more frequently in hot weather, examine the level of the electrolyte in the cells. If necessary add distilled water to bring the electrolyte just level with the separator guards, which can be seen when the vent plugs are removed.

The use of a Lucas battery filler will be found helpful in this topping-up process, as it ensures that the correct electrolyte level is obtained automatically and also prevents distilled water from being spilled over the battery top.

Distilled water should always be used for topping-up. In an emergency however, clean soft rain water collected in an earthenware container may be used.

Note: Never use a naked light when examining a battery, as the mixture of oxygen and hydrogen given off by the battery when on charge, and to a lesser extent when standing idle, can be dangerously explosive.

REMOVAL

Mark the position of the bonnet hinges relative to the bonnet. Remove the four setscrews securing the bonnet to the hinges.

Release the two spring clips and remove the battery cover.

Remove the two securing screws and detach the terminals from the lugs.

Unscrew the two battery securing bolts and detach the retaining band and rubber.

Lift out the battery from the tray.

REFITTING

Refitting is the reverse of the removal procedure. Before refitting the cable connectors, clean the terminals and coat with petroleum jelly.

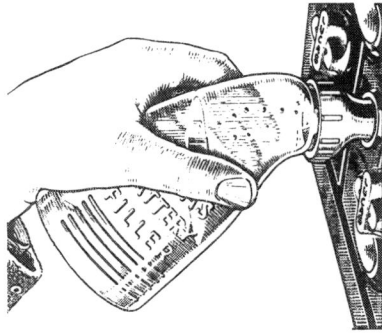

Fig. 1. Lucas battery filler.

PERSISTENT LOW STATE OF CHARGE

First consider the conditions under which the battery is used. If the battery is subjected to long periods of discharge without suitable opportunities for recharging, a low state of charge can be expected. A fault in the generator or regulator, or neglect of the battery during a period of low or zero mileage may also be responsible for the trouble.

Vent Plugs

See that the ventilating holes in each vent plug are clear.

Level of Electrolyte

The surface of the electrolyte should be just level with the tops of the separator guards. If necessary, top up with distilled water. Any loss of acid from spilling or spraying (as opposed to the normal loss of water by evaporation) should be made good by dilute acid of the same specific gravity as that already in the cell.

Cleanliness

See that the top of the battery is free from dirt or moisture which might provide a discharge path. Ensure that the battery connections are clean and tight.

Hydrometer Tests

Measure the specific gravity of the acid in each cell in turn with a hydrometer. To avoid misleading readings, do not take hydrometer readings immediately after topping-up.

The readings given by each cell should be approximately the same. If one cell differs appreciably from the others, an internal fault in the cell is indicated.

The appearance of the electrolyte drawn into the hydrometer when taking a reading gives a useful indication of the state of the plates. If the electrolyte is very dirty, or contains small particles in suspension, it is possible that the plates are in a bad condition.

The specific gravity of the electrolyte varies with the temperature, therefore, for convenience in comparing specific gravities, this is always corrected to 60°F., which is adopted as a reference temperature. The method of correction is as follows:

For every 5°F. below 60°F. deduct ·002 from the observed reading to obtain the true specific gravity at 60°F.

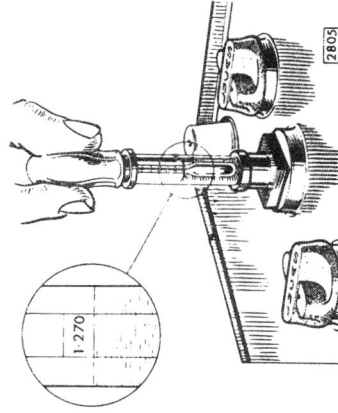

Fig. 2. Testing with a Hydrometer.

State	Home and climates with shade temperature ordinarily below 80 F. (26·6°C.). Specific gravity of electrolyte (corrected to 60°F.)	Climates with shade temperature frequently over 80 F. (26·6°C.). Specific gravity of electrolyte (corrected to 60°F.)
Fully charged	1·270–1·290	1·210–1·230
About half discharged	1·190–1·210	1·130–1·150
Completely discharged	1·110–1·130	1·050–1·070

For every 5 F. above 60 F. add ·002 to the observed reading to obtain the true specific gravity at 60°F.

The temperature must be that indicated by a thermometer actually immersed in the electrolyte, and not the air temperature.

Compare the specific gravity of the electrolyte with the values given in the table and so ascertain the state of charge of the battery.

If the battery is in a discharged state, it should be recharged, either on the vehicle by a period of day-time running or on the bench from an external supply, as described under "Recharging from an External Supply."

Discharge Test

A heavy discharge tester consists of a voltmeter, 2 or 3 volts full scale, across which is connected a shunt resistance capable of carrying a current of 150-160 amperes. It is important to use only a suitable rated instrument. Pointed prongs are provided for making contact with the inter-cell connectors.

Press the contact prongs against the positive and negative terminals of each cell. A good cell will maintain a reading of 1·2—1·5 volts, depending on the state of charge, for 10 seconds. If, however, the reading rapidly falls off, the cell is probably faulty and a new plate assembly may have to be fitted.

RECHARGING FROM AN EXTERNAL SUPPLY

If the battery tests indicate that the battery is merely discharged, and is otherwise in a good condition, it should be recharged, either on the vehicle by a period of day-time running or on the bench from an external supply.

If the latter, the battery should be charged at 5 amperes until the specific gravity and voltage show no increase over three successive hourly

readings. During the charge the electrolyte must be kept level with the tops of the separator guards by the addition of distilled water.

A battery that shows a general falling-off in efficiency common to all cells, will often respond to the process known as "cycling." This process consists of fully charging the battery as described above, and then discharging it by connecting to a lamp board, or other load, taking a current of 5 amperes. The battery should be capable of providing this current for at least 7 hours before it is fully discharged, as indicated by the voltage of each cell falling to 1·8. If the battery discharges in a shorter time, repeat the "cycle" of charge and discharge.

PREPARING NEW UNFILLED, UNCHARGED BATTERIES FOR SERVICE

Preparation of Electrolyte

Batteries should not be filled with acid until required for initial charging.

Electrolyte of the specific gravity required is prepared by mixing distilled water and concentrated sulphuric acid, usually of 1·840 specific gravity. The mixing must be carried out either in a lead-lined tank or in a suitable glass or earthenware vessel. Slowly add the acid to the water, stirring with a glass rod. **Never add the water to the acid**, as the resulting chemical reaction causes violent and dangerous spurting of the concentrated acid. The correct specific gravity for the filling acid and approximate proportions of acid and water are indicated in the following table:

Heat is produced by the mixture of acid and water, and the electrolyte should be allowed to cool before taking hydrometer readings—unless a thermometer is used to measure the actual temperature, and a correction applied to the reading before pouring the electrolyte into the battery.

Specific Gravity of Filling Acid (corrected to 60 F.)

Home and Climates with shade temperature ordinarily below 80 F. (26·6°C.).	Climates with shade temperatures frequently above 80 F. (26·6 C.).
1·260	1·210
Add 1 part by volume of acid (1·840 S.G.) to 3·2 parts of distilled water to mix this electrolyte	Add 1 part by volume of acid (1·840 S.G.) to 4·3 parts of distilled water to mix this electrolyte

Quantity of electrolyte required per cell 1¼ pint approximately (710 c.c.)

Filling the Battery

The temperature of the acid, battery and filling-in must be between 40 F.—100 F. (5·5 C —37·7 C).

Carefully break the seals in the filling holes and fill each cell to the level of the separator guard with electrolyte of the appropriate specific gravity. Allow the battery to stand for twelve hours, in order to dissipate the heat generated by the chemical action of the acid on the plates and separators. Restore levels by adding more acid of the same specific gravity and then proceed with the initial charge.

Initial Charge Rate

Charge at a rate of 4 amperes until the voltage and specific gravity readings show no increase over five successive hourly readings. This may take up to 80 hours, depending on the length of time the battery has been stored before charging.

Keep the current constant by varying the series resistance of the circuit or the generator output.

This charge should not be broken by long rest periods. If, however, the temperature of any cell rises above the permissible maximum (that is, 100°F. (37·7 C.) for batteries filled with 1·260 S.G. acids, 120 F. (48·8 C.) for those with 1·210 S.G. acid), the charge must be interrupted until the temperature has fallen at least 10 F. (—12·0 C.) below that figure. Throughout the charge, the electrolyte must be kept level with the top of the separator guards by the addition of acid solution of the same specific gravity as the original filling-in acid, until the specific gravity and voltage readings have remained constant for five successive hourly readings. If the charge is continued beyond that point, top up with distilled water.

At the end of the charge carefully check the specific gravity in each cell to ensure that, when corrected to 60 F., it lies within the specified fully-charged limits. If any cell requires adjustment, some of the electrolyte must be siphoned

off and replaced either by distilled water or by acid of the strength originally used for filling-in, depending on whether the specific gravity is too high or too low. Continue the charge for an hour or so to ensure adequate mixing of the electrolyte and again check the specific gravity readings. If necessary, repeat the adjustment process until the desired reading is obtained in each cell. Finally, allow the battery to cool, and siphon off any electrolyte above the tops of the separator guards.

PREPARING NEW "DRY-CHARGED" BATTERIES FOR SERVICE

Filling the Cells

Carefully break the seals in the filling holes and fill each cell with correct specific gravity acid as shown in the table above to the top of the separator guards in one operation. The temperatures of the filling room, battery and acid should be maintained at between 60 F. and 100 F. (15·5 C.—37·7 C.). If the battery has been stored in a cool place, it should be allowed to warm up to room temperature before filling.

Freshening Charge

Batteries filled in this way are up to 90% charged and capable of giving a starting discharge one hour after filling. When time permits, however, a short freshening charge will ensure that the battery is fully charged.

Such a freshening charge should be 5 amperes for not more than four hours.

During the charge the electrolyte must be kept level with the top of the separators by the addition of distilled water. Check the specific gravity of the electrolyte at the end of the charge; if 1·260 acid was used to fill the battery, the specific gravity should now be between 1·260 and 1·290; if 1·210 acid, between 1·210 and 1·230.

Maintenance in Service

After filling, a dry-charged battery needs only the attention normally given to all lead-acid type batteries.

DISTRIBUTOR

REMOVAL

Spring back the clips and remove the distributor cap.

Disconnect the low tension wire from the distributor terminal.

Disconnect the vacuum pipe by withdrawing the elbow sleeve junction.

Slacken the distributor plate pinch bolt and withdraw the distributor.

REFITTING

Refitting is the reverse of the removal procedure, but it will be necessary to reset the ignition timing as follows:

Ignition Timing

Set the micrometer adjustment in the centre of the scale.

Enter the distributor into the cylinder block with the vacuum advance unit connection facing the cylinder block.

Rotate the rotor-arm until the driving dog engages with the distributor drive shaft.

Connect the low tension wire to the terminal on the distributor body.

Rotate the rotor until the rotor-arm approaches the No. 6 (front) cylinder segment in the distributor cap.

Slowly rotate the engine until the ignition timing scale on the crankshaft damper is the appropriate number of degrees before the pointer on the sump. (See Data).

Connect a 12 volt test lamp with one lead to the distributor terminal (or the CB terminal of the ignition coil) and the other to a good earth.

Slowly rotate the distributor body until the points are just breaking, that is, when the lamp lights up.

Tighten the distributor plate pinch bolt.

A maximum of six clicks on the vernier adjustment from this setting, to either advance or retard, is allowed.

ROUTINE MAINTENANCE

Distributor Contact Breaker Points

Every 3,000 miles (5,000 km.) (First 500 miles [834 km.] with new contact set) check the gap between the contact points with feeler gauges when the points are fully opened by one of the cams on the distributor shaft. A combined screwdriver and feeler gauge is provided in the tool kit.

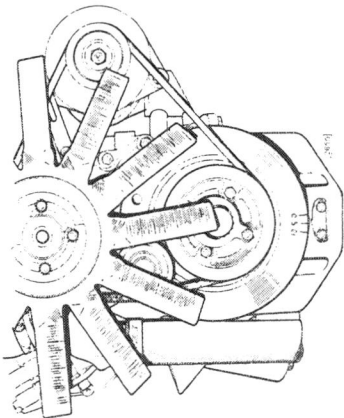

Fig. 3. *Timing scale marked on the crankshaft damper.*

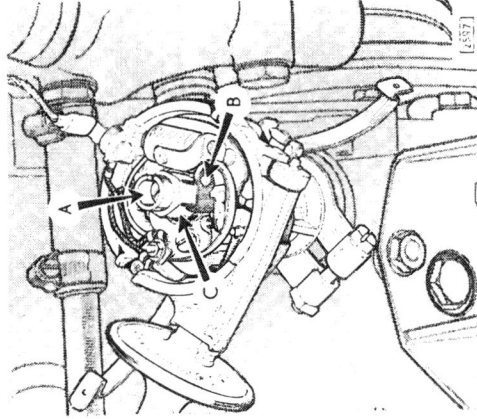

Fig. 5. *Distributor lubrication points.*

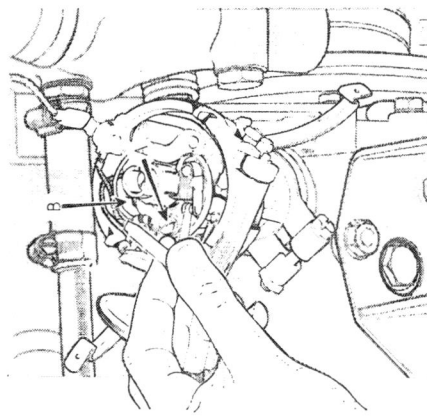

Fig. 4. *Checking the gap between the distributor contact points. The screw 'A' secures the fixed contact point, the contact gap is adjusted by turning a screwdriver in the slot 'B' in the contact plate.*

If the gap is incorrect, slacken (very slightly) the contact plate securing screw (A, Fig. 4) and adjust the gap by turning a screwdriver in the nick in the contact plate and the slot in the base plate. (B). Turn clockwise to decrease the gap and anti-clockwise to increase. Tighten the securing screw and plate.

Lubrication—Every 3,000 miles (5,000 km.)

Remove the moulded cover and withdraw the rotor arm. A tight rotor arm can be withdrawn using a pair of suitable levers carefully applied at opposite points below the rotor moulding—never against the metal electrode.

Important: Do not allow oil or grease on or near the contacts when carrying out the following lubrication.

Cam Bearing

To lubricate the cam bearing, inject a few drops of thin machine oil into the rotor arm spindle (A, Fig. 5). Do not remove or slacken the screw located inside the spindle—a space is provided beneath the screwhead to allow the lubricant to reach the cam bearing.

Cam

Lightly smear the faces of the cam (C, Fig. 5) with Mobilgrease No. 2 or with clean engine oil.

Centrifugal Timing Control

Inject a few drops of thin machine oil through a convenient aperture in the contact breaker base plate.

Cleaning

Clean the moulded cover inside and outside with a soft dry cloth. Pay particular attention to spaces between the terminals. Check that the small carbon brush inside the moulding can move freely in its holder.

Whilst the rotor arm is removed, examine the contact breaker. Rough, burned or blackened contacts can be cleaned with fine carborundum stone or emery cloth. After cleaning remove any grease or metallic dust with a petrol-moistened cloth.

Contact cleaning is facilitated by removing the lever to which the moving contact is attached. To do this, remove the nut, insulating piece and electrical connections from the post to which the contact breaker spring is anchored. The contact breaker lever can then be lifted off the pivot post and the spring from the anchor post.

DISTRIBUTOR DATA

	Compression Ratio	
	7 : 1 – 8 : 1	9 : 1
	22D6	22D6
Lucas Ignition Distributor type	22D6	22D6
Lucas Model No.	40885B/D	40886B/D
Cam dwell angle	34 ± 3°	34 ± 3°
Contact breaker gap	·014" – ·016" (0·36 mm. – 0·41 mm.)	0·14" – ·016" (0·36 mm. – 0·41 mm.)
Contact breaker spring tension (measured at free contact)	18 – 24 ozs. (512 – 682 gms.)	18 – 24 ozs. (512 – 682 gms.)

IGNITION TIMING

	T.D.C.
7 to 1 Compression Ratio	T.D.C.
8 to 1 Compression Ratio	7° BTDC
9 to 1 Compression Ratio	5° BTDC

IGNITION DISTRIBUTOR TEST DATA

VACUUM TIMING ADVANCE TESTS

The distributor must be run immediately below the speed at which the centrifugal advance begins to function to obviate the possibility of an incorrect reading being registered.

Distributor Type	Lucas Service Number	Lucas Vacuum Unit Number	Vacuum in inches of mercury and advance in degrees		No advance in timing below-ins. of mercury
			Inches	Degrees	
22 D6	40885BD	54412145	25 14 10 6	6–8 4½–7½ 1–4½ 0–1	5
22 D6	40886BD	54411969	15 12 6 7	7–9 6–9 0–3 0–¼	2½

CENTRIFUGAL TIMING ADVANCE TESTS

Mount distributor in centrifugal advance test rig and set to spark at zero degrees at 100 r.p.m.

Lucas Advance Springs Number	Accelerate to-RPM and note advance in degrees		Decelerate to-RPM and note advance in degrees		No advance in timing below-RPM
	RPM	Degrees	RPM	Degrees	
5441559	3,400	19	3,000 2,300 1,300 1,000 500	17—19 14—16 10—12 8—10 1—3	200
5441560	2,000	13	1,500 1,100 800 550 400	11—13 8—10 6—8 2—4½ ½—2½	225

2802

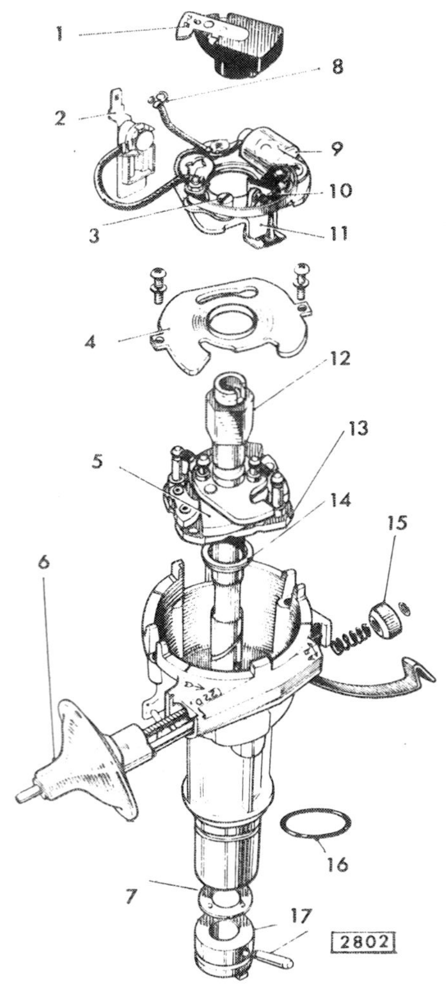

Fig. 6. *Distributor Model 22D6.*

1. Rotor arm
2. L.T. terminal
3. Fixed contact plate securing screw
4. Contact breaker base plate
5. Centrifugal timing control weights
6. Vacuum timing control unit
7. Thrust washer
8. C.B. earth connector
9. Capacitor
10. Contacts
11. Contact breaker moving plate
12. Cam
13. Action plate
14. Distance collar
15. Micrometer adjustment nut
16. Oil seal washer
17. Dog and pin

After cleaning and trimming the contacts, smear the pivot post (B, Fig. 5) with Ragosine Molybdenised Non-creep Oil or with Mobilgrease No. 2. Reassemble the contact breaker and check the setting.

Refit the rotor arm, carefully locating its moulded projection in the spindle keyway and pushing it on as far as it will go.

Refit the moulded cover and spring the two side clips into position.

SERVICING

Dismantling

When dismantling, note carefully the position in which the various components are fitted in order to simplify their reassembly.

Bearing Replacement

The ball bearing at the upper end of the shank can be removed with a shouldered mandrel locating on the inner journal of the bearing.

When fitting a new ball bearing, the shouldered mandrel must locate on both inner and outer journals of the bearing.

The bush is pressed into the shank with a shouldered mandrel. The mandrel should be hardened and polished and approximately 0·0005" greater in diameter than the distributor shaft. To prevent subsequent withdrawal of the bush with the mandrel, a stripping washer should be fitted between the shoulder of the mandrel and the bush.

Under no circumstances should the bush be overbored by reaming or by any other means, since this will impair the porosity and therefore the lubricating quality of the bush.

Reassembly

When reassembling, Ragosine molybdenised non-creep oil or (failing this) clean engine oil, should be smeared on the shaft and, more lightly, on the contact breaker bearing plate.

The bearing bush at the lower end of the shank can be driven out with a suitable punch.

A bearing bush may be prepared for fitting by allowing it to stand completely immersed in medium viscosity (S.A.E. 30–40) engine oil for at least 24 hours. In cases of extreme urgency, this period of soaking may be shortened by heating the oil to 100 C. for two hours and then allowing to cool before removing the bush.

FLASHER UNITS

The flasher unit is housed in a small cylindrical container located behind the instrument panel and is accessible after removing the screen rail (see page P.52) and lowering the instrument panel (page P.52).

The automatic operation of the flasher lamps is controlled by means of a switch, contained in the flasher unit, being operated automatically by the alternative heating and cooling of an actuating wire; also incorporated is a small relay to flash the indicator warning lights when the system is functioning correctly. Failure of either of these lights to flash will indicate a fault.

In the event of trouble occurring the following procedure should be followed:

(i) Check bulbs for broken filaments.

(ii) Refer to the wiring diagram and check all flasher circuit connections.

(iii) Switch on the ignition and check with a voltmeter that flasher unit terminal 'B' is at 12 volts, with respect to earth.

(iv) Connect together flasher unit terminals 'B' and 'L' and operate the direction indicator switch. If the flasher lamps now light the flasher unit is defective and must be replaced.

(v) If after the above checks the bulb still does not light a fault is indicated in the manual flasher switch on the steering column which is best checked by substitution.

Note: It is important that only bulbs of the correct wattage rating (that is, 21 watts) are used in the flasher lamps.

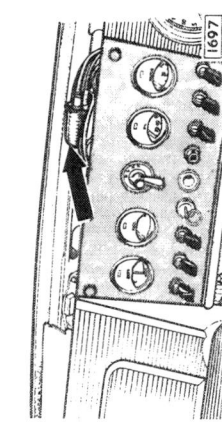

Fig. 7. Showing position of flasher unit behind instrument panel.

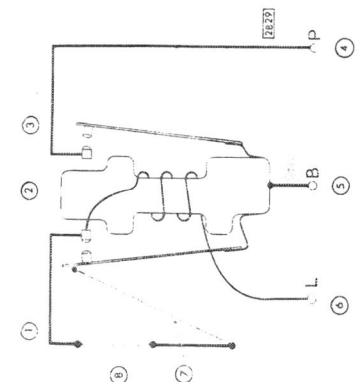

Fig. 8. Flasher unit circuit diagram.

FUSE UNITS

Four Lucas Model 4 FJ fuse units, each carrying two live glass cartridge type fuses (1, 3, Fig. 9) and two spares (4, Fig. 9), are incorporated in the electrical system and are located behind the instrument panel.

Access to the fuses is obtained by removing the two instrument panel retaining screws (top left-hand and top right-hand corners). Withdraw the picnic tray to the full extent and locate the two spring steel retaining clips attached to the tray at the back edge. Press both clips inwards to the centre of the car and complete the withdrawal of the tray. The instrument panel will then hinge downwards exposing the fuses and the fuse indicator panel. The circuits controlled by individual fuses are shown on the indicator panel and it is essential that the blown fuse is replaced by one of the correct value.

Only one end of the spare fuses is visible and they are retained in position by a small spring clip. Always replace the spare fuse as soon as possible.

The fuse for the electrically heated backlight (optional extra) is contained in a plastic fuse holder located in a clip behind the instrument panel.

The in-line fuse for the traffic hazard warning system (U.S.A. market only) is retained in a fuse holder located under the sub-panel.

	Commencing Chassis Nos.	
	R.H. Drive	L.H. Drive
3-4 "S"	1B 1004	1B 25005
3-8 "S"	1B 50321	1B 75053

Commencing at the above chassis numbers, an 8 ampere fuse has been introduced in the intermediate speed hold switch circuit (automatic transmission) and the control switch circuit (overdrive transmission).

The fuse holder is located in a spring clip behind the side facia panel on the steering wheel side.

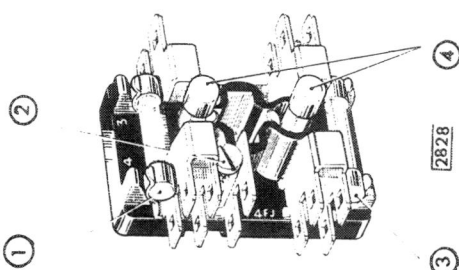

Fig. 9. The 4J fuse unit.

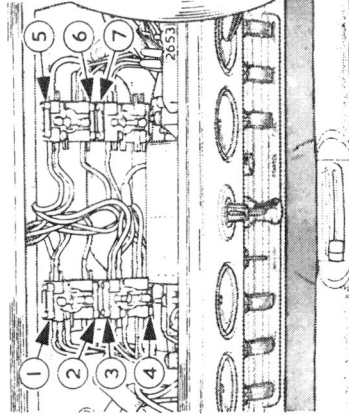

Fig. 10. Location of fuses.

Fuse No.	Circuits	Amps.
1	Headlamp (main beam)	35
2	Headlamp (dip beam)	35
3	Horn relay Screen washer—Stop lamps— Flashers —Reverse lamps— Overdrive solenoid or Anti-creep and Intermediate speed hold solenoids (Automatic Transmission)	35
4	Windscreen wipers Auxiliary starting carburetter, Fuel, Oil and Water gauges, Heater motor	35
5	Horns	50
6	Side, Tail, Panel and Number plate lamps	35
7	Headlamp flashers—Interior lamps— Cigar lighter	35
Separate fuses	Electrically heated backlight (optional extra)	15
	Traffic hazard warning (U.S.A. only)	35
	Overdrive or Automatic Intermediate Speed Hold	8

GENERATOR—MODEL C42

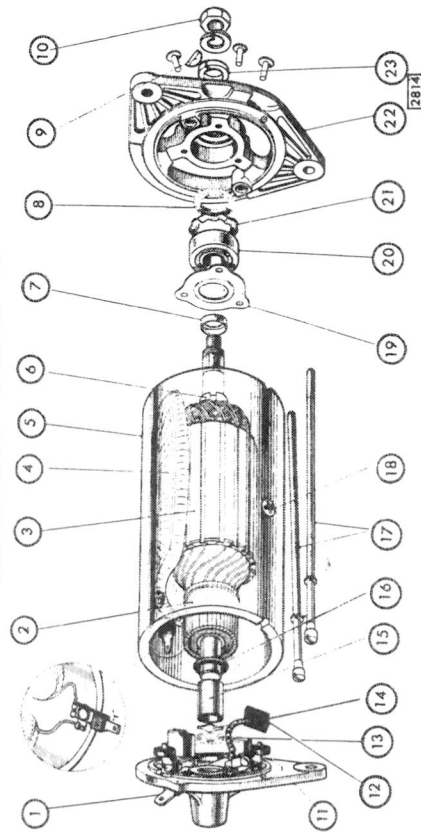

Fig. 11. *Exploded view of Model C42 generator.*

1.	Output terminal 'D'	11.	Commutator end bracket
2.	Commutator	12.	Brushes
3.	Armature	13.	Felt ring
4.	Field coils	14.	Felt ring retainer
5.	Yoke	15.	Porous bronze bush
6.	Shaft collar	16.	Fibre thrust washer
7.	Shaft collar retaining cup	17.	Through bolts
8.	Felt ring	18.	Pole shoe securing screws
9.	Shaft key	19.	Bearing retaining plate
10.	Shaft nut	20.	Ball bearing
		21.	Corrugated washer
		22.	Drive end bracket
		23.	Pulley spacer

GENERAL

The generator is a shunt-wound, two-pole, two-brush machine, arranged to work in conjunction with Lucas regulator unit model RB340. A fan, integral with the driving pulley, draws cooling air through the generator, inlet and outlet holes being provided in the end brackets of the unit.

The output of the generator is controlled by the regulator unit and is dependent on the state of charge of the battery and the loading of the electrical equipment in use. When the battery is in a low state of charge, the generator gives a high output, whereas if the battery is fully charged, the generator gives only sufficient output to keep the battery in good condition without any possibility of over-charging. An increase in output is given to balance the current taken by lamps and other accessories when in use.

ROUTINE MAINTENANCE

(a) Lubrication

Every 6,000 miles, inject a few drops of high quality viscosity (S.A.E. 30) engine oil into the hole marked "OIL" at the end of the C.E. bracket bearing housing.

(b) Inspection of Brushgear

Every 24,000 miles the generator should be removed from the engine and the brushgear checked as detailed in paragraph iv c, page P.18.

PERFORMANCE DATA

Cutting-in Speed	1,250 r.p.m. (max.) at 13·0 generator volts.
Maximum Output	30 amps at 2,200 r.p.m. (max.) at 13·5 generator volts.
Field Resistance	4·5 ohms.

REMOVAL

Disconnect the cables from the two terminals at the rear of the dynamo noting that they are of different sizes.

Remove the nut and bolt securing the adjusting link to the dynamo.

Remove the two nuts and bolts securing the dynamo to the mounting bracket when the dynamo can be lifted out.

Remove the generator belt by pushing the spring loaded jockey pulley inwards and lifting the belt over the generator pulley.

Note: If the car is fitted with power steering disconnect the hoses at the unions on the pump assembly (attached to the rear of the generator) and place the hoses in a raised position to prevent drainage of the oil. Alternatively, allow the oil to drain into a clean container.

REFITTING

Refitting is the reverse of the removal procedure. When replacing the generator belt, hold the spring loaded jockey pulley in towards the block and only release when the belt is sitting securely in the pulleys.

Note: If the car is fitted with power steering reconnect the hoses to the pump and bleed the system as described in Section I (Power-Assisted Steering).

SERVICING

(a) Testing in position to Locate Fault in Charging Circuit

In the event of a fault in the charging circuit, adopt the following procedure to locate the cause of the trouble.

(i) Check the connections on the commutator end bracket. The larger connector carries the main generator output, the smaller connector the field current.

(ii) Pull off the connectors from the terminal blades of the generator and connect the two terminal blades with a short length of wire.

(iii) Start the engine and set to run at normal idling speed.

(iv) Clip the negative lead of a moving coil type voltmeter, calibrated 0—20 volts, to one generator terminal and the positive lead to a good earthing point on the yoke.

(v) Gradually increase the engine speed, when the voltmeter reading should rise rapidly and without fluctuation. Do not allow the voltmeter reading to reach 20 volts and do not race the engine in an attempt to increase the voltage. It is sufficient to run the generator up to a speed of 1,000 r.p.m. If the voltage does not rise rapidly and without fluctuation the unit must be dismantled (see Paragraph iv b) for internal examination.

Excessive sparking at the commutator in the above test indicates a defective armature which should be replaced.

Note: If a radio suppression capacitor is fitted between the output terminal and earth, disconnect this capacitor and re-test the generator before dismantling. If a reading is now given on the voltmeter, the capacitor is defective and must be replaced.

If the generator is in good order, remove the link from between the terminals and restore the original connections.

(b) To Dismantle

(i) Take off the driving pulley.

(ii) Unscrew and withdraw the two through bolts.

(iii) Withdraw the commutator end bracket from the yoke.

(iv) Lift the driving end bracket and armature from the yoke. Take care not to lose the fibre thrust washer or collar from the commutator end of the shaft.

(v) The driving end bracket, which on removal from the yoke has withdrawn with it the armature and armature shaft ball bearing, need not be separated from the shaft unless the bearing is suspected and requires examination, or the armature is to be replaced; in this event the armature should be removed from the end bracket by means of a hand press, having first removed the shaft key.

(c) **Brushgear** (Checking with yoke removed)

(i) Lift the brushes up into the brush boxes and secure them in that position by positioning the brush spring at the side of the brush.

(ii) Fit the commutator end bracket over the commutator and release the brushes.

(iii) Hold back each of the brush springs and move the brush by pulling gently on its flexible connector. If the movement is sluggish, remove the brush from its holder and ease the sides by lightly polishing on a smooth file. Always refit brushes in their original positions. If the brushes are badly worn, new brushes must be fitted and bedded to the commutator. The minimum permissible length of brush is ¼" (6·4 mm).

(iv) Test the brush spring tension utilizing a spring balance. The tension needed to just lift the spring from contact with the brush with a new spring and a new brush is 33 ozs. (936 grams) but with a brush worn to ¼" (6·4 mm) it may reduce to 16 ozs. (453·6 grams). Both pressures should be measured. Renew any brush spring when the tension falls below these values.

(d) **Commutator**

A commutator in good condition will be smooth and free from pits or burned spots. Clean the commutator with a petrol-moistened cloth. If this is ineffective carefully polish with a strip of fine glass paper while rotating the armature. To remedy a badly worn commutator, first rough turn the commutator and then undercut the insulator between the segments to a depth of 5/32" (8 mm). Finally, take a light skim with a very sharp (preferably diamond-tipped) tool. If a non-diamond tipped tool is used for machining, the commutator should be lightly polished with a very fine glass paper. Emery cloth must not be used on the commutator. Finally clean away any dust.

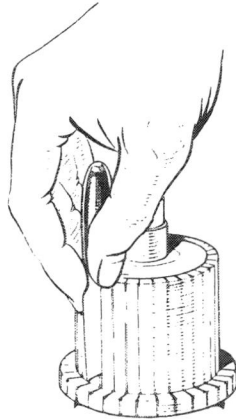

Fig. 12. *Undercutting the commutator insulation.*

(e) **Armature**

Indication of an open-circuited armature winding will be given by burnt commutator segments. If armature testing facilities are not available, an armature can be checked by substitution. To separate the armature shaft from the drive end bracket, press the shaft out of the drive end bracket bearing. When fitting the new armature, support the inner journal of the ball bearing, using a mild steel tube of suitable diameter, whilst pressing the armature shaft firmly home (see also paragraph iv h, Page P21).

Fig. 13. *Showing the correct and incorrect way of undercutting the commutator insulation.*

A — Correct way   B — Incorrect way

(f) **Field Coils**

Measure the resistance of the field coils, without removing them from the generator yoke, by means of an ohm meter connected between the field terminal and the yoke. Field resistance is 4·5 ohms.

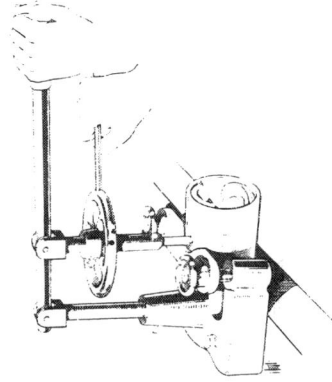

Fig. 14. *Tightening the pole shoe retaining screws.*

If an ohm meter is not available, connect a 12 volt d.c. supply between the field terminal and generator yoke with an ammeter in series. The ammeter reading should be approximately 2·7 amperes. Zero reading on the ammeter or an "Infinity" ohm meter indicates an open circuit in the field winding. If the current reading is much more than 2·7 amperes, or the ohm meter reading much below 4·5 ohms, it is an indication that the insulation of one of the field coils has broken down.

In either event, unless a substitute generator is available, the field coils must be replaced. To do this, carry out the procedure outlined below:

(i) Drill out the rivet securing the field coil terminal assembly to the yoke and remove the insulating sleeve from the terminal block to protect it from the heat of soldering. Unsolder the terminal blade and earthing eyelet.

(ii) Remove the insulation piece which is provided to prevent the junction of the field coils from contacting with the yoke.

(iii) Mark the yoke and pole shoes so that the latter can be refitted in their original positions.

(iv) Unscrew the two pole shoe retaining screws by means of a wheel-operated screwdriver.

(v) Draw the pole shoes and coils out of the yoke and lift off the coils.

(vi) Fit the new field coils over the pole shoes and place them in position inside the yoke. Take care to ensure that the taping of the field coils is not trapped between the pole shoes and the yoke.

(vii) Locate the pole shoes and field coils by lightly tightening the fixing screws.

(viii) Fully tighten the screws by means of the wheel-operated screwdriver.

(ix) Solder the original terminal blade and earthing eyelet to the appropriate coil ends.

(x) Refit the insulating sleeve and re-rivet the terminal assembly to the yoke.

(xi) Refit the insulation piece behind the junction of the two coils.

(g) Bearings

Bearings which are worn to such an extent that they will allow side movement of the armature shaft must be replaced.

To replace the bearing bush in a commutator end bracket, proceed as follows:

(i) Remove the old bearing bush from the end bracket. The bearing can be withdrawn with a suitable extractor or by screwing a $\frac{5}{8}$" tap into the bush for a few turns and pulling out the bush with the tap. Screw the tap squarely into the bush to avoid damage to the bracket.

(ii) Withdraw and clean the felt retainer and felt ring.

(iii) Insert the felt ring and ring retainer in the bearing housing, then press the new bearing bush into the end bracket, using a shouldered, highly polished mandrel of the same diameter as the shaft which is to fit in the bearing, until the visible end of the bearing is flush with the inner face of the bracket.

Porous bronze bushes must not be opened out after fitting, or the porosity of the bush may be impaired.

Note: Before fitting the new bearing bush, it should be allowed to stand for 24 hours completely immersed in a good grade S.A.E. 30 engine oil; this will allow the pores of the bush to be filled with lubricant. The ball bearing at the driving end is replaced as follows:

(i) Drill out the rivets which secure the bearing retaining plate to the end bracket and remove the plate.

(ii) Press the bearing out of the end bracket. Remove and clean the corrugated washer and felt ring.

(iii) Before fitting the replacement bearing, see that it is clean and pack it with high melting point grease such as Energrease RBB3.

(iv) Place the felt ring and corrugated washer in the bearing housing in the end bracket.

(v) Locate the bearing in the housing and press it home.

(vi) Fit the bearing retaining plate. Insert the new rivets from the pulley side of the end bracket and open the rivets by means of a punch to secure the plate rigidly in position.

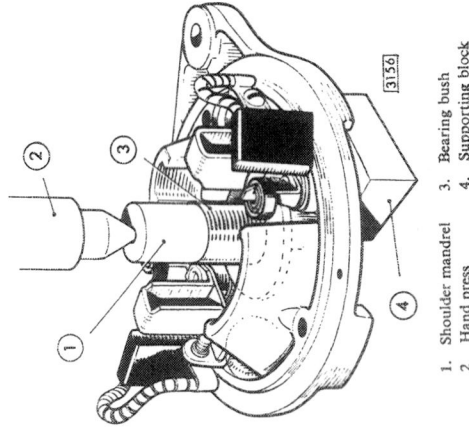

Fig. 15. *Method of fitting the porous bronze bush.*

1. Shoulder mandrel
2. Hand press
3. Bearing bush
4. Supporting block

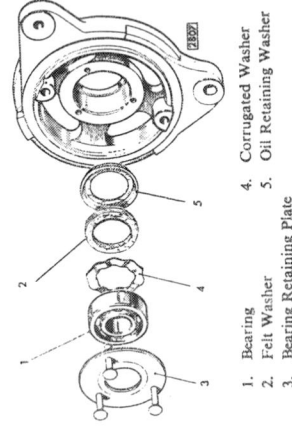

Fig. 16. *Exploded view of drive end bearing.*

1. Bearing
2. Felt Washer
3. Bearing Retaining Plate
4. Corrugated Washer
5. Oil Retaining Washer

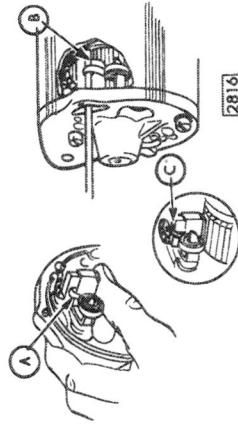

Fig. 17. *Lifting brush spring.*

(h) To Re-assemble

(i) Fit the drive end bracket to the armature shaft. The inner journal of the bearing must be supported by a tube, approximately 4" long $\frac{1}{8}$" thick and internal diameter $\frac{5}{8}$" (101·6 mm × 3·18 mm × 15·88 mm). Do not use the drive end bracket as a support for the bearing whilst fitting an armature.

(ii) Fit the yoke to the drive end bracket.

(iii) Lift the brushes up into the brush boxes and secure them in that position by positioning each brush spring at the side of its brush.

(iv) Fit the fibre thrust washer on the shaft. Fit the commutator end bracket to the yoke, so that the dowel on the bracket locates with the groove on the yoke. Take care not to trap the brush connector pigtails. Insert a thin screwdriver through the ventilator apertures adjacent to the brush boxes and carefully lever up the spring arms until the bushes locate correctly on the commutator.

(v) Refit the two through bolts, pulley spacer and shaft key.

(vi) After reassembly, lubricate the commutator end bearing (see Paragraph 2a).

GENERATOR—MODEL C48

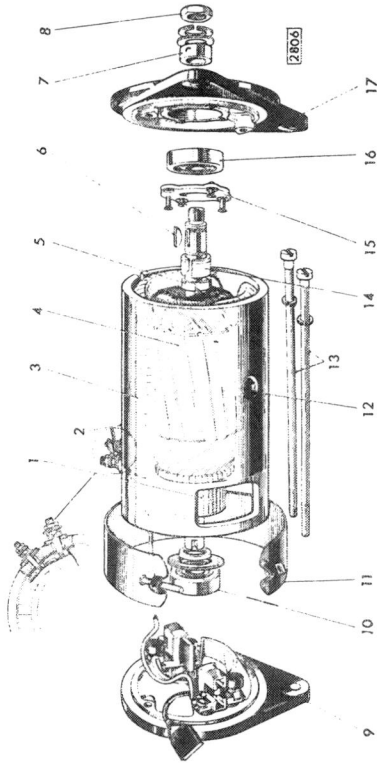

Fig. 18. *Exploded view of Model C48 generator.*

1. Commutator
2. Yoke
3. Field coils
4. Armature
5. Shaft collar
6. Shaft key
7. Pulley spacer
8. Shaft nut
9. Commutator end bracket
10. Ball bearing
11. Cover band
12. Pole shoe securing screws
13. Through bolts
14. Bearing spacer
15. Bearing retaining plate
16. Ball bearing
17. Drive end bracket

GENERAL

The C.48 generator fitted as standard equipment for certain countries and available as an optional extra for all others is similar in construction and operation to the C.42 generator described previously with the exception of the following items:

(i) Larger capacity, giving increased output.

(ii) Cover band fitted to provide access to the brush gear.

(iii) Ball races provided at each end of the generator.

At the commutator end the bearing outer race is a sliding fit in the commutator end bracket while the inner race is a press fit onto the armature shaft. The drive-end bearing is secured with a die-cast retaining plate and four countersunk screws, the end of the screws being caulked over on to the outer face of the bracket.

ROUTINE MAINTENANCE

Inspection of Brushgear

Every 24,000 miles (38,000 km.) the generator should be removed from the engine and the brushgear checked as described under "Servicing - Testing in position" paragraph vi.

PERFORMANCE DATA

Cutting-in speed ..	850 (max.) r.p.m. at 13 generator volts.
Maximum output..	35 amps. at 1,650 (max.) r.p.m. at 13·5 generator volts (on resistance load of 0·385 ohms).
Field resistance ..	6·0 ohms.

REMOVAL.

Proceed as detailed under Generator—Model C.42 (page P.17).

REFITTING

Proceed as detailed under Generator—Model C.42 (page P.17).

SERVICING

Testing in position to locate fault in charging circuit

In the event of a fault in the charging circuit, adopt the following procedure to locate the cause of trouble.

(i) Check that the generator and the control box are connected correctly. The larger generator terminal must be connected to control box terminal "D" and the smaller generator terminal to control box terminal "F".

(ii) Disconnect the cables from the generator terminals and connect the two terminals with a short length of wire.

(iii) Start the engine and set to run at normal idling speed.

(iv) Clip the negative lead of a moving-coil voltmeter, calibrated 0—20 volts, to one generator terminal and the other lead to a good earthing point on the yoke.

(v) Gradually increase the engine speed, when the voltmeter reading should rise rapidly and without fluctuation. Do not allow the voltmeter reading to reach 20 volts and do not race the engine in an attempt to increase the voltage. It is sufficient to run the generator up to a speed of 1,000 r.p.m. If the voltage does not rise rapidly and without fluctuation the unit must be dismantled for internal examination, see "To Dismantle." Excessive sparking at the commutator in the above test indicates a defective armature which should be replaced.

(vi) Remove the cover band and examine the brushes and commutator. Hold back each brush spring and move the brush by pulling gently on its flexible connector. If the movement is sluggish, remove the brush from its holder and ease the sides by gently polishing on a smooth file. Always refit a brush in its original position. Brushes must be replaced when worn to 11/32" (8·73 mm.) in length.

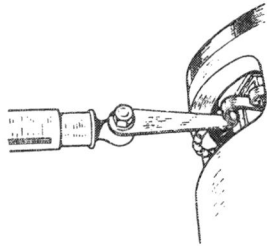

Fig. 19. *Testing the brush spring tension.*

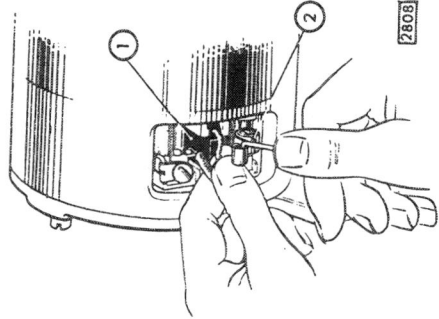

Fig. 20. *Checking the brush gear.*

Test the brush spring tension with a spring balance. The spring force on a new brush should be 25 oz. (720 grms.), and on a brush worn to its minimum length 16 oz. (460 grms.)—both values being measured radially to the commutator. Fit new springs if the tension is low.

If the commutator is blackened or dirty, clean it by holding a petrol-moistened cloth against it while the engine is turned slowly by hand. Re-test the generator as in (vi); if there is still no reading on the voltmeter, there is an internal fault and the complete unit, if a spare is available, should be replaced. Otherwise the unit must be dismantled (see "To Dismantle") for internal examination.

(vii) If the generator is in good order, remove the link from between the terminals and restore the original connections, taking care to connect the larger generator terminal to control box terminal "D" and the smaller generator terminal to control box terminal "F."

To Dismantle

(i) Take off the driving pulley.

(ii) Remove the cover band, hold back the brush springs and remove the brushes from their holders.

(iii) Unscrew and withdraw the two through bolts.

(iv) The commutator end bracket can be withdrawn from the generator yoke, after removing the "D" terminal.

(v) The driving end bracket together with the armature can now be lifted out of the yoke.

(vi) The driving end bracket need not be separated from the shaft unless the bearing is suspected and requires examination, or the armature is to be replaced; in this event the armature should be removed from the end bracket after removal of the bearing retainer plate screws.

Assembly is the reverse of the dismantling procedure.

Commutator

A commutator in good condition will be smooth and free from pits or burned spots. Clean the commutator with a petrol-moistened cloth. If this is ineffective, carefully polish with a strip of fine glass paper while rotating the armature. To remedy a badly worn commutator, mount the armature, with or without drive end bracket, in a lathe, rotate at high speed and take a light cut with a very sharp tool. Do not remove more metal than is necessary. Polish the commutator with very fine glass paper. Undercut the insulators between the segments to a depth of $\frac{5}{32}$" (8 mm.) with a hacksaw blade ground to the thickness of the insulator.

Armature

The testing of the armature winding requires the use of a volt-drop test and growler. If these are not available, the armature should be checked by substitution. No attempt should be made to machine the armature core or to true a distorted armature shaft.

To remove the armature shaft from the drive end bracket and bearing, support the bearing retaining plate firmly and press the shaft out of the drive end bracket. When fitting the new armature, support the inner journal of the ball bearing whilst pressing the armature shaft firmly home.

Field Coils

Measure the resistance of the field coils, without removing them from the generator yoke, by means of an ohm meter connected between the field terminal and yoke. The ohm meter should read 6·0 ohms. If an ohm meter is not available connect a 12 volt D.C. supply with an ammeter in series between the field terminal and generator yoke. The ammeter reading should be approximately 2 amperes.

No reading on the ammeter, or an infinite ohm meter reading, indicates an open circuit in the field winding. If the current reading is much more than 2 amperes, or the ohm meter reading much below 6 ohms, it is an indication that the insulation of one of the field coils has broken down.

In either case, unless a substitute generator is available, the field coils must be replaced. To do this, carry out the procedure outlined below, using a wheel-operated screwdriver.

(i) Remove the field coil terminal from the yoke, and unsolder the field coil connections.

(ii) Remove the insulation piece which is provided to prevent the junction of the field coils from contacting with the yoke.

(iii) Mark the yoke and pole shoes so that the latter can be refitted in their original positions.

(iv) Unscrew the two pole shoe retaining screws by means of the wheel-operated screwdriver.

(v) Draw the pole shoes and coils out of the yoke and lift off the coils.

(vi) Fit the new field coils over the pole shoes and place them in position inside the yoke. Take care to ensure that the taping of the field coils is not trapped between the pole shoes and the yoke.

(vii) Locate the pole shoes and field coils by lightly tightening the fixing screws.

(viii) Fully tighten the screws by means of the wheel-operated screwdriver.

(ix) Replace the insulation piece between the field coil connections and the yoke.

(x) Resolder the field coil connections to the field coil terminal and refit to the yoke.

Bearings

It is extremely unlikely that bearing wear, sufficient to necessitate replacement, will be found to occur during the normal life of the generator. If, however, such wear is found to have taken place, the bearing must be replaced.

The ball bearing at the driving end is replaced as follows:

(i) Withdraw the screws which secure the bearing retaining plate to the end bracket and remove the plate.

(ii) Press the bearing out of the end bracket.

(iii) Before fitting the replacement bearing see that it is clean and pack it with high melting point grease.

(iv) Locate the bearing in the housing and press it home.

(v) Refit the bearing retaining plate.

The ball bearing at the commutator end is replaced as follows:

Withdraw the bearing from the armature shaft by means of a hand press or extractor. The inner race of the replacement bearing must be supported during fitment.

CURRENT VOLTAGE REGULATOR—MODEL RB 340

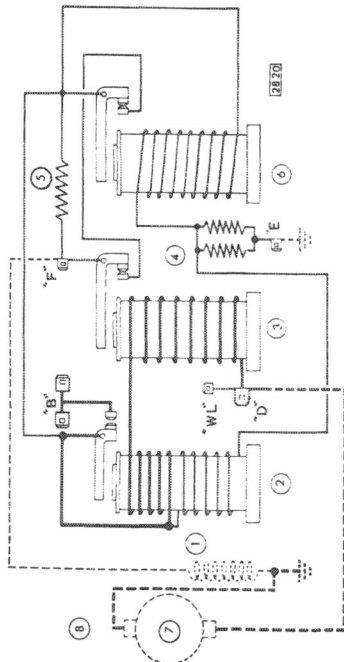

Fig. 21. Circuit diagram of the R.B.340 control box.

1. Field
2. Cut-out Relay
3. Current Regulator
4. Swamp Resistor
5. Field Resistor
6. Voltage Regulator
7. Armature
8. Generator

GENERAL

Preliminary Checking of Charging Circuit

Before disturbing any electrical adjustments, examine as described below to ensure that the fault does not lie outside the control box:

(i) Check the battery by substitution or with an hydrometer and a heavy discharge (150—160A) tester.

(ii) Inspect the generator driving belt. This should just be taut enough to drive without slipping.

(iii) Check the generator by substitution or by withdrawing the cables from the generator terminals and, using a suitable "jumper lead," linking large generator terminal "D" to small terminal "F" and connecting a voltmeter between this link and earth and then running the generator up to about 1,000 r.p.m. (600 engine r.p.m.), when a rising voltage should be shown.

(iv) Inspect the wiring of the charging circuit and carry out continuity tests between the generator, control box and the ammeter.

(v) Check earth connections, particularly that of the control box.

(vi) In the event of reported undercharging, ascertain that this is not due to low mileage.

(vii) The control box terminals are protected with a plastic cover which is a sliding fit over the unit and must be removed before any adjustments can be made.

To ensure the impossibility of making incorrect connections the field cable ter-

minal is insulated with a black plastic sleeve which differs in form and colour from all other terminal insulators.

Note: Should the control box fail to respond correctly to any adjustment given in the following instructions, it should be examined at a Lucas Service Depot or by an official Lucas Agent.

VOLTAGE REGULATOR
Open Circuit Settings

	Ambient Temperature		Voltage
C.42 Generator	10°C.	(50°F.)	14·9—15·5
	20°C.	(68°F.)	14·7—15·3
	30°C.	(86°F.)	14·5—15·1
	40°C.	(104°F.)	14·3—14·9
C.48 Generator	10 C.	(50 F.)	15·0—15·6
	20 C.	(68 F.)	14·8—15·4
	30 C.	(86 F.)	14·6—15·2
	40°C.	(104 F.)	14·4—15·0

Method of Adjustment

Checking and adjusting should be completed as rapidly as possible to avoid errors due to heating of the operating coil.

(i) Withdraw the cable from control box terminal blades "B."

(ii) Connect a first-grade 0—20 moving coil voltmeter between control box terminal "D" and a good earthing point.

Note: A convenient method of making this connection is to withdraw the ignition warning light feed from control box terminal "WL" and to clip the voltmeter lead of appropriate polarity to the small terminal blade thus exposed—this terminal being electrically common with terminal "D".

(iii) Start the engine and run the generator at 3,000 r.p.m. (1,800 engine r.p.m.) C.48 generator, 4,500 r.p.m. (2,700 engine r.p.m.) C.42 generator.

(iv) Observe the voltmeter pointer.
The voltmeter reading should be steady and lie between the appropriate limits (see "Open Circuit Settings"), according to the temperature. An unsteady reading may be due to unclean contacts. If the reading is steady but occurs outside the appropriate limits, an adjustment must be made. In this event, continue as follows:

(v) Stop the engine and remove the control box cover.

(vi) Re-start the engine and run the generator at 3,000 r.p.m. (1,800 engine r.p.m.) C.48 generator, 4,500 r.p.m. (2,700 engine r.p.m.) C.42 generator.

(vii) Using a suitable tool, turn the voltage adjustment cam until the correct setting is obtained—turning the tool clockwise to raise the setting or anti-clockwise to lower it.

(viii) Check the setting by stopping the engine and then again raising the generator speed to 3,000 r.p.m. (1,800 engine r.p.m.) or 4,500 r.p.m. (2,700 engine r.p.m.).

(ix) Restore the original connections and refit the cover.

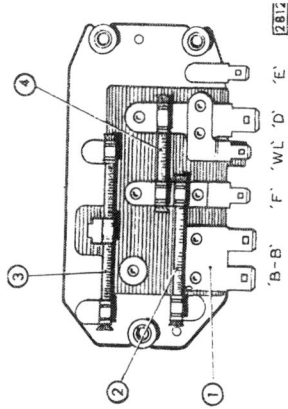

Fig. 22. View of underside of R.B.340 control box.

1. Battery main terminal
2. Swamp resistor
3. Ballast resistors
4. Field resistors

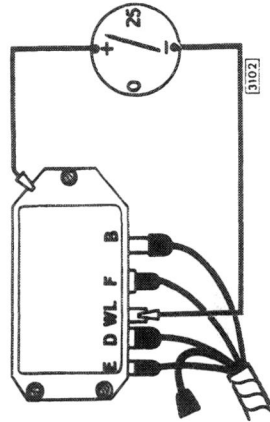

Fig. 23. Checking open circuit setting.

CURRENT REGULATOR

On-Load Setting

The current regulator on-load setting is equal to the maximum rated output of the generator, which is 30 amperes (C.42), 35 amperes (C.48).

Method of Adjustment

The generator must be made to develop its maximum rated output, whatever the state of charge of the battery might be at the time of setting. The voltage regulator must therefore be rendered inoperative, and this is the function of the bulldog clip used in (ii) below in keeping the voltage regulator contacts together.

(i) Remove the control box cover.

(ii) Using a bulldog clip, short out the contacts of the voltage regulator. (see Fig 24).

(iii) Withdraw the cable from control box terminal blades "B".

(iv) Using a suitable "jumper lead", connect the cables removed in (iii) to the load side of a first-grade 0-40A moving coil ammeter.

(v) Connect the other side of the ammeter to one of the control box terminal blades "B".

Note: It is important to ensure that terminal "B" carries only this one connection. All other load connections (including the ignition coil feed) must be made to the battery side of the ammeter.

(vi) Switch on all lights, to ensure that the generator develops its full rated output.

(vii) Start the engine and run the generator at 4,500 r.p.m. (2,700 engine r.p.m.), C.42 generator, 3,000 r.p.m. (1,800 engine r.p.m.), C.48 generator.

(viii) Observe the ammeter pointer.

The ammeter pointer should be steady and indicate a current equal to the maximum rated output of the generator. An unsteady reading (one fluctuating more than $\pm$ 1 ampere) may be due to unclean contacts. If the reading is too high or too low an adjustment must be made. In this event proceed as follows:

(ix) Using a suitable tool, turn the current adjustment cam until the correct setting is obtained—turning the tool clockwise to raise the setting or anti-clockwise to lower it.

(x) Switch off the engine and restore the original connections.

(xi) Refit the control box cover.

1. "Bulldog" clip
2. V.R. Contacts
3. Voltage Regulator
4. Current Regulator
5. Cut-out Relay
6. Setting Tool
7. Adjustment Cams

Fig. 24. The cam adjuster on the R.B.340 control box. Note the bulldog clip closing the voltage regulator contacts.

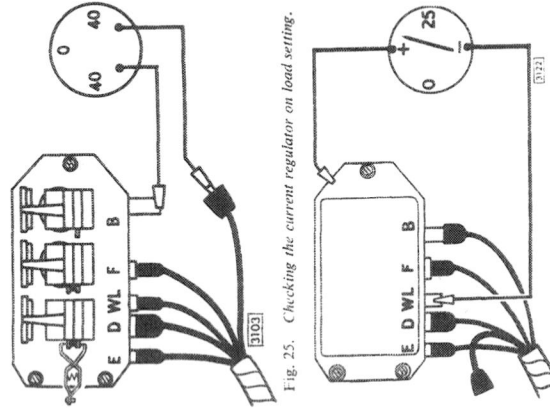

Fig. 25. Checking the current regulator on load setting.

Fig. 26. Checking cut-in voltage.

(iv) Slowly decelerate and observe the voltmeter pointer.

Opening of the contacts, indicated by the voltmeter pointer dropping to zero, should occur between the limits given in "Electrical Settings."

(ii) If the drop-off occurs outside these limits, an adjustment must be made. In this event, continue as follows:

(v) Stop the engine and remove the control box cover.

(vi) Adjust the drop-off voltage by carefully bending the fixed contact bracket. Reducing the contact gap will raise the drop-off voltage; increasing the gap will lower the drop-off voltage.

Retest and if necessary, re-adjust until the correct drop of setting is obtained.

Note: This should result in a contact "follow through" or blade deflection of 0·010"—0·035" (0·25—0·80 mm.).

Refit the connections and cover.

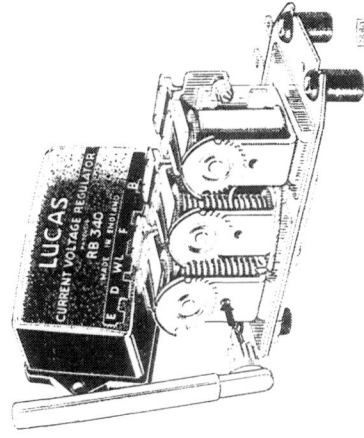

Fig. 27. Adjusting the cut-out.

CUT-OUT RELAY

Electrical Settings

(i) Cut-in Voltage   12·6—13·4

(ii) Drop-off Voltage   9·3—11·2

Method of Cut-in Adjustment

Checking and adjusting should be completed as rapidly as possible to avoid errors due to heating of the operating coil.

(i) Connect a first-grade 0—20 moving-coil voltmeter between control box terminal "D" and a good earthing point, referring to the note in "Voltage Regulator—Method of Adjustment."

(ii) Switch on an electrical load, such as the headlamps.

(iii) Start the engine and slowly increase its speed.

(iv) Observe the voltmeter pointer.

The voltage should rise steadily and then drop slightly at the instant of contact closure. The cut-in voltage is that which is indicated immediately before the pointer drops back. It should occur between the limits given in "Electrical Settings." (i) above. If the cut-in occurs outside those limits, an adjustment must be made. In this event, reduce generator speed to below cut-in value and continue as follows:

(v) Remove the control box cover.

(vi) Using a suitable tool, turn the cut-out relay adjustment cam a small amount in the appropriate direction—turning the tool clockwise to raise the setting or anti-clockwise to lower it.

(vii) Repeat the above checking procedure until the correct setting is obtained.

(viii) Switch off the engine, restore the original connections and refit the cover.

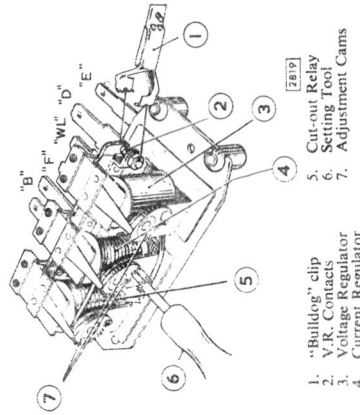

Method of Drop-off Adjustment

(i) Withdraw the cables from control box terminal blades "B".

(ii) Connect a first-grade 0—20 moving-coil voltmeter between control box terminal "B" and earth.

(iii) Start the engine and run up to approximately 3,000 generator r.p.m. (1,800 engine r.p.m.).

ADJUSTMENT OF AIR GAP SETTINGS

Air gap settings are accurately adjusted during production of the control box and should require no further attention. If the original adjustments have been disturbed, it will be necessary to reset as described below.

Armature-to-Bobbin Core Gaps of Voltage and Current Regulators

(i) Using a suitable tool, turn the adjustment cam to the point giving maximum lift to the armature tensioning spring, i.e. by turning the tool to the fullest extent anti-clockwise.

(ii) Slacken the adjustable contact locking nut and screw back the adjustment contact.

(iii) Insert a flat feeler gauge of 0·045" (1·04 mm.) thickness between the armature and the copper separation on the core face, taking care not to turn up or damage the copper shim. The gauge should be inserted as far back as the two rivet heads on the underside of the armature.

(iv) Retaining the gauge in position and pressing squarely down on the armature, screw in the adjustable contact until it just touches the armature contact.

(v) Retighten the locking nut and withdraw the gauge.

(vi) Carry out the electrical setting procedure.

Contact "follow-through" and Armature-to-Bobbin Core Gap of Cut-out Relay

(i) Press the armature squarely down against the copper separation on the core face.

(ii) Adjust the fixed contact bracket to give a "follow-through" or blade deflection of the moving contact of 0·010"—0·035" (0·25—0·89 mm.).

(iii) Release the armature.

(iv) Adjust the armature back stop to give a core gap of 0·035"—0·045" (·89—1·13mm.).

(v) Check the cut-in and drop-off voltage settings.

CLEANING CONTACTS

Regulator Contacts

To clean the voltage or current regulator contacts, use fine carborundum stone or silicon carbide paper followed by methylated spirits (denatured alcohol).

Cut-out Relay Contacts

To clean the cut-out relay contacts, use a strip of fine glass paper—never carborundum stone or emery cloth.

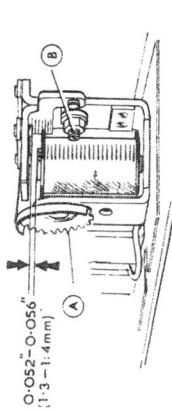

0·052"—0·056" (1·3—1·4 mm.)

A. Turn cam to minimum lift. B. Slacken contact screw.

Fig. 28. Voltage regulator gap setting.

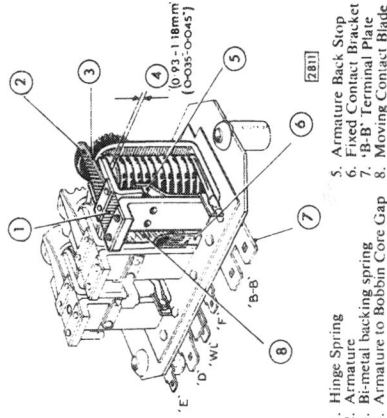

(0·93 - 1·18mm) (0·035—0·0045")

1. Hinge Spring
2. Armature
3. Bi-metal backing spring
4. Armature to Bobbin Core Gap
5. Armature Back Stop
6. Fixed Contact Bracket
7. 'B-B' Terminal Plate
8. Moving Contact Blade

Fig. 29. Cut-out air gap setting.

STARTER MOTOR

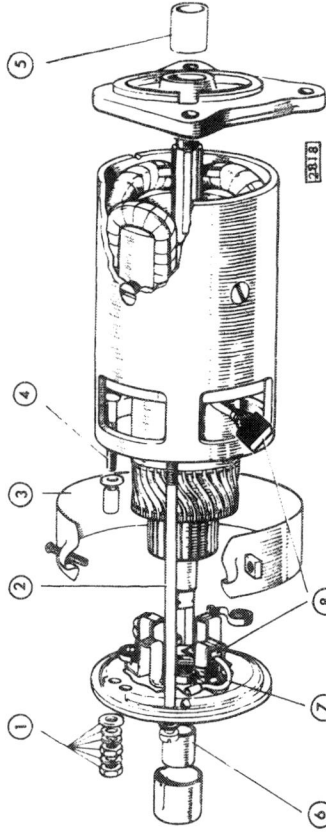

Fig. 30. Exploded view of the starter motor.

1. Terminal nuts and washers
2. Through bolt
3. Cover band
4. Terminal post
5. Bearing bush
6. Bearing bush
7. Brush spring
8. Brushes

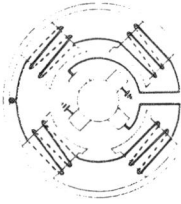

Fig. 31. Internal connections of the starter motor.

PERFORMANCE DATA

Model	M 45 G
Lock Torque	22 lb/ft. with 430-450 amperes at 7·8-7·4 volts. (3·04 kg/m.)
Torque at 1,000 r.p.m.	8·3 lb/ft. with 200-220 amperes at 10·2-9·8 volts. (1·44 kg/m.)
Light running current	45 amperes at 5,800-6,800 r.p.m.

GENERAL

The electric starter motor is a four-pole, four-brush machine having an extended shaft which carries the engine engagement gear, or starter drive as it is more usually named. The diameter of the yoke is 4·5" (11·4 cm.).

The starter motor is of similar construction to the generator except that heavier copper wire is used in the construction of the armature and field coils. The field coils are series parallel connected between the field terminal and the insulated pair of brushes.

REMOVAL.

The starter motor can only be removed from beneath the car.

Detach the earth lead from the battery. Disconnect the cable from the terminal at the end of the starter motor.

Remove the two nuts from the rear ends of the starter motor securing strap. Support starter motor from below by hand and withdraw the strap.

The top bolt is accessible from above the car.

Withdraw starter motor.

Note: On cars fitted with automatic transmission it will be necessary to disconnect the filler tube extension at the rubber joint and blank off the end of the tube to prevent the oil escaping. Release the gland union on the gearbox and swing the tube away.

REFITTING

Refitting is the reverse of the removal procedure.

ROUTINE MAINTENANCE

The only maintenance normally required by the starter motor is the occasional checking of brush-gear and commutator. About every 10,000 miles, remove the metal band cover. Check that the brushes move freely in their holders by holding back the brush springs and pulling gently on the flexible connectors. If a brush is inclined to stick, remove it from its holder and clean its sides with a petrol-moistened cloth. Be careful to replace brushes in their original positions in order to retain "bedding." Brushes which have worn so that they will not "bed" properly on the commutator or have worn less than 5/16" (7.9 mm.) in length must be renewed.

The commutator should be clean, free from oil or dirt and should have a polished appearance. If it is dirty, clean it by pressing a fine dry cloth against it while the starter is turned by hand by means of a spanner applied to the squared extension of the shaft. Access to the squared shaft is gained by removing the thimble-shaped metal cover. If the commutator is very dirty moisten the cloth with petrol.

SERVICING

Testing in Position

Check that the battery is fully charged and terminals are clean and tight. Recharge if necessary.

(i) Switch on the lamps and operate the starter control. If the lights go dim, but the starter motor is not heard to operate, an indication is given that the current is flowing through the starter motor windings but that the armature is not rotating for some reason; possibly the pinion is meshing permanently with the geared ring on the flywheel. In this case the starter motor must be removed from the engine for examination.

(ii) Should the lamps retain their full brilliance when the starter switch is operated, check the circuit for continuity from battery to starter motor via the starter switch, and examine the connections at these units. If the supply voltage is found to be applied to the starter motor when the switch is operated, an internal fault in the motor is indicated and the unit must be removed from the engine for examination.

(iii) Sluggish to slow action of the starter motor is usually due to a loose connection causing a high resistance in the motor circuit. Check as described above.

(iv) If the motor is heard to operate, but does not crank the engine, indication is given of damage to the drive.

Bench Testing and Examination of Brushgear and Commutator

(i) Remove the starter motor from the engine, as described on this page.

(ii) After removing the starter motor from the engine secure the body in a vice and test by connecting it with heavy gauge cables to a battery of the appropriate voltage. One cable must be connected to the starter terminal and the other held against the body or end bracket. Under these light road conditions, the starter should run at a very high speed without excessive noise and without excessive sparking at the commutator.

(iii) If the operation of the starter motor is unsatisfactory, remove the cover band and examine the brushes and commutator. Hold back each of the brush springs and move the brush gently on its flexible connector. If the movement is sluggish, remove the brush from its holder and ease the sides by lightly polishing on a smooth file. Always replace brushes in their original positions. If the brushes are worn so that they will not bear on the commutator, or if the brush flexible is exposed on the running face, they must be replaced.

Check the tension of the brush springs with a spring scale. The correct tension is 30—40 oz. (850·5 gms.—1134 gms.). New springs should be fitted if the tension is low.

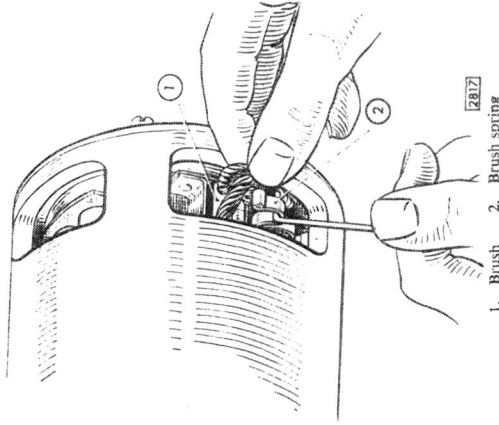

1. Brush   2. Brush spring

Fig. 32.   Checking the brush gear.

(i) If the commutator is blackened or dirty, clean it by holding a petrol-moistened cloth against it while the armature is rotated.

(iv) Re-test the starter as described under (ii). If the operation is still unsatisfactory, the unit can be dismantled for detailed inspection and testing as follows:

TO DISMANTLE (Fig. 30)

(i) Remove the cover band, hold back the brush springs and lift the brushes from their holders.

(ii) Remove the nuts from the terminal post which protrudes from the commutator end bracket.

(iii) Unscrew the two through bolts from the commutator end bracket. Remove the commutator end bracket from the yoke.

(iv) Remove the driving end bracket complete with armature and drive from the starter motor yoke. If it is necessary to remove the armature from the driving end bracket, it can be done by means of a hand press after the drive has been dismantled.

Commutator

A commutator in good condition will be smooth and free from pits and burned spots. Clean the commutator with a petrol-moistened cloth. If this is ineffective, carefully polish with a strip of fine glass paper, while rotating the armature. To remedy a badly worn commutator, dismantle the starter drive and remove the armature from the end bracket. Now mount the armature in a lathe, rotate at a high speed and take a light cut with a very sharp tool. Do not remove any more metal than is necessary. Finally polish with a very fine glass paper.

The insulators between the commutator segments MUST NOT BE UNDERCUT.

Armature

Examination of the armature may reveal the cause of failure, e.g., conductors lifted from the commutator due to the starter motor being engaged while the engine is running and causing the armature to be rotated at an excessive speed. A damaged armature must always be replaced — no attempts should be made to machine the armature core or to true a distorted armature shaft.

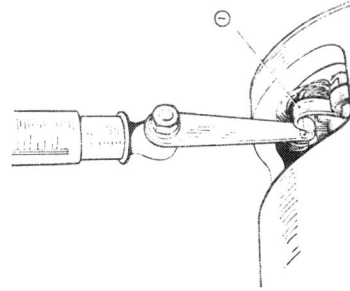

1. Brush spring

Fig. 33.   Testing the brush spring tension.

Take care to ensure that the taping of the field coils is not trapped between the pole shoes and the yoke.

Locate the pole shoes and field coils by lightly tightening the fixing screw. Fully tighten the screws with the wheel-operated screwdriver. Replace the insulation piece between the field coil connections and the yoke.

Bearings

Bearings which are worn to such an extent that they will allow excessive side-play of the armature shaft must be replaced. To replace the bearing bushes proceed as follows:

(i) Press the bearing bush out of the end bracket.

(ii) Press the new bearing bush into the end bracket using a shouldered, highly polished mandrel of the same diameter as the shaft which is to fit in the bearing. Porous bronze bushes must not be opened out after fitting, or the porosity of the bush may be impaired.

Note: Before fitting a new porous bronze bearing bush it must be completely immersed for 24 hours in clean thin engine oil.

REASSEMBLY

The reassembly of the starter motor is a reversal of the dismantling procedure.

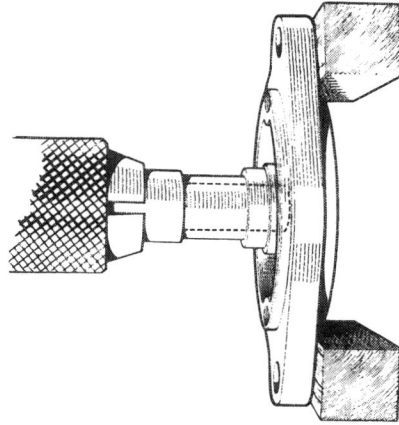

Fig. 35. *Method of fitting bush.*

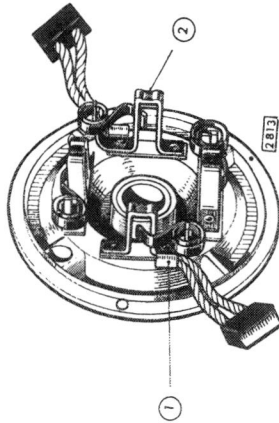

1. Terminal Eyelet   2. Brush Boxes

Fig. 34. *Commutator end bracket connections.*

Replacement of Brushes

If the brushes are worn to less than $\frac{5}{16}$" (7.9 mm.) in length, they must be replaced.

Two of the brushes are connected to terminal eyelets attached to the brush boxes on the commutator end bracket and two are connected to the field coils.

The flexible connectors must be renewed by unsoldering and the connectors of the new brushes secured in their place by soldering. The new brushes are preformed so that the bedding to the commutator is unnecessary.

Field Coils

(i) Test the field coils for continuity by connecting a 12-volt test lamp between the starter motor terminal and to each field brush in turn.

(ii) Lighting of the lamp does not necessarily mean that the field coils are in order, as it is possible that one of them may be earthed to a pole-shoe or to the yoke. This may be checked with a 110-volt test lamp, the test leads being connected between the starter motor terminal and a clean part of the yoke. If the lamp lights, defective insulation of the field coils or of the terminal post is indicated. In this event, see that the insulating band is in position and examine the field coils and terminal connections for any obvious point of contact with the yoke. If from the above tests the coils are shown to be open-circuited or earthed and the point of contact cannot be readily located and rectified, either the complete starter motor or the field coils must be replaced. If the field coils are to be replaced, follow the procedure outlined below, using a wheel-operated screwdriver.

Remove the insulation piece which is provided to prevent the intercoil connectors from contacting with the yoke.

Mark the yoke and pole shoes so that the latter can be refitted in their original positions. Unscrew the four-pole shoe retaining screws with the wheel-operated screwdriver.

Draw the pole shoes and coils out of the yoke and lift off the coils. Fit the new field coils over the pole shoes and place them in position inside the yoke.

STARTER DRIVE

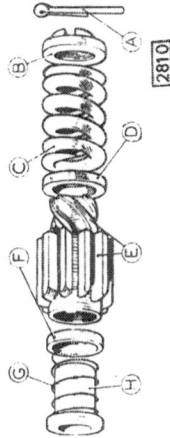

Fig. 36. *Exploded view of the starter drive assembly.*

GENERAL

The pinion is mounted on a threaded sleeve which is carried on splines on the armature shaft, the sleeve being arranged so that it can move along the shaft against a compression spring so as to reduce the shock loading at the moment engagement takes place.

When the starter switch is operated, the shaft and screwed sleeve rotate, and owing to the inertia of the pinion the screwed sleeve turns inside the pinion causing the latter to move along the sleeve into engagement with the flywheel ring. The starter will then turn the engine.

As soon as the engine fires and commences to run under its own power, the flywheel will be driven faster by the engine than by the starter. This will cause the pinion to be screwed back along the sleeve and so thrown out of mesh with the flywheel teeth. In this manner the drive safeguards the starter against damage due to being driven at high speeds by the engine.

A pinion restraining spring is fitted over the starter shaft to prevent the pinion being vibrated into contact with the flywheel when the engine is running.

ROUTINE MAINTENANCE

If any difficulty is experienced with the starting motor not meshing correctly with the flywheel, it may be that the drive requires cleaning. The pinion should move freely on the screwed sleeve; if there is any dirt or other foreign matter on the sleeve it must be washed off with paraffin.

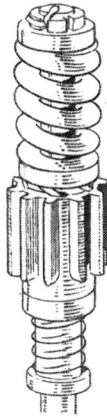

Fig. 37. *Showing the starter drive assembled.*

In the event of the pinion becoming jammed in mesh with the flywheel, it can usually be freed by turning the starter motor armature by means of a spanner applied to the shaft extension at the commutator end.

This is accessible by removing the cap which is a push fit.

DISMANTLING AND REASSEMBLY

Having removed the armature as described in the section dealing with starter motors the drive can be dismantled as follows:

Remove the split pin (A) from the shaft nut (B) at the end of the starter drive. Hold the squared starter shaft extension at the commutator end by means of a spanner and unscrew shaft nut (B). Lift off the main spring (C), washer (D), screwed sleeve with pinion (E), collar (F), pinion restraining spring (G) and restraining spring sleeve (H).

Note: If either the screwed sleeve or pinion are worn or damaged they must be replaced as a pair, not separately.

The reassembly of the drive is a reversal of the dismantling procedure.

LAMPS

LAMP BULBS

LAMP	LUCAS BULB No.	VOLTS	WATTS	APPLICATION
Head	Sealed Beam Unit	12	75/45†	Home and R.H.D. Export Middle East, S. America and U.S.A.
	410	12	60/45* 50/40	
	411	12	45/40	Belgium, Holland, Sweden, Austria and Italy
			45/40 (Yellow)	France
Side	989	12	6W	
Front and Rear Flashing Indicators	382	12	21W	
Rear/Brake	380	12	21/6W	
Number plate and luggage compartment illumination	989	12	6W	
Reversing light	382	12	21W	
Fog lights	323 Phillips No. 683	12 12	48W 48W	Not fitted on cars for U.S.A. Switzerland only
Interior light	254	12	6W	
Glovebox illumination	254	12	6W	
Map light	989	12	6W	
Instrument illumination: Headlamp warning light Ignition warning light Handbrake/Brake Fluid warning light Side lamp warning light	987	12	2.2W	
Switch indicator strip Flasher indicator warning light Automatic Transmission indicator light Overdrive indicator light	281	12	2W	Italy only

* Early cars
† Later cars

HEADLIGHTS

Description

The car is fitted with the two headlight system, the standard light units fitted being of the sealed beam type having aiming pads mounted into the lenses. These pads are for use with an approved mechanical aimer (such as the Lucas Lev-L-Lite).

Headlamps for certain continental countries are fitted with pre-focus bulbs replacing the sealed beam units.

To obtain the best possible results from the headlights, it is essential that they are correctly adjusted. The alignment of the headlight beam is set correctly before the car leaves the factory but if for any reason adjustment becomes necessary and an approved beam setter is not available the following procedure should be carried out.

HEADLIGHT BEAM SETTING

Adjusting the Headlight Beam

Place the car on a level surface in front of a garage door or wall; the car should be at least 25 feet (7·6 m.) away and square to the door or wall. Carry out the work with conditions as dark as possible so that the oval shaped light areas can be clearly seen.

With the headlamps in the full beam positions, that is, not dipped, the beams from the two headlamps should be parallel with the ground and with each other; measurements should be taken from the centres of the headlamps and the horizontal and vertical axes of the oval light areas.

If adjustment is required, remove the headlamp rim by unscrewing the retaining screw. Switch on the headlamps and check that the beams are not in the dipped position.

The setting of the beams is adjusted by two screws, one being located at the top centre and the other one centre right-hand side. The top screw is for vertical adjustment, that is, to raise or lower the beam; turn the screw anti-clockwise to lower the beam and clockwise to raise the beam.

The side screw is for side adjustment, that is, to turn the beam to the right or left. To move the beam to the right turn the screw clockwise; to move the beam to the left turn the screw anti-clockwise.

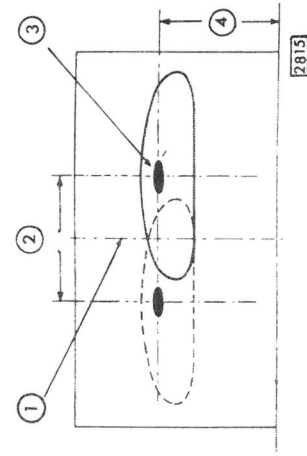

Fig. 38. Headlight beam setting.

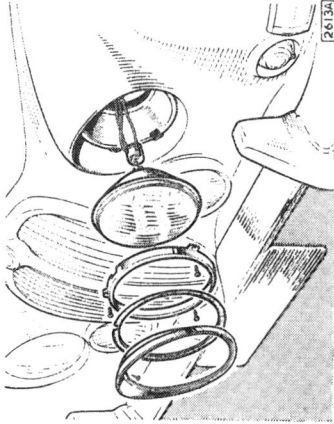

Fig. 39. The headlamp beam setting screws.

Headlight Unit—Replacement

Remove the top retaining screw and withdraw the headlight embellisher, noting the two retaining lugs at the lower edge. Remove the three cross-headed screws and the headlight unit retaining rim. Withdraw the headlight and unplug the adaptor from the rear of the unit, the headlight can now be replaced with a sealed beam unit of the correct type (see "Light Bulbs" page xi).

On cars fitted with non-sealed beam headlights, proceed as described above until the headlight unit is withdrawn, release the bulb retaining spring clips and withdraw the bulb. Replace with a bulb of the correct type (see page xi). When re-assembling note that a groove in the bulb plate must register with a raised portion on the bulb retainer.

Note: Do not turn the two slotted screws or the setting of the headlight will be upset.

Fig. 40. Headlight sealed beam unit removed.

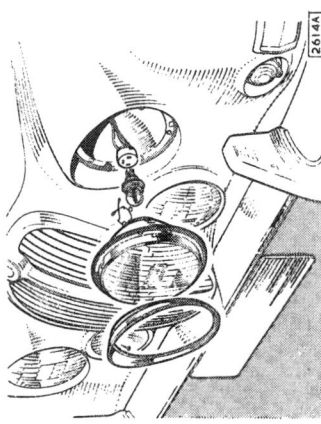

Fig. 41. Headlight bulb removal—non-sealed beam unit.

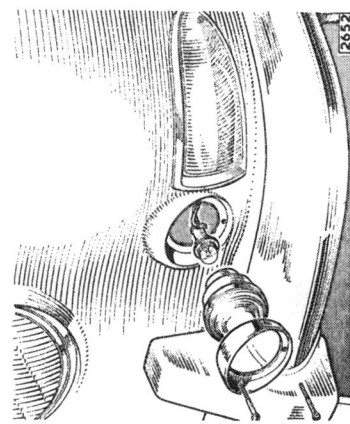

Sidelight Bulb—Replacement

Remove the two screws retaining the light embellisher. Withdraw the embellisher and light unit, remove the bulb holder from the rear of the unit. The bulb may then be removed by pressing in and turning anti-clockwise.

Sidelight Warning Light Bulb—Replacement (Italian market only)

Withdraw the bulb carrier unit only from the holder (accessible from the rear face of the side facia panel). Unscrew the bulb and replace with one of the correct value, that is, 12 volt, 2·2 watts screw cap.

Fig. 42. Side light bulb removal.

Number Plate Light Bulb—Replacement

Remove two screws securing the luggage compartment lid panel to allow access to the bulb holders. The number plate bulb holders are the two in the centre of the group. Press the tag in, lift and withdraw the holder. The bulb may then be removed by pressing in and turning anti-clockwise.

Luggage Compartment Light Bulb—Replacement

The bulb is accessible through an aperture in the luggage compartment lid. Remove the bulb by pressing in and turning anti-clockwise.

Reversing Light Bulb—Replacement

Proceed as for the number plate light bulb. The reversing light bulb holders are those on the outsides of the group.

Interior Light Bulbs—Replacement

Using care to avoid breakages, prise the cover from the appropriate interior light noting the stud fixings. Remove the bulb by pressing in and turning through 90. Replace the bulb with one of the correct value by pressing the bulb into the holder and turning until the notches inside the holder are located. Replace the cover by pressing onto the securing studs.

Glovebox Light Bulb—Replacement

Open the glove box lid and remove the mauve glass from its holder. Care should be taken when removing this glass to avoid breakages. Remove the bulb from between the two contacts and replace with a bulb of the correct value. Replace the glass.

Map Light Bulb—Replacement

Remove the bulb which is situated under the centre of the screen rail in front of the instrument panel. Removal is effected by pressing the bulb inwards, rotating slightly and withdrawing outwards.

Replace the bulb by a reversal of the above procedure.

Indicator Strip Bulbs—Replacement

Three bulbs are provided along the bottom rear edge of the instrument panel. Withdraw the bulbs by pulling out from the sockets provided in the rear panel.

Replace the appropriate bulb with one of the correct value.

Fig. 46. Number plate light bulb removal.

Fig. 47. Reverse light bulb removal.

Fig. 43. Front flasher bulb removal.

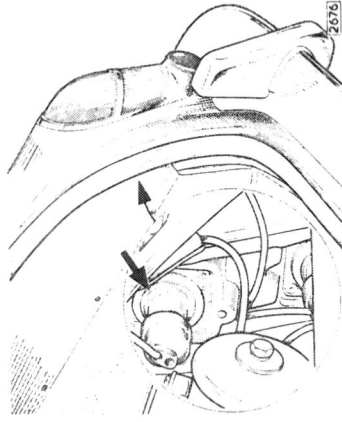

Fig. 44. Rear flasher bulb removal.

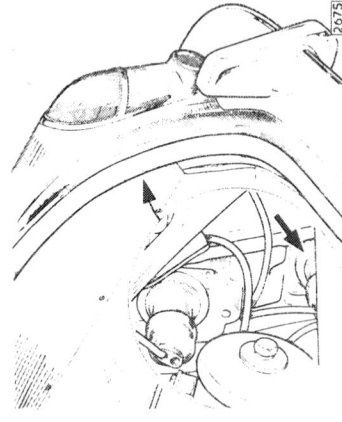

Fig. 45. Rear brake light bulb removal.

Front Flasher Bulb—Replacement

Remove the screw retaining the light glass and disconnect the glass at the three tags under the chrome surround. The bulb may then be removed by pressing in and turning anti-clockwise.

Rear Flasher Bulb—Replacement

Remove the screws securing the appropriate luggage compartment casing and withdraw the upper bulb holder from the rear of the light assembly. The bulb may then be removed by pressing in and turning anti-clockwise.

Flashing Indicator Bulb—Replacement

Disconnect the earth lead at the battery.

Detach the switch cover from above the steering column by withdrawing the two most sunken screws from below. Withdraw one or both flasher indicator warning light bulb holders from the outer sockets of the upper switch cover. Remove the bulb from the holder by applying inward pressure and rotating through 90 in either direction. The bulb is replaced by inserting into the bulb holder and rotating through 90 until the notches inside the bulb holder are located. Replacing the bulb holder and upper switch cover is the reverse of the removal procedure.

Rear Brake Light Bulb—Replacement

Proceed as for rear flasher bulb but withdraw the lower bulb holder. When fitting a replacement bulb note that the pins are offset.

When replacing the bulb holder check that the wide gap between the retaining claws registers with the projection in the lamp body.

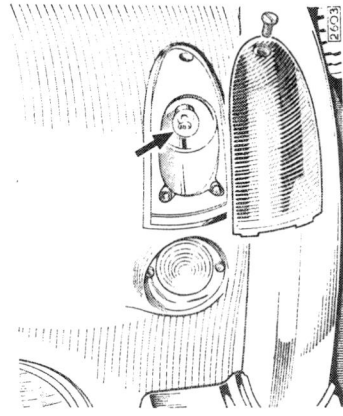

Automatic Selector Bulb—Replacement

Disconnect the earth cable at the battery. Detach the upper switch cover from the steering column by removing the two most sunken screws from below. Remove the bulb holder from the centre socket in the switch cover. Remove the bulb from the holder by pressing in and turning through 90° in either direction. The bulb is replaced by inserting into the bulb holder and turning until the notches inside the bulb holder are located. Replacing the bulb holder and upper switch cover is the reverse of the removal procedure.

Fog-Light Unit—Replacement

Unscrew the screw from the bottom of the lamp and disengage the rim at the top. Disconnect the cables, remove the four spring clips and withdraw the light unit.

Refitting is the reverse of the removal sequence.

Fog Lamp Bulb—Replacement (Switzerland)

Remove the two front securing screws and detach the light unit. Withdraw the bulb and replace with one of the correct type.

Fog Lamp Bulb—Replacement (France)

Unscrew the screw from the bottom of the lamp, disengage the rim at the top and withdraw the light unit from the back shell. Ease back the earth contact and withdraw the bulb. When replacing the bulb align the groove in the bulb plate with its register in the reflector. When refitting the light unit care must be taken to ensure that the contact blade coupled to the red and yellow cable registers with the centre contact on the bulb.

Adjusting Fog Lamp Beams

The beam of the fog lamp can be adjusted by slackening the nut of the attachment bolt, access to which is gained from beneath the car, and moving the lamp into the desired position. The nut is then tightened while a second person holds the lamp steady.

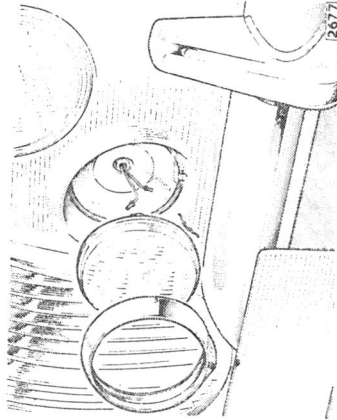

Fig. 48. *Fog light unit removal.*

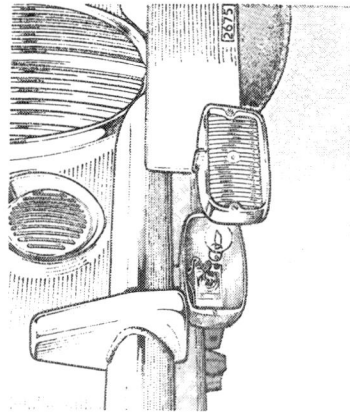

Fig. 49. *Fog light bulb removal (Switzerland).*

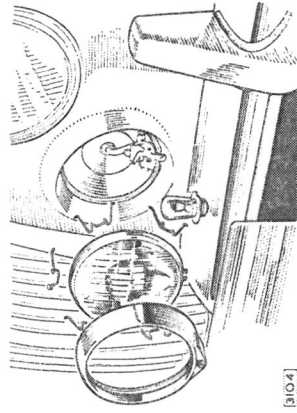

Fig. 50. *Fog light bulb removal (France).*

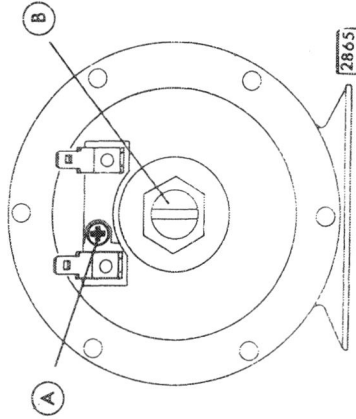

HORNS—MODEL 9H

Description

The Lucas 9H horns are mounted at the front end on either side of the engine compartment immediately below the radiator. The horn circuit operates through a Lucas 6 RA relay, the contacts C1 and C2 closing when the relay coil is energised by depressing the semi-circular ring attached to the steering wheel, or, by pressing the centre button.

Maintenance

In the event of the horn(s) failing to sound or performance becoming uncertain, check that the fault is not due to external causes before any adjustments are made.

Check as follows and rectify as necessary:

(i) Battery condition.

(ii) Loose or broken connections in the horn circuit. Test with voltmeter at cable terminals.

(iii) Loose fixing bolts. It is important to keep the horn mountings tight and to maintain rigid the mountings of any units fitted near the horns.

(iv) Faulty relay. Check by substitution after verifying that current is available at terminal C2 (cable colour—brown and blue) and terminal W (green).

(v) Check that fuse 3 (35 amperes) and fuse 5 (50 amperes) have not blown.

Note: Horns will not operate until the ignition is switched on.

Adjustment

The horns cannot be conveniently adjusted in position. Remove and securely mount on the test fixture.

A small serrated adjusting screw is provided to take up wear of moving parts only in the horn and it is located adjacent to the horn terminals. Turning this screw does not alter the pitch of the note.

Connect an 0—25 moving coil ammeter in series with the horn supply feed. The ammeter should be protected from overload by connecting an ON-OFF switch in parallel with its terminals.

Keep this switch ON except while taking readings, that is, when the horn is sounding.

Turn the adjustment screw anti-clockwise until the horn just fails to sound.

Turn the screw clockwise until the horn operates within the specified current limits of 6·5—7·0 amperes.

Fig. 51. *The Lucas 9H horn.*

A. Contact Breaker Adjustment Screw
B. Slotted Centre Core (do not disturb)

WINDSCREEN WIPER

DATA

Wiping Speed		
Normal	...	45–50 cycles per minute
High	...	60–70 cycles per minute
Light Running		
Normal Speed	...	2.7—3.4 amperes
High Speed	...	2.6 (or less) amperes
Stall Current	...	10—11 amperes (DR3)
Control Switch	...	79.5A.
Pressure of Blades against Windscreen	...	11—13 ounces (312—369 gms.)
Maximum permissible force to move cable rack in protective tubing with motor, arms and blades disconnected	...	6.0 lb. (2.72 kilograms)

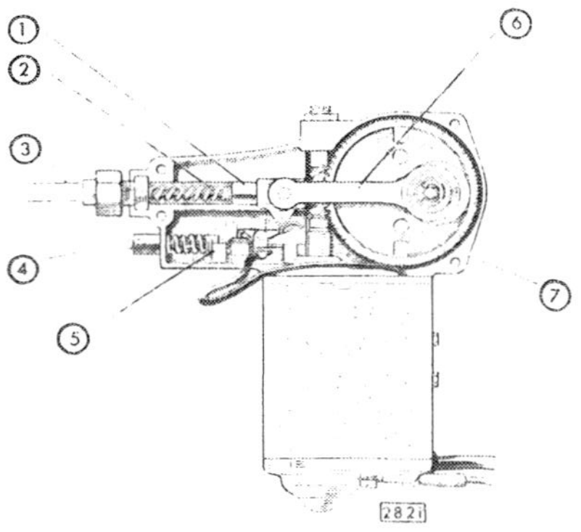

DESCRIPTION

The windscreen wiper assembly consists of a two-speed, thermostatically protected motor coupled by a cable rack drive to two scuttle mounted wheel boxes. The cable rack comprises a flexible inner core of steel wire wound with a wire helix. A reciprocating motion is imparted to the rack by a connection rod in the motor gearbox and transmitted to the wiper arm spindles by engagement of the rack with a gear in each wheelbox.

The wipers are self parking and are controlled by a switch on the instrument panel, giving Park, Slow, and Fast speed operation. The fast speed is intended for use when driving fast through heavy rain or light snow. It should NOT be used with heavy snow or a drying windscreen.

If overloaded, the motor windings will overheat and cause the thermostat to trip and isolate the motor from the supply. Possible causes include: Packed snow or ice on the screen, over-frictional or oil-contaminated blades, damaged drive mechanism or spindle units. Provided the obstruction or other cause of excessive heating is removed, normal working resumes automatically when the temperature falls to a safe level.

MAINTENANCE

Efficient wiping is dependent upon having a clean windscreen and wiper blades in good condition.

Use methylated spirits (denatured alcohol) to remove oil, tar spots and other stains from the windscreen.

Silicone and wax polishes should not be used for this purpose.

Fig. 54. Showing the D.R.3 wiper motor with cover plate removed.

1. Crosshead
2. Cable Rack
3. Protective Tubing
4. Adjusting Nut
5. Self-Parking Switch
6. Connecting Rod
7. Limit Gear

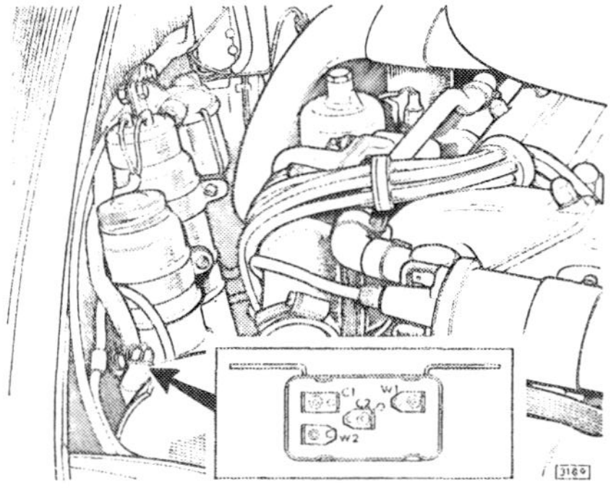

Fig. 52. The 6 RA relay connections.

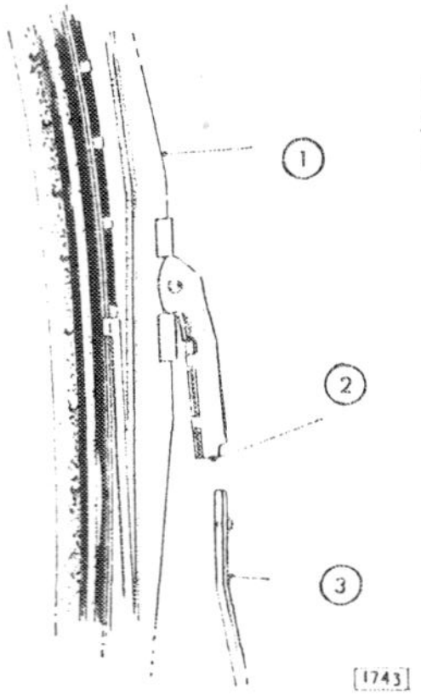

Fig. 53. Wiper blade to arm attachment.

1. Blade
2. Entry Slot
3. Arm

Service Replacements

When fitting replacement horns it is essential that the following procedure be carried out.

(i) Refit the lockwashers in their correct positions, one on each side of the mounting bracket centre fixing.

(ii) Ensure after positioning the horn, that the 3/8" centre fixing bolt is secure but not over-tightened. Over-tightening of this bolt will damage the horn.

(iii) Ensure that, when a centre fixing bolt or washers other than the originals are used, the bolt is not screwed into the horn to a depth greater than 11/16" (17.5 mm.).

Horn Relay—Checking

If the horn relay is suspected, check for the fault by substitution or by the following method:

(i) Check that fuses No. 3 and No. 5 have not blown. Replace if necessary.

(ii) Check with a test lamp that current is present at relay terminals W1 (green) and C2 (brown and purple). Switch on ignition before checking terminal W1.

(iii) Remove cable from terminal W2 (purple and black) and earth the terminal to a clean part of the frame. Relay coil should now operate and close contacts. Reconnect cable.

(iv) Remove cables from terminal C2 (brown and purple). Check for continuity by means of an earthed test lamp when horn button or ring is depressed with the ignition ON. Replace relay if faulty.

Note: The wiper blades fitted to the 'S' model are manufactured with special anti-smear properties. Renew only with Genuine Jaguar replacement parts.

Worn or perished wiper blades are readily removed for replacement.

When necessary, adjustment to the self-parking mechanism can be made by turning the knurled nut located near the cable rack outlet on the wiper motor. Turn the nut only one or two serrations at a time, and test the effect of each setting before proceeding.

REMOVAL OF WIPER MOTOR AND CABLE

Withdraw the wiper arms from the spindles.

Unscrew the large nut connecting cable guide to the wiper motor.

Remove the setscrew securing the earth wire to the motor.

Disconnect the cable harness attached to the motor at the snap connectors noting the cable colours.

From the underneath of the right hand front wing remove the three nuts securing the wiper motor to the wing valance.

The wiper motor and cable can now be removed as an assembly by drawing the cable through the guide tube.

Disconnecting the cable

Remove the four small set-screws from the gear cover.

Lift off the cover, remove the circlip from the post in the gear wheel.

Remove the washer, spring, shaped washer and connecting link from the post. Lift out the connecting link from the crosshead.

Lift out the cable ferrule from the gear casing.

REFITTING

Refitting is the reverse of the removal procedure.

Renew grommets in the wing valance if worn or damaged.

REMOVAL OF WHEELBOXES

Remove the screen rail as detailed on page P.52.

Remove the picnic tray and lower the instrument panel as detailed on page P.52.

Release the main panel harness from the two plastic covered clips behind the instrument panel.

Detach the fabric cover, attached with upholstery solution between the demister ducts and the panels.

Withdraw the wiper blades and arms.

Remove the wiper motor as detailed above and withdraw the drive cable from the wheelboxes.

Remove the two securing nuts and detach the back plates from the wheelboxes and withdraw the conduit tubing.

Unscrew the nuts securing the wheelboxes to the scuttle and remove the chrome distance pieces and rubber seals.

Withdraw the wheelboxes.

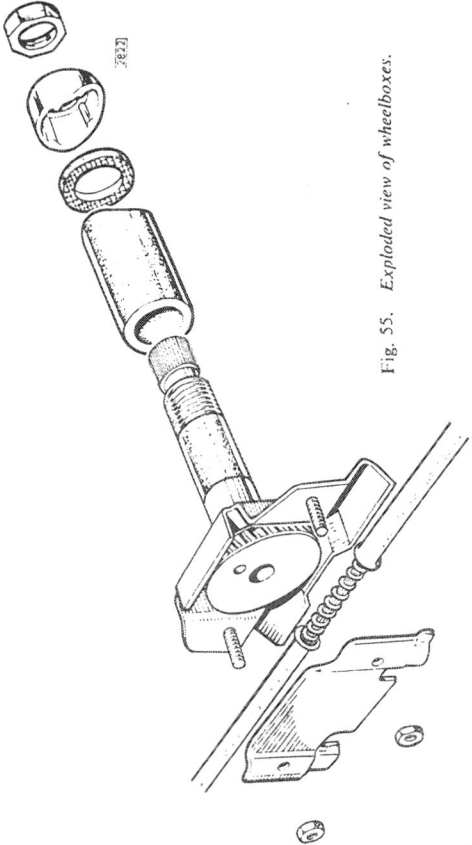

Fig. 55. *Exploded view of wheelboxes.*

---

1. Resistor
2. Field Coil
3. R.H. Terminal
4. Armature
5. 79 SA Switch
6. L.H. Terminal
7. Parking Switch
8. Thermostatic Circuit Breaker

G. Green
GR. Green with Red
GY. Green with Yellow
GW. Green with White
M. Brown
O. Orange
R. Red
U. Blue

2824

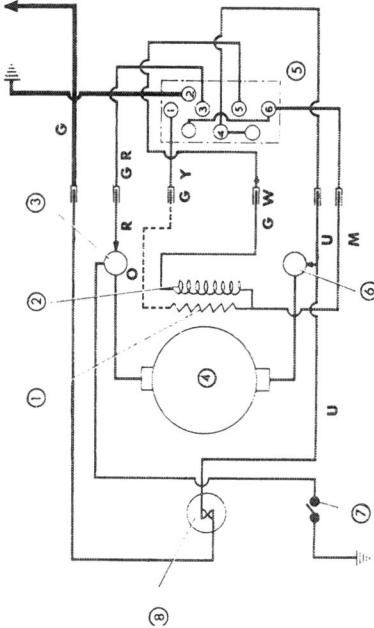

Fig. 56. *Wiring connections switch to wiper.*

REFITTING

Refitting is the reverse of the removal procedure. When refitting ensure that the flared ends of tubes register with the narrow slots in cover plate.

FAULT DIAGNOSIS

Poor performance can be electrical or mechanical in origin and not necessarily due to a faulty motor, for example:

Low voltage at the motor due to poor connections or to a discharged battery.

Cable rack binding in protective tubing;

Excessive loading on the wiper blades;

Wheelboxes loose, out of alignment or spindles binding in the bearing housing.

TESTING

Unless the origin of the fault is apparent, proceed as follows to determine the cause of failure.

Measuring Supply Voltage

Using a first grade moving coil voltmeter, measure the voltage between the motor supply terminal (to which the green cable is connected) and a good earthing point. This should be 11·5 volts with wiper working normally. If the reading is low, check the battery, switch (by substitution), cabling and connections.

To Check the "Fast" Speed Current

Using a fully charged 12-volt battery and two test leads, connect the "GREEN" cable on the wiper motor to the "Negative" battery terminal. Join the "YELLOW" and "RED" cables together and connect to the "Positive" battery terminal. Connect the "BLUE" and "WHITE" cables together. Check the cycles per minute of the wiper spindle.

To Check the "Slow" Speed Current

Connect the "GREEN" cable to the "Negative" battery terminal. Join the "BROWN" and "RED" cables together and connect to the "Positive" battery terminal. Connect the "BLUE" and "WHITE" cables together. Check the cycles per minute of the wiper spindle.

Measuring the Light Running Current

The light running current must not exceed 3·4 amperes at normal speed of 45—50 r.p.m./or t.p.m. of the output motor shaft; also 2·6 amperes at fast speed 60—70 c.p.m./or r.p.m. of output motor shaft.

If the current is in excess of these figures, change the wiper motor. See DATA chart for other information.

Checking the brush gear

Withdraw the two through bolts and detach the end cover.

Expand the retaining spring and lift off the two brush carriers as an assembly.

Note: The two brushes are loose in the carriers and care must be taken that they are not misplaced when removed.

Refitting

Refitting is the reverse of the removal procedure. Ensure that the brushes are refitted in the same way as originally fitted.

Checking Cable Rack and Tubing

The maximum permissible force to move the cable rack in its protective tubing is 6 pounds with the wiper arms, blades and motor disconnected. The measurement can be made by hooking a spring balance in the hole in the cross-head (into which a pin on the connecting rod is normally located) and withdrawing the rack with the balance.

Before checking, disconnect the cable rack from the motor, remove the motor, wiper arms and blades.

When refitting the tubing check that they do not foul the body at any point. Failure to ensure this may result in the transmission of cable rack noise.

Binding of the rack can be due to kinked or flattened tubing or to faulty installation. Minor faults can be cleared with a suitable tested mandrel sold specifically for checking wiper installations. Badly kinked or flattened tubing must be renewed. Any bends of less than 9″ radius must be reformed.

It is ESSENTIAL that all the flared ends of the tubing are registered in the slots provided in the wheelbox plates before tightening the wheelbox cover plate securing nuts.

The cable rack should be well lubricated with Duckham's HBB grease.

Checking Wheelboxes

Check the wheelboxes for misalignment or looseness and rectify as required.

Renew seized wheelboxes.

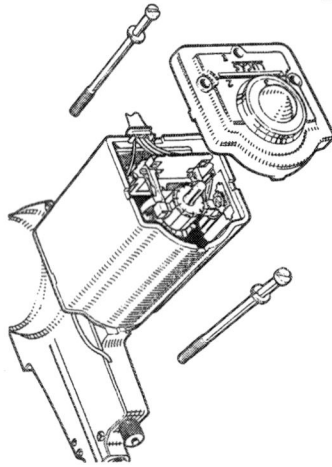

3159

Fig. 57. *Showing access to brush gear.*

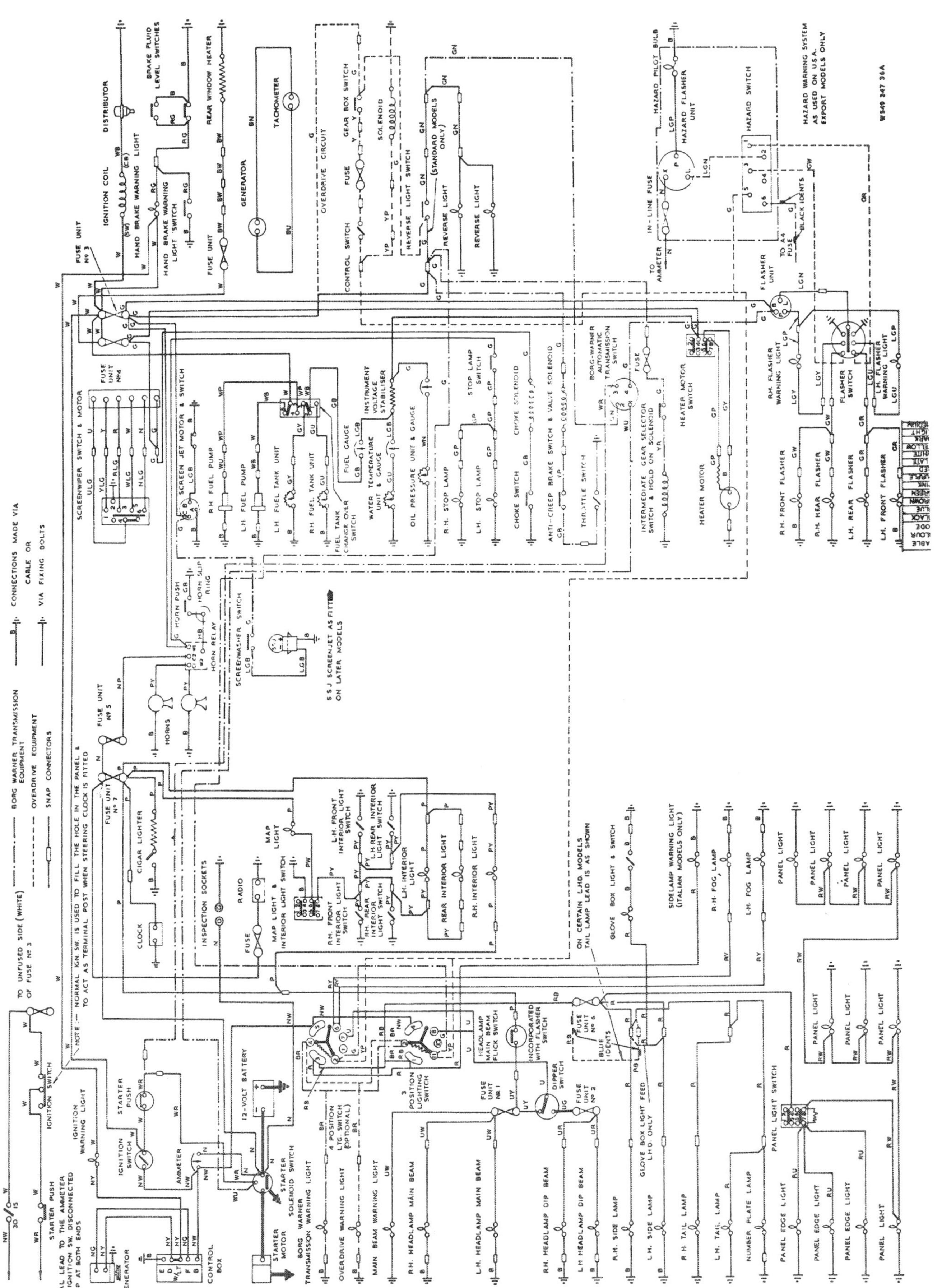

Fig. 58. *Wiring diagram*

WIRING HARNESS REPLACEMENT

The wiring harness consists of four main items, namely, the forward, the panel and right-and left-hand body harnesses.

The junctions between the forward and body harnesses are behind the side facia panel and the glove-box adjacent to the door pillars.

When replacing items, all harnesses must be secured in the clips provided and all grommets must be renewed if worn or damaged.

The body harnesses are routed through the door sills.

Refer to the wiring diagram when making any connections (see page P.49).

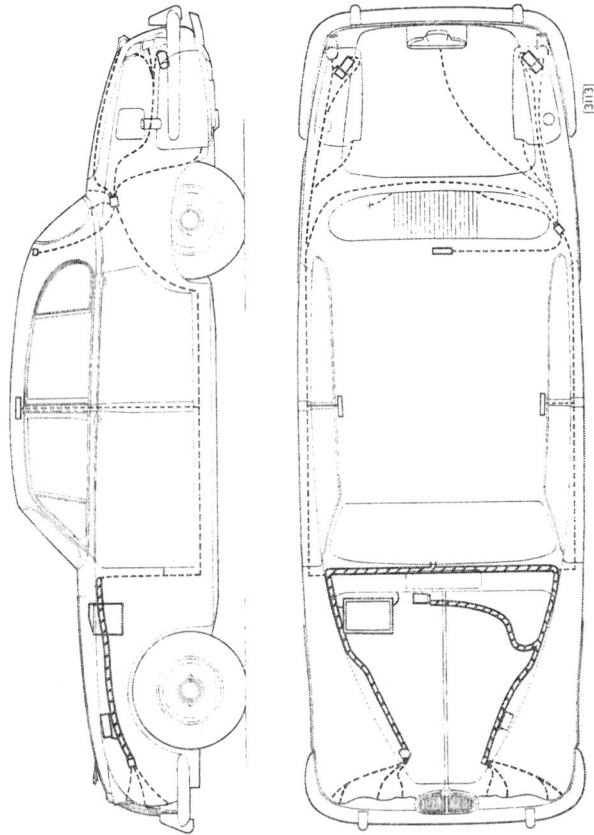

Fig. 59. *Layout of wiring harnesses.*

---

bulb holder ensure that the red transparent window is retained in the bezel by a small circlip position the designation plate on the bulb holder and screw on the bezel.

FLASHING INDICATOR CONTROL

Removal

Disconnect the earth lead at the battery. Detach the upper and lower switch covers from around the steering column by removing the two sunken screws and three screws from below. Disconnect the seven cable harness at the snap connectors at the left-hand side of the steering column. Detach the flasher indicator control from the left-hand side of the steering column by withdrawing two horizontally positioned screws from the right-hand side.

Refitting

Refitting is the reverse of the removal procedure. Insert the wires into the connectors so that similar coloured wires are opposite each other.

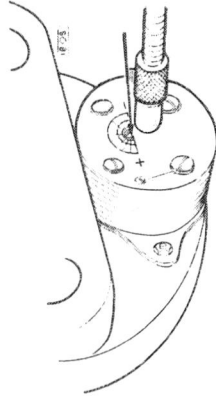

Fig. 60. *Adjustment screw for clock.*

MISCELLANEOUS

ELECTRIC CLOCK

Removal

Detach the earth lead from the battery. Remove the revolution counter from the instrument panel as detailed under "Revolution Counter and Clock Removal". Detach the clock from the rear face of the revolution counter by removing the two nuts. The flexible setting drive can be removed by slackening the knurled nut. Disconnect the cable at the snap connector.

Refitting

Refitting is the reverse of the removal procedure.

It is ESSENTIAL to check that the clock is *always* running when the battery is connected. Failure to ensure this will result in the gold-plated contacts in the clock burning and will effectively shorten the life of the instrument.

Adjustment

Adjustment is effected by means of a small screw surrounded by a semi-circular scale located at the back of the instrument.

If the clock is gaining turn the screw towards the minus (—) sign; if the clock is losing turn the screw towards the positive (+) sign.

Note: The action of resetting the hands automatically restarts the clock.

Warning: The electric clock incorporates a rectifier in the movement.

If at any time the clock is removed for servicing and subsequent testing on the bench IT IS MOST IMPORTANT that the feed terminal on the back of the clock is connected to the negative side of the battery and that the outer casing of the clock is positively earthed. Incorrect connection of a rectified clock to the battery will *instantly destroy the rectifier.*

BRAKE FLUID AND HANDBRAKE WARNING LIGHT

Renewing the Bulb

Unscrew the bezel of the lamp exercising care to control the run of the spring loaded bulb beneath. Feed the bulb into the spring-loaded

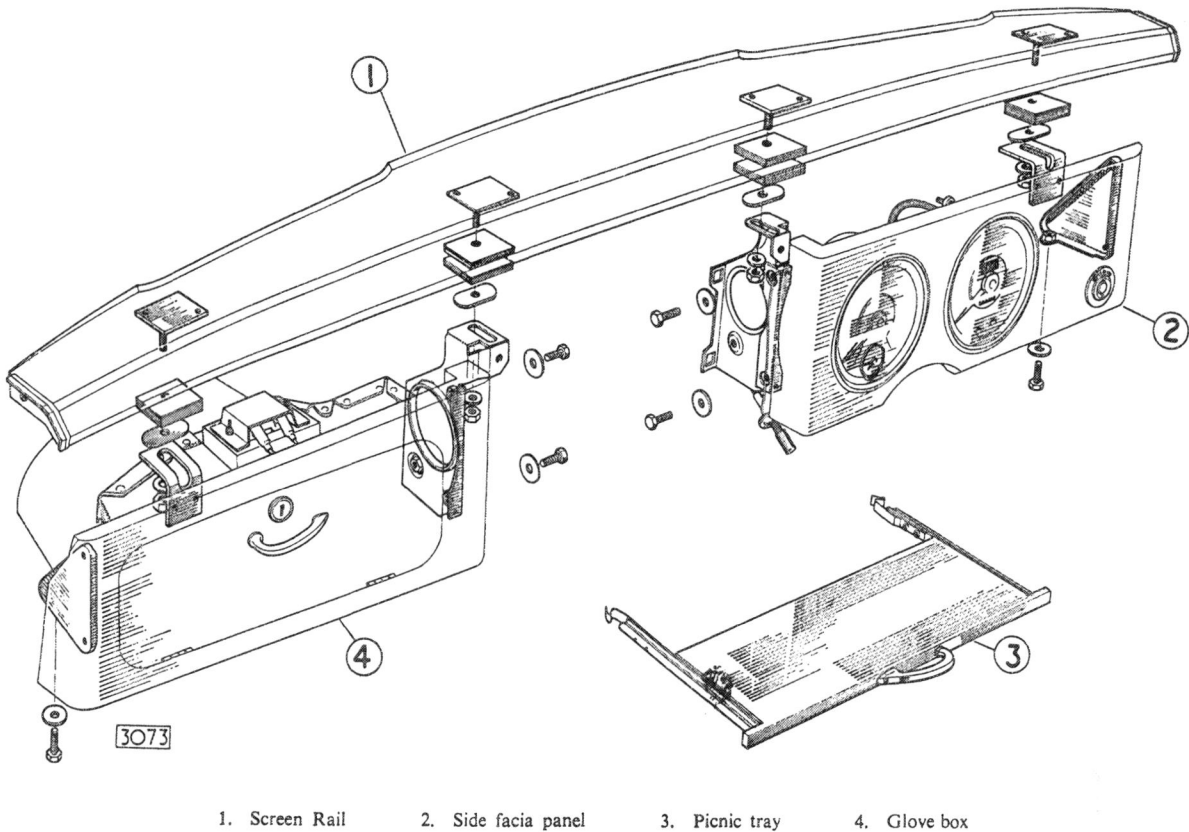

1. Screen Rail 2. Side facia panel 3. Picnic tray 4. Glove box

Fig. 61. *Showing the attachment points for the side facia panel, glove box and screen rail.*

The nut situated above the glove-box is accessible through a hole in the glove-box top panel, the centre nut being exposed when the instrument panel is lowered. Disconnect the two cables connected to the maplight and withdraw the rail.

Refitting

Refitting is the reverse of the removal procedure.

THE SIDE FACIA PANEL (2, Fig. 61)

Removal

Detach the earth lead from the battery.

Withdraw the picnic tray and lower the instrument panel.

Remove the screen rail.

Remove the upper cover from the steering column nacelle by withdrawing the two sunken screws from beneath.

Identify and withdraw the flasher warning light harness through the loop bracket attached to the hidden face of the instrument panel by disconnecting the snap connectors.

Detach the fabric covered steering column cover, located above the parcel tray, by removing the two drive screws.

Release the two nuts, now exposed, securing the steering column assembly to the body to the full extent of the thread and lower the column to the newspaper tray.

Remove the fabric sheeting, attached with upholstery solution, between the demister ducts and the facia panel.

Remove the setscrew securing the facia panel to the bracket attached to the body below the screen pillar.

Withdraw the two facia panel securing screws located in the instrument panel aperture.

Disconnect the speedometer drive cable, warning lights and electrical leads and remove the facia panel.

Refitting

Refitting is the reverse of the removal procedure. Ensure that the flasher warning lights are replaced in their correct holders.

THE INSTRUMENTS

THE INSTRUMENT PANEL

Opening

Detach the earth lead from the battery.

Withdraw the picnic tray to the full extent and locate the two spring steel retaining clips attached to the tray at the back edge. Press both clips inwards to the centre of the car and complete the withdrawal of the tray.

Remove the ignition key and cigar lighter for safe keeping. Hinge the centre instrument panel downwards on its bottom edge, after withdrawing the thumb screws situated in each top corner.

Removal

The instrument panel can be removed completely by detaching the earth lead from the battery, identifying and removing the leads from the instruments, cigar lighter and switches, removing the electrical harness and clips from the instrument panel and withdrawing two bolts from the extended portion of each hinge, accessible through the newspaper tray beneath.

Refitting

Refitting is the reverse of the removal procedure, but particular attention must be given to the following point.

That the leads are refitted in accordance with their colour coding, utilizing the wiring diagram as a reference.

Closing

Closing is the reverse of the opening procedure.

Always ensure that the clips securing the main harness to the instrument panel will in no way foul any of the switch or instrument terminals, otherwise a direct short will occur when the battery is connected.

THE SCREEN RAIL (1, Fig. 61)

Removal

Remove the screen rail from the base of the windscreen after detaching three nuts, two adjacent to each end and one central.

THE GLOVEBOX

Removal

Detach the earth lead from the battery.

Withdraw the picnic tray and lower the instrument panel (see page P.52).

Remove the screen rail (see page P.52).

Remove the fabric sheeting, attached with upholstery solution, between the demister ducts and the glove box.

Remove the setscrew securing the glovebox to the bracket attached to the body below the screen pillar.

Remove the two glovebox securing screws located in the instrument panel aperture.

Disconnect the illumination lamp cables and remove the glovebox.

Refitting

Refitting is the reverse of the removal procedure.

THE SPEEDOMETER

Removal

Detach the earth lead from the battery and raise the steering to the highest position. Detach the speedometer from the facia board by removing the two knurled nuts, earth lead and the two retaining pieces.

Withdraw the flexible drive from the centre of the instrument by slackening the knurled sleeve nut.

Remove the speedometer from the facia board; identify and remove the three warning lamps and the two instrument illumination lamps from the hidden face of the instrument. Remove the odometer trip setting drive by slackening the knurled sleeve nut.

Refitting

Refitting is the reverse of the removal procedure. When inserting the instrument lights, ensure:

(i) That the two instrument illumination lamps are inserted in the apertures, at the side of the instrument.

(ii) That the headlamp warning light is inserted in the right-hand bottom aperture.

(iii) That the fuel warning light is inserted in the centre bottom aperture.

(iv) That the ignition warning light is inserted in the left-hand bottom aperture.

THE REVOLUTION COUNTER AND CLOCK

The revolution counter and clock are of the electrical type and the electrical leads to both are included in the car harness.

The clock is mounted at the bottom of the revolution counter indicator head and to effect its removal it is necessary to remove the revolution counter from the side facia panel.

The revolution counter consists of an A.C. generator fitted to the rear end of the camshaft with an indicator head mounted in the side facia panel

Removal

Remove the speedometer from the side facia panel as previously detailed, this will give the necessary working clearance. Detach the revolution counter from the facia board by removing two knurled nuts, earth lead and retaining pieces, then withdraw the revolution counter after removing the two centre leads and two instrument illumination bulbs from the hidden face of the instrument and from the clock at the snap connector. Detach the clock setting drive by slackening the knurled sleeve and remove the clock from the revolution counter by removing two nuts.

Refitting

Refitting is the reverse of the removal procedure.

Refit the leads in accordance to their colour coding, utilizing the wiring diagram as a reference.

TESTING OPERATION OF REVOLUTION COUNTER

Utilizing an A.C. voltmeter check the current across the terminals of the generator at the rear of the right-hand camshaft while the engine is running; as a rough guide it can be assumed that there is one volt output per 100 r.p.m. When electrical current is evident, check the continuity of the two leads by attaching the terminals to the generator and connecting the voltmeter to the opposite ends of the cables after removal from revolution counter. If when running engine continuity is evident, it can be assumed that the instrument is unserviceable and must be exchanged.

THE REVOLUTION COUNTER DRIVE

The revolution counter drive takes the form of a small A.C. electrical generator fitted at the rear R.H. end of the cylinder head where its tongued driving spindle engages a slotted adaptor screwed in the rear end of the inlet camshaft. Leads included in the electrical harness of the car connect with the Lucar tabs pointing upward in the body of the generator and with similar tabs at the rear of the instrument lead in the side facia panel. The Lucar tabs are of the same size and the leads can be fitted either way round.

Removal

Open the engine compartment and detach the earth lead from the battery. Remove the electrical harness from the two Lucar tabs on the A.C. generator on the rear right-hand end of the cylinder head. Detach the A.C. generator from the rear right-hand end of the cylinder head by withdrawing three Allen screws and a plate washer, remove the generator in a rearward direction and note the position of the tongued driving spindle.

Refitting

Refitting is the reverse of the removal procedure but particular attention must be given to the following point:

That the tongued driving spindle is positioned in the same attitude as it was when it was removed; whenever difficulty is experienced in engaging the tongued spindle do not apply any force but remove the generator, ascertain the position of the slot in the camshaft with a mirror and set the tongued drive in a similar position.

REMOVAL OF THE INSTRUMENT PANEL COMPONENTS

Detach the earth lead from the battery.

Remove the picnic tray as detailed on page P.52.

Remove the ignition key and cigar lighter for safe keeping. Hinge the centre instrument panel downwards on its bottom edge, after withdrawing the thumb screws situated in each top corner.

Ignition Switch

Identify and remove the ignition switch cables. Withdraw the ignition switch from the rear of the instrument panel by removing the chrome ring and fibre washer.

Note the locating washer fitted to the threaded portion of the switch.

The lock barrel can be withdrawn by inserting a thin rod through a hole in the body of the switch and depressing the plunger in the lock. Insert the key and turn to the "ON" position to gain access to the plunger.

Refitting is the reverse of the removal procedure. When refitting a new lock barrel check that the number of the key is the same as that stamped on the lock barrel.

Insert the key in the lock and turn the switch to the "ON" position before inserting the lock barrel. Refit the locating washer over the threaded portion of the switch before inserting in the panel and locate the tag with the cut-out portion in the panel hole.

Cigar Lighter Element

Withdraw the cigar lighter and ensure that it is cold. Place the unit in the palm of the hand, knob first, and hold the sleeve downward against the pressure of the spring. Unscrew the lighter element and fit a replacement. It is important **not** to omit the spring as it ejects the lighter unit when it attains the correct temperature.

Cigar Lighter Unit

Detach the earth lead from the battery. Withdraw the cigar lighter. Identify and remove the cables from the cigar lighter housing. Unscrew the outer casing at the rear of the panel and withdraw the inner section of the cigar lighter unit.

Refitting is the reverse of the removal procedure.

Starter Push Button

Remove the cables from the push button. Withdraw the push button through the face of the instrument panel by removing the nut, washer and spring washer at the rear of the instrument panel.

Refitting is the reverse of the removal procedure.

THE BI-METAL RESISTANCE INSTRUMENTATION

Engine Temperature, Fuel Tank and Oil Pressure Gauges

DESCRIPTION

The Bi-metal Resistance Instrumentation for engine temperature, petrol tank contents and engine oil pressure consists of a gauge unit fitted in the instrument panel, a transmitter unit fitted in the engine unit or petrol tank and connected together to the battery, the oil pressure **gauge** being an exception, through a common voltage regulator. The purpose of the latter is to ensure a constant power supply at a predetermined voltage thus avoiding errors due to a low battery voltage. In the instance of the oil pressure gauge this is not quite so critical to supply voltage.

In all systems the gauge unit operates on the thermal principle utilizing a heater winding wound on a bi-metal strip, while the transmitter units of the engine temperature and petrol tank contents gauge are of the resistance type but in both instances the system is voltage sensitive. The transmitter unit of the oil pressure gauge is of the thermal pressure principle utilizing a heater winding wound on a bi-metal strip having contact at one end with the second contact mounted on a diaphragm which is sensitive to engine oil pressure.

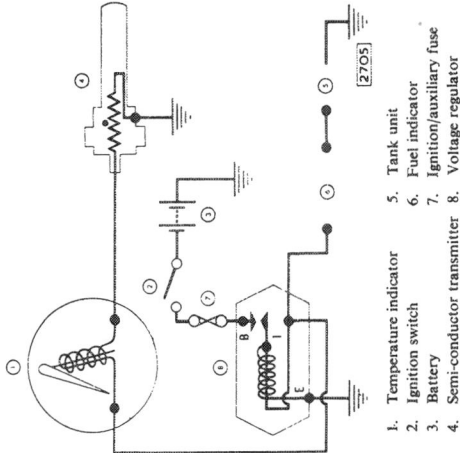

Fig. 62. *The combined wiring diagram of the fuel tank contents and water temperature gauges with the voltage regulator.*

1. Temperature indicator
2. Ignition switch
3. Battery
4. Semi-conductor transmitter
5. Tank unit
6. Fuel indicator
7. Ignition/auxiliary fuse
8. Voltage regulator

OPERATION OF THE ENGINE TEMPERATURE GAUGE

The transmitter unit of the engine temperature gauge is fitted in the water outlet pipe of the engine unit and is a variable resistance and consists of a temperature sensitive resistance element contained in a brass bulb. The resistance element is a semi-conductor which has a high negative temperature co-efficient of resistance and its electrical resistance decreases rapidly with an increase in its temperature. As the temperature of the engine unit rises the resistance of the semi-conductor decreases and increases the flow of current through the transmitter, similarly a decrease in engine temperature reduces the flow of current.

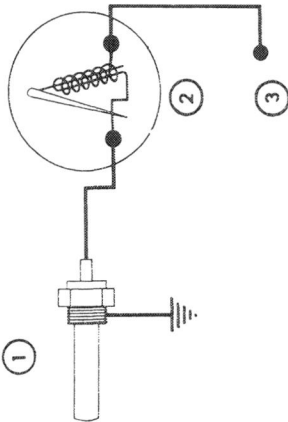

1. Temperature transmitter
2. Rear of indicator
3. Voltage regulator terminal "I"

Fig. 63. *The water temperature gauge circuit.*

Head and Side Light Switch

Detach the earth lead from the battery.

Remove the light switch control lever from the face of the instrument panel by depressing the plunger in the right-hand side.

Identify and remove the leads from the light switch.

Remove the three nuts, shakeproof washers, washers and blade terminal from the switch mounting posts. Withdraw the light switch. The designation plate can be removed from the instrument panel face by detaching the nut on the rear of the panel.

Refitting is the reverse of the removal procedure. Re-position the designation plate on the instrument panel by allowing a flat on the threaded barrel to locate a flat in the panel.

The light switch control lever is pressed onto the light switch so that the plunger locates with a drilling in the hub of the control lever.

Tumbler Switches

Detach the earth lead from the battery.

Identify and remove the leads from the Lucar tags on the body of the switch. Withdraw the tumbler switch from the rear of the instrument panel by holding the switch lever in a horizontal position and removing the screwed chromium ring from the face of the instrument panel.

Refitting is the reverse of the removal procedure. The flat face of the switch lever should be facing downwards.

Ammeter and Oil Pressure Gauges

Withdraw the illumination bulb holder from the rear of the gauge. Remove the cables from the Lucas connectors and terminals. Remove the two knurled nuts and "U" clamp. Withdraw the gauge through the front face of the instrument panel.

When refitting the gauges, check that the "U" clamp does not foul any terminals or the bulb holder.

Fuel and Water Temperature Gauges

Removal and refitting of these gauges is similar to the ammeter and oil pressure gauges. But in this case, the "U" clamp is retained by one knurled nut.

The removal and replacement of the fuel gauge tank unit and water temperature transmitter unit are detailed in the "Fuel System" and "Cooling System" sections respectively.

Voltage Regulator (Fuel and Water Temperature Gauges)

Remove the cables (noting their respective positions) from the voltage regulator situated in the left-hand corner of the instrument panel. Withdraw the voltage regulator by removing one nut, shakeproof washer and blade terminal.

When refitting the voltage regulator, ensure that a good earth is made between the regulator and panel.

Switch Indicator Strip

Remove the indicator strip, chrome finisher and light filter from the bottom edge of the instrument panel by withdrawing the four screws.

Refitting

Refitting is the reverse of the removal procedure.

ANALYSIS OF THE ENGINE TEMPERATURE AND PETROL TANK GAUGE FAULTS

NOTE: THE INSTRUMENT PANEL GAUGES MUST NEVER BE CHECKED BY SHORT-CIRCUITING THE TRANSMITTER UNITS TO EARTH

Symptom	Unit Possibly at Fault	Action
Instrument panel gauge showing a "zero" reading	Voltage regulator	Check that output voltage at terminal 'I' is 10 volts.
	Instrument panel gauge	Check for continuity between the gauge terminals with the leads disconnected.
	Transmitter unit in petrol tank or engine unit	Check for continuity between the terminal and the case with lead disconnected.
	Wiring	Check for continuity between the transmitter and the voltage regulator, also that the transmitter unit is earthed.
Instrument panel gauge showing a high/low reading when ignition switched on	Voltage regulator	Check output voltage at terminal 'I' is 10 volts.
	Instrument panel gauge	Check by substituting another instrument panel gauge.
	Transmitter unit in petrol tank or engine	Check by substituting another transmitter unit in petrol tank or engine unit.
	Wiring	Check for leak to earth.
Instrument panel gauge showing a high reading and overheating	Voltage regulator	Check output voltage at terminal 'I' is 10 volts.
	Wiring	Check for short circuits on wiring to each transmitter unit.
Instrument panel gauge showing an intermittent reading	Voltage regulator	Check by substituting another voltage regulator.
	Instrument panel gauge	Check by substituting another instrument panel gauge.
	Transmitter unit in petrol tank or engine unit	Check by substituting another transmitter unit in petrol tank or engine unit.
	Wiring	Check terminals for security, earthing and wiring continuity.

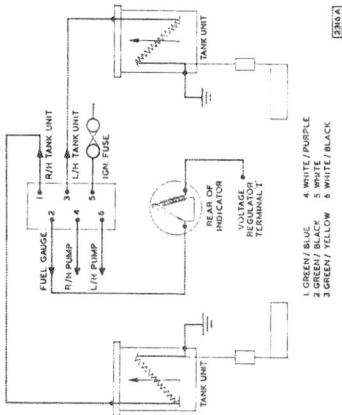

Fig. 64. *The fuel tank contents gauge circuit.*

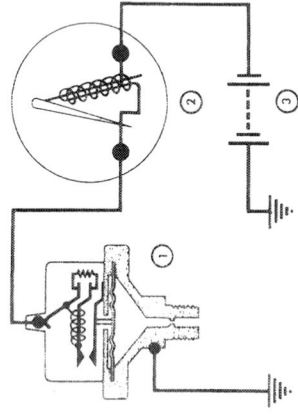

1. Thermal pressure transmitter
2. Rear of Indicator
3. Battery

Fig. 65. *The engine oil pressure gauge circuit.*

The gauge unit fitted in the instrument panel consists of a heater winding, connected at one end to the transmitter unit and at the second end to the "I" terminal of the voltage regulator, wound on a bi-metal strip which is linked to the indicator needle. The heater winding and bi-metal strip assembly is sensitive to the changes in voltage received from the transmitter unit causing the heater winding to heat or cool in the bi-metal strip, resulting in the deflection of the indicator needle over the scale provided. The calibration of the scale is such that the movement of the indicator needle over it is relative to the temperature of the transmitter unit bulb and therefore the temperature of the engine unit.

OPERATION OF THE FUEL TANKS GAUGE

The transmitter units of the petrol gauge are fitted in the petrol tanks and are of variable resistance actuated by floats, the arms of which carry contacts travelling across resistances housed in the transmitter bodies. The float arms take up a position relative to the level of petrol in the tanks and thus vary the amount of current passing through the indicator unit.

The gauge unit in the instrument panel consists of a heater winding, connected at one end to the transmitter units and at the other to the "I" terminal of the voltage regulator, wound on a bi-metal strip which is linked to the indicator needle. The heater winding and bi-metal strip assembly is sensitive to the changes in voltage received from the position of the transmitter float, causing the heater winding to heat or cool the bi-metal strip, resulting in the deflection of the indicator needle over the scale provided. The calibration of the scale is such that the movement of the indicator needle over it is relative to the position of the transmitter float actuated by the level of the contents in the petrol tank.

Exaggerated indicator needle movement due to petrol swirl in the tanks is considerably reduced as there is a delay before current changes from the transmitter unit can heat or cool the bi-metal and heater winding assembly in the indicator unit, which in fact causes the deflection of the needle. Similarly the indicator needle will take a few moments to register the contents of the petrol tank being used when the ignition is first switched on.

ANALYSIS OF THE OIL PRESSURE GAUGE FAULTS

Symptom	Unit Possibly at Fault	Action
Instrument panel gauge showing a "zero" reading	Wiring	Check for continuity between the gauge and the transmitter unit and that the latter is earthed.
	Instrument panel gauge	Check for continuity between the gauge terminals with leads disconnected. If satisfactory replace the transmitter unit.
Instrument panel gauge showing a reading with ignition switched on but engine not running	Transmitter unit on oil filter head	Check by substituting another transmitter unit.
Instrument panel gauge showing a high reading and overheating	Transmitter unit on oil filter head	Check by substituting another transmitter unit.
Instrument panel gauge showing a below "zero" reading with ignition switched off	Instrument panel gauge	Check by substituting another instrument panel gauge.

OPERATION OF THE OIL PRESSURE GAUGE

The transmitter unit of the oil pressure gauge, fitted in the head of the engine oil filter, is a voltage compensated pressure unit and consists of a diaphragm, a bi-metal strip with a heater winding wound thereon, a resistance and a pair of contacts. One contact is attached to the diaphragm while the second is mounted on one end of the bi-metal strip, the second end of which is connected through the resistance and the gauge unit to the battery supply; the heater winding is also connected to the battery supply but not through the resistance. Engine oil pressure will close the contacts causing current to flow through the gauge unit, bi-metal strip and contacts to earth resulting in the heating of the heater winding which will, after a time, open the contacts.

The gauge unit fitted in the instrument panel consists of a winding, connected at one end to the battery supply and at the second to the transmitter unit wound on to a bi-metal strip which is linked to an indicating needle. The heater winding and bi-metal strip assembly is sensitive to the continuity changes received from the thermal pressure unit, fitted in the engine oil filter, causing the heater winding to heat or cool the bi-metal strip resulting in the deflection of the indicating needle over the scale provided.

The changes in continuity of current from the transmitter unit will vary according to the amount of oil pressure for, as the latter rises, the outward travel of the diaphragm contact limits the return travel of the bi-metal strip contact thus allowing a longer continuity period. This results in a greater heating of the heater winding in the gauge unit and increased deflection of the indicating needle over the scale showing a greater oil pressure.

The opening and closing of the transmitter unit contacts is continuous thus the temperature of the heater winding in the gauge unit is kept within close limits and the calibration of the scale is such that the movement of the indicating needle over it is relative to the opening of the transmitter unit contacts and therefore the oil pressure of the engine is recorded.

THE SPEEDOMETER DRIVE CABLE

Removal

Disconnect the flexible cable and remove the speedometer from the side instrument facia as previously detailed. Detach the flexible drive cable from the attachment to the gearbox and release it from the retaining clips.

Refitting

Refitting is the reverse of the removal procedure but particular attention must be given to the following points:

(i) That the run of the flexible drive cable is without any sharp bends.

(ii) That the securing clips are so shaped that they only hold the cable in position without crushing it.

Important: All speedometer drive cables are marked with a white marker tape. This marker MUST be on the engine side of the bulkhead grommet when the cable is fitted.

SPEEDOMETER CABLE

General Instructions

Flexible cable condition to a great extent affects performance of speedometers. Poor installation or damage to the flexible drive will show up as apparent faults. It is most important that the flexible drive should be correctly fitted and maintained as illustrated in the following diagrams.

1. **Smooth Run**
   Run of flexible drive must be smooth. Minimum bend radius 6". No bend within 2" of connections.

2. **Securing**
   Avoid sharp bends at clips. If necessary change their positions. Do not allow flexible drive to flap freely. Clip at suitable points.

3. **Securing**
   Avoid crushing flexible drive by over-tightening clip.

4. **Connection**
   Ensure tightness of outer flex connections. They should be finger tight only. It may be necessary to clean thoroughly the point of drive before the connection can be screwed completely home.

Fig. 66. Clipping of cable (right hand).

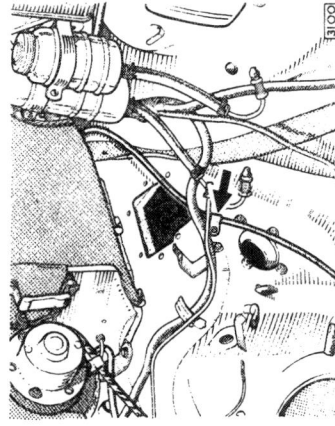

Fig. 67. Clipping of cable (left hand).

SPEEDOMETERS—GENERAL INSTRUCTIONS

Speedometer performance is dependent on the flexible drive, and apparent faults in the instrument may be due to some failure of the drive. Before returning a speedometer for service, the flexible drive should be checked, as described in the previous paragraphs. The following diagrams show you how to check the instrument performance.

15. Instrument Not Operating

Flexible drive not properly connected (see paragraph 5). Broken or damaged inner flexible shaft or fault at point of drive (see paragraphs 12 and 13), in which case remove and replace flex (see paragraphs 6 and 8) or rectify point of drive fault. Insufficient engagement of inner shaft (see paragraph 10). Defective instrument—return for service.

16. Instrument Inaccurate

Incorrect speedometer fitted. Check code number.

17. Speedometer Inaccurate

Check tyre pressures. Inaccuracy can be caused by badly worn tyres. Non-standard tyres fitted, apply to Smiths for specially calibrated instrument.

18. Speedometer Inaccurate

Rear axle non-standard. Drive ratio in vehicle gearbox non-standard. A rapid and simple check is obtained by entering in the formula the figures found in the test (see paragraph 19).

$$\frac{1680 \text{ N}}{\text{R}} = \text{T.P.M. No.}$$

Where N = Number of turns made by the inner shaft for six turns of rear wheel and R = Radius of rear wheel in inches measured from centre of hub to ground.

Example

Cardboard pointer on inner shaft (see 19) rotates $9\frac{1}{8}$ times as vehicle is pushed forward six turns of rear wheel. Rear wheel radius $12\frac{1}{4}''$.

Flex turns per mile:

$$1680 \times 9\frac{1}{8} = 15330 \div 12\frac{1}{4} = 1251 = \text{T.P.M. No.}$$

$$12\frac{1}{4}$$

Fig. 71. *Showing the code number on the face of the instrument.*

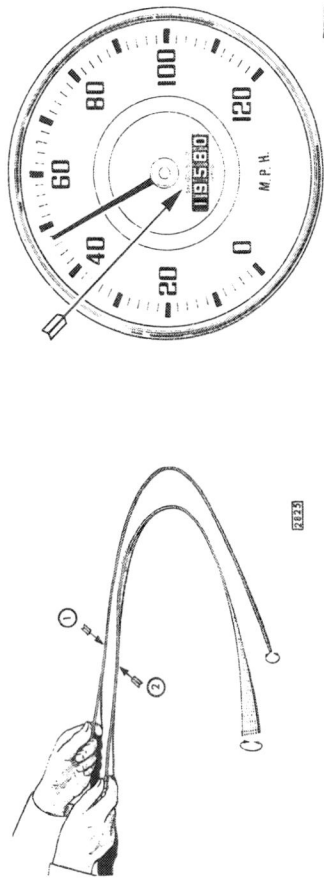

Fig. 68. *Checking the inner flex for kinks.*

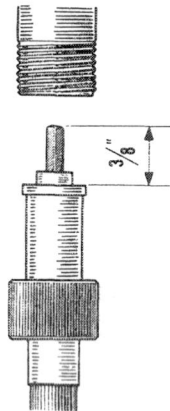

Fig. 69. *Showing the amount the inner flex must protrude from the outer cable.*

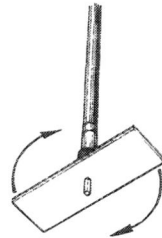

Fig. 72. *Cardboard pointer on the inner flex for checking the number of turns.*

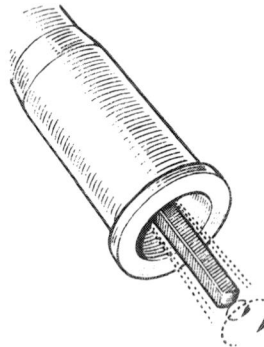

Fig. 70. *Checking the inner flex for "run-out".*

5. Connection of Inner Flexible Shaft

Where possible slightly withdraw inner flex and connect outer first. Then slide inner into engagement.

6. Removal of Inner Shaft

Most inner flexes can be removed by disconnecting instrument end and pulling out flex. Broken inner flex will have to be withdrawn from both ends.

7. Examination of Inner Flexible Shaft

Check for kinked inner flexible shaft by rolling on clean flat surface. Kinks will be seen and felt.

8. Lubrication Every 10,000 Miles

Withdraw inner flexible drive (see paragraph 6). Place blob of grease on end of outer cable and insert flex through it, carrying grease inside. Use Esso T.S.D. 119 or equivalent. Do NOT use oil.

9. Excessive Lubrication

Avoid excessive lubrication. If oil appears in flexible drive, suspect faulty oil-seal at point of drive.

10. Inner Shaft Projection

Check $\frac{3}{8}''$ projection of inner flex beyond outer casing at instrument end. This ensures correct engagement in instrument and point of drive.

11. Concentric Rotation

Check that inner flex rotates in centre of outer cable.

12. Damaged Inner Shaft

Examine inner flex ends for wear or other damage. Before fitting new flex ensure instrument main spindle is free.

13. Damage Drive End Connections

Examine point of drive for damage or slip on gears in gearbox.

14. Ensuring Correct Drive Fitted

When ordering, state Make, Year and Model of vehicle. State also length of drive required when alternatives are shown.

19. **Gearing Test**

Disconnect flexible drive from speedometer. With the gears in neutral, count the number of turns of the inner shaft for six turns of the rear wheels when the vehicle is pushed forward in a straight line. Measure rolling radius of rear wheels—centre of hub to ground. Apply figures in formula (see paragraph 18).

20. **Correct Speedometer**

Number illustrated should correspond within 25 either way with the number obtained from paragraphs 18 and 19. If it does not, apply to Smiths for specially calibrated instrument, giving details of test and vehicle.

21. **Pointer Waver**

Oiled up instrument. Replace oil seal if necessary, clean and lubricate flexible drive (see paragraph 8). Return instrument for replacement.

22. **Pointer Waver**

Inner flexible shaft not engaging fully. Check 10, then try 4. Also check 12.

23. **Pointer Waver**

Kinked or crushed flexible drive. Check 7 and 3. For withdrawal of inner shaft see paragraph 6. Bends of too small radius in flexible drive, check 1.

24. **Pointer Waver**

If 21, 22 and 23 show no sign of trouble, instrument is probably defective. Return for replacement.

25. **Noisy Installation**

Tapping noises. Check 5 and 2. Flexible drive damaged. Check 7 and 12 (also see paragraph 6). Check lubrication is sufficient. Check 10 and 11.

26. **Noisy Installation**

General high noise level. Withdraw inner shaft (see paragraph 6) and reconnect outer flex. If noise continues at lower level then source of noise is in vehicle point of drive. Fitting new P.V.C. covered flexible drive with nylon bush on inner shaft and instrument with rubber mounted movement should overcome this trouble.

27. **Noisy Installation**

Regular ticking in time with speedometer decimal distance counter. Return speedometer for replacement.

28. **Noisy Installation**

Loud screeching noise more prevalent in cold weather, return instrument for replacement.

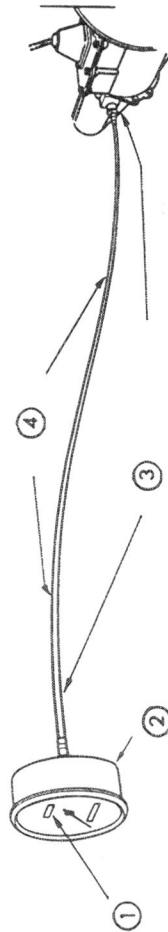

Fig. 73. *Showing the turns per mile on the face of instrument.*

1. Regular ticking
2. Screeching noise
3. General high noise level
4. Tapping noises

Fig. 74. *Diagram showing apparent source and type of noise*

OPTIONAL EXTRAS

This section covers the installation of SMITHS Radiomobile radio sets and Heated Backlight available as optional extras for the 3·4 and 3·8 "S" Models.

RADIO

SMITHS "Radiomobile" radio sets are available in the following models to suit the broadcasting requirements of different countries. Rear extension speakers are also available if required.

920 T—Long and Medium wave band.

922 T—Medium wave band.

530 T—Medium and Short wave band.

This instruction covers both left- and right-hand drive cars.

Warning: Before connection to battery supply is made, it is essential to ensure that the receiver is connected for POSITIVE GROUND. DAMAGE TO TRANSISTORS IS INEVITABLE IF POLARITY IS INCORRECT.

The radio unit and front loudspeaker are fitted in the console situated centrally below the parcel tray.

The Aerial is fitted on the **drive side** front wing.

AERIAL MOUNTING

Warning: When removing trim secured by adhesive, extreme care must be taken.

Disconnect battery.

Remove the drive side trimmed scuttle casing.

Remove the cover plate secured by four screws.

Drill $\frac{7}{8}$" dia. hole in drive side front wing, as shown in Fig. 76.

From inside the car pass aerial mast assembly up through hole in wing and secure base of aerial to the bracket provided in car. See Fig. 77.

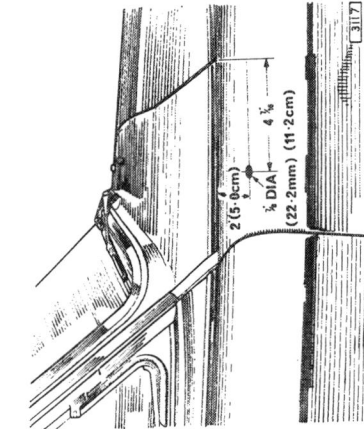

Fig. 75. *Location of hole for aerial mounting.*

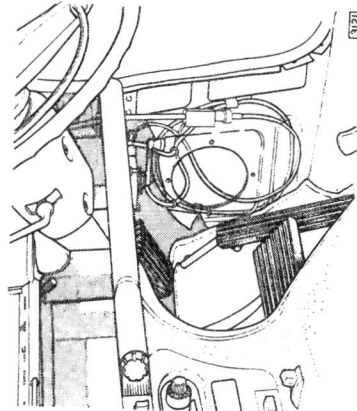

Fig. 76. *Aerial base mounting and run of operating cable.*

Fit aerial grommet to wing.

Wind up aerial mast to maximum height and remove the drive cable from winder mechanism, as shown in Fig. 78.

Route the drive cable via hole inside scuttle; as shown in Fig. 77. Fit the grommet provided and refit cable to winder box.

Secure the winder box to underside of parcel tray, as shown in Fig. 77.

Note: On right-hand drive cars the three fixing holes are provided as shown. On left-hand drive cars two fixing studs and one hole are provided.

Fit grommets provided for aerial lead and reservoir cable.

Route aerial lead and drive cables as shown in Fig. 77. Drive cable should be routed to maximum radius. Clip and tape up cables as shown.

RADIO UNIT AND FRONT LOUDSPEAKER MOUNTING (Fig. 79)

Carefully pull out escutcheon surrounding heater push button controls.

Remove two screws which are covered by escutcheon.

Remove the perforated cover plate in centre tray. This plate is secured to tray with nylon studs and should be eased off.

Carefully pull back parcel tray trim to expose two fixing screws.

Remove fixing screws and pull out veneered wooden radio panel.

Remove the escutcheon secured by wood screws from behind radio aperture in panel. Discard the grille and remove excess material covering scale aperture and holes in escutcheon.

Note: Model 530T: An escutcheon with piercing to suit is supplied in the kit for this model.

Assemble radio unit to panel as shown.

Connect fuse lead to feed side (brown-lead) of fuse panel. No. 7, located on bulkhead behind the centre instrument panel. Connect other end to battery fly lead from radio unit.

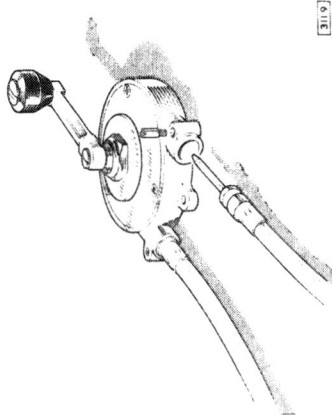

Fig. 77. The aerial winder.

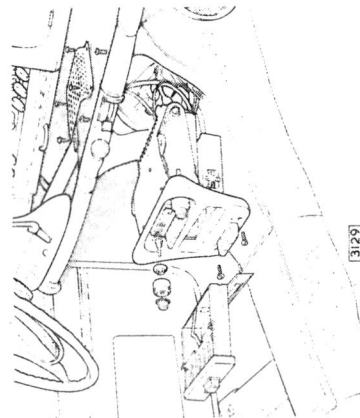

Fig. 78. Fitting the receiver to the console.

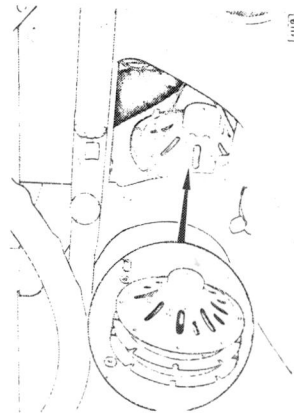

Fig. 79. Fitting the front speaker.

Fit loudspeaker lead to loudspeaker.

Secure loudspeaker to two fixing studs provided inside of console as shown in Fig. 86.

Note: The excess studding securing loudspeaker bezel and grille must be cut flush with the securing nuts.

Connect loudspeaker lead to radio unit.

Connect the aerial lead to radio unit.

Re-connect battery.

Switch on the radio and tune to a weak signal on 1200 Kc s (approx. 250 metres) and adjust aerial trimmer for maximum volume.

Replace wooden radio panel assembly.

Note: It is important that the area underneath and behind the heat sink is well ventilated. Felt should be removed from this area if necessary.

Suppression

It is important to scrape to bare metal all points at which an earth connection is made.

Fit 1 mfd. capacitor to dynamo output terminal. Earth to dynamo fixing bolt.

Fit 1 mfd. capacitor to (SW) terminal on coil. Earth under coil fixing screw.

Fit 1 mfd. capacitor to each petrol pump feed. Earth under pump fixing bolt.

Fit .25 mfd. capacitor to oil pressure indicator transmitter. Earth under suitable filter fixing bolt ensuring that bolt is re-tightened securely.

REAR SPEAKER FITTING

1. Disconnect battery.
2. Remove rear seat.
3. Remove two 1" U.N.F. screws retaining the attached brackets at the bottom of the rear squab to body of car.
4. Remove rear squab in an upward and forward direction disengaging from the retaining clips attached to the front edge of the metal parcel shelf.
5. Break adhesive bond of parcel tray trim to body of car and locate and remove two PK screws retaining trimmed parcel tray to metal parcel shelf.
6. Disengage retaining clips on underside of the combined trimmed parcel tray and lower rear light finisher from metal parcel tray. Remove trimmed parcel tray.
7. Remove the felt packing on top of metal parcel shelf.

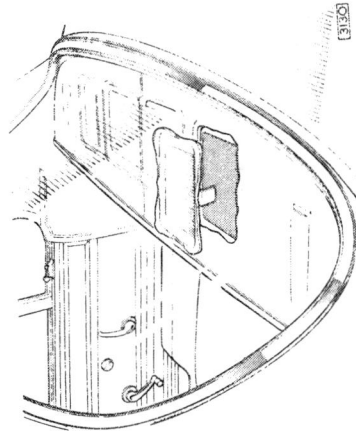

Fig. 80 Cut away "Dedshete" to expose depression in parcel tray.

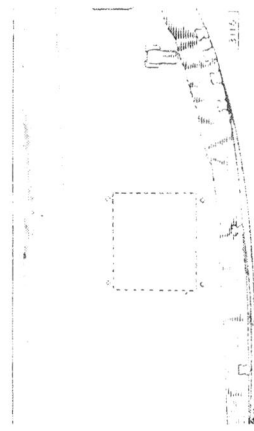

Fig. 81. Aperture to be cut 3 (95 mm.) less all round than bezel outline.

8. Cut away painted "DEDSHETE" in area of left-hand depression in metal parcel shelf to expose metal cover plate. See Fig. 80.

9. From inside boot, remove sufficient felt on underside of metal parcel shelf in the area of metal cover plate to expose welded retaining tabs.

10. Remove and discard metal cover plate.

11. From inside boot, position loudspeaker mounting board symmetrically about the aperture in metal parcel shelf. Using mounting board as a template, mark and drill four ¼" diameter holes in the metal parcel shelves, as shown in Fig. 81.

Note : On later production cars, these fixing holes are already provided in the body.

12. Temporarily replace trimmed parcel tray onto metal parcel shelf. From inside boot, mark onto underside of trimmed parcel tray the outline of aperture in metal parcel shelf.

13. Remove trimmed parcel tray.

14. Secure loudspeaker mounting board, and loudspeaker to metal parcel shalf using fixings provided.

15. Locate in top left-hand corner of boot, twin plastic loudspeaker lead in wiring harness.

16. Bare the ends of the loudspeaker lead in wiring harness and connect to the two way terminal block provided. Connect bare ends of the short loudspeaker lead provided to the two way terminal block. Connect loudspeaker lead to loudspeaker.

17. On the underside of trimmed parcel tray locate centre of marked aperture. About this centre, mark and drill four 3/16" (5 mm.) diameter holes.

Note : On later production cars the bezel is pre-pierced.

18. Temporarily assemble the bezel and the grille to the trimmed parcel tray and mark round the outline of the bezel.

19. Remove bezel from parcel tray. Using outline as guide, reduce size of bezel outline by ⅜" (9.5 mm.) all round.

20. Cut and trim aperture.

21. Assemble bezel and masking cloth to the parcel tray using fixings provided.

22. Replace felt packing and trimmed parcel tray assembly adhesing trim of parcel tray to the body of car with a suitable adhesive.

23. Replace rear squab and rear seat.

24. From inside the boot, on the underside of metal parcel shelf and using suitable adhesive, replace felt, trimming around outline of loudspeaker mounting board as required.

25. From behind centre instrument panel, locate twin plastic loudspeaker lead in weaved cotton covering, routed with main wiring harness above fuses.

26. In vertical rear face of front parcel tray, on drive side of steering column, drill a ⅜" diameter (9.5 mm.) hole to dimensions as shown in Fig. 82 for balance control.

27. Remove the radio from console unit.

28. Remove existing lead from receiver to front loudspeaker. Replace with long loudspeaker lead terminated at one end with small "Lucar" type connectors and bare ends at the other.

29. Connect new loudspeaker lead terminated with two pin plug to appropriate socket in receiver.

30. Connect the twin plastic lead from rear loudspeaker to two-way terminal block provided. Bare the ends of unterminated lead provided and connect to two-way terminal block.

31. Bare other ends of unterminated lead and connect to points "B" and "C" on terminal block of the balance control assembly—See Fig. 83.

32. Connect the loudspeaker lead from front loudspeaker to points "A" and "B" on the terminal block of balance control assembly - See Fig. 83.

33. Connect loudspeaker lead from receiver to points A and C on terminal block of balance control assembly—See Fig. 83.

34. Affix the balance control assembly to vertical rear face of front parcel tray.

35. Replace receiver into console unit.

36. Re-connect battery.

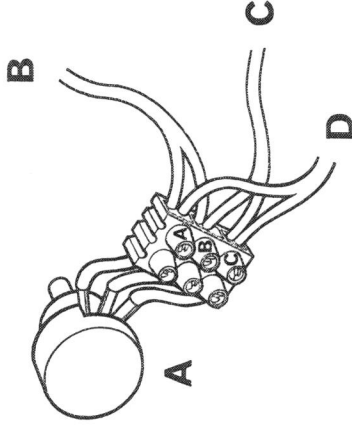

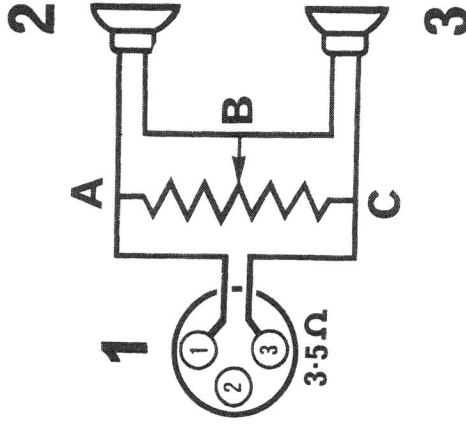

Fig. 83. *Circuit diagram.*

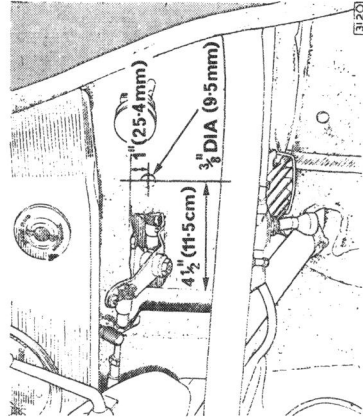

Fig. 82. *Drilling in rear face of front parcel tray.*

ELECTRICALLY-HEATED BACKLIGHT

Description

An electrically-heated backlight to provide demising or defrosting is fitted as an optional extra.

Operation

The heating element consisting of a fine wire mesh between laminations of the glass is connected to the main wiring harness.

The element will come into operation when the ignition is switched on, no separate switch being provided.

The current consumption is approximately 5 amperes.

A 15 amp. fuse contained in a plastic holder, located in a clip behind the instrument panel, is provided in the circuit as a safety precaution.

Fitting Instructions

Remove the backlight as detailed in Section N page N.15.

Remove the rear seat cushion and squab as detailed in Section N page N.23.

Lift up the rear parcel tray trimming where stuck to the rear squab panel and remove the two drive screws, now exposed, securing the parcel tray trim board. Pull the board away from the three retaining clips at the rear edge.

Drill two ¼" (6·4 mm.) holes in the parcel tray 17" (43·2 cm.) from either side of the centre line of the backlight and 6" (15·2 cm.) from the front edge. Fit the two small grommets in the holes.

Fit the backlight as detailed in Section N page N.15.

Feed the two cables attached to the backlight through the grommets.

Connect the left-hand cable to the black and white cable located in the luggage compartment behind the left-hand hinge bracket.

Attach the earth contact to the right-hand hinge bracket and connect the black cable.

Refit the parcel tray trim board.

Refit the squab and seat cushion.

Disconnect the battery.

Remove the picnic tray (see page P.52) and lower the centre instrument panel.

Drill a ⁷⁄₆₄" (2·78 mm.) hole in the top left-hand corner of the metal back plate of the instrument panel.

Important: EXTREME CARE must be taken to ensure that the drill does not penetrate the wooden facia panel. Fit a stop or sleeve over the drill shank to allow the drill to pierce the back plate only.

Attach the spring clip with the small drive screw to the panel and clip in the fuse holder.

Connect the white cable to the vacant A3 terminal on the fuse block along with the existing white cables.

Connect the black/white cable from the fuse-holder to the equivalent cable already situated in the existing body harness.

Refit the instrument panel and picnic tray.

Reconnect the battery, switch on the ignition and test through.

Fault Diagnosis

Check that the fuse has not blown. Replace if necessary by one of the correct value.

Check the rear light element by disconnecting the cable connectors in the luggage compartment and reconnecting the backlight cables to a 12-volt battery with a 0—20 moving coil ammeter in series.

If no reading is apparent on the meter replace the backlight glass.

If a reading is shown on the meter check the feed cables connection in the luggage compartment for continuity with a volt meter. Insert fuse and switch on ignition before checking.

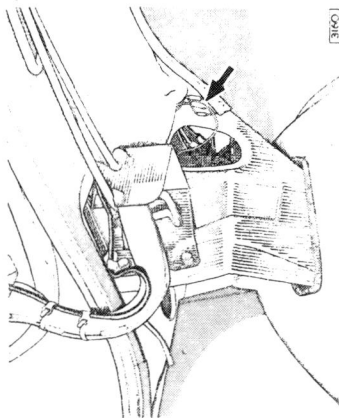

Fig. 84. *Location of B W cable in boot.*

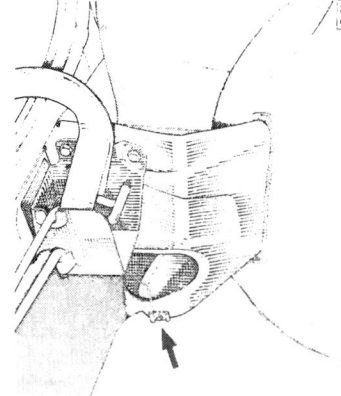

Fig. 85. *Location of earth tab in boot.*

SUPPLEMENTARY INFORMATION TO SECTION P

"ELECTRICAL AND INSTRUMENTS"

ELECTRICALLY HEATED BACKLIGHT

Introduction of Control Switch in Electrical Circuit

	Commencing Chassis No.	
	R.H. Drive	L.H. Drive
3·4 "S"	1B 6438	1B 25850
3·8 "S"	1B 57175	1B 79231

Commencing at the above chassis numbers, the heated backlight has a control switch, warning light and relay with resistance included in the electrical circuit.

The warning light operating through the resistance, is dimmed when the side lights are switched on.

The circuit remains ignition controlled and there is no change in the fuse rating, fuse location and current consumption.

The control switch and warning light is mounted on the facia panel above the brake fluid warning light; the relay is mounted on the reinforcement panel behind the facia.

Fitting Instructions

Fit the backlight and cables as detailed on page P.70. Fit the fuse holder as described on Page 71.

Remove both side facia panels.

Drill two holes $\frac{7}{16}$" (11·1 mm.) in diameter above the brake fluid warning light in the position illustrated in Fig. 86.

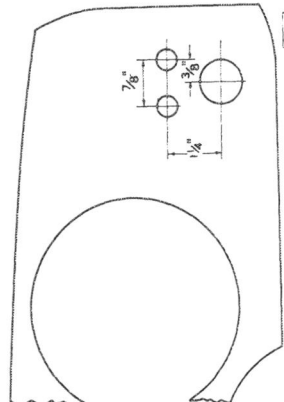

Fig. 86. *Drill two holes $\frac{7}{16}$" (11.1 mm.) diameter.*

Mount the switch escutcheon switch and warning light.

Using the relay as a template drill (No. 32 drill) two holes in the position shown in Fig. 87 to mount the relay. Attach the relay with two No. 8 metal drive screws.

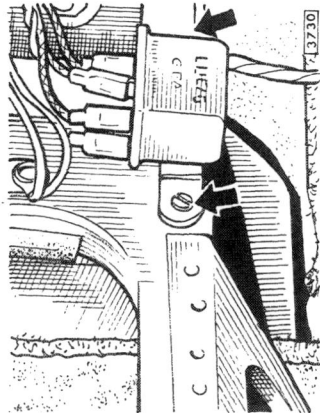

Fig. 87. *Location of relay mounting.*

Route the switch connector harness behind the instrument panel and tape into the panel harness.

Connect the red cable (centre breakout) to the side lamp fuse (No. 6); the black white with green ident to the in-line fuse and the remaining black/white cable to the heated rear window connector already existing in the body harness.

Connect the harness relay junctions to the relay as follows:—

Red (R)	..	Relay terminal (W1)
Black/White (BW)	..	„ „ (C2)
Red Brown (RN)	..	„ „ (C1)
Black (B)	..	„ „ (W2)

Attach the black cables with the eyelet terminals to one of the relay fixing screws.

Connect the harness switch junction B/W and B W with green ident to the control switch and insert the bulb holder in the socket.

Refit the facia panels, picnic tray and instrument panel.

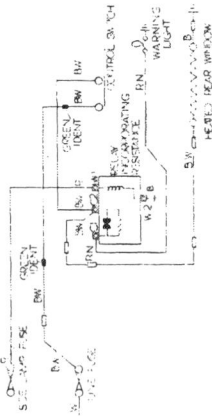

Fig. 88. *Wiring diagram for fitting heated backlight.*

Publication No. E/133/3

OFFICIAL TECHNICAL BOOKS

Brooklands Technical Books has been formed to supply owners,
restorers and professional
repairers with official factory literature.

Workshop Manuals

Jaguar Service Manual 1946-1948		9781855207844
Jaguar XK 120 140 150 150S & Mk 7, 8 & 9		9781870642279
Jaguar Mk 2 (2.4 3.4 3.8 240 340)	E121/7	9781870642958
Jaguar Mk 10 (3.8 & 4.2) & 420G	E136/2	9781855200814
Jaguar 'S' Type 3.4 & 3.8	E133/3	9781870642095
Jaguar E-Type 3.8 & 4.2 Series 1 & 2		
	E123/8, E123 B/3 & E156/1	9781855200203
Jaguar E-Type V12 Series 3	E165/3	9781855200012
Jaguar 420	E143/2	9781855201712
Jaguar XJ6 2.8 & 4.2 Series 1		9781855200562
Jaguar XJ6 3.4 & 4.2 Series	E188/4	9781855200302
Jaguar XJ12 Series 1		9781783180417
Jaguar XJ12 Series 2 / DD6 Series 2	E190/4	9781855201408
Jaguar XJ6 & XJ12 Series 3	AKM9006	9781855204010
Jaguar XJ6 OWM (XJ40) 1986-94		9781855207851
Jaguar XJS V12 5.3 & 6.0 Litre	AKM3455	9781855202627
Jaguar XJS 6 Cylinder 3.6 & 4.0 Litre	AKM9063	9781855204638

Owners Workshop Manuals

Jaguar E-Type V12 1971-1974	9781783181162
Jaguar XJ, Sovereign 1968-1982	9781783811179
Jaguar XJ6 Workshop Manual 1986-1994	9781855207851
Jaguar XJ12, XJ5.3 Double Six 1972-1979	9781783181186

Parts Catalogues

Jaguar Mk 2 3.4	J20	9781855201569
Jaguar Mk 2 (3.4, 3.8 & 340)	J34	9781855209084
Jaguar Series 3 12 Cyl. Saloons		9781783180592
Jaguar E-Type 3.8	J30	9781869826314
Jaguar E-Type 4.2 Series 1	J37	9781870642118
Jaguar E-Type Series 2	J37 & J38	9781855201705
Jaguar E-Type V12 Ser. 3 Open 2 Seater	RTC9014	9781869826840
Jaguar XJ6 Series 1		9781855200043
Jaguar XJ6 & Daimler Sovereign Ser. 2	RTC9883CA	9781855200579
Jaguar XJ6 & Daimler Sovereign Ser. 3	RTC9885CF	9781855202771
Jaguar XJ12 Series 2 / DD6 Series 2		9781783180585
Jaguar 2.9 & 3.6 Litre Saloons 1986-89	RTC9893CB	9781855202993
Jaguar XJ-S 3.6 & 5.3 Jan 1987 on	RTC9900CA	9781855204003

Owners Handbooks

Jaguar XK120		9781855200432
Jaguar XK140	E101/2	9781855200401
Jaguar XK150	E111/2	9781855200395
Jaguar Mk 2 (3.4)	E116/10	9781855201682
Jaguar Mk 2 (3.8)	E115/10	9781869826765
Jaguar E-Type (Tuning & prep. for competition)		9781855207905
Jaguar E-Type 3.8 Series 1	E122/7	9781870642927
Jaguar E-Type 4.2 2+2 Series 1	E131/6	9781869826383
Jaguar E-Type 4.2 Series	E154/5	9781869826499
Jaguar E-Type V12 Series 3	E160/2	9781855200029
Jaguar E-Type V12 Series 3 (US)	A181/2	9781855200036
Jaguar XJ (3.4 & 4.2) Series 2	E200/8	9781855201200
Jaguar XJ6C Series 2	E184/1	9781855207875
Jaguar XJ12 Series 3	AKM4181	9781855207868

Carburetters

SU Carburetters Tuning Tips & Techniques	9781855202559
Solex Carburetters Tuning Tips & Techniques	9781855209770
Weber Carburettors Tuning Tips and Techniques	9781855207592

Jaguar - Road Test Books

Jaguar and SS Gold Portfolio 1931-1951	9781855200630
Jaguar XK120 XK140 XK150 Gold Port. 1948-60	9781870642415
Jaguar Mk 7, 8, 9, 10 & 420G	9781855208674
Jaguar Mk 1 & Mk 2 1955-1969	9781855208599
Jaguar E-Type	9781855208360
Jaguar XJ6 1968-79 (Series 1 & 2)	9781855202641
Jaguar XJ12 XJ5.3 V12 Gold Portfolio 1972-1990	9781855200838
Jaguar XJS Gold Portfolio 1975-1988	9781855202719
Jaguar XJ-S V12 1988-1996	9781855204249
Jaguar XK8 & XKR 1996-2005	9781855207578
Road & Track on Jaguar 1950-1960	9780946489695
Road & Track on Jaguar 1968-1974	9780946489374
Road & Track On Jaguar XJ-S-XK8-XK	9781855206298

Brooklands Books Ltd., PO Box 146,
Cobham, Surrey KT11 1LG, England

Ref: J180WH   Part No. E.133/3   ISBN 9781870642095

www.brooklandsbooks.com

Printed in Great Britain
by Amazon